Christian Alexander Steinle

A first level trigger approach for the CBM experiment

disserta Verlag

Steinle, Christian Alexander: A first level trigger approach for the CBM experiment, Hamburg, disserta Verlag, 2012

ISBN: 978-3-95425-008-0
Druck: disserta Verlag, ein Imprint der Diplomica® Verlag GmbH, Hamburg, 2012

Bibliografische Information der Deutschen Nationalbibliothek
Die Deutsche Nationalbibliothek verzeichnet diese Publikation in der Deutschen Nationalbibliografie; detaillierte bibliografische Daten sind im Internet über http://dnb.d-nb.de abrufbar.

Die digitale Ausgabe (eBook-Ausgabe) dieses Titels trägt die ISBN 978-3-95425-009-7 und kann über den Handel oder den Verlag bezogen werden.

Universität Heidelberg
Fakultät für Mathematik und Informatik
ziti – Lehrstuhl für Informatik V

A first level trigger approach for the CBM experiment

Inaugural dissertation
to obtain one's doctorate of Natural Sciences in the
Combined Faculties for the Natural Sciences and for Mathematics
of the Ruperto-Carola University of Heidelberg, Germany

submitted by:

Diplom-Informatiker Christian Steinle

born in Heidelberg

Date of oral examination: January, 27th 2012

1. Referee: Prof. Dr. Reinhard Männer
2. Referee: Prof. Dr. Peter Fischer

Corrected and Extended by Curriculum Vitae

Thanks

First of all, I want to thank my dissertation adviser Prof. Dr. Reinhard Männer for the possibility to work on the very interesting topic of massively parallelized processing.

Additional thanks to all my colleagues at the department of computer engineering at the University of Heidelberg and at the the GSI in Darmstadt, especially Andreas Wurz, Joachim Gläss, Andreas Kugel, Volker Friese, Johann Heuser and Walter Müller.

Moreover special thanks for Lou Scagnetti, who was always patient with me and serves good ideas allegorizing initiatives for different steps of my work.

In particular I want to thank my parents Karl-Heinz Willhelm Steinle and Ursel Steinle as well as my brother Michael Willhelm Steinle and his family for supporting me and giving me the chance to do my diploma and dissertation.

This work has been funded by the DFG Graduates Support Program, the BMBF DEPFET project Hadronen & Kernphysik 06MN229I and the GSI University Programs MAMAEN and HDMAEN as well as the Chair for Computer Science V at the University of Heidelberg.

> I do not agree with math. I think that an accumulation of zeros is a dangerous number.

(Citation of Stanisław Jercy Lec, see [Sta])

Declaration of Authorship

I certify that the work presented here is, to the best of my knowledge and belief, original and the result of my investigations, except as acknowledged, and has not been submitted, either in part or whole, for a degree at this or any other University.

Christian Alexander Steinle
March 11, 2012

Kurzfassung

Schon im Altertum studierte die Menschheit bekanntermaßen Mutter Natur mit der Zielsetzung fundamentale Naturgesetze zu entdecken und zu verstehen. Bei diesem Vorhaben kam allerdings schon frühzeitig zum Vorschein, daß unterstützende Ausrüstung vollkommen unverzichtbar ist, um den zu inspizierenden Versuchsaufbau, die Durchführung des Versuchs und die erworbenen Ergebnisse präzise zu beschreiben und zu analysieren, auch wenn erste Bestrebungen gänzlich ohne Instrumente unternommen wurden. Resultierend ging daraus nun eine einfache aber mächtige Wechselbeziehung hervor, denn die Entwicklung von immer weiter verbesserten Maschinen brachte natürlich wiederum steigendes Wissen mit sich.
Begründet auf diesem Umstand führte die wachsende Komplexität der Kenntnisse im Laufe der Zeit zu der wohlbekannten Aufteilung der allgemeinen Studie der Natur in eine Vielzahl an Fachrichtungen wie zum Beispiel Physik, Biologie und Chemie, während die verwendeten Maschinen zu modernen Hochleistungscomputern weiterentwickelt wurden. Darüber hinaus unterteilt sich jede einzelne Fachrichtung heutzutage in viele Teilfelder wie beispielsweise Hochenergiephysik, die verständlicherweise in höchstem Maße Gebrauch von der neuesten Computerspitzentechnologie macht und deshalb das Betätigungsfeld dieser Arbeit festlegt.

Ein verbreitetes Problem in der Hochenergiephysik besteht darin, daß Experimente eine Datenmenge produzieren, die viel zu groß ist, um sie mit irgendeiner Gerätetechnologie für spätere Analysen aufzuzeichnen. Aus diesem Grund werden Algorithmen benötigt, welche in der Lage sind, diese Daten direkt zu verarbeiten indem interessante Fragmente einfach aus der Masse herausgefiltert werden. Somit definiert sich die Hauptaufgabe dieser sogenannten Trigger-Algorithmen durch die Reduzierung der auftretenden Daten auf eine Größe, die entweder aufgenommen oder unmittelbar im Anschluß von langsameren Algorithmen analysiert werden kann. In diesem Zusammenhang muß natürlich zusätzlich eingeräumt werden, daß der verminderte Anteil im günstigsten Fall lediglich alle zu betrachtenden Objekte umfasst.

Da das geplante Komprimierte Baryonische Materie (CBM) „fixed-target“ (Zielscheiben) Experiment an der Einrichtung für Antiprotonen- und Ionenforschung (FAIR) ausdrücklich bis zu 10^7 Nukleon-Nukleon Kollisionen pro Sekunde mit einer Vielzahl von bis zu 1 000 Partikeln generiert, was offensichtlich eine Datenflut bisher ohnegleichen zur Folge hat, ist es nicht verwunderlich, daß keiner der bekannten Trigger, die in anderen Experimenten eingesetzt werden, entworfen wurde, um eine solch enorme Rate bewältigen zu können.

Um die effektive Verarbeitung von Datenmengen diesen Ausmaßes zu gewährleisten, habe ich daher die allgemeine Hough-Transformation angepaßt und sie überdies auf das CBM Experiment zugeschnitten.
Zu diesem Zweck entwickelte ich zunächst eine analytische Formel für die Annäherung der Differentialgleichung, welche gemeinhin die Flugbahn eines geladenen Teilchens im inhomogenen Magnetfeld beschreibt. Diese Formel präzisiert ferner eine parametrische Beschreibung und ist daher als Berechnungskern für die Hough-Transformation, die einen allgemeinen Ansatz in der Mustererkennung und Merkmalsextraktion darstellt, verwendbar.
Im Anfangsstadium entsprang die Gestaltung meiner Formel der geläufigen Physik für die Ablenkung eines geladenen Teilchens im homogenen Magnetfeld. Um diese dann an die Beschaffenheit eines inhomogenen Magnetfeldes anzupassen, fügte ich einfach Faktoren für geeignete Abtastpunkten hinzu, die ein homogenes Magnetfeld so vertreten, daß dieses theoretisch unterstellte homogene Magnetfeld die gleiche Ablenkungskraft wie das echte inhomogene Magnetfeld an jedem dieser festgelegten Abtastpunkte verursacht.

Nebenbei entwarf ich ebenfalls ein Softwarepaket zu Simulationszwecken, welches den Trigger als Modul im CBMROOT Framework der Gesellschaft für Schwerionenforschung (GSI) realisiert. Dieses Framework ist zudem nicht nur in der Lage die Hough-Transformation anzuwenden, sondern kann auch genutzt werden, um vorher Pakete mit Eingangsdaten, welche aus einer Reihe von Partikelinteraktionen mit den Detektorlagen (Hits) im Silizium Tracking System (STS) bestehen, mit Hilfe eines anderen schon installierten Moduls zu generieren. In Anlehnung an die Eingangsdaten können dann anschließend die resultierenden Ausgangsdatenpakete, welche selbstverständlich durch eine Gruppe einzelner Partikelspuren jeweils beschrieben durch ihren rekonstruierten Impuls und im Debug-Modus zusätzlich durch die beitragenden Hits bestimmt werden, wie erwartet an das nachfolgend verwendete Framework-Modul übergeben werden. Ergänzend baute ich sogar einige Analysen für die Optimierung der Parameter des Algorithmus, einige Analysen für die Prüfung der Leistung des Algorithmus und einige Standardanalysen für die Abschätzung der notwendigen Anforderungen an die Hardware in den Quellcode des

Softwaremoduls ein.
Da der Algorithmus, den ich entwickelt habe, darüber hinaus sehr gut für die Umsetzung in Hardware geeignet ist, verspricht die Verwendung von parallelen Architekturen wie der Cell BE oder vor allem der Einsatz des umfassenden Parallelisierungspotenzials von Feld programmierbaren (Logik-)Gatter-Anordnungen (FPGAs) eine einfachere Realisierung der erforderlichen Verarbeitungsgeschwindigkeit. In diesem Zusammenhang, soll außerdem ausdrücklich hervorgehoben werden, daß jede realisierte funktionale Einheit des Algorithmus eingehend in einem eigenen Abschnitt vorgestellt wird. Weiterhin ersetzen Look-Up-Tabellen die analytische Formel im Berechnungskern der Hough-Transformation, was viel Zeit einspart, die ansonsten für die notwendigen aufwändigen Rechenoperationen verbraucht worden wäre. Parallel dazu ermöglicht die Aufteilung des großen 3D Hough-Raums in getrennte autonome 2D Lagen die serielle Verarbeitung dieser Lagen völlig unabhängig voneinander. Nebenbei bemerkt stellen beide Konzepte Schlüsselelemente für eine schnelle Verarbeitung dar. Überdies bietet die Herausnahme der aufwändigen Rechenoperationen aus der Online-Verarbeitung begleitend die Möglichkeit an, jede Art von modernen Berechnungsalgorithmen für die Look-Up-Tabellen zu verwenden, was insbesondere das Runge-Kutta Verfahren einschließt, ohne Auswirkung auf die benötigte Online-Rechenzeit.

Bei näherer Untersuchung des CBM Experiments stellt sich sofort heraus, daß die ausgewählten Faktoren für die analytische Formel, welche die Differentialgleichung annähert, recht gut für das installierte Magnetfeld, das durch vernachlässigbare Komponenten in Abhängigkeit jeder Koordinatenrichtung außer für die y Komponente in z Richtung (Strahlachse) geprägt ist, abschneiden. Darüber hinaus reduziert sich der riesige Speicherbedarf, der üblicherweise Voraussetzung für die Realisierung des Hough-Raumes ist, faktisch durch eine spezielle Dekomposition in eine beliebige Anzahl einer seriell verarbeitbaren Menge an parallelen Lagen, die ferner nur die unbedingt erforderlichen Quantisierungsstufen in Bezug auf die benötigte Auflösung für den Impuls der Partikel beinhalten. Zudem sind natürlich alle Probleme, die aus dieser Dekomposition hervorgehen, erfolgreich beseitigt worden. Die Einführung des Signatur-Konzeptes in Kombination mit der Fenstertechnik wirkt obendrein dem herkömmlichen Problem der verrauschten Maxima im Hough-Raum, die gewöhnlich häufiger durch die Verwendung von Zählern und zu großen Quantisierungsstufen auftreten, stark entgegen.
Folgerichtig zeigen CBMROOT Framework Simulationen des vorgestellten Algorithmus ein sehr gutes Leistungsverhalten für zentrale Gold+Gold Kollisionsmusterereignisse, welches über den gesamte Impulsbereich nahezu stabil ist und einen Impulsfehler von weniger als 10 % beinhaltet. Des Weiteren

sind Effizienzen um die 86 % in Kombination mit einer Trackingreduktionsrate von rund 80 % und einer Trackingidentifizierungsrate von etwa 67 % mühelos erreichbar, während die unerwünschten Seiteneffekte, abhängig von den allgemein bekannten Einflussfaktoren Detektoraufbau, Strahlenergie und Ereignisfülle, unter 10 % bleiben. In diesem Zusammenhang veranschaulicht eine kurze aber gründliche Analyse zusätzlich, daß die Monolithische-Aktive-Pixel-Sensor (MAPS)-Technologie für alle geplanten Detektorlagen wegen der hohen Abweichung zwischen der Auslese- und der Reaktionsrate ungeeignet ist, wohingegen die Mikro-Strip-Technologie für bis zu 40 % der wenigstens fünf benötigten Detektorlagen einsetzbar ist, beginnend bei der, die am weitesten vom „Target“ (von der Zielscheibe) entfernt ist. Im Gegensatz dazu ist die Hybrid-Pixel-Technologie allerdings total unproblematisch. Weiterhin muß außerdem noch angemerkt werden, daß der Einfluß der Strahlenergie auf die Leistung durch die Anpassung des Detektoraufbaus abgeschwächt werden kann.

Bedeutender ist allerdings, dasß die Berechnung eines der oben erwähnten typischen Musterereignisses, welches aus Sicht des Algorithmus durch sieben Detektorlagen, $15{,}66 \cdot 10^3$ Hits, drei Signaturen für die Enkodierung, drei Prioritätsklassen, neun Elemente im Fenster und einer Quantisierung des Histogramms von $95 \cdot 31 \cdot 191$ Zellen geprägt ist, auf einem durchschnittlichen Computer ungefähr 2,36 s dauert, ohne daß nennenswert Speicher erforderlich ist. Daher führt die Verknüpfung dieses Ergebnisses mit dem CBM Umfeld, welches durch $25 \cdot 10^5 \frac{\text{Kollisionensereignisse}}{\text{s}}$ mit einer Vielzahl von bis zu 1 000 Partikeln pro zentraler Kollision abgesteckt ist, offenkundig zu der Anforderung von $5{,}9 \cdot 10^6$ parallel eingesetzten Systemen in einer Rechenanlage, die rund $\$\,2{,}8 \cdot 10^9$ kostet.

Im Gegensatz zu diesen Berechnungselementen zeichnet sich das Parallelverarbeitungspotenzial einer Cell BE in einem Sony Playstation III System durch die Bearbeitung des gleichen Musterereignisses in weniger als 21,5 ms aus, was natürlich dann lediglich $53{,}8 \cdot 10^3$ parallele Systeme und somit nur $\$\,21{,}4 \cdot 10^6$ zur Folge hat.

Darüber hinaus bringt das noch flexiblere Parallelisierungspotenzial der FPGA-Technologie sogar mehrere nennenswerte Umsetzungsstrategien für eine betriebsfähige Plattform mit sich, die durch die Verwendung eines einzigen Chips für den gesamten Algorithmus, durch das Erreichen der schnellstmöglichen Verarbeitungsgeschwindigkeit mit mehreren Chips oder durch das Realisieren der kostengünstigsten Version mit mehreren Chips gekennzeichnet sind. Aus diesem Grund wird selbstverständlich die beste Laufzeit für den Durchsatz und die Latenz jeder Strategie jeweils bezogen auf das Musterereignis getrennt ausgewiesen. Diese Rahmenbedingung legt sodann für die Version mit einem einzigen Chip den Durchsatz auf 62,80 μs und die Latenz auf 126,02 μs, für die schnellste Version mit mehreren Chips den Durchsatz auf 62,80 μs und die Latenz auf 126,02 μs als auch zu guter Letzt für die kostengünstigste Version mit mehreren Chips den Durchsatz auf 88,46 μs und die Latenz auf 237,81 μs fest. Demnach kann man auf der Grundlage dieses Zeitverhaltens die entsprechende Anzahl an notwendigen parallelen Plattformen unmittelbar zu $1{,}9 \cdot 10^3$ mit $1 \frac{\text{Baustein}}{\text{Plattform}}$, 157 mit $15 \frac{\text{Bausteinen}}{\text{Plattform}}$ und 222 mit $3 \frac{\text{Bausteinen}}{\text{Plattform}}$ ermitteln. Weiterhin kann man somit die verursachenden Kosten für die unterschiedliche Stückzahl an Bauelementen auf $\$\,3{,}0 \cdot 10^6$ für die Version mit einem einzigen Chip, auf $\$\,3{,}8 \cdot 10^6$ für die schnellste Version mit mehreren Chips und auf $\$\,1{,}1 \cdot 10^6$ für die kostengünstigste Version mit mehreren Chips berechnen.

Schlussendlich untermauern diese Fakten, daß die Plattformarchitektur mit drei FPGA-Chips am Besten für das Trigger-System des CBM Experiments geeignet ist. Zudem besitzt dieser Ansatz, dank der Möglichkeit zusätzliche aufeinander folgende Verbesserungsalgorithmen zu nutzen, sogar noch mehr Raum für Optimierungen. Da ferner die Notwendigkeit für eine Beschränkung der Anzahl an Berechnugseinheiten allgemeinhin für Experimente aufkommen wird, wenn man sich mit solch enormen Datenraten auseinandersetzen muß, liefert die vorgeschlagene Umsetzung erwiesenermaßen hervorragende Ergebnisse bezüglich Verarbeitungsgeschwindigkeit als auch Leistung.

Abstract

Since ancient and medieval times mankind studied, as everybody knows, Mother Nature with the objective of discovering and understanding fundamental laws. Admittedly this intention revealed very early that supporting equipment is absolutely essential to precisely describe and analyze the examined experimental setup, the performance of the tests and the gained results, even though first attempts are made entirely without tools. Consequentially a simple but powerful interrelation arised now, because the development of more and more improved machines entailed obviously in turn increasing knowledge.
So originating in this circumstance, the growing complexity of knowledge lead over the time to the well-known split of the universal study of nature into varying fields like physics, biology and chemistry for example, whereas the involved machines were enhanced to modern high-performance computers. Moreover each single field is in these days classified into many subfields like high energy physics for instance, which makes naturally maximum use of todays leading edge of computer technology and defines thus the field of activity for this thesis.

A common problem in high energy physics is that experiments produce an amount of data far too large to be recorded by any hardware technology for later analyses. For that reason, algorithms are required, which are able to process those data on-line by simply filtering interesting fractions out of the total bulk. Consequently the defining task of these so-called trigger algorithms is to reduce the occurring data to a quantity, which can be either captured or directly analyzed in detail by slower algorithms. In this context, it has to be of course granted in addition that the diminished quantum merely contains at best all objects to study.
As the planned Compressed Baryonic Matter (CBM) fixed-target experiment at the Facility for Antiprotons and Ions Research (FAIR) produces particularly up to 10^7 nucleus-nucleus collisions per second with multiplicities of up to 1,000 particles, which implicates evidently an unprecedented flood of data

up to now, it is not astonishing that none of the known triggers utilized in other experiments have been designed to cope with such an enormous rate.

To ensure the effective processing of data on such a scale, I adapted therefore the general Hough transform and customized it further to the CBM experiment.
For this purpose, I initially evolved an analytic formula approximating the differential equation, which commonly describes the motion trajectory of a charged particle in an inhomogeneous magnetic field. This formula specifies furthermore a parametric description and is thence usable as computational kernel in the Hough transform, which constitutes a general approach in pattern recognition and feature extraction.
In the early stages, the design of my formula issued from the common physics for the deflection of a charged particle in a homogeneous magnetic field. To adjust for the conditions in an inhomogeneous magnetic field, I simply attached factors at appropriate sample points, which represent a homogeneous magnetic field in the manner that this theoretically assumed homogeneous field causes the same deflection force as the real inhomogeneous magnetic field at each of these given sample points.

Besides this, I designed also a software package for simulation purposes, which implements the trigger as module of the CBMROOT framework established by the Gesellschaft für Schwerionenforschung (GSI). This framework is moreover not only able to apply the Hough transform, but can be as well used to previously produce bundles of input data, which consists of a set of particle interactions with the detector stations (hits) in the Silicon Tracking System (STS), with the help of another already installed module. Just as with the input, the resulting packets of output data, which are surely defined by a group of individual particle tracks with each described by its reconstructed momentum and in debug mode accessorily by the contributing hits, can be then subsequently delivered as expected to the consecutively applied framework module. Supplementary I engineered even some analyses to optimize the algorithm's parameters, some analyses to study the algorithm's performance and some standard analyses to estimate the necessary hardware requirements into the source code of the software module.
Since the algorithm I have developed is beyond that highly suitable for an implementation in hardware, an easier realization of the required processing speed is evidently promised by the utilization of parallel architectures like the Cell BE or in particular by the application of the full parallelism capability of Field Programmable Gate Arrays (FPGAs). In this connection, it should be also explicitly highlighted that every realized functional unit of the algorithm is presented in detail, each in its own section. Further on, Look-up-tables

(LUTs) replace the analytic formula in the computational kernel of the Hough transform, which saves much time that would have been spent otherwise for the indispensable complicated arithmetic operations. In addition to this, the fragmentation of the large 3D Hough space into separate autonomous 2D layers enables the serial processing of these layers completely independent of each other. By the way, both concepts are key essentials to a fast processing. Anyway the extraction of the complicated arithmetic operations in the computational kernel from the on-line processing offers collaterally the option to use any kind of state-of-the-art LUT generation algorithm, which implies especially the Runge-Kutta approach, with no effect on the attended on-line computing time.

Within a more detailed study of the CBM experiment, it turns immediately out that the selected factors for the analytic formula approximating the differential equation perform quite well for the setup of the magnetic field, which is shaped by negligible components in dependency of each coordinate direction except for the y component in the z direction (beam axis). Over and above, the huge amount of memory, which is normally requisite to implement the entire Hough space, is effectively reduced by a specialized decomposition into any number of serially processable parallel layer volumes, which include also just the essential quantization steps with regard to the needed particle momentum resolution. Moreover all problems following this decomposition of the Hough space have been of course eliminated successfully. Into the bargain, the introduction of the signature concept in combination with the windowing technique strongly counteracts the conventional problem of noisy maximums in the Hough space, which appear usually more frequent when using counters and too many quantization steps.
Consequentially CBMROOT framework simulations of the introduced algorithm show a very good performance for central Au+Au collision sample events, which is almost stable over the entire momentum range including a momentum error less than 10 %. Furthermore efficiencies around 86 % can be easily achieved in combination with a tracking reduction rate about 80 % and a tracking identification rate of approximately 67 %, while the unwanted side-effects are kept below 10 % dependent on the familiar influencing factors detector setup, beam energy and event multiplicity. In this connection, a brief but detailed analysis demonstrates additively that the Monolithic Active Pixel Sensor (MAPS) technology is unsuitable for all planned detector stations of the experiment because of the high discrepancy between the readout and the reaction rate, whereas the micro-strip technology is applicable for up to 40 % of the at least required five detector stations starting from the one, which is farthermost from the target. In contrast to this, the hybrid

pixel technology is however unproblematic at all. Further on, it has to be noted as well that the impact of the beam energy to the performance can be attenuated by the adaption of the detector setup.
More important is indeed that the evaluation of one of the above mentioned typical sample events, which is characterized from the algorithm point of view by seven detector stations, $15.66 \cdot 10^3$ hits, three signatures for the Encoding, three priority classes, nine elements in the window and a histogram quantization of $95 \cdot 31 \cdot 191$ cells, takes nearly 2.36 s on an ordinary personal computer without requiring a considerable size of memory. Thence the combination of this score with the CBM environment, which is defined by $25 \cdot 10^5 \frac{\text{events}}{\text{s}}$ with multiplicities of up to 1,000 particles per central collision, leads obviously to the requirement for $5.9 \cdot 10^6$ parallel used systems in a computation factory, which costs roughly $\$\,2.8 \cdot 10^9$.
Unlike these computational elements, the parallel execution capabilities of the Cell BE in a Sony Playstation III system features the processing of the same sample event in less than 21.5 ms, which certainly results then in only $53.8 \cdot 10^3$ parallel systems and thus in just $\$\,21.4 \cdot 10^6$.
Moreover the even more flexible parallelism capability of the FPGA technology entails actually several considerable implementation strategies for an operational platform, which are identified by utilizing only a single chip for the entire algorithm, by achieving the fastest processing speed with multiple chips or by realizing the cheapest version with multiple chips. For that reason, the best timing for the throughput and the latency is of course reported separately for each strategy respectively related to the sample event. This prevailing circumstance determines then the throughput to $752.18\,\mu s$ and the latency to $1.57 \cdot 10^3\,\mu s$ for the single-chip version, the throughput to $62.80\,\mu s$ and the latency to $126.02\,\mu s$ for the fastest multi-chip version as well as finally the throughput to $88.46\,\mu s$ and the latency to $237.81\,\mu s$ for the cheapest multi-chip version. So based on these timings, the corresponding number of necessary parallel platforms can be directly calculated to $1.9 \cdot 10^3$ with $1 \frac{\text{device}}{\text{platform}}$, 157 with $15 \frac{\text{devices}}{\text{platform}}$ and 222 with $3 \frac{\text{devices}}{\text{platform}}$. Further on, the inducing costs for the different quantity of devices can be therefore computed to $\$\,3.0 \cdot 10^6$ for the single-chip version, to $\$\,3.8 \cdot 10^6$ for the fastest multi-chip version and to $\$\,1.1 \cdot 10^6$ for the cheapest multi-chip version.

In conclusion to these substantiating facts, the platform architecture with three FPGA chips implements the most suitable trigger system for the CBM experiment. This approach has furthermore even more space for optimizations due to the possibility to utilize additional successive improvement algorithms. Since the need for a restriction to the amount of computational units will universally arise for experiments when dealing with such enormous data rates, the proposed implementation yields evidentially excellent results in speed as well as in performance.

Contents

List of Figures

List of Tables

Chapter 1

Introduction

This chapter presents an introduction into the subject of this thesis. For this purpose, the historical background of the Compressed Baryonic Matter (CBM) experiment is exhibited, before the demanding physics are introduced in detail. As furthermore this thesis examines just a small but important part of this experiment, the subsequent section features precise information about the measurement mechanism, while the final section informs about the accurate definition of a trigger, which has to be developed.

1.1 Historical Background of the CBM experiment

The idea, that all matter is composed of elementary particles, can be backdated to at least the 6th century Before Christ. In these days, the philosophical doctrine of atomism and the nature of elementary particles were founded in abstract reasoning rather than empirical observation by ancient Greek philosophers such as Leucippus, Democritus and Epicurus. Since that time, these subjects are further cogitated by human cultures until today. For example, ancient Indian philosophers such as Kanada, Dignāga and Dharmakirti and medieval scientists such as Alhazen, Avicenna and Algazel are engaged in these topics as well as early modern European physicists such as Pierre Gassendi, Robert Boyle and Sir Isaac Newton. But it lasts till the 19th century until the name atom, which represents since then a single and unique type of particle composing each element in nature, is born. Although this idea was inferred by John Dalton from his recent work on stoichiometry, the roots of the word atom originate from the Greek word 'atomos', which means 'indivisible'. However in the early 20th century, Otto Hahn, Fritz Straßmann, Lise Meitner and Otto Robert Frisch discovered and explained the nuclear fission, which is an absolute proof that atoms are not the fundamental particles of nature, but conglomerates of even smaller ones.

Besides this, the discovery and explanation of the nuclear fusion in different disciplines by Ernest Rutherford, 1. Baron Rutherford of Nelson, George Anthony Gamow, Sir Marcus Laurence Elwin Oliphant and Hans Albrecht Bethe lead to an active industry of generating one atom from another, which enables even the not profitable nuclear transmutation of lead into gold. Moreover the invention of bubble chambers and spark chambers in the 1950s enables scattering experiments to exhibit a large and ever-growing number of strongly interacting particles, which are named hadrons. As these experiments result in the so-called 'particle zoo', it suggests further that such an amount of particles can not be fundamental at all. Hence the search for a particle classification induces the introduction of the three flavors up, down and strange for smaller particles called quarks inside the hadrons. This first model of these smaller particles, which was independently proposed by Murray Gell-Mann and George Zweig in the 1960s, is then adapted in the following years to the recent model, which contains six types of quarks known as up, down, strange, charm, top and bottom. In addition to the flavor, quarks have also intrinsic properties, which include electric charge, color charge, spin and mass. Furthermore these particles are the only elementary particles in the Standard Model of particle physics (SM), which is introduced during the 1970s, experiencing all four fundamental interactions or forces, which are determined by gravitation, electromagnetism, weak interaction and strong interaction. As this SM explains further the large number of particles in the 'particle zoo' to be combinations of a relatively small number of fundamental particles, it contains also force carrying particles called gluons, which mediate the strong interaction. Consequently it has to be finally mentioned here that the theory of the strong interaction, which specifies the interaction of quarks and gluons to form hadrons, is described by the Quantum ChromoDynamics (QCD).

Since the entire mentioned physics are about particles, the discipline is of course called particle physics and studies the elementary constituents of matter and radiation in combination with their interactions. In contrast to the early years, in which the focus has been restricted to molecules, atoms and nucleons, it is in todays scientific work changed to the elementary particles. In modern particle physics, the experimental examination of physical models is mainly realized by particle accelerators and detector systems, because they offer the possibility to produce and measure energetic particle collisions up to the range of multiple GeV or TeV. For this purpose, the accelerators use electric fields to propel ions or charged subatomic particles to high speeds, which are even close to the velocity of light, and to contain them in well-defined beams. Due to the particle collisions, the newly created particles and interactions, which do often not occur in nature under normal circumstances, can be then measured by detector systems and thus analyzed and identified. As such experiments run at very high energies, it is nowadays also referred to as high energy physics. However it has to be noticed here that heavy-ion physics experiments at high energies are also included in this part of physics. Besides this, the information, which is gained from such experiments, is important for the understanding of the composition of today's world and also the early universe in an extremely hot phase where particle interactions occurred quite different than known today. Hence the biggest particle accelerators are applied in basic research, while their importance in medicine and other fields of industry is growing.

Coming now to more details about particle accelerators, it is interesting that there are principally only two basic types, which are characterized and named by linear and circular. Further on, both types can be applied in just two different experimental setups called collider, which involves directed beams against each other, and fixed-target, which implies only a single beam aligned to a target.
Besides this, a very brief description of a linear accelerator is defined by the application of linear arrays of plates or drift tubes, which offer an alternating high energy field, to accelerate particles in a straight line with a target of interest at one end.
In contrast to this, circular accelerators use electromagnets to enforce particles to move in an accelerating circle until they reach sufficient energy. When comparing both types now, the circular accelerator combines the advantage of being smaller at comparable power with the disadvantage of emitting the so-called synchrotron radiation, which limits the functionality, because the energy loss caused by the radiation can be as big as the added energy by the acceleration.

The earliest circular accelerators are determined by cyclotrons, which are invented in 1929 by Ernest Orlando Lawrence and at first built with a dimension of 9 inch at the University of California's Lawrence Berkeley National Laboratory also known today as Berkeley Radiation Laboratory. These cyclotrons accelerate particles by the application of an electric field with constant frequency to kinetic energies in the range of MeV, while using a single, large, uniform and constant dipole magnet to bend their path into a circular orbit, which obviously increases as faster the particle becomes. So the functionality is easily defined by continuously injecting the particles in the center

of the magnet and extracting them at the outer edge where they feature their maximum energy and velocity. Although there are possible modifications due to either the electric or magnetic field, the accelerated particles are still limited in their energy, which leads thus to the next generation of circular particle accelerators.
These ones keep the particles in a ring of constant radius by synchronizing the bending power of the magnetic field to the frequency-varying electric field, which is responsible for the acceleration. An immediate advantage of these circular accelerators, which are called synchrotrons, is constituted by the circumstance that the narrow beam pipe can be surrounded by much smaller and more tightly focusing magnets. Moreover such accelerator types are commonly established by the usage of some straight acceleration sections between some bending sections, which result thus in a round polygon shape instead of a torus. So compared to the cyclotrons, the synchrotrons cannot accelerate particles continuously, but must operate cyclically by supplying particles in bunches.
More complex and modern synchrotrons deliver these particle bunches even further into storage rings, which are realized by magnets with a constant field, because the particles can continue to orbit there for long periods until experimentation or further acceleration. The highest-energy machines are actually accelerator complexes consisting of cascaded specialized elements, which include linear accelerators or high voltage power supplies like the Cockcroft-Walton generator for initial beam creation, one or more low energy synchrotrons to reach intermediate energy, storage rings to accumulate or cool beams and a last large ring for final acceleration and experimentation.

Beyond that, a variety of such accelerators exist due to their related duty. But as each Cathode Ray Tube (CRT) of old televisions or old computer monitors contains a linear accelerator, the only one, which should be noticed here, is the longest one in the world. The so-called 'SLAC Linac' is located at the Stanford Linear accelerator center since 1966, is 3 km long and is able to accelerate electron-positron particles to kinetic energies of 50 Gev.
Moreover, one of the first large synchrotrons is determined by the Bevatron, which is operating at the Berkeley Radiation Laboratory since 1954. It is specifically designed to accelerate protons to sufficient energy in the range of 6.2 GeV to create anti-protons and verify the particle-antiparticle symmetry of nature.
Another interesting synchrotron is the Alternating Gradient Synchrotron (AGS), which is located at the Brookhaven National Laboratory in Long Island, New York, because it is the first large synchrotron with alternating gradient and strong-focusing magnets. This innovative breakthrough con-

cept in accelerator design allows scientists to accelerate protons to kinetic energies of 33 GeV, while the beam aperture is reduced and therefore the size and costs of the bending magnets.
The first major European particle accelerator is the Proton Synchrotron, which is in general similar to the AGS and operates at the Conseil Européen pour la Recherche Nucléaire (CERN) since 1959. Actually this one is used to feed for example the Super Proton Synchrotron (SPS) and the Large Hadron Collider (LHC) with protons of a kinetic energy about 28 GeV.
For a quite long time, the Tevatron, which operates at the Fermi National Accelerator Laboratory in Batavia near Chicago, Illinois since the early 1990s, has been the highest energy synchrotron in the world. It is able to accelerate protons and anti-protons to kinetic energies slightly less than 1 TeV and collides them together to further realize observations in the Collider Detector at Fermilab (CDF) experiment and the DØ experiment.
But in recent days, the LHC, which is built at the CERN, has roughly seven times more energy. It is housed in the 27 km tunnel, which has formerly housed the Large Electron-positron Collider (LEP) collider. So the LEP experiments, which are Apparatus for LEP PHysics (ALEPH), DEtector with Lepton, Photon and Hadron Identification (DELPHI), Omni-Purpose Apparatus for Lep (OPAL) and L3, are replaced by the new LHC experiments, which are A Large Ion Collider Experiment (ALICE), A Toroidal LHC ApparatuS (ATLAS), Compact Muon Solenoid (CMS), Large Hadron Collider beauty (LHCb) and TOTal Elastic and diffractive cross section Measurement (TOTEM). For these experiments, it is interesting that some of them observe hadrons, while others analyze ion collisions.

Besides these particle accelerators, there are other noticeable ones like the Relativistic Heavy Ion Collider (RHIC) and the Hadron Elektron Ring Anlage (HERA). The RHIC, which is able to collide hadrons as well as ions with kinetic energies of up to 100 GeV, is operating at the Brookhaven National Laboratory in Long Island, New York since the year 2000 and employs a circular ring of 3.8 km for particle acceleration. At this facility, there are five experiments established, which are named Solenoidal Tracker at RHIC (STAR) and Pioneering High Energy Nuclear Interactions eXperiment (PHENIX) in conjunction with the less known Broad RAnge Hadron Magnetic Spectrometers (BRAHMS), Phobos and pp2pp.
The HERA with its four experiments H1, ZEUS, HERMES and Hadron Electron Ring Accelerator B hadrons (HERAB) is located at the Deutsches Elektron SYnchrotron (DESY) in Hamburg, Germany since the early 1990s. It utilizes a circular ring of 6.3 km to collide electrons and protons with kinetic energies of 27.5 GeV for the electron and 920 GeV for the proton.

The last but not least accelerator complex mentioned here is the Facility for Antiprotons and Ions Research (FAIR), which is actually under construction and located at the Gesellschaft für Schwerionenforschung (GSI) in Darmstadt, Germany. This international accelerator facility of the next generation adopts the already established UNIversal Linear ACcelerator (UNILAC) and Schwer-Ionen-Synchrotron (SIS) and combines them with a new double-ring synchrotron, some new storage rings and a new Super-Fragment-Separator to feature experiments, which offer novel insights into the structure of matter and the evolution of the universe as well as possibilities to work in other practical disciplines. Therefore the accelerator should provide a high energetic heavy ion beam to offer an observation possibility for the synthesis of heavy elements and the strong interaction between elementary particles. Aside from this, the creation of Quark-Gluon Plasma (QGP) should enable the analysis of a state of matter, which has existed in a very short span of time after the big bang. In addition to this, slowed down anti-protons may draw conclusions about symmetry violations between the laws of nature of our world and an antimatter world, because they offer a totally new field of research, which is given by the exact determination of anti-atoms.

As almost all of these particle accelerators and experiments are setup during the last 60 years, it is obvious that many developments in this discipline as well as in the fields of detectors and computing technologies have made life easier or have offered the possibility for research in new areas. At the same time, the data rates of the experiments have generally increased to permit the investigation of rare and thus interesting interactions of matter, which are covered by a huge background of well-known physics. For this purpose, the purely manual study of event images in the early experiments has changed to on-line electronics, which offer a data reduction in many orders of magnitude, before the information is recorded on a mass storage. In addition to this, detectors have changed from bubble chambers to gaseous and and solid-state electronic detectors. Moreover the applied computational electronics have changed from large mainframes to small modern desktops or special integrated circuits offering a computational power, which is several orders of magnitude higher. So as the data analysis techniques in high-energy physics experiments has been performed over the years in a continuously changing environment, the applied methods have evolved accordingly to cope with these changes.

In the following all presented concepts are developed to be suitable for the CBM experiment, which allows the study of strongly interacting matter at high baryon densities where the QCD phase diagram is poorly known. In this experiment the applied detector concept is new to heavy-ion physics, because all charged particles as well as secondary vertices from heavy-flavor decays are exclusively reconstructed in a high-performance Silicon Tracking System (STS), which will be installed in a magnetic dipole field between the fixed target and further detection systems for particle identification and calorimetry. In addition to this, the high track densities and high collision rates require the application of most advanced silicon detectors. The technological challenges include high position resolution in thinnest possible hybrid pixel and micro-strip sensors combined with extreme radiation hardness, fast self-triggered read-out and ultra low-mass mechanical supports.

1.2 CBM Physics

The CBM experiment (see [CBM04], [CBM05], [CBM07] and [HMS$^+$05b]) is a fixed-target heavy-ion experiment, which is planned at the SIS of the FAIR, to study nucleus-nucleus collisions for the further exploration of the QCD phase diagram, which is depicted in figure 1.1. For this purpose, the SIS will provide high intensity beams of protons and antiprotons, nuclei up to uranium as well as radioactive beams.

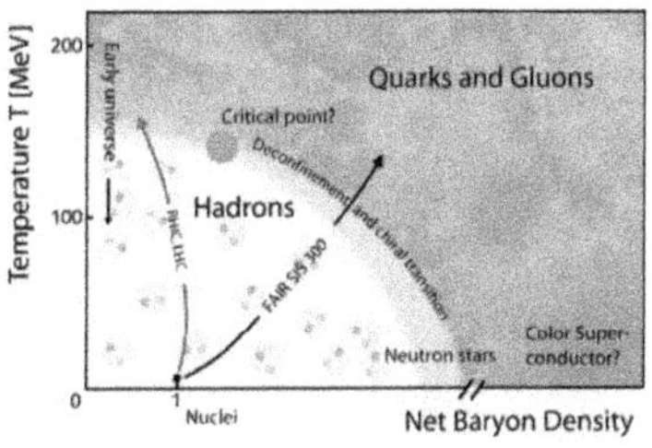

(a) Source: [CBM08]

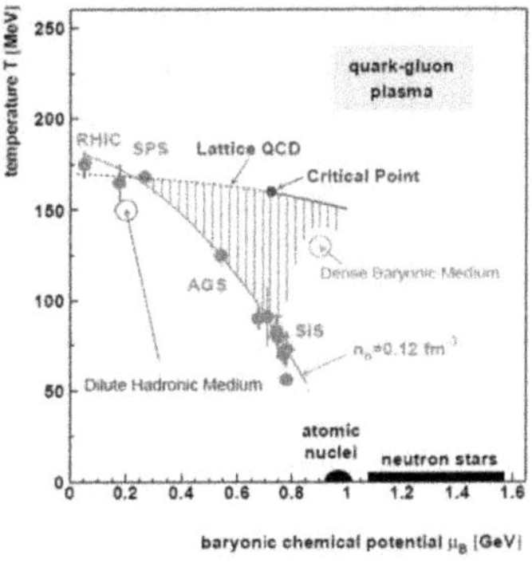

(b) Source: [Sen02]

Figure 1.1: Phase diagram of strongly interacting matter

The key to the physics of the CBM experiment is identified by some measurements and Lattice Quantum ChromoDynamics (LQCD) calculations, which predict that high energy densities cause quarks and gluons to be not anymore confined in ordinary hadrons, but form a phase called QGP. Furthermore this phase defines a state of matter, which was the substance of the Universe in the first microseconds after the Big Bang and is considered to exist in the interior of neutron stars. For that reason, the CBM experiment will address such predictions based on precision, large acceptance and high-statistics measurements of 'common' hadrons (π, K, p) as well as of so-called 'rare probes', like low-mass di-leptons (ρ, ϕ), charmed hadrons (D mesons, J/ψ) or multi-strange baryons (Ξ, Ω). Besides this, it has to be noticed that many of these observables will be measured for the first time in heavy-ion collisions at these energies.

The main goal of the CBM research program is defined by the simultaneous measurement of observables, which are sensitive to high density effects and phase transitions, because these measurements are considered to provide indications on the chiral symmetry restoration at high baryon densities ρ_B. For that purpose, in-medium modifications of hadrons in dense matter, the deconfinement phase transition at high baryon densities ρ_B and the equation of state of strongly interacting matter at high baryon densities ρ_B is taken into account as well as the first order phase transition and its critical endpoint as direct evidence for a phase boundary. In this connection, the major challenge is obviously determined by the discovery of diagnostic probes, which are connected to chiral symmetry restoration and to the deconfinement phase transition. The observation of in-medium modifications of hadrons might be a signature for the onset of chiral symmetry restoration. The in-medium spectral function of short-lived vector mesons (ρ, ω, φ) can be measured directly via their decay into di-lepton pairs. Since leptons are essentially unaffected by the passage through the high-density matter, they provide, as a penetrating probe, almost undistorted information on the conditions in the interior of the collision zone. The anomalous suppression of charmonium due to screening effects in the QGP was predicted to be an experimental signal of the QGP. Particles containing heavy quarks like charm are produced in the early stage of the collision.
At FAIR, open and hidden charm production will be studied at beam energies close to the kinematic threshold, while the production mechanisms of D, J/ψ, ψ' and Λ_c mesons and baryons will be sensitive to the conditions inside the early fireball. One of the early predictions for a QGP signal was the increased production of strangeness in the deconfined phase resulting in an enhanced yield of strange particles after hadronization. This effect was expected to be even more pronounced for multi-strange hyperons (Λ, Ξ and Ω).
Recently, data on the excitation function of strangeness production measured

by NA49 (see [Thee]) have revived the discussion on the role of strangeness as a signature for a deconfinement phase transition. Many signals for the the QGP have been proposed and are still under discussion, even though the hope for the discovery of a 'smoking gun' has not yet become true. Furthermore the detection of the critical endpoint of the deconfinement phase transition would be such a direct indication for the existence of a new phase. Theoretical investigations suggest that particle density fluctuations occur in the vicinity of the critical endpoint, which might be observed experimentally as non-statistical event-by-event fluctuations of observables.

Coming now to more numerical facts, the dense matter, which should be analyzed by the CBM experiment, will be created in collisions of intense heavy-ion beams extracted from SIS-100/SIS-300 at energies up to 90 GeV for protons, 45 AGeV for light ions and 35 AGeV for ions as heavy as gold and uranium with nuclear targets at up to $10^7 \frac{\text{interactions}}{\text{s}}$. Within these conditions, the explored region of the phase diagram corresponds to the intervals in temperatures of T $\backsimeq$ 100 - 160 MeV and baryon-chemical potential $\mu_b \backsimeq$ 400 - 600 MeV. So the CBM experiment will explore the phase diagram at high baryon densities, which determines a complementary region compared to the experiments at the AGS, the SPS and the RHIC with collision systems at high temperatures and low baryon-chemical potentials. However future experiments at the LHC will probe collisions at even higher temperatures and lower baryon-chemical potentials.

Beyond that, the therefor required CBM detector is optimized for the detection of rare and penetrating probes as open charm and di-lepton decays of low-mass vector mesons. Those are important observables for the initial energetic and dense phase of nuclear collisions. The experiment combines a STS, which is installed in a compact magnetic dipole field directly behind the target, for exclusive and precise charged-particle tracking and vertexing in high-multiplicity events with detectors for particle identification (Ring Imaging CHerenkov (RICH), Transition Radiation Detector (TRD)), Time Of Flight (TOF) measurement based on Resistive Plate Chambers (RPCs) and Electromagnetic CALorimeter (ECAL) downstream. A muon identification system (Muon Chamber (MUCH)) and other specific detector configurations are under study. More detailed information about each detector subsystem can be found in the following section 1.3 on page 10. More specific information can be found in [HMS+05b], [HMS+05c], [And06], [Sen06], [CBM08], [Ht08], [Fri06a], [Cha08], [Sen02], [Höh07], [Sen07], [CBM04], [CBM05], [GSI06], [CBM07] and [Fri06c].

1.3 CBM Detector System

The CBM detector system requires the adoption of a novel concept for charged particle tracking and vertexing because of the complex collision environment, which is necessary for the physics program. This concept is realized with a STS of unprecedented performance concerning rate capabilities and radiation hardness, which is similar to a system used in the NA60 experiment at the SPS (see [Thef] and [HBD+06]).

The measurement of open charm is one of the prime interests in the CBM experiment and at the same time one of the most difficult tasks. So the benchmark for the performance of the STS is the reconstruction of D-mesons through their hadronic decays $D^0 \longrightarrow K^-\pi^+$ ($c\tau \approx 124\,\mu$m) and $D^\pm \longrightarrow K^\mp\pi^\pm\pi^\pm$ ($c\tau \approx 317\,\mu$m), which have to be identified by the measurement of displaced vertices with an accuracy of about $50\,\mu$m along the beam axis to suppress the combinatorial background of prompt pions and kaons. (see [HMS+05c]).
Further on, the study of the very different CBM observables calls for a modular setup and different running conditions. For instance, the measurement of Jψ mesons via their electron-positron decay has three requirements. The first one is a high reaction rate of up to $10^7 \frac{\text{events}}{\text{second}}$ to compensate for the low production cross sections at the threshold. The next one is an on-line event selection to reduce the data flow down to an archiving rate of 25 kHz. And the final requirement is a detector system for electron identification and pion suppression (electron to pion ratio of 10^{-4} required).
Moreover the measurement of low-mass vector mesons requires the rejection of close pairs (i.e. the detection of soft electrons) to reduce the combinatorial background caused by electrons/positrons from π^0 Dalitz decays and gamma conversions. Further on, the study of observables event-by-event requires particle identification and a large acceptance, which does not vary with beam energy (see [Sen06]).
Coming now to the point, the experimental setup has to fulfill all following requirements (see [CBM04]):

- Identification of electrons, which require a pion suppression factor in the order of 10^4
- Identification of hadrons with large acceptance
- Determination of primary and secondary vertices (accuracy $\approx 50\,\mu$m)
- High granularity of the detectors

- Fast detector response and read-out
- Very small detector dead time
- High-speed trigger and data acquisition
- Radiation hard detectors and electronics
- Tolerance towards delta-electrons

So with regard to these issues, figure 1.2 depicts schematically the proposed CBM detector system, which is able to cover the needs.

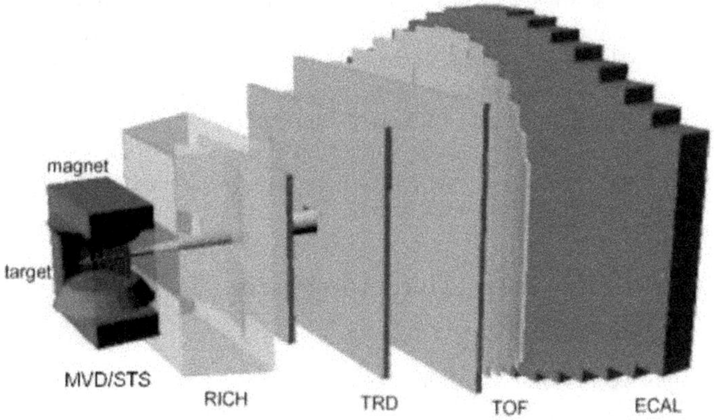

Figure 1.2: Sketch of the proposed experimental setup for CBM (Source: [GSM05])

Since the CBM experiment is a fixed-target one, it is clear that the beam enters this experimental setup from the left hand side, because the target is located there. Hence the beam hits at first upon the CBM target at the entrance of the dipole magnet (red). Afterwards the collided particles move through different sequenced detector systems, which begin with the Micro Vertex Detector (MVD) system and the STS (**blue**) inside the magnetic field. These systems are then followed by a RICH detector system (yellow), a TRD system (green), a TOF detector system (turquoise) and an ECAL detector system (violet). For that reason, the total length of the entire detector setup is about 12 m. More information about this setup can be found in [Heu06].

Besides this, it is important to realize that all of these diverse detector systems are really necessary, because each one features the measurement of separate information. For instance, the STS performs charged particle tracking, high-resolution momentum measurement and vertexing combined with the MVD system. In addition to this, the RICH detector system offers identification of electron pairs from low-mass vector meson decays, and the TRD system provides charged particle tracking as well as identification of high energy electrons. Further on, Hadron identification is realized in the TOF detector system, while identification and measurement of the energies of electrons and photons is exhibited by the ECAL detector system.

Although these systems are able to measure sufficient information, alternative approaches are under study anyway. For example, the electron pair measurement and the feasibility of detecting vector mesons via their decays into muon pairs is possibly implemented by a MUCH detector system. So if the RICH detector system is replaced by an active absorber, muons would be identified in a sandwich of absorber and detection layers (see [Heu07]). More ongoing information can be found in [CBM04], [CBM05], [GSI06] and [CBM07]. Nevertheless the subsequent sections present a very brief description of each introduced detector system.

1.3.1 Micro Vertex Detector System & Silicon Tracking System

The combined MVD and STS appoints the central component of the CBM experiment, because its task is determined by the reconstruction of the motion trajectories for the charged particles, which are created in a central heavy-ion collision. The complexity of this task can be simply imagined by a closer look to the primary and secondary vertex resolution of about $30\,\mu$m as well as the momentum resolution of around $\frac{\Delta p}{p} \approx 1\,\%$. Furthermore it has to be mentioned that such a precision is needed, because it dictates a prerequisite for high-resolution mass measurements. Thus a thin detection system is required, which combines high-resolution space point measurement of about $10\,\mu$m with read-out rates of about 25 ns to accommodate reaction rates of 10 MHz. In this connection, a benchmark observable is defined by the D meson ('open charm'), which has to be identified via its hadronic decays $D^0 \rightarrow K^-\pi^+$ and $D^\pm \rightarrow K^\mp\pi^\pm\pi^\pm$ in typically around several $100\,\mu$m distance from the collision vertex. For that reason, the presence of the MVD in addition to the STS is also immediately clear, because it offers a particularly high position resolution with a very low material budget, which is below $0.3\,\%$ radiation length per layer to reduce multiple scattering. In addition to this, the MVD possesses a high radiation tolerance up to a dose of 50 MRad, which corresponds to the dose accumulated in ten years of running. Coming now to the point, the summarized requirements are fixed by:

- Material budget below 0.3 % of the radiation length per station to reduce multiple scattering
- Hit resolution of about 10 μm for the pixel sensors to achieve a vertex resolution of about 50 μm along the beam axis
- Radiation hardness of up to 50 MRad corresponding to the dose accumulated in ten years of running
- Read-out times of about 25 ns to accommodate reaction rates of 10 MHz

Further on, the challenge of this detector system is determined by the high track densities of hundreds of charged particles, which are exemplary caused by Au+Au collisions with 25 $\frac{\text{GeV}}{\text{nucleon}}$. To get a simple imagination about this challenge, figure 1.3 depicts the occurring track densities by the visualization of such a typical collision.

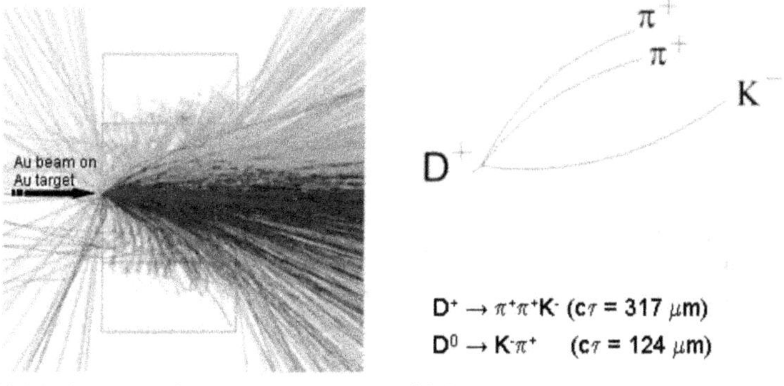

(a) Simulated particle motion trajectories for the entire collision

(b) Simulated particle motion trajectories for a rare "open charm" decay

Figure 1.3: Illustration of the track densities in a central Au+Au collision at 25 $\frac{\text{GeV}}{\text{nucleon}}$ in the STS (Source: [Sil])

Furthermore a closer look to the left figure 1.3a shows the simulation result, while the right figure 1.3b illustrates a rare 'open charm' decay, which has to

be identified as detached vertex in about 100 μm distance from the primary vertex.
Nevertheless a detector setup, which is suggested to meet the requirements of the CBM experiment, is depicted in figure 1.4.

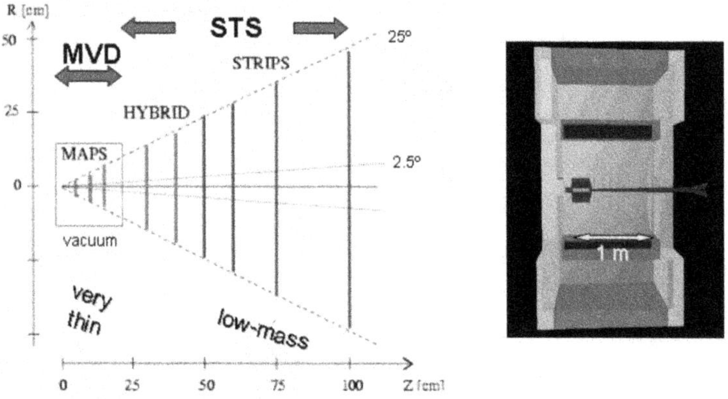

(a) Structure (b) Visualization

Figure 1.4: Layout of the MVD and STS (Source: [Sil])

So with regard to this figure, the proposed detector setup comprises actually nine detector stations covering laboratory polar angles from 2.5° to 25°. Moreover all stations are arranged in the 1 m long gap of a superconducting dipole magnet with 1 Tm bending power surrounding the beam pipe. In the innermost region, three pixel detector stations with a very high spatial resolution and an active area of about 1.5 m^2 are actually used. These stations form the MVD and are located at 5 cm, 10 cm and 20 cm downstream of the target inside a vacuum vessel. In this connection, it should be mentioned that a different detector setup, which applies only two pixel stations, is also considered. Further on, the successive six stations form the STS, which defines the largest part of the tracking setup. These stations are located at 30 cm, 40 cm, 50 cm, 60 cm, 75 cm and 100 cm downstream of the target inside the dipole magnet but outside of the vacuum vessel. Actually the first two stations are hybrid pixel stations, while the last four are built from thin double-sided micro-strip detectors, which are possibly complemented by one or two more hybrid pixel stations. At this juncture, it should be also mentioned that different setups are considered, which apply more micro-strip stations with or without hybrid pixel stations. Furthermore another remarkable setup is determined by the usage of two very close and rotated double-sided micro-strip planes, which build then a single detector station.

Coming now to the direct implementation of the pixel stations, which are used for the MVD system, it is quite hard to ascertain that current technologies combine only two out of four performance requirements, which are essential to CBM. So the actual selection can be made between technologies, which support either a very small pixel size and a very low total thickness (Complementary Metal Oxide Semiconductor (CMOS) sensors, Monolithic Active Pixel Sensors (MAPSs)) or a fast read-out and radiation hardness (hybrid pixel detectors). For that reason, the development of a pixel detector prototype has been started, which bases on the MAPS technology, because its already excellent position resolution of $1.5\,\mu$m and the small thickness of ultimately around $50\,\mu$m is very attractive. In this connection, a particular challenge is faced by the development of a very thin module structure, which includes cooling for the placement in the vacuum vessel to make use of those advantages. Furthermore the R&D on the sensors focuses on an improvement of the radiation tolerance to $10^{13}\,\mathrm{n_{equiv.}}$, while achieving the read-out of a sufficiently large array within $10\,\mu$s. However, the harsh radiation conditions at the first pixel station, which is $4 \cdot 10^{15}\,\mathrm{n_{equiv.}}$ for one year of $10^{7}\,\frac{\mathrm{interactions}}{\mathrm{s}}$, suggests to apply MAPSs only at lower beam intensities in the initial phase of CBM. Hence alternative sensor technologies like new developments of thin hybrid pixel or SOI-monolithic pixel detectors with small sensor cells have to be considered for the future.

Focusing yet the implementation of the STS, there are optionally two generic hybrid pixel detector stations followed by six micro-strip stations.
Further on the hybrid pixel stations are generally segmented into squared pixels of $50\,\mu$m yielding a spatial resolution of about $15\,\mu$m, while the corresponding sensors and matching read-out electronics are placed on two separate chips, which are interconnected pixel by pixel through microscopic solder balls. Even though hybrid pixel detectors are relatively thick and require presumably active cooling in the acceptance, the stations contribute in the current STS concept with unambiguous space points to the tracking in regions where the track densities are high.
Besides this, the micro-strip stations are built from sensor modules with $50\,\mu$m strip pitch in combination with a 15° stereo angle. Furthermore the modules base on a ladder structure of double-sided micro-strip sensors, which are read out at the periphery of a station. Thus this approach avoids material budget from the read-out of hybrid pixel stations to cooling lines and so on

in the acceptance of the tracking system. So this detector concept has been assessed in terms of sensor occupancies, tracking efficiencies with momentum resolution and the vertex detection capabilities.

Finally it has to be noticed that simulations are taken into account to optimize the number of detector stations, their positions, their layout and their material budget with regard to efficient tracking, high-resolution momentum and vertex reconstruction. Moreover it is obvious that these simulations are closely linked to the detector R&D activities [HMS+05b], [HMS05a]. More information can be found in [Heu07], [Sen06], [HDMS06] and [CBM08].

Monolithic Active Pixel Sensor Stations

The very thin and highly-granulated detector stations of the MVD, which are located in the upstream section close to the target, are uniquely composed of MAPSs featuring a pixel size of $40 \cdot 40\,\mu m^2$. In addition to this, these MAPSs have the the properties of being radiation hard in the dimension of some MRad (see [DBD+05]), being fast in the order of (10 μs and very thin in the scale of 150 μm. Moreover these silicon pixel sensors, which offer also a high position resolution of typically 5 μm, are combined with the individual preamplifiers on the same CMOS chip. For that reason, this technology determines the best choice to fulfill the essential prerequisites for the identification of D mesons, which exhibit typically a decay length of 100 μm next to the target. So the D mesons could be only reconstructed with regard to the very high combinatorics, if the trajectory of the decay products can be identified by a common secondary vertex. Hence the resolution of this vertex reconstruction along the beam line must be much smaller than the decay length of the D mesons. Regarding now all these issues, table 1.1 summarizes now the prerequisites of the MVD and the properties of the MAPSs (see [AYBD+]).

	Prerequisite	MAPS
Vertex resolution	$< 50\,\mu$m	-
MIPS detection eff.	-	$>> 99\,\%$
Position resolution	$< 5\,\mu$m	$1.5\,\mu$m - $5\,\mu$m
Material budget	few $0.1\,\%$ x_0 (few $100\,\mu$m $si_{equivalent}$)	-
Read-out	$< 10\,\mu$s	test: $20\,\mu$s
Radiation hardness	$> 10^{13}\,\frac{\text{n}}{\text{cm}^2}$, 2.5 MRad	$\sim 10^{12}\,\frac{\text{n}}{\text{cm}^2}$
Thickness	-	$120\,\mu$m (test: $50\,\mu$m)
Sensor area	-	$3.5\,\text{cm}^2$ (reticle size)

Table 1.1: Comparing outline for the prerequisites of the MVD and the properties of the MAPS

With regard to this table, it is obvious that not all prerequisites can be satisfied so far. However this circumstance is negligible here, because the MAPS technology is still under development. More information about the MAPS can be found in [BCC+], [GCC+05], [Heu07] and [CBM08].

Hybrid pixel Stations

The first and major important issue concerning the hybrid pixel detector stations is certainly determined by the fact that they have no detailed layout yet. Hence the simulation treats them simply like generic silicon discs, which are segmented into a pixel grid. So each active pixel creates just a hit, which holds the coordinates equal to the pixel center. Figure 1.5 depicts an example for such a hit generation. Since the x and y coordinate of a hit are set according to the pixel center of the grid, it is immediately clear that the z coordinate of the hit corresponds simply to the mid-plane of the station. Thus the errors in x and y are defined by the respective pixel size divided by sqrt(12), while the error in z is zero. However there is obviously no covariance between the errors.

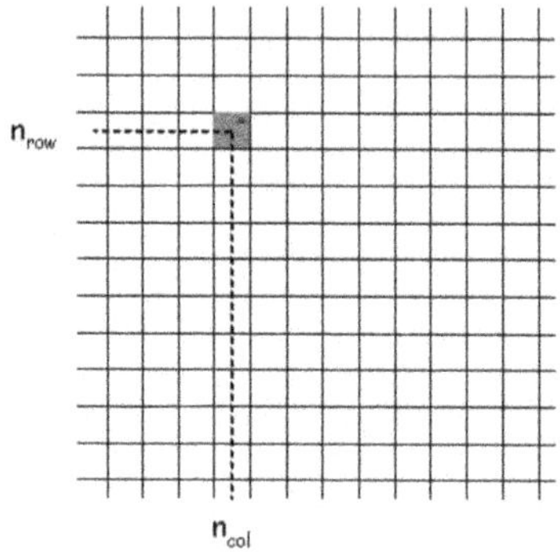

Figure 1.5: Schematic for the generation of a hit representation in a hybrid pixel detector station (Source: [Det])

Moreover the applicable detector station thicknesses range from 200 μm up to 1,600 μm with 680 sectors and 885,760,000 channels. Coming now to the visualization of an example, figure 1.6 shows the distribution of the hits in such a hybrid pixel station during a simulation of a central Pb+Pb event at 25 AGeV with a pitch of $50 \cdot 50\,\mu m^2$.

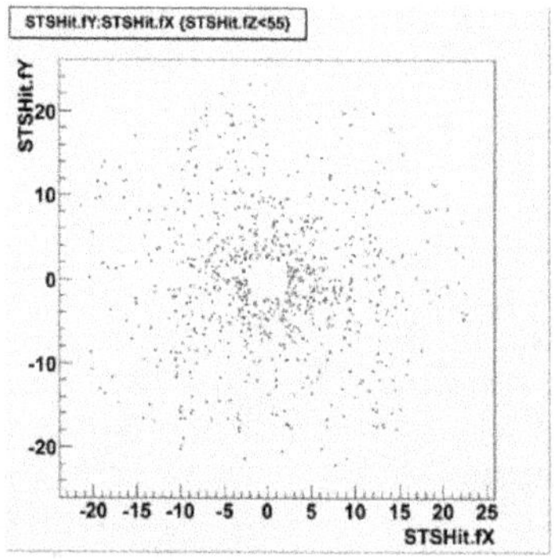

Figure 1.6: Occupancy of a hybrid pixel detector station due to the simulation of a central Pb+Pb event at 25 AGeV: $50 \cdot 50\,\mu m^2$ pitch, 830 active channels $\Rightarrow$ 830 hits

Micro-strip Stations

As the aim of this type is defined by a low-mass detector, the double-sided rotated micro-strip stations base on modules, which are made from several sectors of micro-strip sensors on the front side and the back side. Furthermore each station is planned to be read out in individual segments of single or several chained sensors through thin and flat multi-line cables, which route the analog signals to the electronics at the periphery where all services including cooling can be provided. So based on this concept, the hit coordinates in x and y are defined by the center of the overlap area of the front side and the rotated back side micro-strip. Hence this obviously resulting parallelogram, which is shown in figure 1.7a, narrows the measurement of the correct hit from a simple and commonly long micro-strip to a smaller and thus more applicable area. However the rotation of the back side micro-strips entails a not negligible problem, because if two hits occur almost close to each other, the two correct ones could be impossibly found out of the four existing ones due to the overlapping area of the active micro-strips. Furthermore a closer look to the visualization of figure 1.7b shows the reason for that behavior, because the circumstance that the active micro-strips of two very close correct hits, which are identified by blue dots, generate two extra hits indicated by red dots via the active yellow and green micro-strips. These two additional hits are so-called fake hits due to the fact that they do not correspond to real hits of the experiment. So based on this context, it is clear that the number of fake hits depends directly on the length of the micro-strips and the rotation angle between the front and the back side.

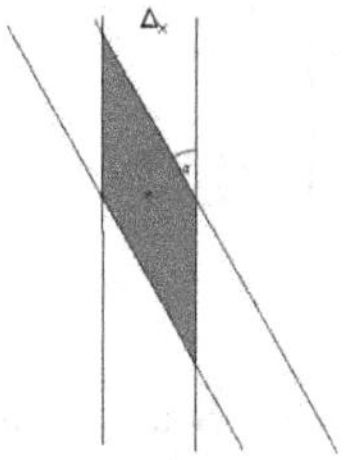

(a) Two active intersecting micro-strips

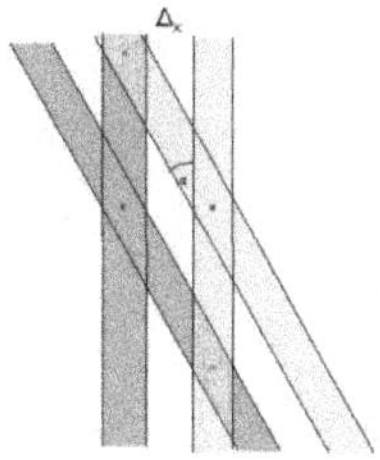

(b) Four active intersecting micro-strips

Figure 1.7: Schematic for the generation of a hit and a fake hit representation in a micro-strip detector station due to the overlapping area of active front and back side micro-strips

Since the x and y coordinate of a hit are set according to the center of the overlapping parallelogram, it is obvious that the z coordinate of the hit corresponds simply to the mid-plane of the station. Thus the errors in x and y are defined by the micro-strip pitch respectively divided by sqrt(12) and the absolute value of the tangent of the stereo angle, while the error in z is zero. Moreover it has to be noted that the errors in x and y are generally correlated.

Besides this, the first concept for the construction of such stations arranges the modules of a station 'vertically' with the micro-strip sensors oriented perpendicular to the main bending plane in the magnetic field. Furthermore as the track particle densities are evidently higher in the inner part around the beam axis, it is appropriate to use smaller modules there, which lead then to a flattened overall module occupancy below 2 % for a central Au+Au collision at 25 AGeV. Nevertheless it is clear that all modules base on the same construction scheme just adapting the micro-strip sensor length in combination with the number of read-out channels. So based on this concept, figure 1.8 illustrates the layout of such a station including the application of modules with different length.

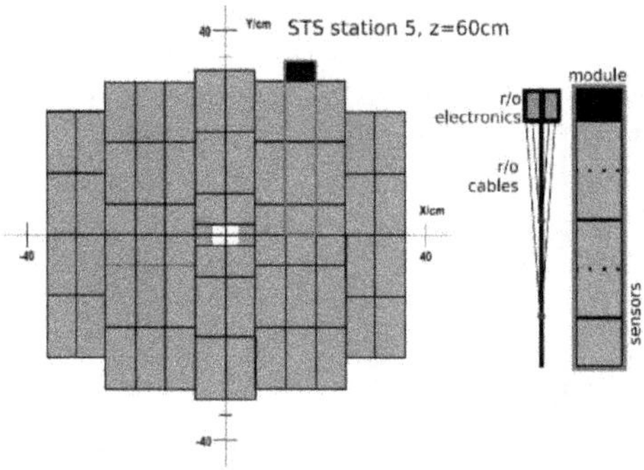

Figure 1.8: Layout of 'vertical' arranged micro-strips (Source: [Sil])

Coming now to the visualization of an example, figure 1.9 shows the distribution of the hits in such a micro-strip station during a simulation of a central Pb+Pb event at 25 AGeV with a pitch of 50 μm.

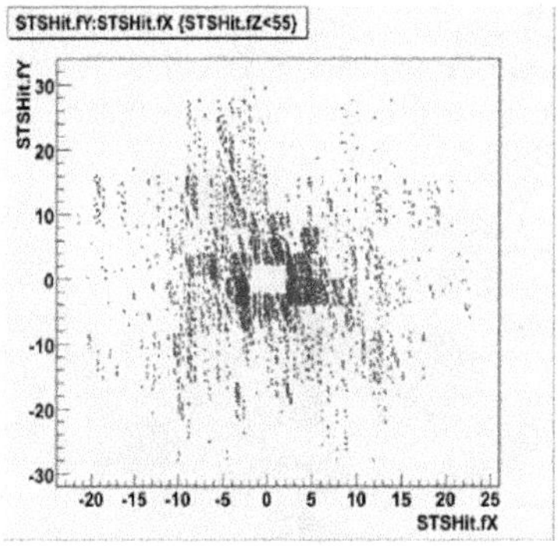

Figure 1.9: Occupancy of a 'vertical' oriented micro-strip detector station due to the simulation of a central Pb+Pb event at 25 AGeV: 50 μm pitch, $814 + 820$ active channels $\Rightarrow$ 5,762 hits

Beyond that, an alternative micro-strip station concept arranges the detector modules 'radially' in quadrants. These modules are then further divided into several sectors of different effective micro-strip sensor lengths consisting of either single or groups of chained sensors. Having now the first concept with the 'vertically' arranged modules in mind, the reason for this layout is immediately clear, because the micro-strips should be repeatedly shorter close to the beam line than at larger radii due to the track particle densities. So based on this second concept, figure 1.10 illustrates the layout of such a station including the application of modules with different size.

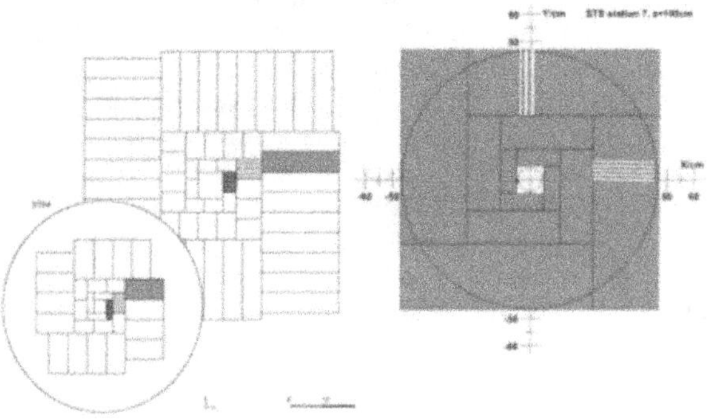

Figure 1.10: Layout of 'radial' arranged micro-strips (Source: [Det])

The visualization of the same example than used to give an imagination for the first concept leads now to figure 1.11, which shows the distribution of the hits in such a micro-strip station during a simulation of a central Pb+Pb event at 25 AGeV with a pitch of 50 μm.

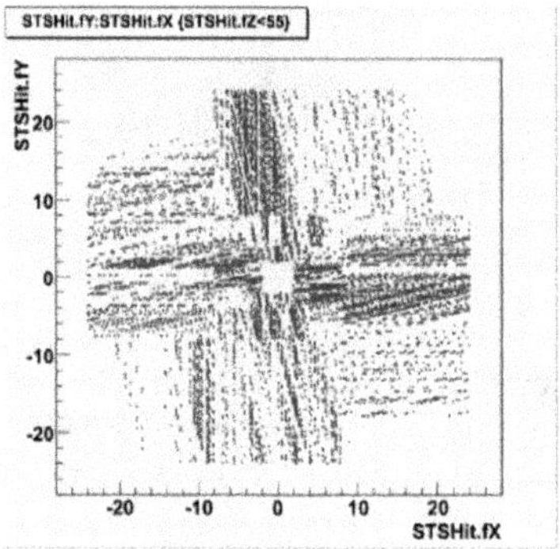

Figure 1.11: Occupancy of a 'radial' oriented micro-strip detector station due to the simulation of a central Pb+Pb event at 25 AGeV: 50 μm pitch, $822 + 827$ active channels $\Rightarrow$ 10,865 hits

With regard to the figures 1.9 and 1.11, an interesting circumstance can be directly deduced, because it is obvious that the arrangement of the modules effects the result significantly and is thus important. In addition to this, such an effect might be also caused by the different applied read-out strategies for the hits, which are generated for each combination of an active front and back side micro-strip crossing within a sector, because if the read-out is assumed at the bottom edge of the sector, the corresponding hit coordinates would be projected to this edge along the micro-strips. For this purpose, a visualization is used at first to get an easier access to both available read-out strategies. Therefore figure 1.12 depicts the common read-out strategy for the front side of the double-sided micro-strip stations including dotted lines indicating the read-out caused projection, while figure 1.12b and figure 1.12c shows respectively the actually used read-out strategy for the back side of the 'radial' and the 'vertical' version.

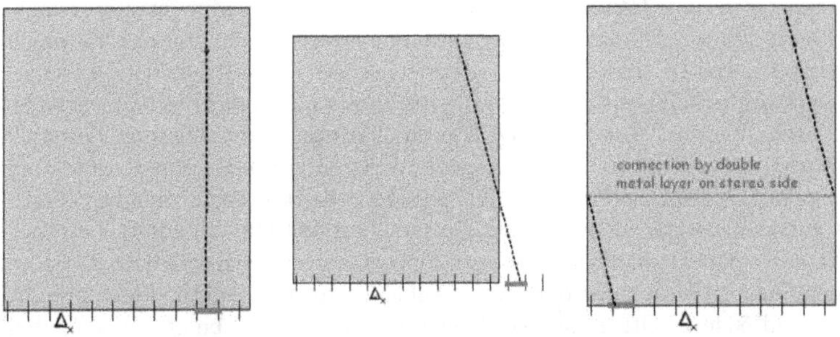

(a) Common read-out strategy for the front side of both versions

(b) Actual strategy for the back side of the 'radial' version

(c) Actual strategy for the back side of the 'vertical' version

Figure 1.12: Assembly scheme for the read-out of the micro-strip sector sides (Source: [Det])

So with regard to the figures 1.12a and 1.12b, the read-out strategy, which is actually applied for the 'radial' version, involves obviously a double-sided sensor with the hit coordinates vertically projected at the front side and with a stereo angle at the back side. Hence it is important to realize that the number of channels on the back side is larger than at the front side, because micro-strips, which end at the vertical edge of a sector, are also projected to the bottom edge. In contrast to this, figure 1.12c shows that the read-out strategy of the back side, which is actually used for the vertical version, just connects the corresponding micro-strips on each vertical edge of a sector horizontally by the usage of a double metal layer. Furthermore this transverse joint causes then the possibility for two active micro-strips to exhibit more than one crossing, which leads further evidently to the ability of an additional hit generation for each of these crossings. Thus this functionality introduces a new type of fake hits, which originates from a correct combination of front and back side micro-strip, but is vertically displaced from the correct hit coordinate. So within this context, it is obvious that a single read-out channel can correspond to several micro-strip segments, while the number of channels is consequently equal for the front and the back side, which features an easier chaining.

Coming now to more implementation specific data, the first micro-strip sensor prototypes have a thickness of $200\,\mu$m - $300\,\mu$m, are double-sided with AC-coupled micro-strips of $50\,\mu$m micro-strip pitch and feature a 15° stereo angle between its front and back side micro-strips. Moreover this stereo angle determines a compromise between the number of combinatorial hits, which are produced by the projective sensor topology in high-track density conditions, and a sufficiently high spatial resolution achieved in two coordinates. At this juncture, it is has to be noticed that even if the measurement error of a space point coordinate can be reduced by larger stereo angles, they yield also a larger fraction of fake hits. Besides this, the actual number of sectors is 680 , while the number of channels is 1,083,440 . In addition to this, actually planed micro-strip sensors foresee a micro-strip pitch of $25\,\mu$m in combination with three different possible micro-strip length, which are 20 , 40 and 60 mm. Other aspects of the sensor design will be of course related to its radiation hardness, the minimization of inactive area near its edges and the touch robustness, which may be advanced by the biasing technique and the layout of the guard rings structure. The development of the readout cables, which are made from aluminum traces on polyamide material for minimum material budget, is a particular important task, because it includes the realizable fabrication of fine-pitch traces matching the micro-strip pitch of the sensors over lengths of up to about 50 cm. Moreover the mechanical and electrical attachment to both sides of thin double-sided sensors have to be managed. Further on, the full functionality of the micro-strip detector modules require the total capacitance of sensor and cable at the input of the Front-End Electronic (FEE) to be as low as possible in order to achieve a sufficiently large signal-to-noise ratio of the measurements. More information can be found in [Heu07] and [CBM05].

1.3.2 Ring Imaging CHerenkov Detector System

The RICH detector is designed to provide the identification of electrons with a momentum below $8\,\frac{\text{GeV}}{\text{c}}$ in combination with a suppression factor of more than 500 for pions in the momentum range of electrons decayed from low-mass vector-mesons. Because of the high rate and high track density environment, the considered RICH detector is conventional gaseous with a vertical geometry thus shielding the photodetector planes with the magnet yoke. Commonly such a detector uses a radiator with a low refractive index of $\gamma_{th} \geq 30$. As a rather large material budget is located in front of the RICH detector, most of the detected Cherenkov rings stem from secondary electrons produced upstream. Together with the numerous charged tracks and high ring densities in the inner part of the RICH detector, these conditions lead to a certain probability of ring-track mismatches and the creation of fake rings from random combinations of close hits. This circumstance challenges obviously any di-electron measurement, even more considering the fact that di-electrons from π^0-Dalitz decays and γ-conversion are difficult to suppress. In addition to this, the magnetic field opens the close pairs before the electron identification is performed. Detailed simulations have been carried out in order to investigate the influence of these conditions on the feasibility of the foreseen physics measurement.

Coming now to more implementation specific data, the actual layout of the RICH detector contains a radiator, a mirror and a photodetector. Furthermore the radiator has a length of 2.2 m and is filled with gas of either 100 % nitrogen (N_2), 40 % helium (He) and 60 % methane (CH_4) or 50 % nitrogen (N_2) and 50 % methane (CH_4). The mirror has a radius of 4.5 m and is composed of 3 mm thick beryllium (Be) covered with 0.5 mm glass, while the reflectivity of this mirror is comparable to the measurements for the High Acceptance Di-Electron Spectrometer (HADES) (see [HAD]) mirror, which is made up of aluminium (Al) coated with magnesium fluoride (MgF_2). The photodetector is composed of PhotoMultiplier Tubes (PMTs) with a diameter of 6 mm and a quantum efficiency of 20 %, which leads to about 100,000 channels. Further on, the geometry and quantum efficiencies are originated from the Hamamatsu H8500 Multiple Anode PhotoMultiplier Tube (MAPMT), which includes UV windows and is used in the B physics experiment at 2-TeV proton-antiproton accelerator (BTeV) (see [BTe]). However the glass window of the PMTs is here covered with wavelength shifter films (compare WLSF) in order to increase the absorption of Cherenkov photons. Depending on the detection threshold at small wavelengths λ_{min}, the number of measured photons per ring ranges from 15 ($\lambda_{min} = 250\,\text{nm}$) to 33 ($\lambda_{min} = 120\,\text{nm}$). The corresponding figures of merit are $N_0 = 138$ and 292 , respectively. More information can be found in [Sen06], [CBM08], [HDD+08] and [CBM07].

1.3.3 Muon Chamber Detector System

For the complementary measurement of low mass vector mesons and charmonium through their decay into muon pairs, the RICH detector system will be replaced by a MUCH detector system, which is regrettably not able to measure hadrons. Thus the TRD will serve in this alternative setup for pure tracking towards the TOF detector system. Furthermore this TOF detector system will be used for background suppression. So the idea of the MUCH detector system is to continuously track all charged particles through the complete setup, starting with the tracks measured by the STS. For this purpose, it is clear that this concept requires highly granulated and fast tracking detectors.

Coming again to more implementation specific issues, the present design of the MUCH detector system foresees a compact active absorber system to reduce the number of muons from pion and kaon decays behind the magnet. This system combines several iron absorber layers of varying thickness for hadron suppression with 15 to 18 intermediate active tracking stations, which are implemented with high rate gas detectors based on Micropattern technology. For the measurement of muons from low mass vector mesons (ρ, ω, ϕ), the total iron absorber thickness is 125 cm (7.5 λ_I), while 1 m of iron is added (total thickness of 13.4 λ_I) for muons from charmonium. Besides this, it is useful to subdivide each gas detector chamber into ring-like regions with an appropriate pad size, because the hit density, which can reach up to 1 $\frac{\text{hit}}{\text{cm}}$, decreases dependent on the distance from the beam. Moreover the radii of these rings are chosen in the way that the mean hit density is doubled from ring to ring, while the pad area varies inversely. Further on, each region is mapped onto rectangular sectors with a sector being a group of 128 channels, which are connected to a FEE chip. Moreover the identical sector sizes within a ring simplify the design of the printed circuit boards for the detector read-out. Beyond that, another constraint is determined by the spatial resolution, which should match the tracking requirements of about 0.4 mm for the central area and decreases obviously towards the periphery. Thus the required resolution defines the maximal pad size.
For all that reasons, the current design proposes to implement the active stations with Gas Electron Multiplier (GEM) or THick Gas Electron Multiplier (THGEM) detectors.
GEM detectors are made of 50 μm thick polymer foil, which is coated with a thin layer of metal on both sides. Furthermore a regular field of holes is chemically pierced with a fraction of a millimeter distant. By applying a high voltage of about 500 V across the two conducting surfaces, a primary electron produces an avalanche of electrons and ions inside the holes which can then be read-out by using pads or strips.
Besides this, THGEM detectors are an augmented version of GEM detectors. Here the holes are mechanically drilled on thick FR4 plates which have a thickness of 500 μm and more. The size of the holes increases also to a dimension about 300 μm, which leads then to a higher operation voltage. Although the THGEM detectors have a slightly inferior position resolution compared to the GEM detectors, they are more robust and easier to manufacture.
More information can be found in [Fri06b], [CBM08] and [CBM07].

1.3.4 Transition Radiation Detector System

The TRD serves for charged particle tracking and for identification of high energy electrons and positrons with $\gamma > 2{,}000$ to reconstruct J/ψ mesons. In combination with the RICH and the ECAL detector systems, it has to provide sufficient electron identification capability for the measurements of charmonium and low-mass vector mesons. Hence the required pion suppression is in the range of 100 - 200 for electrons with a momentum above 1.5 $\frac{\text{GeV}}{\text{c}}$. Additionally the position resolution has to be in the order of 200 μm to 300 μm.

Coming now to more implementation specific issues, the entire detector system is currently envisaged to be divided into three stations, which are positioned at distances of 4 , 6 and 8 m from the target. Furthermore each station is composed of at least three or better four layers. The total thickness of the detector in terms of radiation length has to be kept as small as possible to minimize multiple scattering and conversions. Further on, the gas mixture of the read-out detectors has to be based on Xe to maximize the absorption of Transition Radiation (TR), which is produced by the radiator. In addition to this, a fast read-out within less than 100 ns has to be provided because of the high interaction rate environment in particular for the inner part of the detector planes covering forward emission angles. To ensure the speed and also to minimize possible space charge effects expected at high rates, it is clear that the detector has to have a thickness smaller than about 1 cm.
For such a detector, there are two solutions well known from other high energy experiments. The first one is a Multiwire Proportional Chamber (MWPC) with pad read-out (ALICE-TRD: see [ALI]), and the second one is formed by straw tubes (ATLAS-TRT: see [ATL]). By the way, both types are currently investigated for alternative designs, while a combination of the two would determine a possible and feasible solution as well.
Besides this, the radiator can be similarly chosen from two possibilities, which are defined by the regular or irregular type.
In that case, the regular radiator, which is composed of foils (polypropylene) and gaps of equal size, determines the choice, which provides the highest TR yield. However the costs may be high due to a complicated construction procedure.
In contrast to this, the irregular radiator, which is built of fibers and/or foams, would have a reduced TR yield, if the material budget is equal, but eases the manufacturing.
Beyond that, the main characteristics of the established TRD are shown in table 1.2.

Characteristic	Information
Cell sizes	1 - 10 cm^2 depending on the polar angle and tuned for an occupancy less than 10 %
Material budget	$\frac{X}{X_0} \simeq 15 - 20\,\%$
Rates	Up to 100 $\frac{\mathrm{kHz}}{\mathrm{cm}^2}$
Doses for charged particles	Up to 16 $\frac{\mathrm{krad}}{\mathrm{year}}$ corresponding to $26 - 40\,\frac{\frac{\mathrm{mC}}{\mathrm{cm}}}{\mathrm{year}}$ charge on chamber wires

Table 1.2: Summary about the main characteristics of the TRD

Finally, it should be said that the actual developed TRD bases on the MWPC technology. By using nine to twelve layers, the requested area of the detector is in the range of 485 - 646 m^2, while the necessary amount of electronic channels is between 562,000 and 749,000. More information can be found in [Sen06], [CBM08] and [And06].

1.3.5 Time Of Flight Detector System

Since the TOF detector system is mainly used for the identification of hadrons, a key task is obviously determined by the separation of pions and kaons. Thus flight paths of the order of 10 m are necessary in order to achieve a two to three sigma separation with time resolutions in the order of 80 ps. For this purpose, the full coverage of the same solid angle as the STS requires an area of the TOF detector system in the scale of 120 m^2 (see [CBM05]). As furthermore such an area can not be equipped with traditional scintillator/photomultiplier counters at an affordable cost, the only choice is determined by the development of a high resolution and high rate version of multi-gap timing RPCs. Besides this, it should be noticed that a system of similar size is currently being installed in the ALICE experiment (see [ALI]), although the requirements imposed by the running conditions are very different.

Coming actually to more implementation specific issues, the equipped TOF detector system in the CBM experiment consists of an array of RPCs with an area of 120 m^2 located 10 m downstream of the target. Further on, 10 MHz minimum bias Au+Au collisions require the innermost part of the detector to

work at rates up to 20 $\frac{\text{kHz}}{\text{cm}^2}$. Moreover the pad size at small deflection angles is about 5 cm^2 corresponding to an occupancy of below 5 % for central Au+Au collisions at 25 AGeV.
Besides this, a RPC consists of a stack of planar electrode plates of high resistivity, which are kept by spacers at fixed distances between 0.2 and 0.35 mm typically. In addition to this, an anode is normally placed in the center of the stack, while two cathodes enclose it at its outer surfaces. So if such a design is operated under a uniform electric field of about 10 $\frac{\text{kV}}{\text{mm}}$ between the electrodes and the right gas mixture, the counters would deliver fast avalanche signals, which could be derived as single pulses from the anode or in differential mode between anode and the cathodes.
But due to the requirements of the CBM experiment, the TOF detector system has to be equipped with different module types like single-cell counters, multi-pad counters, single-strip counters and multi-strip counters.
Within this context, such single-cell counters with four or more gaps and surfaces of typically 10 cm^2 deliver easily time resolutions down to 40 ps at efficiencies above 99 %. If these counters are further adopted in size and shape to the needed granularity, they would be also shielded perfectly, which features thus a minimum cross talk to neighbors. For that reason, this type is one of the choices for the innermost angular region of the planned TOF detector system. In this connection, it has to be supplementary mentioned, that the envisaged time resolutions in the high-rate environment is not that problematic, because suitable electrode material can be found.
Beyond that, a somewhat more economical realization is determined by a multi-pad counter. Such a device exhibits a much larger surface of the electrode stack and an anode, which is subdivided into single pads including a separate read-out. This circumstance leads then obviously to an easier coverage of larger areas with the side effect of enhanced cross talk between the pads. For that reason, such modules could be envisaged for the larger polar regions of the TOF detector system.
Above all, a single-strip counter bases on a thin metallic shielding of rectangular profile, which houses a double gap aluminum strip. Furthermore this strip defines an high-voltage middle electrode, which is surrounded by a 0.3 mm gap of float glass plates on both sides. These glass plates are further coated with a series of narrow long strips on the outside. So a fast timing signal can be read-out at both ends of the aluminum electrode, which can be then used to form the sum and difference of two measured times delivering a time-of-flight measurement (see [FFG03]).
Besides this, the transition from a single-cell to a multi-pad counter suggests that an analog replacement of the single anode strip by a multi-strip anode is also possible. Such an anode possesses then a segmented strip/gap config-

uration and is read-out at both ends of each strip. The advantage of such a design in comparison to a pad like structure is apparently defined by the reduction of electronic channels and a potentially better spatial resolution. Of course the limit for this detector type is determined by the number of hits, which can be detected within an event, but such detectors equipped with normal floating glass electrodes offer an ideal solution for the outer 75 % of the total TOF detector system where the rates stay below 1 $\frac{\text{kHz}}{\text{cm}^2}$. Further on, such multi-strip detectors need to be arranged in a staggered configuration in order to guarantee a full geometrical efficiency as well as single-strip detectors. Moreover the number of strips and the pitch are of course depending on the required spatial resolution and the multi-hit capability. Detectors of this type are presently being developed for the TOF upgrade project of the Four Pi (FOPI) experiment (see [Wel]).

To generate the start signal for the time-of-flight measurement, a polycrystalline Chemical Vapor Deposition of Diamond (CVDD) strip detector is used. These detector types are fast particle sensors, which are radiation hard and can operate in primary heavy ion beams of particle intensities of up to $10^9\,\frac{\text{ions}}{\text{s}}$. The developed detectors have a size of $20 \cdot 20\,\text{mm}^2$ and a thickness between $100\,\mu$m and $300\,\mu$m. More information can be found in [CBM08], [Sen06] and [CBM05].

1.3.6 Electromagnetic CALorimeter Detector System

The ECAL will be used for the identification of electrons and photons, providing unique information for leptons (electrons and muons) and performing a rather accurate time-of-flight analysis. So in combination with the RICH and the TRD, the ECAL will significantly contribute to the particle identification due to its high capability to discriminate photons, electrons and hadrons. Furthermore the large angular acceptance of the CBM experiment and the distance needed for the time-of-flight measurement result in a large area calorimeter system located immediately after the TOF, which implies at least 12 m away from the target.

Coming now to more implementation specific issues, the optimization of the cost-to-performance ratio leads to the proposed ECAL design, which bases on the 'shashlik' technology of sampling scintillator-lead structures read out by plastic Wavelength Shifting Fibers (WLSFs), because this technology is already developed and successfully used in the calorimeters of the PHENIX experiment (see [PHE]), the HERAB experiment (see [Thec]), the LHCb experiment (see [LHC]) and the K zero Pi zero: $K^0_\perp \longrightarrow \pi^0 \nu \bar{\nu}$ (KOPIO)

experiment (see [Thed]).
So based on this concept, it is natural to adopt a variable lateral segmentation in three different cell-size zones following the HERAB and LHCb experiences, because the hit density varies significantly by two orders of magnitude over the calorimeter surface. In addition to this, the 'shashlik' technology allows an easy way to build the whole system with equal sized basic modules, while the required lateral segmentation is achieved by dividing the scintillator plastic plates into the corresponding number of light isolated pieces and grouping their fibers onto the separate photodetectors. Hence single cell modules should be applied in the outer region, 2×2 cells per module in the middle region and 3×3 or even 4×4 cells per module in the inner region.
Besides this, the longitudinal segmentation of the read-out within the calorimeter towers constitutes another method, which features the discrimination between leptons and hadrons and helps to improve the time resolution. But the decision on the practicality of longitudinal segmentation depends on the complexity and cost of the additional photodetectors and electronic channels.
Beyond that, custom photomultipliers have demonstrated perfect performance with 'shashlik' calorimeters during the last 15 years. However the relatively high price of these photodetectors seriously limits the granularity of the system, which is particularly important for CBM. Therefore it would be very interesting to consider the new cheaper photodetectors like Avalanche PhotoDiodes (APDs), multianode PMTs or Silicon PhotoMultipliers (SiPMs)) especially for the preshower and hadron catcher, because the requirements for energy resolution, dynamic range and linearity are not very stringent there.

Actually the ECAL consists of modules composed of 380 alternating layers of lead and plastic scintillator. Furthermore each of these modules combine a preshower part and calorimeter parts, which are physically integrated into a single module box, while their read-out is realized from the front and rear sides respectively. Further on, the preshower elements are composed of 8.4 mm thick lead ($1.5\,\chi_0$) and scintillator pads. In addition to this, a groove in the scintillator pad houses the helical WLSF to collect light, which is then sent from both WLSF ends to the photodetector located at the front surface of the calorimeter module.
Moreover the calorimeter part of the module has a periodical structure along the z-axis of metal absorber and scintillator tiles, which are made of polystyrene doped by 1.5 % para-TerPhenyl (pTP) and 0.04 % POPOP. The scintillator tiles are wrapped by $60\,\mu$m-thick tyvek from both sides for better diffusive light reflection. So scintillating light is collected by optic

WLSFs, which define a type of fibers offering one of the best light attenuation length and elasticity. The fibers penetrate all the layers with the surface density about 0.9 $\frac{\text{fibers}}{\text{cm}^2}$, which provides a good homogeneity of the light collection. WLSFs are looped at the front of the cell and the ends of all fibers are collected into one bundle at the end of the cell to guide the light to a photodetector. The tower of lead-tyvek-scintillator layers is compressed together and held by steel strings in four corners of the cell. More information can be found in [CBM08], [CBM04] and [CBM05].

1.3.7 Projectile Spectator Detector System

The Projectile Spectator Detector (PSD) is planned to measure the number of non-interacting nucleons from a projectile nucleus in nucleus-nucleus collisions. Thus this detector will be used in the CBM experiment for the precise determination of the nuclear collision centrality with the number of participants, the reconstruction of the reaction plane and the beam intensity monitor by detecting the electromagnetic dissociated neutrons. Further on, the experimental extraction of event-by-event fluctuations requires a very precise control over the fluctuations caused by the variation of the number of interacting nucleons due to event-by-event changes in the collision geometry.

Coming again to more implementation specific issues, the need for an excellent energy resolution ($\frac{50\,\%}{\sqrt{\text{E(GeV)}}}$), good transverse uniformity of this resolution and fine granularity as well as the linearity in a wide range of detected energies (10 - 10,000 GeV) cause the developed PSD to be a full compensating modular lead-scintillator hadron calorimeter, which consists of 12×9 individual modules at a distance of about 15 m downstream from the target. Furthermore each single module, which possesses the dimensions of $10 \cdot 10 \cdot 120\,\text{cm}^3$, contains 60 lead/scintillator layers with a thickness of 16 mm and 4 mm.
Further on, the light read-out is provided by WLSFs, which are embedded in round grooves in the scintillator plates, to ensure a high efficiency and uniformity of light collection over the scintillator tile within a very few percent. In addition to this, WLSFs from each six consecutive scintillator tiles are collected together and viewed by a single Micropixel Avalanche PhotoDiode (MAPD) at the end of the module. Moreover the longitudinal segmentation in ten sections ensures the uniformity of light collection along the module as well as the rejection of secondary particles from interaction in the target. Besides this, MAPDs seem to be an optimal choice due to their remarkable properties of high internal gain, compactness, low cost and immunity to the

nuclear counter effect. So ten MAPDs per module are placed at the rear side of the module together with the FEE. In this connection, it should be noticed that a single MAPD exhibits an active area of $3 \times 3\,\text{mm}^2$ and a pixel density of $\frac{10^4}{\text{mm}^2}$. More information can be found in [CBM08], [GSI06] and [CBM07].

1.4 CBM Trigger Principles

At the beginning of this section, the meaning of the word trigger has to be defined, because there are commonly many understandings, which are specific to the field in which this word is used. For instance, if one speaks about firearms, a trigger would mean the mechanism that actuates the firing sequence. If the field is electronics, the Schmitt trigger would represent a comparator circuit, that incorporates positive feedback. Another example is given in psychology. Here a Trauma trigger denotes some form of experience that triggers a traumatic memory in someone who has experienced trauma. So because of these examples, it is obvious that many other fields make use of the word trigger.
But as this thesis is about an application in particle physics, a trigger in this case means a system, which uses a simple criteria to rapidly decide whether the actual event in an experiment should be recorded or ignored. Furthermore in this context, an event means a single set of particle interactions, which are measured by detector stations in a brief span of time.

Since experiments in particle physics are typically searching for 'interesting' events, trigger systems are obviously used to identify and enable the recording of these events for later analysis. Hence it is clear that such systems are only necessary due to the huge amount of occurring data and the limitations in data storage capacity and data rates. For instance, current accelerators have event rates greater than 1 MHz in combination with trigger rates below 10 Hz. Moreover this ratio of the trigger rate to the event rate, which is evaluated to $\frac{1\,\text{MHz}}{10\,\text{Hz}} = 100.000$ in the example, is often referred to as the selectivity of the trigger and can be imaginably very high. For this purpose, it is common to cascade multiple levels of triggers.
The ATLAS experiment (see [ATL]), which is located at CERN, uses for instance three different trigger levels. The first one is based in electronics on the detector, while the other two run primarily on a large computer cluster near the detector. In this connection it is important to realize that the higher the level of the trigger is, the lower is the data rate. Thus the first level trigger must be always the fastest one, is always located next to the detector read-out and the input consists of raw particle interaction data.

The challenges for such a trigger and Data Acquisition (DAQ) system, which can be used in the CBM experiment, are posed by the high reaction rates of up to 10^7 events per second without a fixed bunch crossing and by the high number of multiplicities of up to 1,000 particles per central collision, because the trigger system has to detect efficiently rare probes like open-charm, J/Ψ mesons, or low-mass di-lepton pairs in this harsh environment. Furthermore a trigger for these signals can not be based on a fast preselection by using only a small subset of detector stations, but requires the evaluation of complex signatures for each interaction, which have to be derived from tracking in the STS and vertexing. At this juncture, the most demanding task concerning trigger processing is an open-charm trigger on D mesons in the first level, because D mesons are detected via their weak decay into charged pions and kaons. The difficulty of this measurement lies apparently in the very low multiplicity of D mesons, which is about 10^{-3} per central event, within an environment of about 1,000 charged hardrons produced in such a collision. Evidently, the combinatorial background stemming from directly produced particles has to be suppressed by many orders of magnitude.

In general, tracking out of raw hit data, which is commonly composed of measurements by detector stations, is the hardest part of the trigger generation, because no data reduction can be done so far. In contrast to this, all subsequent parts can take advantage of the results derived from this first level tracking like an assumption of the momentum or the charge of the particles.

Traditionally the tracking, which is usually the predecessor of the vertexing, can be divided into two separate tasks, which are called track finding and track fitting. Furthermore the track finding can be characterized in this context to be a pattern recognition process, that groups subsets of measured detector station hits, which are believed to originate from the same particle, together to form so-called track candidates. Afterwards the track fitting is applied to analyze each track candidate for being an acceptable track or not, to adjust further all acceptable tracks to the chosen track model and finally to estimate a set of parameters describing them. So based on these functionalities, it is obvious that an efficient and reliable track finding algorithm is of major importance with regard to the global efficiency, because a track, which is not found, can be also not fitted and is therefore irrevocably lost. Besides this, the vertexing follows ordinarily a successful tracking to project each fitted track to its origin. Hence this process would be able to decide for each represented particle, if it arises from the primary vertex, or if it is perhaps created in a secondary decay. So if the fact that the relevant tracks must originate in the primary or so-called common vertex is taken into account, it would be possible to further improve the tracking quality by simply removing the tracks, which are identified to represent a particle created in a secondary decay. Figure 1.13 depicts now a simple example to gain a deeper insight by showing the result of an elementary particle collision including the vertex, two marked particle motion trajectories and possible interactions with the detector stations.

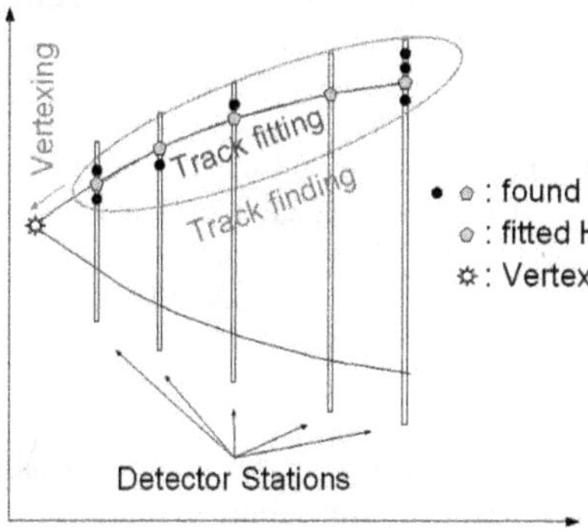

Figure 1.13: Concept details of a trigger including all diverse steps

With regard to this figure, it is obvious that the entire process is not trivial. In the early bubble chamber experiments, the track finding was even done purely manual by inspecting the event images, which contain measured positions from an operator guided machine, on a scanning table (see [Mer80], [JS72]). Although significant effort has been put into the automation of the pattern recognition process, full automation remains still a highly nontrivial task with the computing power up to the late 1960s.
In the following years the boundary between track finding and track fitting becomes also more and more fuzzy. This consolidation process was mainly driven by the invention of the Kalman filter, because due to its recursive nature, optimal predictions of the track parameters into the next detector station are made from already accepted measurements. So this algorithm can be used for finding and fitting concurrently, because the presence or absence of a measurement, which is compatible to the predicted parameters, determines whether the track candidate is further considered or discarded. Nevertheless the Kalman filter is pioneered for the purpose of track finding in the DELPHI experiment at the CERN (see [Bil89]) and is applied to real data for the first time in the ZEUS experiment at the DESY (see [BQ90b], [BQ90a]).

Beyond this, an extension of this method, which is today often called combinatorial Kalman filter, is developed within the context of the HERAB experiment at the DESY (see [Man97]). This extension offers further the possibility to concurrently propagate several track candidates, which originate from the same track seed, by the usage of a combinatorial tree creating new branches each time a measurement is compatible to the prediction. Furthermore such a branch is then obviously discarded as soon as the quality drops below a certain threshold for χ^2 or too many compatible measurements are absent for that track candidate. Thus the optimal end is defined by the survival of only the best track candidate.
Nowadays the Kalman filter is by far the most common method for track finding and track fitting. But a not negligible problem is determined by the possible combinatorial explosion, which can easily occur by upgrading a particle accelerator to higher luminosities. Hence alternatives or adaptations have to be considered especially to cope with the harsh collision environment of the CBM experiment.

Beyond that, the tracking algorithms can be generally classified into three generic methods, which are named the combinatorial, the local and the global method (see [Bes93]).
Here the simplest and most time consuming one is the exhaustive combinatorial method, because it generates all possible hit combinations via the detector stations and checks them against a given track model. So it is obvious that this method can only be used with very few hits, because the number of combinatorial tracks is approximately proportional to N^s, where N represents the number of hits per detector station and s denotes the number of stations. For example, if 1,000 hits in six stations are considered, such an environment could easily result in $1000^6 = 10^{3 \cdot 6} = 10^{18}$ combinations. Hence there is a need for the reduction of the number of combinations by the usage of additional knowledge like the approximated vertex position or possibly the influence of a magnetic field. This circumstance leads then to the more appropriate tracking algorithms, which are categorized by the local and global methods.
Local methods like track following, track road or track element are characterized by processing the tracks independent of each other. For this purpose, some few points generate naturally an initial track candidate, which is further improved by collecting additional hits via interpolation or extrapolation. During this process it is obvious that such a track candidate would be only kept, if more appropriate hits can be found. Otherwise the track candidate is discarded. As thus local methods contain always incomplete track candidates, the same hits have to be used in several combinations. So the

complexity in time rises faster than linear with the number of hits and depends on the ordering of the hits, the initial track candidate and the progress of the tracking obtained so far.
In contrast to this, global methods like conformal mapping, histogramming or template matching are distinguished by processing all hits identically. This means more precisely that such an algorithm typically transforms all hits into a representation, which contains a likelihood information for the possible track candidates. So the result can then be determined by searching with regard to a specific criteria. Therefore the processing complexity of a global method is in principle proportional to the number of hits. Additionally, it should be noticed here that in contrast to local methods, global methods are normally not dependent on the processing order of the hits or something similar.
More detailed and general information about tracking and even vertexing in high energy physics can be found in [FRB+00]. In the following, the focus is changed to special tracking algorithms, which are developed to be applicable for the CBM experiment. Therefore all of them have to fulfill the subsequent requirements, which are introduced in [CBM05] on page 299:

- Tracking efficiency better than 90 %; also for secondary particles with low momenta
- Momentum resolution better than $\frac{\Delta p}{p} = 1\,\%$ for $p > 1\,\frac{\text{GeV}}{\text{c}}$
- Secondary vertex reconstruction with high efficiency and high resolution
- Fast on-line high resolution trigger algorithms for displayed vertices of D mesons

Comparably [Str06] summarizes the tracking algorithms for the LHC experiments ALICE, ATLAS, CMS and LHCb.

1.4.1 Conformal Mapping

One of the standard global methods for tracking in a magnetic field is the conformal mapping. In general, such a method applies a map $\omega = f_{(z)}$, which preserves for example the standard euclidean angles, to transform the hits into another space simplifying the problem of pattern recognition. Further on, the conformal property of such a map is commonly described in terms of the Jacobian derivative matrix of a coordinate transformation, which must consist everywhere of a scalar times a rotation matrix. Before adapting this general approach now to the CBM experiment, another simplification related to the magnetic field has to be introduced. So even if it is not happening in reality, the magnetic field would be at first assumed to be homogeneous, because under such conditions, the particle motion could be described by a circular trajectory in the bending x-z-plane, which is obviously perpendicular to the magnetic field. So the coordinates of such a circular motion trajectory can be easily described by the following well-known equation:

$$r^2 = (x-a)^2 + (z-b)^2 \tag{1.1}$$

Since fitting straight lines is easier and faster than fitting circles, the conformal mapping approach can be used to transform such trajectories into straight lines with the prescriptions:

$$u = \frac{x-a}{r^2} \quad and \quad v = \frac{z-b}{r^2} \tag{1.2}$$

However a more detailed look to this transformation shows a problem, because it requires a particle position (a, b), which lies on the trajectory. But this problem is not crucial, because it can be easily solved by applying the vertex constraint. This means more precisely that the common vertex, which is located at the coordinates $(0, 0)$, is simply assumed to lay invariably on the motion trajectory. Thus the equations turn into:

$$\begin{aligned} r^2 &= x^2 + z^2 \\ u &= \frac{x}{x^2+z^2} \\ v &= \frac{z}{x^2+z^2} \end{aligned} \tag{1.3}$$

Moreover these equations lead then to a straight line of the form:

$$v = -\frac{a}{b} \cdot u + \frac{1}{2b} \tag{1.4}$$

This equation can be then obviously used for pattern recognition by establishing a search road. Further on, the adaption of this formula to the inhomogeneous magnetic field is described in [RJ07] and [Jer05]. In addition to this, the motion trajectory in the y-z-plane is almost a straight line, because there is no bending caused by the magnetic field. Hence the tracking is equivalent to finding the parameters in those two planes.

Finally it should be mentioned here, that simulations show a varying track finding efficiency of this algorithm for central Au+Au collisions at energies of 25 AGeV between 70 % and 90 % for momenta greater than 1 $\frac{\text{GeV}}{\text{c}}$. By the way, the required processing time is about 5 s to 8 s on a common Pentium4 Central Processing Unit (CPU) at 2 GHz. More information about this approach can be found in [CBM05], [GSI06], [Jer05], [RJ07] and [Yep96].

1.4.2 3D Track-Following Method

As mentioned earlier, the track following is a standard local method for tracking. With regard to the CBM experiment, some approaches, which are introduced in [OPP02], can be used to implement such an algorithm in the 3D coordinate space by operating two projected 2D planes simultaneously. For this purpose, the same reasons than already explained in the previous section 1.4.1 on page 37 counts here, which leads naturally to the circumstance that these planes are also spanned by the x-z-plane and y-z-plane combination.

So the resulting algorithm is quite easy, because it simply predicts the next position of an initial track candidate into the consecutive detector station by assuming a parabolic prolongation for the x-z-plane and a straight line extension for the y-z-plane. Moreover this prediction is then checked by searching projected hits in the vicinity of the position in both planes. For this searching process, it is obvious that the region of interest is defined by an asymmetric corridor around the predicted position. Besides this, it is clear that the size of these corridors has to be suitable with regard to the amount of hits, which have to be considered for a single prediction. Hence these limits are calculated carefully and separately for each detector station by the evaluation of statistics, which are obtained from many Monte Carlo (MC) simulations.
Further on, it is self-evident that hits would only be assigned to a track candidate, if their coordinates are simultaneously located in the region of interest of both projections. But as such an approach can not totally avoid the attaching of multiple hits in the same detector station, this problem can be easily solved by splitting the original track candidate into an equivalent number of new ones, which contain only a single hit in the station. These cloned track candidates do not constitute a problem in the following predictions, because they would eliminate themselves, if no hit falls into the corridors of the next detector station.

So up to now the only weak point of this method is the definition of the initial track candidates. But as the y-z-plane extension is a straight line and the common vertex, which is located at the coordinates $(0, 0, 0)$, should be the origin of all relevant tracks, all initial track candidates can be simply determined by connecting the common vertex with all hits of the first detector station in the y-z-plane via a straight line, because two well-known positions are able to configure such lines. Based on this concept, it is finally clear that the selection of track candidates, which contain hits only exclusively, leads to an appropriate tracking algorithm.

Although the track finding efficiency of this algorithm has been quite good for the initial STS setup, this approach is not longer developed for the CBM experiment, because the introduction of the micro-strip detector stations causes a dramatical efficiency felt down from initially 91 %-96 % for momenta greater than 1 $\frac{\text{GeV}}{\text{c}}$, which can only be coped by a more elaborated prediction method. But nevertheless, more information can be found in [CBM05] and [AAB^{+}06].

1.4.3 Cellular Automaton

Another popular and fast implementation of a local method is defined by the Cellular Automaton. In principle, a Cellular Automaton is a discrete, dynamic and deterministic mathematical system of homogeneous cells, which represent a finite number of states combined with only local interaction possibilities, in a multidimensional grid (see [Ila01]). The basic components of a Cellular Automaton are of course cells, their neighborhood relation, the rules of evolution and the time evolution.

For the purpose of tracking, two different kinds are generally described in literature. While the first group, which is similar to Conway's Game of Life, matches detector interactions to cells, the second group, which is preferred here, pairs track segments and cells.
So a cell is further defined as short track segment, which connects in the first approach two hits and afterwards three hits on neighboring and afterwards also overlapping detector stations. In order to improve the speed of the algorithm, such cells are only allowed to receive integer states, which represent simple meanings related to the position in an optimization sequence in the sense of χ^2. Hence all states are required to be initially set to unity, because each track segment is in principle able to initiate the optimal sequence at startup.
In addition to this, the neighborhood relation between two cells is given by sharing at least a single hit of the track segments. So for that reason a combination of cells form a particle motion trajectory, which is also approximated by a parabolic model in the bending plane x-z and a straight line in the y-z

plane.
Besides this, the entire evolution process is divided into three steps, which are named forward evolution, backward pass and quality rejection.
During the forward evolution, the applied rule is determined by increasing the state of a cell by one, only if the existing left neighbor contains the same state. So with regard to this rule, it is obvious that it is executed for all cells iteratively as long as no update in states happens any more. Hence the results of this process are cells, which contain a state representing the length of a leftwards traceable optimization sequence.
Afterwards the backward pass iteratively investigates all cells containing the highest states, because these ones are most likely to be the first segments of a possible track candidate. For this purpose, the earlier mentioned trace is used to check whether all consecutive states in the optimization sequence decrease exactly by a single state. If this prerequisite is then further proven until the initial state is reached, a track candidate would be successfully accepted and the corresponding cells would be marked as 'used'.
Subsequently this backward pass is followed by the quality rejection, which has to be performed to almost avoid the existence of wrong reconstructed tracks. Therefore an estimator is used to favor long and smooth track candidates against all others.

Finally it should be noticed here, that the resulting track finding efficiency of this algorithm for central Au+Au collisions at energies of 25 AGeV is around 97 % for momenta greater than 1 $\frac{\mathrm{GeV}}{\mathrm{c}}$. Since this algorithm is implemented on different computational platforms, like the Cell BE from the Sony-Toshiba-IBM joint venture (STI) (see section 3.3.2 on page 185), the GTX 280 from nVIDIA or the Core 2 or Core i7 from the INTegrated ELectronics corporation (INTEL), it is obvious that the required processing time depends on the platform. Nevertheless the requirement to compute the tracking result as fast as possible highlights the best available combination of a platform and its corresponding implementation by consuming time in the region of 25 ms up to now. Compared to the initial tracking algorithm, which is described earlier, it has to be supplementary noticed that such a speed can be only reached by the application of an analytic formula, which approximates the magnetic field, by a vectorization of the algorithm and by the usage of multi-threading. More information about the Cellular Automaton method can be found in [Kis03], [Kis04], [GK05], [AEGK02b], [FGI+03], [CBM05], [AAB+06], [Kis06], [GK08] and [GKK+09].

1.4.4 Hough Transform

In the following chapters, the focus of this thesis is entirely set to the Hough Transform, which offers another global method for tracking, including all necessary aspects for the CBM experiment. In principle, such an algorithm transforms the described problem into another representation, which eases the evaluation of a solution, while this solution can be afterwards simply transformed reversely, because the prerequisite of this transformation is reversibility.

So with regard to the tracking intention, it is obvious that the problem is described by the reconstruction of the particle motion trajectories based on the given hits, and the application of a parametric description for these trajectories offers a transformation possibility for the hits into a parameter representation. Hence it is clear that this representation has to simplify the tracking. And a closer look to this circumstance immediately shows that the complex task of tracking is realized by a much easier search for local maximums in a complex density distribution, which is commonly implemented by a discrete histogram and particularly a multidimensional accumulator array. Furthermore it is self-evident that such a successful search is followed by the reutilization of the same parametric description to transform the found maximums reversely. So based on this concept, the result is characterized by an amount of grouped parameters, which describe each reconstructed motion trajectory of a particle. Besides this brief outline, an exhaustive introduction to this method is given in section 2 on page 45.

Finally it should be also noticed here that my department is not new into the business of tracking with the Hough Transform in high energy physics experiments, because there are already several applications, which are published in [GBM90] and [HKM+99].

1.4.5 CBM Related Qualification of the Algorithms

At a first glance, the general aim of a tracking algorithm for an experiment in high energy physics is the achievement of an efficiency, which is almost 100 %. But increasing interaction rates require another fact to be taken into account, because a tracking algorithm, which offers such an efficiency, would be obviously also not applicable, if the processing time is so slow that the required hardware is not practicable or too expensive.

So with regard to the demanding environment of the CBM experiment, the Conformal Mapping method and the 3D track-Following algorithm disqual-

ify themselves, because they are both slower and less efficient than the Cellular Automaton method, which is for instance already used in the Analyser of Rare EventS (ARES) experiment (see [GKKO93]), the Neutrino Ettore Majorana Observatory (NEMO) experiment (see [NKK+97]), the LHCb experiment (see [Sch07]) and the HERAB experiment (see [AEKM02] and [AEGK02a]). Since the implemented Cellular Automaton method features furthermore another two advantages in addition to its overwhelming tracking efficiency and good speed, it seems to be at a first guess the best method for the CBM experiment as well.
In this connection, it has to be mentioned that the first supplementary advantage is determined by the avoidance of the exhaustive combinatorial searches, which define a common problem of such an algorithm, while the second one is identified by the significant reduction of the processing information due to the smart definition of the neighborhood relation and the highly structured data. Keeping these facts now in mind, figure 1.14 depicts a typical dependency distribution of the required reconstruction time per event and the number of related MC Tracks.

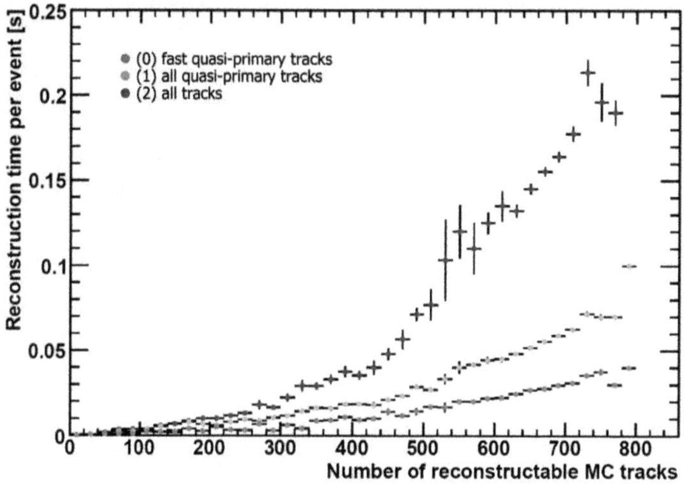

Figure 1.14: Required processing time of the Cellular Automaton method in dependency of the number of related MC Tracks (Source: [GK08] and [GKK+09])

However this figure shows immediately that the Cellular Automaton method is not linear, but probably polynomial in time with regard to the number of related MC Tracks. As this amount is furthermore proportional to the number of MC Points, the Cellular Automaton is also suggested to be at least polynomial in time with regard to the number of hits. Here it should be additionally noticed that this circumstance is considered without thinking about the efficiency of the detectors, because fake hits for example do not improve the situation for instance.

In contrast to this, the Hough Transform, and especially the Field Programmable Gate Array (FPGA) implementation, which is presented in this thesis, is indeed linear in time with regard to the number of hits. Moreover the fuzziness of the algorithm can be explicitly used to adapt the implementation in the range of exactness to almost streamlining. Of course it is clear, that the more exact the algorithm should work, the more processing time is required, and the less time should be required, the more inexactness must be accepted. Hence the Hough Transform method is studied in the following in order to investigate its hardware applicability for the first level tracking in the CBM experiment with regard to the consumed processing time and the acceptable inexactness.

Chapter 2

Hough Transform

This section presents all necessary adaptations of the Hough transform theory in combination with all required customizations for the CBM experiment.

2.1 Hough Transform Theory

The historical background of the Hough transform invention can be summarized very quickly, because there are only three major mentionable steps, which lasts about ten years. The original Hough transform was developed in 1959 by P.V.C. Hough in the field of automated pattern recognition in bubble chamber experiments ([Hou59]) and can be characterized by a very strong relationship to the Radon transform introduced in 1917 by Johann Radon, because it is a special case of this transformation (see [Dea81]). In the following P.V.C. Hough patented his transformation three years later, in 1962, in combination with an electronic solution in cooperation with IBM ([Hou62] and [BB91]). Afterwards A. Rosenfeld proposed the Hough transform another seven years later, in 1969, to be generally applicable in image processing (see [Ros69]), which enables the algorithm to play an important role for pattern recognition or feature extraction (complex lines) since today.

Clarifying now the first technical term, the abstract task of pattern recognition denotes in its simplicity the detection of collinear points in an image, which can be further on easily widened to other more complex patterns. Furthermore it is in general possible to detect every pattern, which can be defined by a parametric description. In the following, the initial example of collinear points is taken to explain the basic idea of the Hough transform, because the parametric description represents a simple straight line, which is usually defined by its slope m and the y axis interception c (see figure 2.1a) in the cartesian coordinate system:

$$y = mx + c \tag{2.1}$$

Based on this equation, the basic idea of the Hough transform (see [Jäh05]) is to change the point of view, which means that the variables x and y are simply swapped with the parameters m and c in the way that m and c are now the variables and x and y the parameters. So this means metaphorically spoken that the x and y values of the collinear points are now taken for given parameters of a transforming function, which leads to many new possible parametric descriptions (see figure 2.1b):

$$c = -mx + y \tag{2.2}$$

The result of this concept is obviously a new space, which owns the characteristic that the pattern of the original space described by equation 2.1 is located in a single point, because it is originally defined by the combination of the parameters m and c. Hence the position of this point determines the feature variables m and c (see feature extraction) of the pattern in the so-called Hough space. Based on this circumstance, the operating mode of the algorithm is now quite obvious, because each collinear point produces due to the transforming function 2.2 a curve in the Hough space, which determines all possible m-c-combinations of straight lines to which the actual collinear point can belong.
So within this context, the algorithm scheme is now self-explanatory, because the Hough curves of all collinear points, which are located on a concrete straight line, must obviously intersect at the corresponding feature point in the Hough space. Thence pattern recognition has turned into the easier task of computing an intersection of curves. Further on it is additionally clear that a point, which does not lay on this straight line, does also not intersect the other Hough curves in the feature point. Moreover another noticeable fact, which can be seen in figure 2.1b, is defined by the circumstance that the corresponding Hough curves are also straight lines. However this situation depends evidently on the transforming function and may vary as shown later in this chapter.

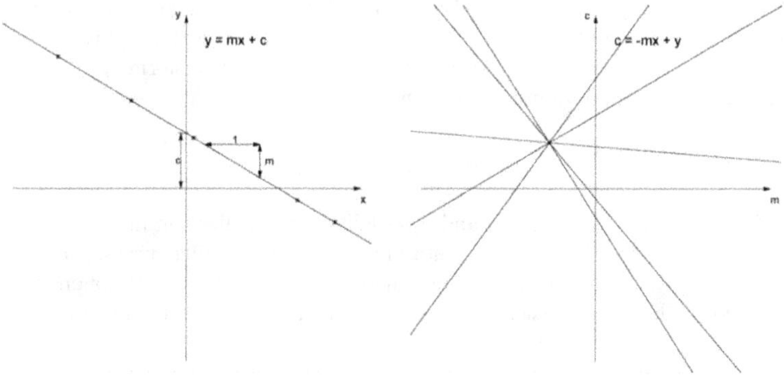

(a) Coordinate space (b) Hough space

Figure 2.1: Illustration of the Hough transform based on a straight line modeled with $f_{(m,c)}$

Since the generic algorithm is explained in detail, it is time to introduce a big problem of this parametric description, which belongs to the range of the slope dimension m in the Hough space. More precisely, a significant problem is caused by the circumstance that the more vertical the straight line gets, the more grows the magnitude of the slope m towards infinity, because this regularity leads evidently to the necessarily covered range of (-∞; +∞), which is naturally very hard to handle.

For that reason R.O. Duda and P.E. Hart introduced in 1972 the so-called Hesse form, whose name belongs to the inventor Ludwig Otto Hesse (1811-1874), as parametric description of a straight line, which induces the availability of a new transforming function ([DH72]). Henceforward the new form bases on the orthogonal normal vector $\vec{n}$ with length one and the orthogonal distance vector $\vec{a} = p \cdot \vec{n}$ with p as distance from the origin (see [BSMM05] pages 200-201):

$$
\begin{aligned}
\vec{n} \cdot (\vec{x} - \vec{a}) &= 0 \\
\vec{n} \cdot (\vec{x} - p \cdot \vec{n}) &= 0
\end{aligned}
$$

With $\vec{n} = \binom{cos(\phi)}{sin(\phi)}$, this formula turns into:

$$x \cdot cos\,(\phi) + y \cdot sin\,(\phi) - p = 0$$

So now a straight line can be also described with their orthogonal distance p to the origin and the angle ϕ, which is located between the normal vector $\vec{n}$ and the x axis (see figure 2.2a). Hence the resulting transforming function for the Hesse description is defined by:

$$p = x \cdot cos\,(\phi) + y \cdot sin\,(\phi) \tag{2.3}$$

With regard to this equation and especially the visualization in figure 2.2b, which depicts the Hough space containing the corresponding curves, it is now evident the Hough curves are not longer straight lines, but trigonometric functions. Thus the earlier mentioned problem of the infinity range for the slope dimension m can be avoided.

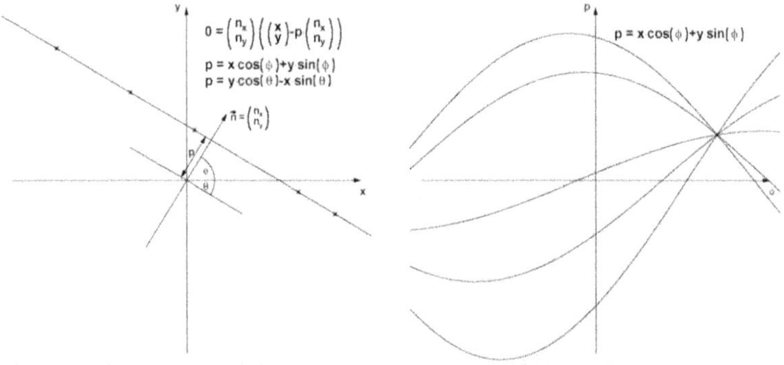

(a) Coordinate space (b) Hough space

Figure 2.2: Illustration of the Hough transform based on a straight line modeled with $f_{(p,\phi)}$

Beyond the solution of the mentioned problem, the feature turns from $(m,\ c)$ to $(p,\ \phi)$, even though the same original pattern is searched. Furthermore it is an interesting circumstance that there are as many several Hough transforms possible as parametric descriptions for the pattern is imaginable. However it has to be noticed that different transformations might have also different advantages and disadvantages. Thus the best one has to be chosen with regard to the given requirements, because if the feature $(\pm\inf,\ c)$ is remembered in the $(m,\ c)$ space for instance, it could surely exist, but could be obviously not found. In contrast to this, the feature $(p,\ \pm\frac{\pi}{2})$ in the $(p,\ \phi)$ space is findable, whether it represents the same feature than $(\pm\inf,\ c)$ in the $(m,\ c)$

space. According to this, it is clear that the Hough space dimension ranges of this new transformation are finite. Moreover a more detailed look to the p dimension shows that the range is determined by the negative and positive maximal distance of the straight line to the origin, while the ϕ dimension range has to cover values from $-\frac{\pi}{2}$ to $+\frac{\pi}{2}$. So since each possible straight line with distance p and angle ϕ is now generally defined in the Hough space by a single point, such a transformation is called line-to-point transformation. Moreover each imaginable transformation of a pattern, which is describable by a parametric description, allows such a to-point transformation to be supposable ([BB91]).

Besides this, another more common transforming function, which bases surely on the introduced (p, ϕ) description, spans the Hough space by the feature (p, θ). Obviously the only difference to the previous function is the substitution of the angle ϕ with the angle θ by using the the following relation:

$$|\phi| + |\theta| = \frac{\pi}{2}$$

In this equation the absolute values of the angles are used, because there are some problems with the angle orientation. This means in detail that if the common angle orientation is used, the angle ϕ would have a positive sign, while the angle θ would have a negative sign. So taking this fact into account, the absolute values can be removed by modifying the equation in the following way:

$$\begin{aligned} \phi - \theta &= \tfrac{\pi}{2} \\ \Rightarrow \quad \phi &= \tfrac{\pi}{2} + \theta \end{aligned}$$

By using this relation in combination with the trigonometric reduction equations (see [BSMM05] on page 79), the function turns into:

$$\begin{aligned} p &= x \cdot cos\,(\phi) + y \cdot sin\,(\phi) \\ \Rightarrow \quad p &= x \cdot cos\left(\tfrac{\pi}{2} + \theta\right) + y \cdot sin\left(\tfrac{\pi}{2} + \theta\right) \\ \Rightarrow \quad p &= -x \cdot sin\,(\theta) + y \cdot cos\,(\theta) \end{aligned}$$

So finally the new formula is given by:

$$p = y \cdot cos\,(\theta) - x \cdot sin\,(\theta) \tag{2.4}$$

As the comparison of the function $f_{(p,\phi)}$ to the new function $f_{(p,\theta)}$ is less meaningful, it is much more easy to compare figure 2.2b to figure 2.3b.

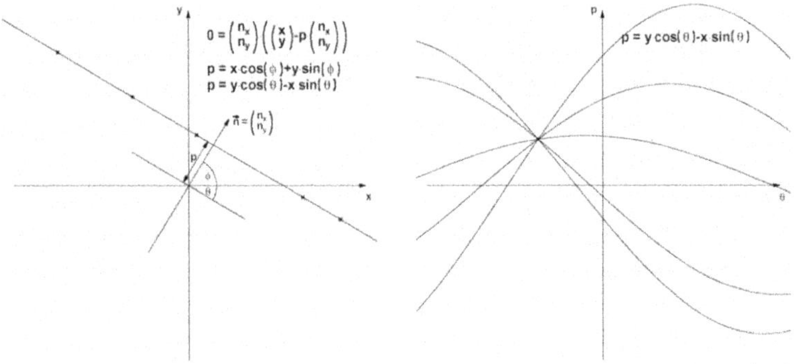

(a) Coordinate space (b) Hough space

Figure 2.3: Illustration of the Hough transform based on a straight line modeled with $f_{(p,\theta)}$

Even if the range of the angle θ is obviously also defined by $[-\frac{\pi}{2}; +\frac{\pi}{2}]$, the main advantage of this description would be determined by the easy transformation of the feature $(p,\ \theta)$ into the original $f_{(m,c)}$ model, because the interest is almost limited to the slope m of the straight line. Therefore the following relations can be used:

$$\begin{aligned} m &= tan\,(\theta) \\ y &= tan\,(\theta)\cdot x + c \\ \Rightarrow\quad c &= y - tan\,(\theta)\cdot x, \quad with \quad (x,y) \in \{line\} \end{aligned}$$

Beyond that, it is clear that many developments take place during the years, which improve the original Hough transform further. Some of them can be found in [Eis99], [Mai85], [IK88], [Lea93] and [KHXO95]. In the following, there is just a short summary presented, which lists possible areas of improvement:

- Improving the usage of the parameter space
  - Muff transform does not need sin and cos ([Wal85])
  - Spreading of parameter space axis ([FS89])
- Minimizing the memory usage
  - adaptive Hough transform with a refining accumulator by iteration ([IK87])
- Minimizing the computation time
  - Combinatorial Hough transform accumulating just parts of Hough curves ([BTS90] and [dFBTS90])
  - randomized Hough transform accumulating random parts of the source image ([XOK90], [XO92] and [XO93])
  - Hierarchical Hough transform works on segmented pyramid-ordered source images ([PIK89] and [BJ95])
- Increasing exactness
  - Hough transform containing gray-values ([LT95])
- Increasing exactness and the robustness against noise
  - Hough transform containing brightness information ([IND92])
  - Fuzzy Hough transform accumulating a band of weighted Hough curves ([HKP93])
- Increasing exactness and the robustness against quantization errors
  - Hough transform containing Mahalonobis distance ([XV94b] and [XV94a])
- Algorithm adapted to special hardware
  - Hough processor ([KOM95])

Although the listed examples describe possible improvements in some areas, the charges are of course often for other areas. So the optimal trade off has to be found by the adaption to the special requirements of the actual task of pattern recognition. For this purpose, a comparison of different Hough transforms can be found in [PYIK89] or [Eis99].

Remembering the already mentioned circumstance that the searched pattern for the common Hough transform must have a parametric description (see [GW00]), some simple examples are straight lines (2D Hough space), circles (3D Hough space) (see [KBS75]) and ellipses (5D Hough space) (see [YWXT04]). However there is also a transformation, which does not need this prerequisite at all. This so-called generalized Hough transform, introduced by D. H. Ballard in 1981, modifies the original Hough transform by using the principle of template matching ([Bal81]). So with such a new transformation, it is evidently possible to detect arbitrary patterns in a picture, because it allows a pattern to vote for a particular Hough space cell by using a predefined Look-up-table (LUT). Moreover it is clear that this LUT depends at first on the pattern, but can be also upgraded to cover rotated or scaled pattern versions by the application of some simple calculations to the LUT.

Beyond that, the main advantage of the Hough transform used for pattern recognition is that partly covered or fragmented patterns can be also detected. So this transformation is very robust against noise and systematic errors. In addition to this, it is also clear that just patterns, which match the parametric description, can be found and nothing else.
Further on, the major important advantage related to the very demanding task of tracking in the CBM experiment is the complexity of the algorithm with regard to the computing time, because it is proportional to the number of transformed elements, which is commonly called linear and represented by the notation $T = O(n)$. As the general Hough transform algorithm is exhaustively presented now, the next paragraphs show more details about different aspects.

In image processing the input to the Hough transform is commonly a binary image or a gradient image, which are both obviously discrete. Accordingly the Hough space is usually implemented by a discrete n-dimensional array field including one dimension for each feature variable (see feature extraction). This circumstance leads then to the fact that each cell of this array field represents quantized values for a possible combination of the feature parameters.
Based on this concept, the startup of the algorithm is quite clear, because each Hough curve has to be computed and inserted into this array field at

first. Normally this process is realized by the computation of the parameters for each curve, which can be further used to increment the value of all corresponding cells in the array field, that the parameters fall into, by one. So obviously an important prerequisite of this process is that the initial value of each cell is zero. Furthermore such an array field is often called histogram or accumulator, because each cell accumulates the number of Hough curves, which intersect in the representing combination of feature parameters.
Beyond that, this context implies also the major important advantage of the array field implementation, because the computation of the intersections, which determines the next part of the algorithm, alters to a search of cells containing local maxima in the array field instead of solving mathematical equations. Furthermore these identified cells are then called peaks and are most likely to be recognized patterns. For that reason, the two universal steps of the Hough transform are called Histogramming and Peak-finding.

Since the process of Histogramming is straight forward, the focus is set in the following to the Peak-finding without any restriction to the generality. Although the simplest Peak-finding process is described by the application of some form of threshold to all accumulator cells independently, other techniques, which introduce relationships to neighboring cells like windowing or something else, may yield better results in different circumstances.
For instance, a not negligible problem of the Peak-finding process is that rounding errors and inexactnesses in the transformation lead in reality often to so-called clusters, which represent fuzzy peaks covering more than just a single cell in the accumulator ([BB91]). But since only the peak with the highest value should commonly survive the Peak-finding process for one cluster, this correct peak has to be obviously identified, while the noisy rest has to be of course removed. Therefor the windowing technique can be applied in combination with a sufficiently sized window to involve the corresponding neighborhood. For that reason, the original Peak-finding process has to be expanded in the way that it turns from just a simple search into searching and morphing or filtering.
Further on, it is obvious that this problem implies also an important limitation of the Hough transform, because the cell, which contains the correct peak, must be observable at all. So usually, the new required Peak-finding process would be more and more efficient, if a higher number of votes fall into the same or better the correct cell, because then this cell can be much easier detected amid the background noise.

Besides this, another big issue, which influences also the cluster building problem or simply clustering, is formed by the determination of the number of accumulator cells for each dimension (quantization), because there is ob-

viously a trade off between computation exactness and computation time or memory usage.
For instance, if too many cells are taken, the feature resolution is very good, but the computation time would be exhaustive, the memory usage enormous and clustering would aggravate the Peak-finding. However, if too few cells are taken contrarily, too many peaks would be lost, because they correspond to the same cell, whether the computation time and the amount of memory would be acceptable. So based on this two desperate conditions, it is instantly clear that the applied quantization of the dimensions has to be adapted to the needed requirements ([BB91]).
Moreover the selection of an applicable quantization implies a further complicating aspect, because a highly occupied accumulator results in found peaks, which does not correspond to real existing feature objects, because many Hough curves intersect in wrong cells. So a more precise look to such a situation shows that the occupancy of the accumulator depends immediately on the relation between the quantization and the number of features, which should be extracted (see feature extraction), because this number corresponds directly to the number of inserted Hough curves in the accumulator and thus to the occupancy. Thus a common solution for this problem is determined by using simply the same quantization for the Hough space than already used for the image space, because the occupancy of the Hough space can be then directly estimated to be in the order of the image space.

Beyond that, it should be also noticed here that there are some other lacks of the Hough transform with regard to their application.
For instance, the starting point and the ending point of a straight line, which is often of special interest in digital image processing, can not be evaluated, because such features are not determined in the parametric description. So additional algorithms have to be applied after the Hough transform to cope with such a lack.
Nevertheless the Hough transform is a widely spread algorithm, which is used for example in digital image processing to analyze camera pictures for robot cars ([Faz98]), to read bar code labels on a carton driving past on a band-conveyor ([Wen99]) or just to find picture edges ([OC76]).
Further on, more hardware related examples are given by the description of an architecture for a chip set ([AAFN96]), a complete Hough processor from the company Innovasic Semiconductor ([Inn]), which is compatible to the product of LSI Logic Corporation ([LSI]), or another special parallel Hough processor from the department of computer engineering of the university of Mannheim ([KOM95]).
Hough transform examples in high energy physics can be found for digitized

streamer chamber pictures in [PRK88], for layered track chambers in [NO86] or for drift chambers in [PSSW86].

Since many different universal aspects of the Hough transform have been successfully inspected, the following sections present dedicated adaptations and customizations for the CBM experiment like the parametric description of the pattern, which should be detected, because it is similar to a special form of a parabola and not to the recently discussed straight lines, circles or ellipses.
In addition to this, a hardware realization has to be considered in each step of development, because the CBM experiment requires a very challenging processing speed.

2.2 Customizing the Hough Transform

At first, it is evident that the Hough transform has to be customized to be able to process the input from the CBM experiment, which is imaginably not a picture in the common sense of digital image processing, because the delivered information has to be composed of the particle interactions with the detector stations. Furthermore it is clear that the major important characteristic of such an information element, which is called hit, is obviously defined by a three dimensional coordinate including a single dimension for each degree of freedom of the possible particle motion trajectories. So within this context, it is obvious that such a picture, which is naturally called coordinate or track space, contains only the hits instead of the total three dimensional picture space, because this circumstance offers the possibility to reduce the enormous amount of data to just the important one.

Besides this, the chosen Hough transform is a generalized one, because a map φ, which evaluates the Hough curve based on the hit information ($\varphi_{(hit)} \rightarrow$ Hough curve), is used. In correlation to this, it is of further interest that the thesis presents two different algorithms, which are able to build the map φ. At this juncture, the first and easier one is characterized by an analytic formula, which approximates the common motion trajectory of a charged particle in an inhomogeneous magnetic field. For this purpose, the basic idea originates from the common equation for the deflection of a charged particle in a homogeneous magnetic field, which is further expanded to the inhomogeneous one by the application of some simple adaptations to compensate the errors and to receive an approximating formula. At this point, it should be additionally noticed that such an analytic formula could be also used in

combination with a standard Hough transform, because the equation represents a parametric description.
Moreover the second algorithm, which features the generation of the map, is distinguished by the application of the Runge-Kutta approach. As however this approach does explicitly not form a usable parametric description, which is applicable for a standard Hough transform, it is certainly indispensable to use a generalized Hough transform to be able to directly compare both maps without any effect to the rest of the transformation process, because they are simply interchangeable.
Having now a more precise look to the map φ, it is clear that the map has to transform the existing hits into the corresponding feature parameter space (see feature extraction) by applying the following rules:

- The map φ has to be uniquely reversible
- The feature parameter space has as many dimensions as the motion trajectory degrees of freedom
- Each point in the parameter space represents one possible motion trajectory
- The value of each point in the parameter space represents the probability of a possible motion trajectory
- The target in the experimental setup is assumed to be the primary interaction point (common vertex)
- The common vertex is the point of origin

By accomplishing now these rules, the three dimensional Hough space, which is customized for the CBM experiment, can be configured with $\theta \backsim \frac{p_x}{p_z}, -\frac{q}{p_{xz}}$ and $\gamma \backsim \frac{p_y}{p_z}$ (see [GSM05]), because the common feature parameters are determined by the three dimensional momentum p_x, p_y and p_z due to the three degrees of freedom for the motion trajectory of a charged particle. In addition to this, it should be noticed here that the common feature parameters are not taken itself, because some limitations exist for the transformation, which are comparable to the problem introduced and solved by the $f_{(m,c)}$ and the $f_{(p,\theta)}$ model in section 2.1 on page 45. Moreover figure 2.4 depicts the orientation of the Hough space dimensions, which denotes a very important aspect in the following implementations.

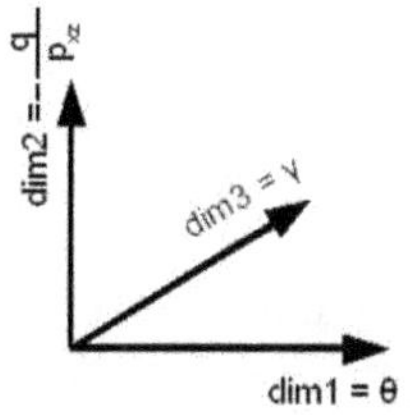

Figure 2.4: Sketch of the Hough space

By taking further the detector setup constraints of the CBM experiment, which can be found in section 1.3.1 on page 12, into account, the ranges of the dimensions for the coordinate space are fixed to:

$$
\begin{aligned}
& z \in [+30; +100] \\
& \frac{|x^{max}_{layer}|}{z_{layer}} = const = 0.5 \\
\Rightarrow\quad & x \in [-50; +50] \\
& \frac{|y^{max}_{layer}|}{z_{layer}} = const = 0.5 \\
\Rightarrow\quad & y \in [-50; +50]
\end{aligned}
$$

Applying these constraints additionally to the Hough space, the ranges of the corresponding dimensions are determined by:

$$
\begin{array}{rrcccccc}
 & \frac{|x^{max}_{layer}|}{z_{layer}} & = & \frac{|y^{max}_{layer}|}{z_{layer}} & = & const & = & 0.5 \\
\Rightarrow & \tan(\gamma^{max}) & = & \tan(\theta^{max}) & = & const & = & 0.5 \\
\Rightarrow & \gamma^{max} & = & \theta^{max} & = & \arctan(0.5) & & \\
 & & \approx & 27^\circ & = & \frac{3}{20}\pi & = & 0.47 \\
\Rightarrow & \theta & \in & [-0.47; +0.47] & & & & \\
\Rightarrow & \gamma & \in & [-0.47; +0.47] & & & & \\
 & & & & & & & \\
 & q & \in & \{-1; +1\} & & & & \\
 & p_{xz} & \geq & 0.9\,\frac{\mathrm{GeV}}{\mathrm{c}} & & & & \\
\Rightarrow & -\frac{q}{p_{xz}} & = & [-1.11; +1.11] & & & &
\end{array}
$$

So based on these ranges, the Hough space is finally defined by:

$$\begin{aligned} dim1 &= \theta, \quad \theta \in [-0.47; +0.47] \\ dim2 &= -\tfrac{q}{p_{xz}}, \quad p_{xz} \in [-1.11; +1.11], \quad q \in \{-1; +1\} \\ dim3 &= \gamma, \quad \gamma \in [-0.47; +0.47] \end{aligned} \tag{2.5}$$

2.3 Basic Hough Transform Formulas

Since the major essential prerequisite of the applied Hough transform is the uniquely reversibility, the basic formulas of this transformation have to determine the correlation of the coordinate or track space to the Hough space and vice versa. So for this purpose, it is obvious that at least two different formulas have to be available, which feature the computation of the representation of a single element of one space in the other space.
More precisely, one formula has to determine the corresponding feature parameters (feature extraction) of a track, which are originally identified by the momentum $\vec{p} = \begin{pmatrix} p_x \\ p_y \\ p_z \end{pmatrix}$, for a given histogram cell, while another formula has to designate accordingly the histogram cell for a given track, which is again defined by the momentum (feature parameters). Moreover these connections are very useful, because they allow for instance the computation of the histogram cell, which should accommodate the peak of a given MC Track.
Besides this, the Hough space dimensions suggest that the feature parameters, which are originally identified by the momentum, are not unique. So another more popular representation, which is called track parameter set or just track parameters, should be also taken into account. For this purpose, it is clear that the two required basic formulas have to be adapted to be suitable for these feature parameters, which are devised by $\begin{pmatrix} t_x \\ t_y \\ \mathbf{qp} \end{pmatrix}$. But within this context, it is important to realize that the variable **qp** is just a name, which is set and used by the CBMROOT framework. So this fact defines **qp** to explicitly not represent the multiplication of the charge q with the momentum p ($\mathbf{qp} \neq q \cdot p$).

Further on, it is also obvious that it might be good to be able to transform the different feature representations into each other. But before presenting now all basic formulas including their derivation, it has to be noticed that the usage of the momentum representation is preferred in this thesis, because it is more conceivable for me. Nevertheless the track parameter representation is additionally presented, because it is mainly used in the software framework CBMROOT. More information about this framework can be found in section 3.2.1 on page 102.

2.3.1 Evaluating the Track Parameter Set for a Given Momentum

Since the introduction of this section indicates that a feature can be defined by two different representations, it might be good to be able to transform one representation into the other one. So for that reason, this section presents the largely known formulas, which allow the computation of the track parameter set for a given momentum.

$$\boxed{\begin{aligned} t_x &= \tfrac{p_x}{p_z} \\ t_y &= \tfrac{p_y}{p_z} \\ \mathrm{qp} &= \sqrt{p_x^2 + p_y^2 + p_z^2} \end{aligned}} \tag{2.6}$$

2.3.2 Evaluating the Momentum for a Given Track Parameter Set

The formulas to compute the momentum for a given track parameter set can be accordingly found by reforming the equations 2.7 of section 2.3.1 on page 60:

$$\begin{aligned} p_x &= t_x p_z \\ p_y &= t_y p_z \\ p_z &= \tfrac{\mathrm{qp}}{\sqrt{t_x^2+t_y^2+1}} \end{aligned}$$

to:

$$\boxed{\begin{aligned} p_x &= t_x \frac{qp}{\sqrt{t_x^2+t_y^2+1}} \\ p_y &= t_y \frac{qp}{\sqrt{t_x^2+t_y^2+1}} \\ p_z &= \frac{qp}{\sqrt{t_x^2+t_y^2+1}} \end{aligned}} \tag{2.7}$$

2.3.3 Evaluating the Momentum for a Given Histogram Cell

As the conversions of both feature representations into each other are successfully presented, the focus is now set to the formulas, which offer the possibility to evaluate the momentum for a given histogram cell. Here it should be additionally noticed that the benefit of these formulas is simply identified by the reverse transformation of the Hough transform results, which mean the determination of the momentum for each found peak characterized by its histogram cell.
Further on, the basic concept, which is used here to derive the formulas, is distinguished by the circumstance that the angles between $\vec{s_x}$, $\vec{s_z}$ and $\vec{p_x}$, $\vec{p_z}$ and between $\vec{s_y}$, $\vec{s_z}$ and $\vec{p_y}$, $\vec{p_z}$ are equal because of the well-known equation $\vec{p} = m \cdot \vec{v} = m \cdot \frac{\vec{s}}{t}$ (see [TMP06] page 213 and [Vog95] page 10). So based on this, the following simple formulas result:

$$\begin{aligned} & \vec{p_{xz}} &=& \; m\vec{v}_{xz} = m\frac{\vec{s}_{xz}}{t} \\ \Rightarrow & \angle\left(|\vec{p_x}| = p_x; |\vec{p_z}| = p_z\right) &=& \; \angle\left(|\vec{s_x}| = x; |\vec{s_z}| = z\right) \\ \Rightarrow & \tan(\theta) &=& \; \frac{p_x}{p_z} = \frac{x}{z} \\ \Rightarrow & p_x &=& \; p_z \tan(\theta) \end{aligned}$$

and

$$\begin{aligned} & \vec{p_{yz}} &=& \; m\vec{v}_{yz} = m\frac{\vec{s}_{yz}}{t} \\ \Rightarrow & \angle\left(|\vec{p_y}| = p_y; |\vec{p_z}| = p_z\right) &=& \; \angle\left(|\vec{s_y}| = y; |\vec{s_z}| = z\right) \\ \Rightarrow & \tan(\gamma) &=& \; \frac{p_y}{p_z} = \frac{y}{z} \\ \Rightarrow & p_y &=& \; p_z \tan(\gamma) \end{aligned}$$

Moreover the momentum in z direction can be evaluated by the equation:

$$\begin{array}{rrclcl} & p_{xz} & = & \sqrt{p_x^2+p_z^2} & = & p_z\sqrt{\left(\frac{p_x}{p_z}\right)^2+1} \\ \Rightarrow & -\frac{q}{p_{xz}} & = & dim2 & = & -\frac{q}{p_z\sqrt{\left(\frac{p_x}{p_z}\right)^2+1}} \\ \Rightarrow & p_z & = & -\frac{q}{dim2\sqrt{\left(\frac{p_x}{p_z}\right)^2+1}} & = & -\frac{q}{dim2\sqrt{(\tan(\theta))^2+1}} \end{array}$$

By applying now the familiar Hough space dimensions, which are determined by $\theta = dim1$, $-\frac{q}{p_{xz}} = dim2$ and $\gamma = dim3$, the final result is given by:

$$\boxed{\left.\begin{array}{rcl} p_x & = & -\frac{q}{dim2\sqrt{(\tan(dim1))^2+1}}\tan(dim1) \\ p_y & = & -\frac{q}{dim2\sqrt{(\tan(dim1))^2+1}}\tan(dim3) \\ p_z & = & -\frac{q}{dim2\sqrt{(\tan(dim1))^2+1}} \end{array}\right\} p_z \in (0;\infty) \Rightarrow q=-1} \quad (2.8)$$

2.3.4 Evaluating the Histogram Cell for a Given Momentum

The actual section contains in analogousness the formulas, which offer the possibility to evaluate the histogram cell for a given momentum. Further on, it should be again mentioned that the employment of these formulas can be found in several algorithms, which analyze the Hough transform to optimize their parameters, because the cell containing the peak identifying a given MC Track can be evaluated.
Moreover the basic concept, which is applied here to derive the formulas, is again determined by the fact that the angles between $\vec{s_x}$, $\vec{s_z}$ and $\vec{p_x}$, $\vec{p_z}$ and between $\vec{s_y}$, $\vec{s_z}$ and $\vec{p_y}$, $\vec{p_z}$ are equal because of the well-known equation $\vec{p} = m \cdot \vec{v} = m \cdot \frac{\vec{s}}{t}$ (see [TMP06] page 213 and [Vog95] page 10). So based on

this, the following simple formulas result:

$$
\begin{aligned}
& \vec{p_{xz}} &=& \quad m\vec{v}_{xz} = m\frac{\vec{s}_{xz}}{t} \\
\Rightarrow & \angle\left(|\vec{p_x}| = p_x; |\vec{p_z}| = p_z\right) &=& \quad \angle\left(|\vec{s_x}| = x; |\vec{s_z}| = z\right) \\
\Rightarrow & \tan(\theta) &=& \quad \frac{p_x}{p_z} = \frac{x}{z} \\
\Rightarrow & \theta &=& \quad \arctan\left(\frac{p_x}{p_z}\right) = \arctan\left(\frac{x}{z}\right)
\end{aligned}
$$

and

$$
\begin{aligned}
& \vec{p_{yz}} &=& \quad m\vec{v}_{yz} = m\frac{\vec{s}_{yz}}{t} \\
\Rightarrow & \angle\left(|\vec{p_y}| = p_y; |\vec{p_z}| = p_z\right) &=& \quad \angle\left(|\vec{s_y}| = y; |\vec{s_z}| = z\right) \\
\Rightarrow & \tan(\gamma) &=& \quad \frac{p_y}{p_z} = \frac{y}{z} \\
\Rightarrow & \gamma &=& \quad \arctan\left(\frac{p_y}{p_z}\right) = \arctan\left(\frac{y}{z}\right)
\end{aligned}
$$

In addition to this, the third parameter is defined by the equations:

$$
\begin{aligned}
& p_{xz} &=& \sqrt{p_x^2 + p_z^2} &=& \; p_z\sqrt{\left(\frac{p_x}{p_z}\right)^2 + 1} \\
\Rightarrow & -\frac{q}{p_{xz}} &=& -\frac{q}{p_z\sqrt{\left(\frac{p_x}{p_z}\right)^2 + 1}} &&
\end{aligned}
$$

By applying now the familiar Hough space dimensions, which are defined by $dim1 = \theta$, $dim2 = -\frac{q}{p_{xz}}$ and $dim3 = \gamma$, the final result turns into:

$$
\boxed{
\begin{aligned}
dim1 &= \arctan\left(\frac{p_x}{p_z}\right) \\
dim2 &= -\frac{q}{p_z\sqrt{\left(\frac{p_x}{p_z}\right)^2 + 1}} \\
dim3 &= \arctan\left(\frac{p_y}{p_z}\right)
\end{aligned}
} \tag{2.9}
$$

2.3.5 Evaluating the Track Parameter Set for a Given Histogram Cell

As the two basic formulas for the momentum representation are presented, the focus is now set to the track parameter representation. So for that purpose, the actual section contains the formulas, which offer the possibility to evaluate the track parameters for a given histogram cell. These formulas are commonly used to compute the feature parameters of a found track for the interface of the CBMROOT framework (see section 3.2.1 on page 102).
Further on, the basic concept, which is applied here to derive the formulas, is again characterized as expected by the fact that the angles between $\vec{s_x}$, $\vec{s_z}$ and $\vec{p_x}$, $\vec{p_z}$ and between $\vec{s_y}$, $\vec{s_z}$ and $\vec{p_y}$, $\vec{p_z}$ are equal because of the well-known equation $\vec{p} = m \cdot \vec{v} = m \cdot \frac{\vec{s}}{t}$ (see [TMP06] page 213 and [Vog95] page 10). So based on this, the following simple formulas result:

$$\begin{array}{rrcl} & \vec{p_{xz}} & = & m\vec{v}_{xz} = m\frac{\vec{s}_{xz}}{t} \\ \Rightarrow & \angle\left(|\vec{p_x}| = p_x; |\vec{p_z}| = p_z\right) & = & \angle\left(|\vec{s_x}| = x; |\vec{s_z}| = z\right) \\ \Rightarrow & \tan\left(\theta\right) & = & \frac{p_x}{p_z} = \frac{x}{z} \\ \Rightarrow & t_x = \frac{p_x}{p_z} = \tan\left(\theta\right) & & \end{array}$$

and

$$\begin{array}{rrcl} & \vec{p_{yz}} & = & m\vec{v}_{yz} = m\frac{\vec{s}_{yz}}{t} \\ \Rightarrow & \angle\left(|\vec{p_y}| = p_y; |\vec{p_z}| = p_z\right) & = & \angle\left(|\vec{s_y}| = y; |\vec{s_z}| = z\right) \\ \Rightarrow & \tan\left(\gamma\right) & = & \frac{p_y}{p_z} = \frac{y}{z} \\ \Rightarrow & t_y = \frac{p_y}{p_z} = \tan\left(\gamma\right) & & \end{array}$$

In addition to this, the third parameter is defined by the equations:

$$\begin{array}{rrcccl} & p_{xz} & = & \sqrt{p_x^2 + p_z^2} & = & p_z\sqrt{\left(\frac{p_x}{p_z}\right)^2 + 1} \\ \Rightarrow & -\frac{q}{p_{xz}} & = & dim2 & = & -\frac{q}{p_z\sqrt{\left(\frac{p_x}{p_z}\right)^2+1}} \\ \Rightarrow & p_z & = & -\frac{q}{dim2\sqrt{\left(\frac{p_x}{p_z}\right)^2+1}} & = & -\frac{q}{dim2\sqrt{(\tan(\theta))^2+1}} \\ & & = & \left.-\frac{q}{dim2\sqrt{(\tan(dim1))^2+1}}\right\} & & p_z \in (0;\infty) \Rightarrow q = -1 \end{array}$$

followed by:

$$\begin{aligned}\mathbf{qp} &= \sqrt{p_x^2 + p_y^2 + p_z^2} &&= \sqrt{\left(\frac{p_x^2}{p_z^2} + \frac{p_y^2}{p_z^2} + 1\right) p_z^2} \\ &= p_z \sqrt{\left(\frac{p_x}{p_z}\right)^2 + \left(\frac{p_y}{p_z}\right)^2 + 1} &&= p_z \sqrt{t_x^2 + t_y^2 + 1}\end{aligned}$$

By applying now the familiar Hough space dimensions, which are determined by $\theta = dim1$, $-\frac{q}{p_{xz}} = dim2$ and $\gamma = dim3$, the recent result turns into:

$$\begin{aligned}\Rightarrow \mathbf{qp} &= -\frac{q}{dim2\sqrt{(\tan(\theta))^2+1}} \sqrt{t_x^2 + t_y^2 + 1} \\ \mathbf{qp} &= -\frac{q}{dim2\sqrt{(\tan(\theta))^2+1}} \sqrt{(\tan(\theta))^2 + (\tan(\gamma))^2 + 1}\end{aligned}$$

In combination with the definition of the track parameter set, which is introduced in the equations 2.6 of section 2.3.1 on page 59, the final result looks like:

$$\boxed{\begin{aligned} t_x &= \tan(dim1) \\ t_y &= \tan(dim3) \\ \mathbf{qp} &= -\frac{q}{dim2\sqrt{(\tan(dim1))^2+1}} \sqrt{(\tan(dim1))^2 + (\tan(dim3))^2 + 1} \end{aligned} \Bigg\} q = -1} \quad (2.10)$$

2.3.6 Evaluating the Histogram Cell for a Given Track Parameter Set

The actual section contains in analogousness the formulas, which offer the possibility to evaluate the histogram cell for a given track parameter set. But in contrast to the equations 2.10 of section 2.3.5 on page 64, these formulas are just shown here to complete the formulary, because they are not required in any occurring case.
Besides this the basic concept, which is applied here to derive the formulas, is again expectedly designed by the fact that the angles between $\vec{s_x}$, $\vec{s_z}$ and $\vec{p_x}$, $\vec{p_z}$ and between $\vec{s_y}$, $\vec{s_z}$ and $\vec{p_y}$, $\vec{p_z}$ are equal because of the well-known equation $\vec{p} = m \cdot \vec{v} = m \cdot \frac{\vec{s}}{t}$ (see [TMP06] page 213 and [Vog95] page 10). So combining this with the definition of the track parameter set, which is introduced in the equations 2.6 of section 2.3.1 on page 59, the following simple formulas result:

$$\begin{array}{rrcl} & \vec{p_{xz}} & = & m\vec{v}_{xz} = m\frac{\vec{s}_{xz}}{t} \\ \Rightarrow & \angle(p_x; p_z) & = & \angle(s_x = x; s_z = z) \\ \Rightarrow & \tan(\theta) & = & \frac{p_x}{p_z} = \frac{x}{z} \\ \Rightarrow & \theta = \arctan\left(\frac{p_x}{p_z}\right) & = & \arctan(t_x) \end{array}$$

and

$$\begin{array}{rrcl} & \vec{p_{yz}} & = & m\vec{v}_{yz} = m\frac{\vec{s}_{yz}}{t} \\ \Rightarrow & \angle(p_y; p_z) & = & \angle(s_y = y; s_z = z) \\ \Rightarrow & \tan(\gamma) & = & \frac{p_y}{p_z} = \frac{y}{z} \\ \Rightarrow & \gamma = \arctan\left(\frac{p_y}{p_z}\right) & = & \arctan(t_y) \end{array}$$

In addition to this, the third parameter is defined by the equations:

$$\begin{array}{rrcccl} & \mathbf{qp} & = & \sqrt{p_x^2 + p_y^2 + p_z^2} & = & \sqrt{\left(\frac{p_x^2}{p_z^2} + \frac{p_y^2}{p_z^2} + 1\right) p_z^2} \\ & & = & p_z\sqrt{\left(\frac{p_x}{p_z}\right)^2 + \left(\frac{p_y}{p_z}\right)^2 + 1} & = & p_z\sqrt{t_x^2 + t_y^2 + 1} \\ \Rightarrow & p_z & = & \frac{\mathbf{qp}}{\sqrt{t_x^2+t_y^2+1}} & & \end{array}$$

followed by:

$$\begin{array}{rrcccl} & p_{xz} & = & \sqrt{p_x^2 + p_z^2} & = & p_z\sqrt{\left(\frac{p_x}{p_z}\right)^2 + 1} \\ \Rightarrow & -\frac{q}{p_{xz}} & = & -\frac{q}{p_z\sqrt{\left(\frac{p_x}{p_z}\right)^2+1}} & = & -\frac{q}{p_z\sqrt{t_x{}^2+1}} \\ & & = & -\frac{q}{\frac{\mathbf{qp}}{\sqrt{t_x^2+t_y^2+1}}\sqrt{t_x{}^2+1}} & & \\ & & = & -\frac{q}{\mathbf{qp}}\sqrt{\frac{t_x^2+t_y^2+1}{t_x^2+1}} & & \end{array}$$

By applying now the familiar Hough space dimensions, which are defined by $dim1 = \theta$, $dim2 = -\frac{q}{p_{xz}}$ and $dim3 = \gamma$, the final result turns into:

$$\boxed{\begin{array}{lcc} dim1 & = & \arctan(t_x) \\ dim2 & = & -\frac{q}{\mathbf{qp}}\sqrt{\frac{t_x^2+t_y^2+1}{t_x^2+1}} \quad \left.\begin{array}{l} q=+1, \text{ if } \mathbf{qp} \geqq 0 \\ q=-1, \quad \text{else} \end{array}\right\} \\ dim3 & = & \arctan(t_y) \end{array}} \tag{2.11}$$

2.4 Analytic Formula Usable for the Hough Transform

Before having a closer look to the analytic formula, it is required to remember the basic concept of the general Hough transform, which means the application of a map to transform all hits into the track parameter space. Further on, the customization of this map φ causes the content to represent two independent but superposing models for the motion trajectory of a charged particle in an inhomogeneous magnetic field. Beyond that, this superposition principle can be further easily used to enable the three dimensional Hough transform to be separable into a two dimensional transformation (the bending x-z-plane projection of the trajectory with a parabola), which is expandable in the third dimension (the y-z-plane projection of the trajectory with a slightly curved line). So based on this concept, it is immediately clear that a single three dimensional Hough curve, which is obviously a plane at the end, is composed of two two dimensional curve projections.
Coming now to the point, it is self-evident that a single track with the parameters p_x and p_z is represented by a single point in this projection of the Hough space, which is of course two dimensional (see figure 2.5).

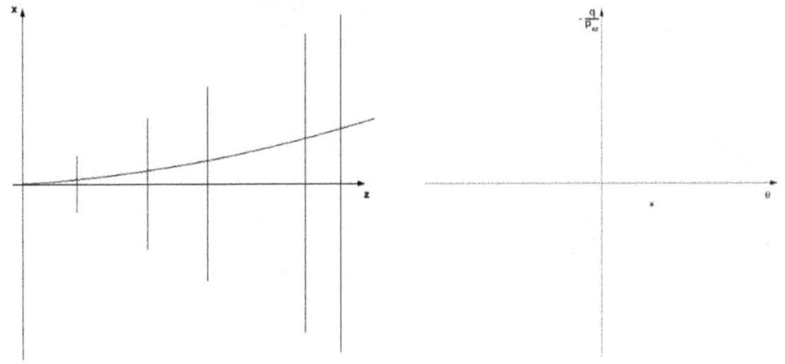

(a) One track in coordinate space (b) One track in Hough space

Figure 2.5: Illustration of the basic Hough transform concept

As furthermore a single measured hit can contribute to several tracks, the model for the x-z-plane projection has to take care about the appropriateness of the curve in the two dimensional Hough space (see fig. 2.6), which means that the x and z coordinates have to obey the corresponding equation.

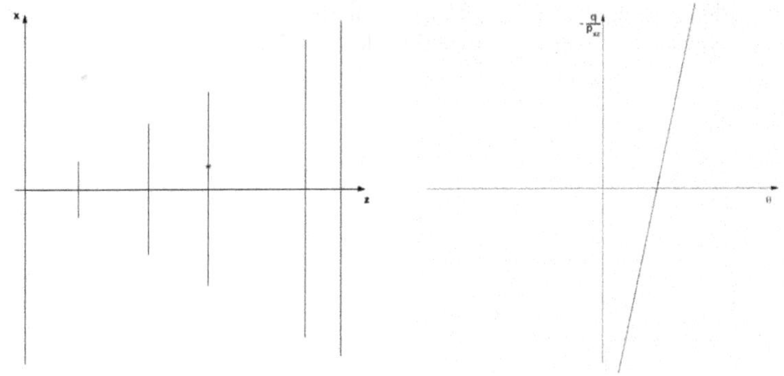

(a) One hit in coordinate space (b) One hit in Hough space

Figure 2.6: Illustration of the basic Hough transform process

So the enhancement of the previous assertion implies also that several sampling hits of the motion trajectory of a track lead to several curves in the two dimensional Hough space, which intersect all in the point that describes the feature parameters of exactly this track (see fig. 2.7).

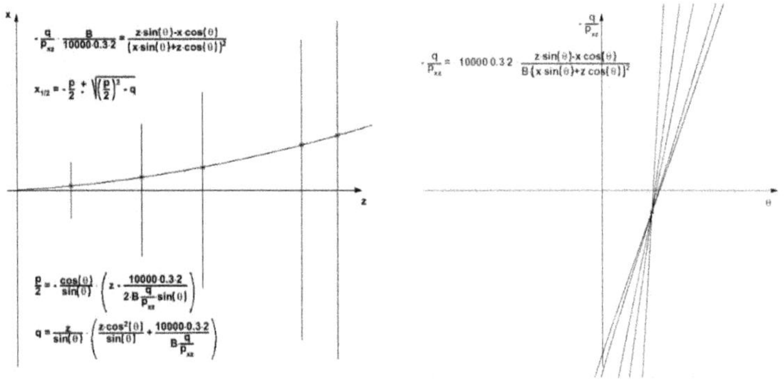

(a) Five hits in the coordinate space (b) Five hits in the Hough space

Figure 2.7: Illustration of the x-z-plane projection of a single track

Moreover the already mentioned clue for the expansion into the third dimension is obviously defined by the superposed application of the y-z-plane projection, which implies the usage of another equation to determine the index range of the x-z-plane projections for the Hough curve of each hit.

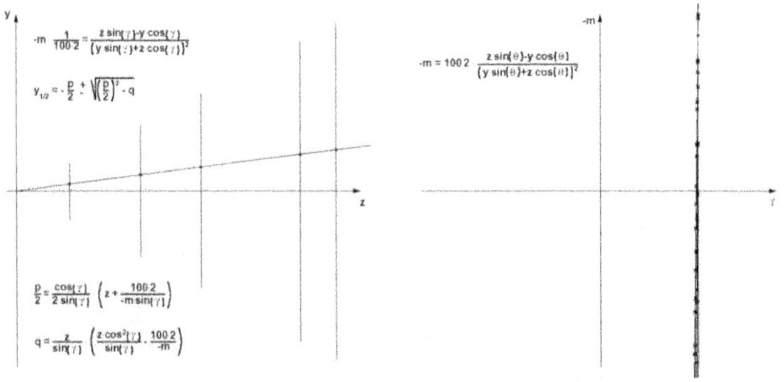

(a) Five hits in the coordinate space (b) Five hits in the Hough space

Figure 2.8: Illustration of the y-z-plane projection of a single track

So within this context, it is clear that the principals of both models are equal (see figure 2.8), which implies that a single track with the parameters p_y and p_z is also represented by a single point in this projection of the Hough space. However it has to be noticed here that the major important element of this model is defined by the index range, because it is varying dependent on the hit and should be kept as small as possible. Returning now to the global point of view, it is finally clear that the result of the composite algorithm is just a single peak for each track in the overlapping projection planes. Hence the two projection models lead to a three dimensional Hough transform, which maps the detector hits in accordance to the track parameters θ, $-\frac{q}{p_{xz}}$ and γ in the Hough space.

2.4.1 Basic Analytic Formula Containing a Homogeneous Magnetic Field

This section presents the derivation of the analytic formula, which originates at the beginning from the homogeneous magnetic field, because simple well-known physics can be applied therefore.

y-z-Plane Projection

The first chosen projection to introduce is the one in the y-z-plane, because the negligible influence of the magnetic field (see section 4.4 on page 223) leads to a quite simple transformation model, which is assumed to be a straight line at first.
So with regard to the starting angle of the particle motion trajectory (see figure 2.9), the following simple equations can be deduced:

$$
\begin{aligned}
m &= \frac{p_y}{p_z} \\
y &= mz \\
\Rightarrow m &= \frac{y}{z} = \frac{p_y}{p_z} \\
m &= tan\left(\gamma\right) \\
\Rightarrow \gamma &= arctan\left(\frac{p_y}{p_z}\right) = arctan\left(\frac{y}{z}\right)
\end{aligned}
$$

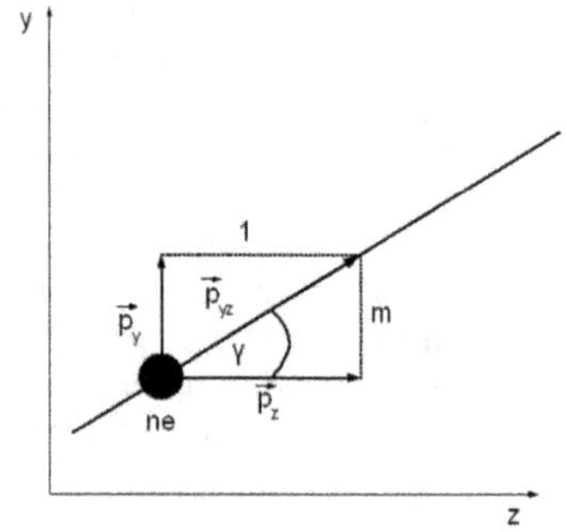

Figure 2.9: Illustration of the charged particle motion influenced by an insignificant magnetic field

However the real shape of the trajectory differs from an exact straight line due to inexactness in the computation because of side effects like multiple scattering or quantization issues. Hence a slightly curved line is preferred including a security region for the γ dimension of the Hough space. For this purpose, the usual computed γ entry is characterized by a range $[\gamma_{min}; \gamma_{max}]$ instead of just a single value. Moreover it is self-evident that this security region or range has to get bigger with an increasing distance from the target, which corresponds to an increasing z coordinate. Thus the transformation model is now changed and further distinguished by the following equation:

$$\begin{aligned} y &= \tfrac{1}{2} m z^2 \\ \Rightarrow m &= 2\tfrac{y}{z^2} \end{aligned}$$

But as the common particle motion trajectory contains a starting angle, this fact has to be taken into account. Hence the simple approach of a rotated coordinate system, which is depicted in figure 2.10, is applied.

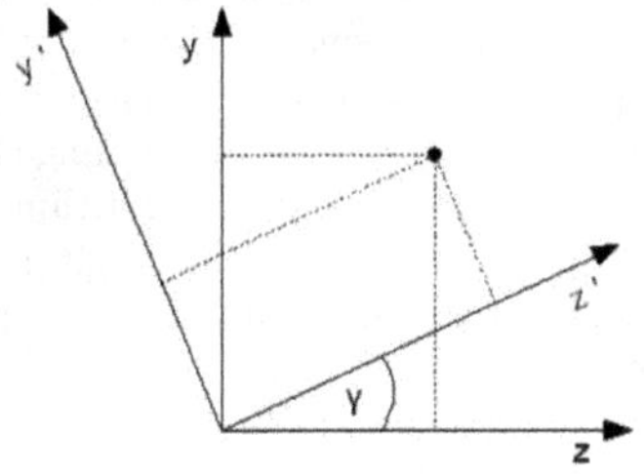

Figure 2.10: Relating sketch for the original and the rotated coordinate system

So rotating the coordinate system by the starting angle γ and the rotation matrix $D = \begin{pmatrix} \cos(\gamma) & -\sin(\gamma) \\ \sin(\gamma) & \cos(\gamma) \end{pmatrix}$ leads to the new coordinates (see [BSMM05] page 196):

$$\begin{aligned} y' &= -z\sin(\gamma) + y\cos(\gamma) \\ z' &= y\sin(\gamma) + z\cos(\gamma) \end{aligned} \tag{2.12}$$

Applying this knowledge now to the recent introduced parabola transformation model, the equation turns into:

$$m = 2\frac{-z\sin(\gamma)+y\cos(\gamma)}{(y\sin(\gamma)+z\cos(\gamma))^2}$$

Further on, the multiplication of both equation sides with -1 leads to a more readable formula in the programming source code. Therefore the equation changes further to:

$$-m = 2\frac{z\sin(\gamma)-y\cos(\gamma)}{(y\sin(\gamma)+z\cos(\gamma))^2} \quad \left[\frac{1}{\text{m}}\right]$$

This resulting formula offers now the possibility to define a varying security region for the γ dimension of the Hough space in dependency of the fix basis range for m and the varying y and z coordinates of the hits. Thence a γ_{min} and a γ_{max} value has to be computed for each hit by this equation, which determine a boundary for accepted γ values. Hence the range of m arranges the overlapping area, which is obviously strongly dependent on the input data, because if the range for m is too big, a unique peak would be found in multiple consecutive layers of the histogram increasing the tracking ghost rate. And if the range for m is too small, the peak could be not high enough to be found. For this purpose, section 4.8.2 on page 268 presents an analysis, which features the quality evaluation of a chosen range for m with regard to a given data set. Furthermore there is also an algorithm described, which tries to automatically determine this range based on given MC data. Returning now to the major topic, the actual units of this basic formula are summarized in the following table 2.1.

Parameter	**Unit**
m	$\frac{1}{\mathrm{m}}$
y	m
z	m
γ	rad

Table 2.1: Units of the basic formula for the y-z-plane projection

But as other units are required for the direct usage of the equation in the environment of the CBM experiment, the equation hast to be customized. Therefore the necessary units are shown in table 2.2 at first.

Parameter	**Unit**
m	$\frac{1}{\mathrm{m}}$
y	cm
z	cm
γ	rad

Table 2.2: Units of the customized formula for the y-z-plane projection

While transforming now the equation to accept these units, the following results:

$$
\begin{aligned}
-m &= 2\frac{\frac{1}{100}z\sin(\gamma)-\frac{1}{100}y\cos(\gamma)}{\left(\frac{1}{100}y\sin(\gamma)+\frac{1}{100}z\cos(\gamma)\right)^2} \\
&= 2\frac{\frac{1}{100}(z\sin(\gamma)-y\cos(\gamma))}{\left(\frac{1}{100}\right)^2(y\sin(\gamma)+z\cos(\gamma))^2}
\end{aligned}
$$

So finally the implementable formula for the first LUT looks like:

$$
\boxed{-m = 100\cdot 2\frac{z\sin(\gamma)-y\cos(\gamma)}{(y\sin(\gamma)+z\cos(\gamma))^2}\quad\left[\frac{1}{\mathrm{m}}\right]} \tag{2.13}
$$

x-z-Plane Projection

In contrast to the y-z-plane projection, which is presented in section 2.4.1 on page 69, the x-z-plane projection is modeled by the common physics theory for the motion of a charged particle in a homogeneous magnetic field at the beginning (see [TMP06] page 830 and [Vog95] pages 451-452). So for that reason, the startup is given by the following equation:

$$
\begin{aligned}
\vec{F}_{Lorentz} &= \vec{F}_{Zentripetal} \\
\Rightarrow\quad |\vec{F}_{Lorentz}| &= |\vec{F}_{Zentripetal}| \\
\Rightarrow\quad nevB\sin(\alpha) &= m\frac{v^2}{r}
\end{aligned}
$$

By assuming now the velocity to be perpendicular to the magnetic field ($\vec{v}\perp\vec{B}\Rightarrow\sin(\alpha)=\sin(90°)=1$) in combination with the well-known equation $p=mv$ (see [TMP06] page 213), the formula can be rewritten as:

$$
\begin{aligned}
nevB\sin(\alpha) &= m\frac{v^2}{r} \\
\Rightarrow\quad nevB &= p\frac{v}{r}
\end{aligned}
$$

and then finally as:

$$
\boxed{\frac{neB}{p} = \frac{1}{r}} \tag{2.14}
$$

A closer look to this formula or its visualization in figure 2.11 shows at present that the motion trajectory of a charged particle in a homogeneous magnetic field is generally described by a circle with radius r. Although this figure depicts such a common experimental setup, it visualizes also the reason for the later proved inapplicability to the CBM experiment.

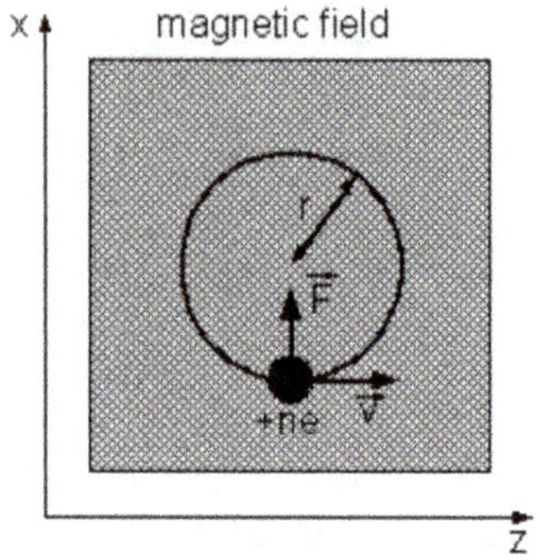

Figure 2.11: Illustration of the charged particle motion inside a significant homogeneous magnetic field

Since the particles, which should be detected, are characterized by $p_z \geq 1\,\frac{\mathrm{GeV}}{\mathrm{c}}$ and the homogeneous magnetic field is assumed to be 10 kGauss (see unit transformations in equation 2.16), the following minimal radius of the circle results:

$$\begin{aligned} r &= \frac{p}{neB} \\ \Rightarrow r &\geq \frac{1\cdot\frac{1.602\cdot10^{-19}}{0.3}\,\mathrm{Ns}}{1\cdot1.602\cdot10^{-19}\,\mathrm{C}\cdot1\,\mathrm{T}} \\ \Rightarrow r &\geq 3\tfrac{1}{3}m \end{aligned}$$

So based on this result, it is immediately obvious that this formula is not applicable for the CBM experiment, because the size of the magnetic field encompasses just one meter in each direction, and the particles enter additionally in the middle of the left side. Thence only circles with a radius smaller than half a meter would be possible for the environmental setup of the CBM experiment. For that reason, the transformation model is changed to the also well-known physics theory for the motion of a charged particle, which is deflected by a perpendicular magnetic field. For that purpose, figure 2.12 gives a good imagination about this model.

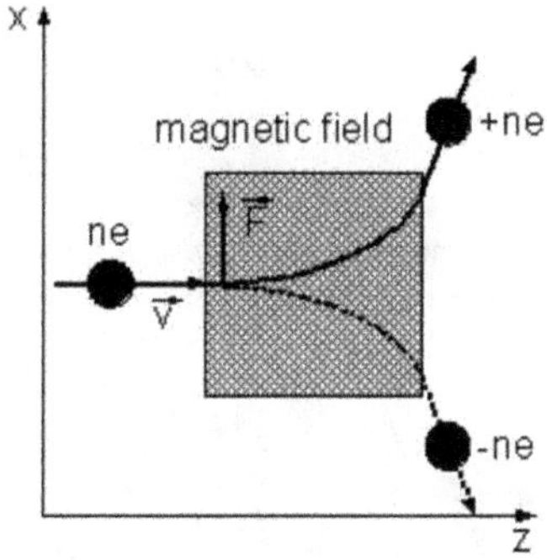

Figure 2.12: Illustration of the charged particle motion deflected by a significant homogeneous magnetic field

Within this context, it is clear that the deflection is modeled by a uniform accelerated motion (see [TMP06] page 29):

$$x = \frac{1}{2}at^2$$

Further on, the time of the particle being in the magnetic field is defined by (see [TMP06] page 29):

$$t = \frac{z}{v}$$

Besides this, the uniform acceleration can be computed by using the second axiom of Sir Isaac Newton (1643-1727) (see [TMP06] page 78) $F = m \cdot a$ in combination with the Lorentz Force (see [Vog95] page 451-452) $F_{Lorentz} = n \cdot e \cdot v \cdot B$:

$$\begin{aligned} F &= ma &= nevB \\ \Rightarrow a &= \tfrac{F}{m} &= \tfrac{nevB}{m} \end{aligned}$$

So by putting these two rudiments now together, the following equation results:

$$\begin{aligned} x &= \tfrac{1}{2}at^2 \\ \Rightarrow\quad x &= \tfrac{1}{2}\cdot\tfrac{nevB}{m}\cdot\left(\tfrac{z}{v}\right)^2 \\ \Rightarrow\quad x &= \tfrac{1}{2}\cdot\tfrac{neB}{mv}\cdot z^2 \end{aligned}$$

Using additionally the well-known equation $p = mv$ (see [TMP06] page 213), the formula can be further rewritten as:

$$\begin{aligned} x &= \tfrac{1}{2}\cdot\tfrac{neB}{mv}\cdot z^2 \\ \Rightarrow\quad x &= \tfrac{1}{2}\cdot\tfrac{neB}{p}\cdot z^2 \\ \Rightarrow\quad \tfrac{neB}{p} &= 2\cdot\tfrac{x}{z^2} \end{aligned} \tag{2.15}$$

Remembering then the equation 2.14, it is not surprising that both formulas are quite similar. The only difference between them is evidently manifested at the right side of the equation, which defines the particle motion trajectory to be changed from a circle into a parabola. Further on, the starting angle of the particle motion, which is of course defined at the position where the particle enters the magnetic field, is again accounted to receive a more realistic and thus more applicable transformation model (see equation 2.12 of section 2.4.1 on page 71). In addition to this, the multiplication of both equation sides with -1 leads to a more readable formula in the programming source code. So the equation changes further to:

$$\begin{aligned} \tfrac{ne}{p} &= 2\tfrac{x}{Bz^2} \\ \Rightarrow\quad -\tfrac{ne}{p} &= 2\tfrac{z\sin(\theta)-x\cos(\theta)}{B(x\sin(\theta)+z\cos(\theta))^2} \end{aligned}$$

Since only the y component of the magnetic field has to be incorporated due to the order of magnitude based on the experimental setup (see section 4.4 on page 223), the effect is obviously limited to the x and z component of the momentum. Hence the formula can be transformed into:

$$-\tfrac{ne}{p_{xz}} = 2\tfrac{z\sin(\theta)-x\cos(\theta)}{B_y(x\sin(\theta)+z\cos(\theta))^2} \quad \left[\tfrac{\mathrm{C}}{\mathrm{Ns}}\right]$$

Coming now to the point, the actual units of this basic formula are summarized in the following table:

Parameter	**Meaning**	**Unit**
n	number of elementary charges	-
e	elementary charge	C
p_{xz}	momentum	Ns
x	coordinate in x-dimension	m
z	coordinate in z-dimension	m
θ	starting angle	rad
B_y	component of the magnetic field	T

Table 2.3: Units of the basic formula for the x-z-plane projection

But as other units are required for the direct usage of the equation in the environment of the CBM experiment, the equation hast to be customized. Therefore the necessary units are shown in table 2.4 at first.

Parameter	**Meaning**	**Unit**
q	number of elementary charge	C
p_{xz}	momentum	$\frac{\text{GeV}}{\text{c}}$
x	coordinate in x-dimension	cm
z	coordinate in z-dimension	cm
θ	starting angle	rad
B_y	component of the magnetic field	kG

Table 2.4: Units of the customized formula for the x-z-plane projection

With regard to the pages 140, 643, 782, 825, 966 and 1,279 in [TMP06], the necessary transformations to support these units are given by the following equations:

$$\begin{aligned}
ne & & &= 1.602 \cdot 10^{-19} \cdot q \\
\mathrm{Ns} &= \frac{\mathrm{Nm}}{\frac{\mathrm{m}}{\mathrm{s}}} &&= \frac{0.3{\cdot}10^{9}\,\mathrm{Nm}}{0.3{\cdot}10^{9}\,\frac{\mathrm{m}}{\mathrm{s}}} \\
&= \frac{0.3{\cdot}10^{9}\,\mathrm{Nm}}{\mathrm{c}} &&= \frac{0.3{\cdot}10^{9}\,\mathrm{J}}{\mathrm{c}} \\
&= \frac{0.3{\cdot}10^{9}{\cdot}1.602{\cdot}10^{-19}\,\mathrm{J}}{1.602{\cdot}10^{-19}\,\mathrm{c}} &&= \frac{0.3}{1.602{\cdot}10^{-19}}\frac{\mathrm{GeV}}{\mathrm{c}} \\
\mathrm{m} &= 100\,\mathrm{cm} \\
\mathrm{T} &= \frac{\mathrm{N}}{\mathrm{Am}} &&= 10\,\mathrm{kG}
\end{aligned} \tag{2.16}$$

Further on, the transformation of the equation to accept the units for the momentum and the elementary charge will lead to:

$$\begin{aligned}
\frac{ne}{p_{xz[\mathrm{Ns}]}} &= \frac{1.602{\cdot}10^{-19}{\cdot}q}{p_{xz[\mathrm{Ns}]}} \\
&= \frac{1.602{\cdot}10^{-19}{\cdot}q}{\frac{1.602{\cdot}10^{-19}}{0.3}p_{xz\left[\frac{\mathrm{GeV}}{\mathrm{c}}\right]}} = \frac{0.3{\cdot}1.602{\cdot}10^{-19}{\cdot}q}{1.602{\cdot}10^{-19}p_{xz\left[\frac{\mathrm{GeV}}{\mathrm{c}}\right]}} \\
&= 0.3\frac{q}{p_{xz\left[\frac{\mathrm{GeV}}{\mathrm{c}}\right]}}
\end{aligned}$$

In combination with the adaption of the coordinates and the magnetic field, the formula turns additionally into:

$$\begin{aligned}
-\frac{q}{p_{xz}} &= \frac{1}{0.3}\frac{2(z\sin(\theta)-x\cos(\theta))}{B_y(x\sin(\theta)+z\cos(\theta))^2} \\
&= \frac{1}{0.3}\frac{2\left(\frac{1}{100}z\sin(\theta)-\frac{1}{100}x\cos(\theta)\right)}{\frac{1}{10}B_y\left(\frac{1}{100}x\sin(\theta)+\frac{1}{100}z\cos(\theta)\right)^2}
\end{aligned}$$

So finally the implementable formula for the second LUT looks like:

$$\boxed{-\frac{q}{p_{xz}} = \frac{1{,}000}{0.3}\frac{2(z\sin(\theta)-x\cos(\theta))}{B_y(x\sin(\theta)+z\cos(\theta))^2}\left[\frac{\mathrm{C}}{\frac{\mathrm{GeV}}{\mathrm{c}}}\right]} \tag{2.17}$$

2.4.2 Adapted Analytic Formula Containing an Inhomogeneous Magnetic Field

As the magnetic field was for the present assumed to be homogeneous in the introduction of section 2.4.1 on page 69, it is obvious that this circumstance has to be adjusted to fit the real experimental setup. For that purpose, the equation 2.13 of section 2.4.1 on page 73 and the equation 2.17 of section 2.4.1 on page 78, which are both required to build the LUTs, have to be generally adapted to suit the real occurring inhomogeneous magnetic field.

However the equation 2.13 of section 2.4.1 on page 73 can be excluded from this process, because the magnetic field is not included at all.
But in contrast to this, an adaption of equation 2.17 of section 2.4.1 on page 78 is necessary. So the next two sections present two separate available and imaginable adjustments of this formula, which feature different advantages. At this juncture, it has to be also noticed here that the basic concept of both adaptations concerns only the factor B_y in the formula, which represents just the magnetic field. That means in detail that the total rest of the formula is still kept and this factor typifies now an imaginary homogeneous magnetic field at the actual coordinate position in the way that the deflection of the particle at this coordinate position is equal to the deflection, which will be realized by the real existing inhomogeneous magnetic field.

Constant Factor

By summarizing now at first the task, it is clear that the factor B_y, which represents universally a homogeneous magnetic field in equation 2.17 of section 2.4.1 on page 78, has to be adapted as stated above to fit the real occurring inhomogeneous magnetic field due to the experimental setup already introduced in section 1.3 on page 10.
So within this context, a practicable solution seems to be determined at a first glance by keeping the equation and adjust only the factor for the discrete evaluation positions so that the deflection of the particle at this coordinate is equal to the deflection, which will be realized by the real existing inhomogeneous magnetic field. Furthermore such an implementation is evidently very easy, because just a simple LUT is required, which consists of the corresponding factors B_y for each detector station. However, in contrast to the service, the computation of these constant factors is quite complex, because they have to represent the best value related to the side-condition that most transformed hits encounter the correct histogram cell. Hence the generation algorithm is certainly determined by the application of a sample MC data

set, because then each MC Track fixes the correct histogram cell, which has to be encountered by all corresponding hits at its best. This means more precisely that each MC Track of the data set is used to compute the correct representing histogram cell by the usage of equation 2.9 of section 2.3.4 on page 62 for the defined momentum, while the corresponding hits can be afterwards inserted in combination with this evaluated histogram cell in equation 2.17 of section 2.4.1 on page 78 to receive the factor B_y as unique existing variable in the formula. So based on this concept, it is clear that the optimal factor of each detector station is finally determined by the best solution for the amount of such equations.
Of course it is needless to say that enough statistics have to be used for the foundation to fix the values for the factor B_y of each detector station. Besides this, it should be also noticed here that an important disadvantage of this approach is obviously determined by the requirement of MC information. Hence the next section presents a different proposal, which avoids this disadvantage by the direct inclusion of the inhomogeneous magnetic field.

Averaged Integral

In contrast to the previous section 2.4.2 on page 79, which introduces the approach comprising constant factors for B_y in the formula, this section presents the adaptation of equation 2.15 of section 2.4.1 on page 76 to the real occurring inhomogeneous magnetic field by involving the field itself to approximate the deflection of the particle.
For this purpose, it is clear that the basic principle is characterized by the evaluation of the physical deflection force of the real inhomogeneous magnetic field, which can be simply computed by averaging the integrated field up to the actual coordinate positions of the particle (see figure 2.13). So based on this concept, the major advantage of this approach is obviously that B_y must not be constant for the whole detector station, but is additionally dependent of the coordinate dimensions x and y. Thus each hit of a station causes a different factor B_y due to its variable coordinate position.

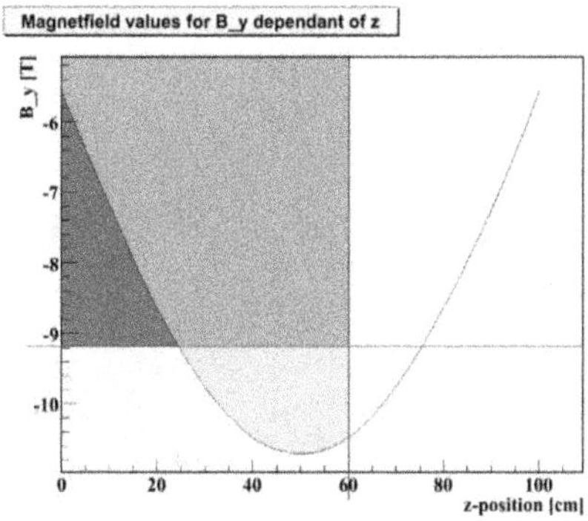

Figure 2.13: Sketch of the averaged integrated magnetic field

Having now a more detailed look to figure 2.13, it is obvious that it depicts a simple example for the recent approach. In this connection, the parabola represents the concrete inhomogeneous magnetic field, while the detector station containing the example hit is located at the position $z = 60\,\text{cm}$. Further on, the green and the yellow marked area show the integral of the magnetic field from the target to this position, whereas the average of this integral is visualized by the green and the blue area. So based on the actual concept, the result of this approach is a factor B_y, which occupies the same integral area from the target to the detector station, because the deflection caused by this homogeneous magnetic field is equal to the real inhomogeneous one. In addition to this, it is clear that the variation of the magnetic field in the x and y dimension is skipped here only for simplicity.
Besides this, a palpable problem of this approach is naturally determined by the fact that the correct motion trajectory of a charged particle through the magnetic field is unknown. However the assumption of a straight line for this trajectory from the target position ($x = 0\,\text{cm}$, $y = 0\,\text{cm}$, $z = 0\,\text{cm}$) to the actual hit coordinate introduces a very small error, which is negligible, because the real trajectory is almost straight and the magnetic field does not change that rapidly (see [GK05] page 6).

For all that reasons, the resulting factor B_y can be expressed by:

$$
\begin{aligned}
B_y &= \frac{\int\limits_{(0;0;0)}^{(x;y;z)} field_y d\vec{s}}{|\vec{s}|} \\
\vec{s} &= \begin{pmatrix} x \\ y \\ z \end{pmatrix} \\
\Rightarrow \quad B_y &= \frac{\left[\int field_y\right]_{(0;0;0)}^{(x;y;z)}}{\sqrt[2]{x^2+y^2+z^2}}
\end{aligned}
$$

This circumstance leads then to the final implementable formula for the second LUT:

$$
\boxed{-\frac{q}{p_{xz}} = \frac{1{,}000}{0.3} \frac{2(z\sin(\theta)-x\cos(\theta))}{\left(\frac{\left[\int field_y\right]_{(0;0;0)}^{(x;y;z)}}{\sqrt[2]{x^2+y^2+z^2}}\right)(x\sin(\theta)+z\cos(\theta))^2}} \tag{2.18}
$$

2.5 Runge-Kutta Approach Applicable for the Hough Transform

Since the customized Hough transform, which is introduced in section 2.2 on page 55, applies off-line calculated LUTs to transform the hits into the Hough space, it is at first important to realize that this circumstance allows obviously each imaginable algorithm to be used to generate these LUTs without consequences to the complexity of the Hough transform. As it is further on common to model the necessary motion trajectory of a charged particle in an inhomogeneous magnetic field by a differential equation, the Runge-Kutta method constitutes a good approach to solve this equation. For that reason, an approach, which combines this model with that solution, is under study for the generation of the LUTs and thus presented in this section. However it has to be noticed that the the usage of a differential equation causes the required map φ to be in no case directly deducible, because the basic principle of the revealed approach is quite different from the one applied to create the analytic formula introduced in section 2.4.1 on page 69 and section 2.4.2 on page 79. Nevertheless the fundamentals of the customized Hough transform, which are summarized in the following, are uniform:

- The track with the parameters p_x, p_y and p_z is represented by a unique point in the three dimensional Hough space
- One measured coordinate point (x, y, z) of a hit contributes to several tracks, which obey the differential equation of the model, to become an appropriate curve in the Hough space
- Several hits of one track lead to several curves in the Hough space, which intersect all in the point that describes the feature parameters of the track

Furthermore, the basic principle of the algorithm is characterized by a recursion, because the particle interaction point with a detector station is propagated from one station to the next by the extrapolation of the track parameters p_x, p_y and p_z from station z_i to station z_{i+1} (see [GK05]). The following figure 2.14 depicts a good imagination about a single propagated track.

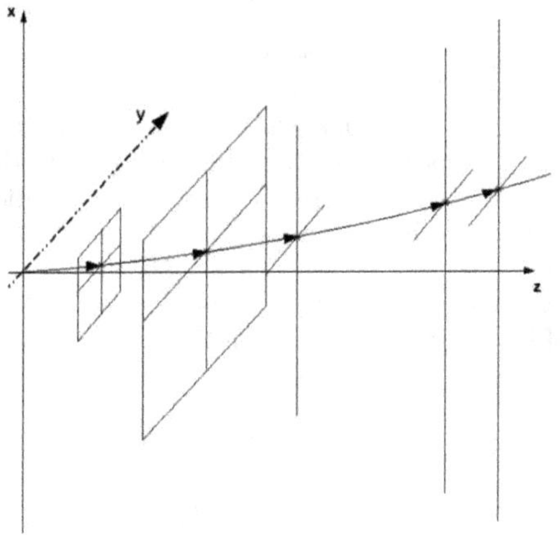

Figure 2.14: Illustration of the Runge-Kutta approach

Besides this, it is clear that the Runge-Kutta approach determines a separate map $\chi_{(Houghparameter)} \rightarrow$ hits instead of the required map φ. But this circumstance is evidently noncritical, because this map correlates of course to the map φ^{-1}. So if an inversion is imaginable and realizable, the needed

map would be received. For this purpose, a quite simple solution, which goes hand in hand with the application of the LUTs in the general Hough transform, is determined by the evaluation of a discrete map χ, whose entry pairs $(address \rightarrow value)$ have to be just swapped afterwards $(value \rightarrow address)$. However it is clear that the real implemented inversion has to face some problems like missing values, repeatedly frequented values for different addresses, the separation into two LUTs and so on.
Nevertheless this simple concept features apparently also the evaluation of the required LUTs for the three dimensional Hough transform, which maps the detector hits according to the track parameters θ, $-\frac{q}{p_{xz}}$ and γ into the Hough space.

2.5.1 Runge-Kutta Formulas

This section presents the derivation of all necessary Runge-Kutta formulas in detail.

Description of the Particle Motion

In contrast to the analytic formula, whose derivation is introduced in section 2.4.1 on page 69 and especially section 2.4.2 on page 79, the Runge-Kutta approach requires at first a more exact model for the motion trajectory of a charged particle in an inhomogeneous magnetic field. For this purpose, the second axiom of Sir Isaac Newton (1643-1727) (see [TMP06] page 78) is combined with the Lorentz Force (see [Vog95] page 451-452), which models the deflection of the particle, to receive the following equation:

$$\vec{F} = m \cdot \vec{a} = ne \cdot \vec{v} \times \vec{B}$$

Besides this, the axiom offers also the possibility to involve the momentum p of a particle in connection to the force F by the equation $\vec{a} = \frac{d\vec{v}}{dt}$ and the equation $p = mv$ (see [TMP06] page 14 and [Vog95] page 213), which leads to the successive changed description:

$$\vec{F} = m \cdot \vec{a} = m \cdot \frac{d\vec{v}}{dt} = \frac{d\,(m\vec{v})}{dt} = \frac{d\vec{p}}{dt}$$

Further on, the employment of this modified axiom results in the next more complex equation:

$$\begin{aligned} \vec{F} &= \tfrac{d\vec{p}}{dt} = ne \cdot \vec{v} \times \vec{B} \\ \Rightarrow \quad & \tfrac{d\vec{p}}{dt} = ne \cdot \tfrac{\vec{p}}{m} \times \vec{B} \end{aligned}$$

At this point, it should be mentioned that an exhaustive documentation of this formula in conjunction with its usage can be found in [GK05], because the Cellular Automaton (see section 1.4.3 on page 40), which constitutes a separate tracking algorithm utilizing exactly this background, is presented there in detail. Beyond that, the following table 2.5 summarizes the units of this basic formula.

Parameter	**Meaning**	**Unit**
$\vec{p}$	momentum	Ns
t	time	s
n	number of elementary charges	-
e	elementary charge	C
$\vec{v}$	velocity	$\frac{\mathrm{m}}{\mathrm{s}}$
$\vec{B}$	magnetic field	T

Table 2.5: Units of the basic formula for the general motion trajectory

But as other units are required for the direct usage of the equation in the environment of the CBM experiment, the equation hast to be customized. Therefore the necessary units are shown in table 2.6 at first.

Parameter	**Meaning**	**Unit**
$\vec{p}$	momentum	$\frac{\text{GeV}}{\text{c}}$
t	time	s
q	number of elementary charge	C
$\vec{v}$	velocity	$\frac{\text{cm}}{\text{s}}$
$\vec{B}$	magnetic field	kG

Table 2.6: Units of the customized formula for the general motion trajectory

With regard to the pages 140, 643, 782, 825, 966 and 1,279 in [TMP06], the necessary transformations to support these units are given by the following equations:

$$\begin{aligned}
ne & & &= 1.602 \cdot 10^{-19} \cdot q \\
\text{Ns} &= \frac{\text{Nm}}{\frac{\text{m}}{\text{s}}} &&= \frac{0.3{\cdot}10^{9}\,\text{Nm}}{0.3{\cdot}10^{9}\,\frac{\text{m}}{\text{s}}} \\
&= \frac{0.3{\cdot}10^{9}\,\text{Nm}}{\text{c}} &&= \frac{0.3{\cdot}10^{9}\,\text{J}}{\text{c}} \\
&= \frac{0.3{\cdot}10^{9}{\cdot}1.602{\cdot}10^{-19}\,\text{J}}{1.602{\cdot}10^{-19}\,\text{c}} &&= \frac{0.3}{1.602{\cdot}10^{-19}}\,\frac{\text{GeV}}{\text{c}} \\
\text{m} &= 100\,\text{cm} \\
\text{T} &= \frac{\text{N}}{\text{Am}} &&= 10\,\text{kG}
\end{aligned} \tag{2.19}$$

Further on, the transformation of the equation to accept the units for the momentum and the elementary charge will lead to:

$$\begin{aligned} \frac{\vec{p}_{[\mathrm{Ns}]}}{ne} &= \frac{\vec{p}_{[\mathrm{Ns}]}}{1.602 \cdot 10^{-19} \cdot q} \\ &= \frac{\frac{1.602 \cdot 10^{-19}}{0.3} \vec{p}_{\left[\frac{\mathrm{GeV}}{c}\right]}}{1.602 \cdot 10^{-19} \cdot q} = \frac{1.602 \cdot 10^{-19} \vec{p}_{\left[\frac{\mathrm{GeV}}{c}\right]}}{0.3 \cdot 1.602 \cdot 10^{-19} \cdot q} \\ &= \frac{1}{0.3} \frac{\vec{p}_{\left[\frac{\mathrm{GeV}}{c}\right]}}{q} \end{aligned}$$

In combination with the adaption of the velocity and the magnetic field, the formula turns additionally into:

$$\vec{F} = \frac{d\vec{p}}{dt} = 0.3 \cdot q \cdot \left(\frac{1}{100} \cdot \vec{v}\right) \times \left(\frac{1}{10} \cdot \vec{B}\right)$$

So finally the formula looks like:

$$\boxed{\begin{aligned} \frac{d\vec{p}}{dt} &= \frac{0.3}{1{,}000} \cdot q \cdot \vec{v} \times \vec{B} \\ \frac{d\vec{p}}{dt} &= \frac{0.3}{1{,}000} \cdot q \cdot \frac{\vec{p}}{m} \times \vec{B} \end{aligned}} \tag{2.20}$$

By taking now the always occurring orthogonality of the Lorentz Force to the direction of the particle motion into account, it is immediately clear that the velocity $\vec{v}$ and thus the momentum $\vec{p} = m \cdot \vec{v}$ is constant (see [GK05]) under the obviously appearing circumstance of a non-varying mass. So within this context, it is applicable to introduce the following unit vector:

$$\vec{u} = \frac{\vec{p}}{|\,\vec{p}\,|} = \frac{m \cdot \vec{v}}{|\,m \cdot \vec{v}\,|} = \frac{\vec{v}}{|\,\vec{v}\,|} \tag{2.21}$$

In addition to this, the time t can be further replaced by the trajectory length s due to the subsequent equation:

$$dt = \frac{ds}{|\,\vec{v}\,|}$$

Returning now to equation 2.20, the employment of these two modulations results in the successive formulas:

$$\begin{aligned}
\tfrac{d\vec{p}}{dt} &= \tfrac{0.3}{1{,}000} \cdot q \cdot \tfrac{\vec{p}}{m} \times \vec{B} \\
\Rightarrow \; d\vec{p} &= \tfrac{0.3}{1{,}000} \cdot q \cdot \left(\tfrac{\vec{p}}{m} \times \vec{B}\right) dt \\
\Rightarrow \; d\vec{p} &= \tfrac{0.3}{1{,}000} \cdot q \cdot \left(\tfrac{\vec{p}}{m} \times \vec{B}\right) \tfrac{ds}{|\vec{v}|} \\
\Rightarrow \; \tfrac{d\vec{p}}{|\vec{p}|} &= \tfrac{0.3}{1{,}000} \cdot \tfrac{q}{|\vec{p}|} \cdot \left(\tfrac{\vec{p}}{m \cdot |\vec{v}|} \times \vec{B}\right) ds \\
\Rightarrow \; d\vec{u} &= \tfrac{0.3}{1{,}000} \cdot \tfrac{q}{|\vec{p}|} \cdot \left(\vec{u} \times \vec{B}\right) ds
\end{aligned}$$

So while finally denoting $|\ \vec{p}\ |$ as p, the equations for the dimensions are determined by:

$$\boxed{\begin{aligned}
du_x &= \tfrac{0.3}{1{,}000} \cdot \tfrac{q}{p} \cdot (u_y \cdot B_z - u_z \cdot B_y)\, ds \\
du_y &= \tfrac{0.3}{1{,}000} \cdot \tfrac{q}{p} \cdot (u_z \cdot B_x - u_x \cdot B_z)\, ds \\
du_z &= \tfrac{0.3}{1{,}000} \cdot \tfrac{q}{p} \cdot (u_x \cdot B_y - u_y \cdot B_x)\, ds
\end{aligned}} \tag{2.22}$$

Description of the Motion Trajectory

Since it is more common within this approach to use the track parameter set for feature representation, equation 2.6 of section 2.3.1 on page 59 has to be remembered, because the motion trajectory of a charged particle is thus determined by the parameters $t_{x(z)} = \frac{dx_{(z)}}{dz}$, $t_{y(z)} = \frac{dy_{(z)}}{dz}$ and $\frac{q}{p} = \frac{1}{\mathbf{qp}}$ with $q \in \{-1; +1\}$. Moreover it should be again explicitly pointed out here that the variable **qp** is just a name, which is set and used by the CBMROOT framework (see section 3.2.1 on page 102). Hence this variable does not implicate the multiplication of the charge q with the momentum p ($\mathbf{qp} \neq q{\cdot}p$). Taking further section 2.5 on page 82 into account, which introduces the basic principle of this approach given by the parameter extrapolation from station to station, it is obvious that one state of this extrapolation process is defined by the coordinates x and y at the z position of the actual detector station in combination with such a track parameter set. So based on this concept, such a state is identified by the so-called state vector, which is described in dependency of z by the following equation (see [GK05]):

$$\hat{r}_{(z)} = \begin{pmatrix} x_{(z)} \\ y_{(z)} \\ t_{x(z)} \\ t_{y(z)} \\ \frac{q}{p} \end{pmatrix}$$

By applying this state vector now to the extrapolation process from position z_i to z_{i+1}, the subsequent formulas result:

$$\frac{d\hat{r}_{(z)}}{dz} = \begin{pmatrix} \frac{dx_{(z)}}{dz} \\ \frac{dy_{(z)}}{dz} \\ \frac{dt_{x(z)}}{dz} \\ \frac{dt_{y(z)}}{dz} \\ \frac{d\frac{q}{p}}{dz} \end{pmatrix}$$

Moreover the usage of equation 2.6 of section 2.3.1 on page 59 and equation 2.21 of section 2.5.1 on page 87 ascertain the next equations:

$$t_x = \frac{p_x}{p_z} = \frac{p_x}{p} \cdot \frac{p}{p_z} = \frac{p_x}{|\vec{p}|} \cdot \frac{|\vec{p}|}{p_z} = \frac{u_x}{u_z}$$

$$t_y = \frac{p_y}{p_z} = \frac{p_y}{p} \cdot \frac{p}{p_z} = \frac{p_y}{|\vec{p}|} \cdot \frac{|\vec{p}|}{p_z} = \frac{u_y}{u_z}$$

So based on this concept, the actual equation for $\frac{d\hat{r}_{(z)}}{dz}$ turns into:

$$\frac{d\hat{r}_{(z)}}{dz} = \begin{pmatrix} \frac{dx_{(z)}}{dz} \\ \frac{dy_{(z)}}{dz} \\ \frac{dt_{x(z)}}{dz} \\ \frac{dt_{y(z)}}{dz} \\ \frac{d\frac{q}{p}}{dz} \end{pmatrix} = \begin{pmatrix} t_{x(z)} \\ t_{y(z)} \\ \frac{d\left(\frac{u_{x(z)}}{u_{z(z)}}\right)}{dz} \\ \frac{d\left(\frac{u_{y(z)}}{u_{z(z)}}\right)}{dz} \\ 0 \end{pmatrix}$$

By remembering further the two formulas $t_{x(z)} = \frac{dx}{dz}$ and $t_{y(z)} = \frac{dy}{dz}$, ds can be expressed by the following equations:

$$\begin{aligned} ds &= \sqrt{dx^2 + dy^2 + dz^2} \\ ds &= dz\sqrt{\left(\frac{dx}{dz}\right)^2 + \left(\frac{dy}{dz}\right)^2 + 1} \\ ds &= dz\sqrt{t^2_{x(z)} + t^2_{y(z)} + 1} \end{aligned}$$

Beyond that, the combination of the quotient rule for derivation ([BSMM05] on page 397) with equation 2.21 of section 2.5.1 on page 87 and equation 2.22 of section 2.5.1 on page 88 leads to the necessary derivations, which are determined by the successive formulas:

$$\begin{aligned} \frac{d\left(\frac{u_{x(z)}}{u_{z(z)}}\right)}{dz} &= \frac{\frac{du_{x(z)}}{dz} \cdot u_{z(z)} - u_{x(z)} \cdot \frac{du_{z(z)}}{dz}}{u^2_{z(z)}} \\ &= k \cdot \left(\frac{u_{y(z)} \cdot u_{z(z)} \cdot B_z - u^2_{z(z)} \cdot B_y - u^2_{x(z)} \cdot B_y + u_{x(z)} \cdot u_{y(z)} \cdot B_x}{u^2_{z(z)}}\right) \\ &= k \cdot \left(t_{y(z)} \cdot B_z - B_y - t^2_{x(z)} \cdot B_y + t_{x(z)} \cdot t_{y(z)} \cdot B_x\right) \\ &= k \cdot \left(t_{x(z)} \cdot t_{y(z)} \cdot B_x - \left(1 + t^2_{x(z)}\right) \cdot B_y + t_{y(z)} \cdot B_z\right) \end{aligned}$$

and

$$
\begin{aligned}
\frac{d\left(\frac{u_{y(z)}}{u_{z(z)}}\right)}{dz} &= \frac{\frac{du_{y(z)}}{dz}\cdot u_{z(z)} - u_{y(z)}\cdot\frac{du_{z(z)}}{dz}}{u_{z(z)}^2} \\
&= k\cdot\left(\frac{u_{z(z)}^2\cdot B_x - u_{x(z)}\cdot u_{z(z)}\cdot B_z - u_{x(z)}\cdot u_{y(z)}\cdot By + u_{y(z)}^2\cdot Bx}{u_{z(z)}^2}\right) \\
&= k\cdot\left(B_x - t_{x(z)}\cdot B_z - t_{x(z)}\cdot t_{y(z)}\cdot B_y + t_{y(z)}^2\cdot B_x\right) \\
&= k\cdot\left(\left(1+t_{y(z)}^2\right)\cdot B_x - t_{x(z)}\cdot t_{y(z)}\cdot B_y - t_{x(z)}\cdot B_z\right)
\end{aligned}
$$

with $k = \frac{0.3}{1{,}000}\cdot\frac{q}{p}\cdot\sqrt{t_{x(z)}^2 + t_{y(z)}^2 + 1}$.

So finally the implementable formula looks like:

$$
\boxed{
\begin{aligned}
\frac{d\vec{r}_{(z)}}{dz} &= \begin{pmatrix} t_{x(z)} \\ t_{y(z)} \\ k\cdot\left(t_{x(z)}t_{y(z)}B_x - \left(1+t_{x(z)}^2\right)B_y + t_{y(z)}B_z\right) \\ k\cdot\left(\left(1+t_{y(z)}^2\right)B_x - t_{x(z)}t_{y(z)}B_y - t_{x(z)}B_z\right) \\ 0 \end{pmatrix} \\
\text{with } k &= \frac{0.3}{1{,}000}\cdot\frac{q}{p}\cdot\sqrt{t_{x(z)}^2 + t_{y(z)}^2 + 1}
\end{aligned}
} \tag{2.23}
$$

Since this formula 2.23 denotes now the characteristic differential equation, the Runge-Kutta method can be applied to find a solution. However it is just a short introduction of this method presented in the following, because it is well known and an exhaustive explanation can be found in [BSMM05] on the pages 931-932 or [PFTV92].
Nevertheless it is noticeable that the necessary exactness of the solution for such a numerically solved equation determines the order, which has to be applied. And as furthermore this method is conventionally called n^{th} order if its error term is $O(h^{n+1})$, the 4th order Runge-Kutta method, which is appropriate here (see [GK05]), leads to an error term of $O(\Delta z^5)$.

Description of the Runge-Kutta Method

The Runge-Kutta method, which is introduced by C. D. T. Runge (1856-1927) in 1895 and further developed by W. M. Kutta (1867-1944) in 1901, is an approach in the field of numerical mathematics to solve differential equations of the form $\frac{d\hat{r}_{(z)}}{dz} = f_{(r,z)}$ with an initial condition $r_{(z_i)} = r_i$ for a given length $\Delta z = z_{i+1} - z_i$. It is mainly characterized by taking r_i and z_i into account to calculate an approximation for r_{i+1} at a brief position $z_{i+1} = z_i + \Delta z$. Furthermore this method, as mostly all numerical solutions for differential equations, attempts to use a finite difference quotient instead of the exact differential quotient, which is given by the derivation. So within this concept, the simplest form of this difference quotient is built by $f_{(r_i,z_i)}$, because the slope m, which enables the linear extrapolation to the position z_{i+1}, is defined at the position z_i by the following equation:

$$r_{i+1} = r_{(z_{i+1})} = r_{(z_i+\Delta z)} = r_i + m \cdot \Delta z \quad with \quad m = f_{(r_i,z_i)} \tag{2.24}$$

However as the correct extrapolation is imaginably almost not describable by a straight line, such a plan is very inaccurate and characterized by rapidly increasing errors, which are directly boosted by many and/or big extrapolation steps Δz. Nevertheless the Runge-Kutta method is able to partly compensate the occurring error for non-linear functions by the adequate combination of multiple different difference quotients. Hence the formula 2.24 is improved by the usage of a slope, which is dependent of the forward behavior of the original differential equation, instead of the slope $f_{(r_i,z_i)}$, which cares only about the initial position z_i. For this purpose, there are some additional steps for the slope computation introduced, which divide the extrapolation length Δz. So the Runge-Kutta method uses a weighted average of approximated values for $f_{(r,z)}$ at several positions within the interval $[z_i; z_{i+1}]$.

Since the order of the Runge-Kutta method entails the error estimation as mentioned earlier, it is obvious that the chosen value is strongly connected to the selection of the additional steps. Thus the 4th order defines the usage of so-called half-steps, which are defined by the positions $z_A = z_i, z_B = z_i + \frac{\Delta z}{2}, z_C = z_i + \frac{\Delta z}{2}$ and $z_D = z_i + \Delta z = z_{i+1}$. Further on, the four resulting slopes m_A, m_B, m_C and m_D at these positions can be then combined to form a unique slope for the linear extrapolation by applying a weighted arithmetical mean (see [BSMM05] page 19), which is determined by the next equation:

$$m_{mean} = \frac{1}{6} \cdot (m_A + 2 \cdot m_B + 2 \cdot m_C + m_D)$$

Further on, figure 2.15 gives a good imagination about this circumstance.

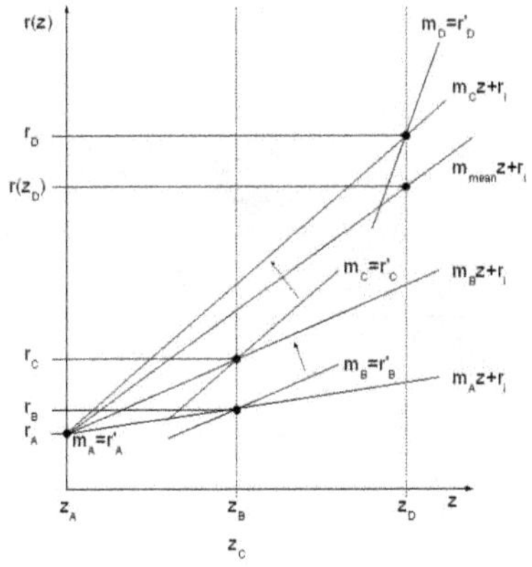

Figure 2.15: Illustration of the Runge-Kutta method

With regard to this figure, it is very important to realize that the computation of the slope m_j depends on the previous slope m_{j-1}. This fact is easily expressed in the following equations:

$$m_j = f_{(r_i + \boldsymbol{\Delta r}_j, z_j)}$$

and

$$\begin{aligned} r_{(z_j)} &= r_i + m_{j-1} \cdot (z_j - z_i) \\ \Rightarrow \quad r_{(z_j)} - r_i &= m_{j-1} \cdot (z_j - z_i) \\ \Rightarrow \quad \boldsymbol{\Delta r_j} &= m_{j-1} \cdot (z_j - z_i) \end{aligned}$$

So finally the implementable formula looks like:

$$\begin{array}{|lcl|}\hline r_{i+1} & = & r_i + \frac{\Delta z}{6} \cdot (m_A + 2 \cdot m_B + 2 \cdot m_C + m_D) \\ \text{with:} & & \\ \quad m_A & = & f_{(r_i, z_i)} \\ \quad m_B & = & f_{(r_i + \Delta r_B, z_i + \frac{\Delta z}{2})} = f_{(r_i + m_A \cdot \frac{\Delta z}{2}, z_i + \frac{\Delta z}{2})} \\ \quad m_C & = & f_{(r_i + \Delta r_C, z_i + \frac{\Delta z}{2})} = f_{(r_i + m_B \cdot \frac{\Delta z}{2}, z_i + \frac{\Delta z}{2})} \\ \quad m_D & = & f_{(r_i + \Delta r_D, z_i + \Delta z)} = f_{(r_i + m_C \cdot \Delta z, z_i + \Delta z)} \\ \hline \end{array} \tag{2.25}$$

Finally a closer look to the application of this concept entails a quite easy algorithm, because the computation of a motion trajectory simply starts with r_0 (common vertex) and r_1 (particle interaction with the first detector station) is evaluated by using the equation 2.25. Afterwards r_1 can be naturally used to calculate r_2 (particle interaction with the second detector station) and so on.
Hence the differential equation 2.23 of section 2.5.1 on page 91, which extrapolates the particle interactions with the detector stations through the whole experimental setup, can be solved under the side conditions of a given momentum and the initial starting point at the target position (0, 0, 0).

Chapter 3

Implementation

This section presents all developed implementations of the adapted Hough transform, which are additionally customized to form a first level tracking system in the STS of the CBM experiment.
Therefore the first section introduces a software implementation, which can be used to customize the parameters of the algorithm, simulate it and finally approve the applicability. In this connection, it should be additionally mentioned that this implementation is not optimized in any case, because it has to be able to run on each imaginable computational platform featuring a compiler for the programming language C++ (see [GCC]) in conjunction with the GNU Make (see [mak] and [GNUb]).
Afterwards the following two sections exhibit respectively a hardware implementation, which differ only in the complexity level. This means more precisely that the first one employs the Sony Playstation III as common embedded system, which is capable of heavy multiprocessing, while the second one utilizes a FPGA platform to realize the highest available parallelism level with especially developed hardware.
But before starting with tangible implementations, it is clear that the basic concepts have to be illustrated.

3.1 Basic Concept of the Implementation

Since the primary aim of this thesis is defined by tracking with maximum speed, the basic concepts focus evidently on the realization of a process pipeline, which is able to cope with one detector station hit per clock cycle. Therefore the complicated calculations of the map $\varphi_{(hit)} \rightarrow$ Hough curve according to the real detector setup and the real magnetic field are implemented with LUTs instead of complex arithmetics. So with regard to all conceivable hits, a first LUT is applied to evaluate the start index and the stop index of the two dimensional histogram layer range, while a second LUT provides the two dimensional Hough curve, which has to be inserted in such a range.
However it has to be additionally noticed here that an important prerequisite of this concept is identified by the requirement of a digitized form for all possible hits, because a LUT can be commonly just indexed with a discrete value.
Moreover it is clear that this concept implements exactly the functionality of a generalized Hough transform, which offers thus the possibility to compare any sufficiently precise LUT generator algorithm with each other. In addition to this, it is also very interesting that the time, which is required to produce the LUTs by the applied generator algorithm, is insignificant, because the calculation of the LUTs is of course realized off-line. For that reason, this thesis discusses the usage of either the developed analytic formula or the Runge-Kutta approach to generate the LUTs.
Besides this, it has to be also mentioned here that each LUT generation algorithm requires usually the hits to be represented in the digitized form as well as in the three dimensional floating point coordinate form, because the applied equations need surely both information.

Coming now to more implementation specific topics, the first noticeable aspect belongs to the LUTs. As the direct implementation of the three dimensional Hough space requires unfortunately a huge amount of memory, the two introduced LUTs feature additionally the decomposition of this Hough space into several two dimensional layers, which offer an adjustable parallelism level with regard to the available hardware resources. So based on this concept, it is clear that the total amount of layers, which have to be computed, is decomposed into a serial computation of a parallel subset. For example, a three dimensional Hough space would require ten million registered bits ($10 \cdot 100^3$) just for the accumulator, if a quantization of one hundred in each Hough space dimension is assumed in combination with ten detector stations, while the decomposition leads exemplary to only one hundred thousand registered bits $(10 \cdot 100^2)$, which are one hundred times used consecutively. Further on, a closer look to the red area in figure 3.1 shows that this decomposition is realized in the way that each detector slice contributes only to a certain amount of two dimensional layers in the Hough accumulator.

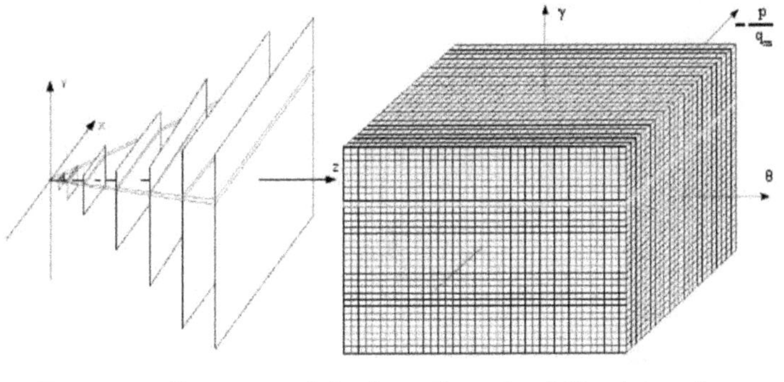

Figure 3.1: Conception of the three dimensional Hough transform

However the adjustment of the transformation process induces the Peak-finding to be also changed, because the neighborhood relation in the third dimension is not any longer directly accessible. Hence two different steps of Peak-finding must be applied, which process at first each two dimensional layer and combine afterwards the results of neighborhood layers. Within this context, it is not surprising that the first step is called 2D Peak-finding, while the subsequent step is named 3D Peak-finding. In addition to this, there is obviously also a need for a supplementary process, which is expectedly called Serialization, because it has to serialize the peaks of each layer after the 2D Peak-finding into a form, which suits for the 3D Peak-finding.

Beyond that, figure 3.2 depicts the processing concept for the Hough curves in the histogram, which involves a systolic array consisting of simple elements. Furthermore the accumulation of the Hough curves in such a simple element requires a counter to store the histogram value, a flip-flop for the short-term storage of the Hough curve and a multiplexer to select the Hough curve shape from two neighboring elements.

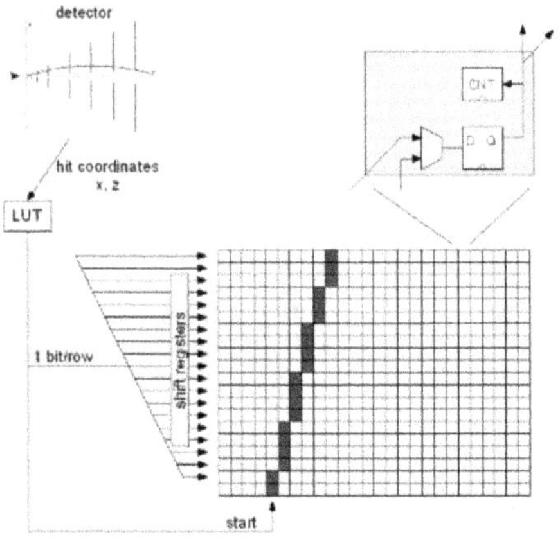

Figure 3.2: Sketch of a single systolic processing histogram layer

So within this context, the consequential usage of this system is obviously defined by applying a new start value to the first row in conjunction with a new selection command to the input of the array in every clock cycle. Afterwards the next clock cycle provokes then these two committed values to be evidently processed by the next straight or diagonal neighbored element in the second row. It is of course needless to say that this process continues up to the end of the histogram, while subsequent parallel input is surely supplied in every clock cycle.

Further on, a more detailed look to such a simple element shows that the multiplexers, which selects the particular input from the two corresponding geographically connected cells, is controlled by a single bit per row delayed by shift registers to the correct clock cycle. So based on this concept, the slope of a curve, which is apparently encoded in the selection command, is restricted to lie arbitrary between the minimum and maximum slope of 45 and 90 degrees.

Besides this, the counter of each simple element stores commonly the number of hits from the different detector stations that lay on a trajectory of a track with the associated feature parameters (see feature extraction). However the implemented counter is modified, because it does not merely count the number of hits, but each hit of a certain detector station sets a certain bit in the counter. Furthermore this leads to the circumstance that each counter is now able to document, which detector station has contributed to its peak. For that reason, the content of these counters is called signature in the following of this thesis. For example, figure 3.3 offers an imagination about such a hit-signature of a peak, which is located in a histogram cell.

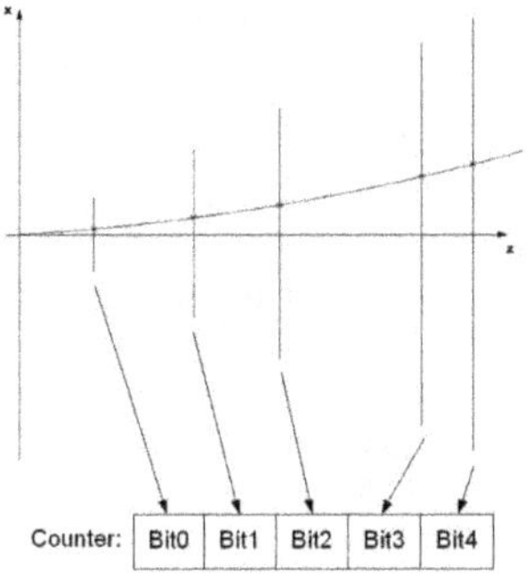

Figure 3.3: Sketch of a counter in the Hough layer representing a signature

Regardless of this documentation advantage, it is for sure that a disadvantage of such an implementation is determined by the auxiliary amount of bits, which are necessary for the signature in contrast to a counter. But this fact would be negligible, if it is compared to the supplementary signature information, because multiple hits in the same detector station, which fall into the same histogram cell, will increment the counter, but just set the same bit in the signature. Hence this concept will lead apparently to less misinterpreted peaks.

However it is nevertheless clear that real tracks, which do not pass all detector stations, or missing hits due to detector inefficiencies cause the definition of signatures, which should be accepted to possibly represent a real track with the feature parameters defined by the actual histogram cell.
For this purpose, an extra step is added to the algorithm, which is called Encoding. So it is evident that this step simply analyzes each histogram cell whether the signature represents a possible track by using an additionally implemented small LUT containing a zero for non-acceptable and another value for acceptable signatures.
Moreover it has to be noticed here that it might be very useful to have not just such a binary information in the LUT, but multiple values representing so-called priority classes. Furthermore the location of this step is of course identified between the accumulation of the Hough curves, which is called Histogramming, and the 2D Peak-finding. For this reason, a peak, which

survives this Encoding step, is afterwards called track candidate and is represented by a priority class instead of a signature.
Besides that, the diverse priority classes can be additionally used in the Peak-finding step to get a better result due to the morphing of the clusters. So it is obvious that the Encoding and the Peak-finding processes are applied to the histogram to get better peaks in the way of removing background noise. Hence these two steps are included in a general process, which is called filtering, because both steps apply a special method to filter the values of the histogram cells.

Since this is still not the total story, another step is needed, which is called Diagonalization. This next step is apparently necessary to cope with the diagonal orientation of the clusters, which originates from the allowed slopes of the accumulated Hough curves, and causes diagonal histogram cells to possess the same priority class. As such a circumstance is obviously a problem in the Peak-finding, the value of these cells except one have to be removed. So the overall result can be simply improved by analyzing the relationship in clusters and increase the priority class of one of the diagonal cells by one, because there remains then just one local peak holding the highest priority class. Further on, it is self-evident that this step has to be inserted between the Encoding and the 2D Peak-finding.

Coming now to the point, the whole algorithm can be summarized in three major steps including a loop in the second step, which serially processes a subset of parallel histogram layers.

1. Create the Hough curves for all hits based on the two LUTs and store them sorted to the start index of the histogram layer
    - Implementing map $\varphi_{(hit)} \rightarrow$ Hough curve
2. Serially process a subset of parallel histogram layers starting with index zero (The subset can also consist just of a single layer)
    (a) Accumulate the Hough curves of the actual layers (Histogramming)
        - Hough curves $\rightarrow$ Peaks (Signatures)

    (b) Encode the actual layers (Encoding)
        - Peaks (Signatures) $\rightarrow$ Encoded track candidates (Priority classes)

(c) Diagonalize the actual layers (Diagonalization)

- Encoded track candidates (Priority classes) $\rightarrow$ Diagonalized track candidates (Priority classes)

(d) Filter the actual layers (2D Peak-finding)

- Diagonalized track candidates (Priority classes) $\rightarrow$ Filtered track candidates (Priority classes)

(e) Serialize the actual layers (Serialization)

- Filtered track candidates (Priority classes) $\rightarrow$ Serialized track candidates (Priority classes)

3. Filter the histogram (3D Peak-finding)

- Serialized track candidates (Priority classes) $\rightarrow$ Found tracks defining feature parameters (see feature extraction)

Finally, there are two more negligible disadvantages of this implementation, which have to be mentioned nonetheless.
The first one is induced by the decomposition of the three dimensional histogram into two dimensional layers, because Hough curves, which are slanted with regard to the histogram layers, have to be inscribed as planes starting with the smallest indexed layer of the curve up to the highest. However this circumstance defines not a crucial problem, because the already mentioned security regions cause each Hough curve to be anyway brightened to a plane to compensate side-effects like multiple scattering. Nevertheless this fact has to be recognized, because some LUT generation algorithms, like the Runge-Kutta approach for instance, have to deal with this situation, whether it is directly implied in the analytic formula.
Besides this, the second disadvantage originates from the combination of the feature parameter $-\frac{q}{p_{xz}}$ with the application of an accumulator, because if the quantization is too small for this feature variable with regard to a real track possessing a very high momentum p_z, such a track would be found in the accumulator coordinates representing the value zero in this Hough space dimension. Even if this situation defines at a first glance not a crucial problem, the circumstance would rapidly change for the extraction of the feature parameters, because the corresponding denominator in the mandatory equations would be then also zero. Hence such a found track receives the assigned momentum $p_z = \infty$. But as such a track is of course absurd, the denominator is simply set to the half of the smallest not quantized value instead of the value zero. Thus such a found track represents a real track, which exhibits a momentum p_z that is at least as high as this maximal value features. So

it is needless to say that the correct momentum p_z can be even higher, but could be only detected correctly, if the quantization of the accumulator is adjusted.

3.2 Software

Since the entire software is divided into twelve libraries and one Application Programming Interface (API), which contains the main object to execute the task in the CBMROOT framework, this chapter exhibits detailed information about each library after a short overview about the framework itself.

3.2.1 Framework Overview

This section presents a short introduction of the framework, which offers the feature to run the tracking software in a well defined environment. For this purpose, the different fragments of the framework are illustrated before the data input, the algorithm implementation and finally the not less important error and warning concept is exhibited.

The packages of the CBMROOT framework

CBMROOT Package: The CBMROOT package is built around the Virtual Monte Carlo (VMC) concept (see section 3.2.1 on page 107) of the ROOT package (see section 3.2.1 on page 106). It offers furthermore the possibility to use the same framework for event simulation and data analysis by applying the task concept of the ROOT system (see [Thea] and [AATB+04]). Hence one task for the simulation and one task for the data analysis, which is called reconstruction, exist. Further on, the schematic design of the simulation part of the framework is depicted in figure 3.4a, while the reconstruction part is shown in figure 3.4b.

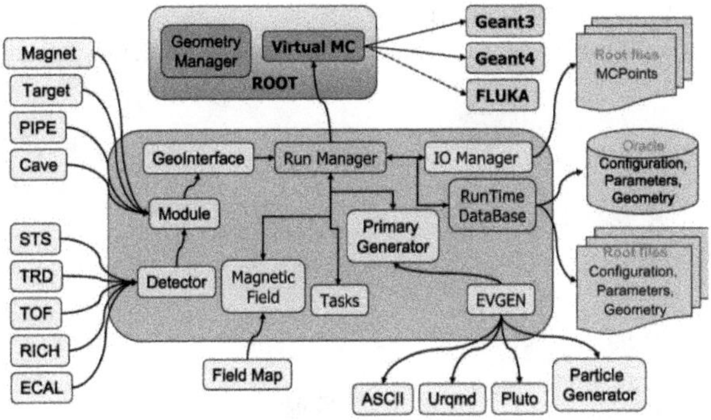

(a) Simulation part

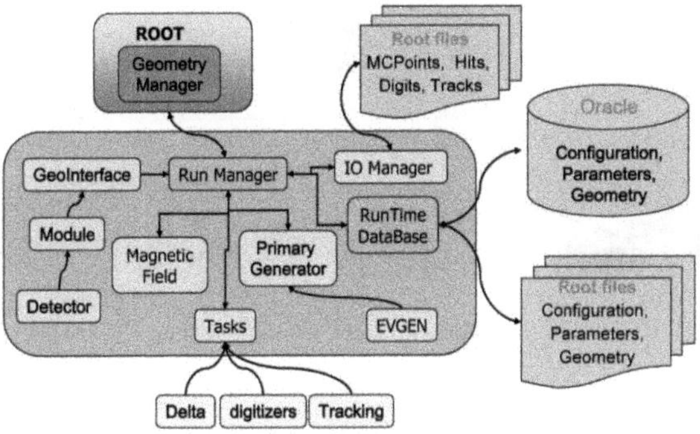

(b) Reconstruction part

Figure 3.4: Schematic design of the CBMROOT framework (Source: [AT06])

With regard to these figures, it has to be noticed that an oracle database, which includes a built-in version management, is used to efficiently store all simulation parameters like the detector geometry or detector station materials for instance.

Besides this, a growing interest in this framework leads to the circumstance that it is split into base classes, which are called the FAIRROOT framework,

and some specific classes for the CBM experiment. So other experiments can be easily added by building another group of specific classes. Moreover it is obvious that the distribution of the actual implemented framework is realized via the tool Subversion (see [Thei]).
Coming now to the simulation chain of the framework, figure 3.5 depicts the actual one.

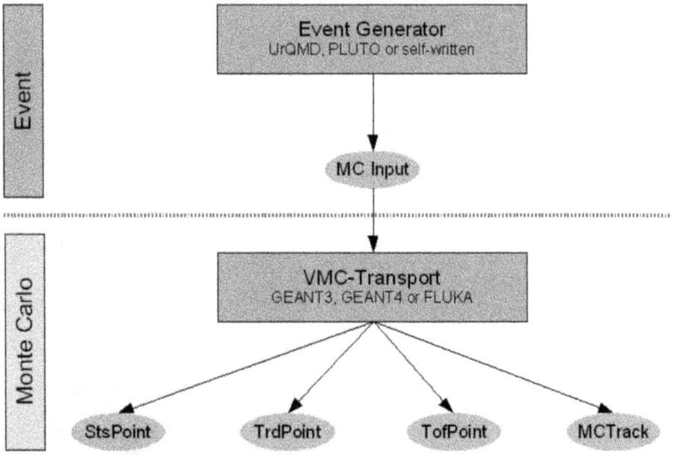

Figure 3.5: Illustration of the simulation chain in the CBMROOT framework

With regard to this figure, the general VMC concept is obviously applied to hide the information of the used particle transport engine. Further on, the input to the VMC is a set of particles, which are produced by an particle event generator like Ultrarelativistic Quantum Molecular Dynamics (UrQMD) or PLUTO, whereas the output of the simulation establishes the so-called MC data, which consists of all motion trajectories or tracks for the transported particles in combination with the points of particle interactions with the detector stations. Furthermore these tracks are represented by their parameters, while the points are mainly determined by their coordinates and their flight time. In addition to this, these two objectives are evidently combined by a link from each point to its corresponding track. Since the reconstruction chain of the framework is also of special interest, figure 3.6 shows the actual one.

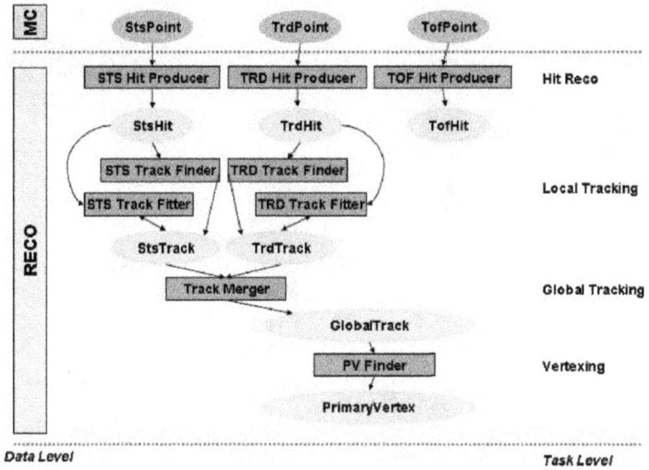

Figure 3.6: Illustration of the reconstruction chain in the CBMROOT framework (Source: [Recb])

With regard to this figure, it is self-evident that the input of the reconstruction is determined by the produced MC data from the simulation. But as the full digitization scheme for the MC data is currently not available due to the still varying detector and read out technology, the digitization and hit reconstruction are both built within the different hit producers. So the major input of the tracking is defined by so-called *StsHit* objects, which base on *StsPoint* objects at the MC level. By the way it should be reminded here that the *StsPoint* objects represent the particle interactions with the detector stations in the STS.

ROOT Package: The ROOT project is started in the context of the NA49 experiment at CERN (see [Thee] and [BBF$^+$97]) and has yielded the ROOT system, which is an object oriented framework for large scale data analyses (see [Rad] and [BRb]). Furthermore this framework is written in C++ and contains an efficient hierarchical object oriented database, a C++ interpreter, advanced statistical analyses like multi dimensional histogramming, fitting, minimization and cluster finding algorithms as well as visualization tools. Further on, user interaction with the ROOT system is possible via a graphical user interface, the command line or batch scripts.
Besides this, large scripts can be additionally compiled and dynamically linked during runtime. At this juncture, the command and scripting language is evidently also C++. Beyond that, the object oriented database design has been optimized for efficient parallel access by multiple processes (see [BRa]). In addition to this ROOT system, there exists also a parallel version of this framework, which is called PROOF.
Finally it has to be mentioned that ROOT is distributed in binary form for the most supported platforms and as source code via anonymous File Transfer Protocol (FTP).
More information can be found in [Theg].

Ultrarelativistic Quantum Molecular Dynamics Package: The Ultrarelativistic Quantum Molecular Dynamics (UrQMD) package uses a microscopic model to simulate (ultra)relativistic heavy ion collisions in the energy range from the Bevalac (combination of Bevatron and SuperHILAC: see [Lis] and [Sea]) and the SIS (see [Uni]) up to the AGS (see [Altb]), the SPS (see [Sup] and [Sea]) and the RHIC (see [RHI]). In this connection, the main goal is the improvement of the understanding about the following physical phenomena within a single model (see particle event generator):

- Creation of dense hadronic matter at high temperatures
- Properties of nuclear matter, Delta & Resonance matter
- Creation of mesonic matter and of anti-matter
- Creation and transport of rare particles in hadronic matter
- Creation, modification and destruction of strangeness in matter
- Emission of electromagnetic probes

Besides this, UrQMD has also been used as a component of various hybrid transport approaches like GEANT4 for example (see section 3.2.1 on page 109).
More information can be found in [Theh].

PLUTO Package: The PLUTO package is a ROOT based framework, which implements a customized streamlining particle event generator for hadronic and electromagnetic decays in C++ (see [Frö07]). Furthermore this package provides tools to set up and manipulate particles, reaction channels and complex reactions. In addition to this, experimental filters on such reaction products like geometrical acceptance or kinematic conditions can be applied. Further on, the package comprises models for resonance and Dalitz decays, resonance spectral functions with mass-dependent widths and anisotropic angular distributions for selected channels. Moreover a decay manager interface enables multi-step or so-called 'cocktail' calculations. Besides this, an extensive particle data base, which covers also particle properties and decay modes, is available with capabilities to support up to 999 particles including user-defined ones. Supplementary thermal distributions are implemented enabling multi-hadron decays of hot fireballs.
More information can be found in [Plu].

Virtual Monte Carlo Concept: The Virtual Monte Carlo (VMC) concept offers the possibility to run different particle transport engines without changing the user code and therefore the input and output format as well as the geometry and detector response definition. For that purpose, the core of the VMC is the category of classes *vmc* in the ROOT package, which provides a set of interfaces to completely decouple the dependencies between the user code and the concrete MC engine. Moreover these interfaces are defined by:

- *TVirtualMC* as interface to the concrete MC program
- *TVirtualMCApplication* as interface to the user's MC application
- *TVirtualMCStack* as interface to the particle stack
- *TVirtualMCDecayer* as interface to the external decayer

So based on this concept, a user has to implement just two mandatory classes determined by the MC application, which has to be derived from *TVirtualM-*

CApplication, and the MC stack, which must be derived from *TVirtualMCStack*. In addition to this, an external decayer, which has to be obviously derived from *TVirtualMCDecayer*, can be evidently also used optionally. Besides this, the concrete MC particle transport engine, like GEANT3, GEANT4 or FLUKA for example, is then selected at run time, while processing a ROOT macro with the instantiation.
More information about the VMC concept can be found in [HAB+03] and [BCHM].

FLUKA Package: The FLUKA package is a tool for particle transport (see particle transport engine) and interactions with matter, which adopts microscopic models whenever possible, while consistency among all the reaction steps and/or reaction types is ensured. Further on, conservation laws are enforced at each step and the results are checked against experimental data at single interaction levels. For this purpose, final predictions are obtained with a minimal set of free parameters fixed for all energy/target/projectile combinations.
Besides this, FLUKA can simulate the interaction and propagation in matter of about 60 different particles including photons, electrons, neutrinos, muons, hadrons, all the corresponding antiparticles, neutrons and heavy ions with high accuracy. Furthermore the package can also transport polarized photons, like synchrotron radiation for instance, and optical photons. Moreover time evolution and tracking of emitted radiation from unstable residual nuclei can be performed on-line.
Beyond that, the FLUKA package can cope with very complex geometries by the usage of an improved version of the well-known combinatorial geometry package, which has been designed to correctly track also charged particles even in the presence of magnetic or electric fields. In addition to all this, various visualization and debugging tools are also available.
More information about the FLUKA package can be found in [FFRS05], [FFR+03] and [FLU].

GEANT3 And GEANT4 Package: The GEometry ANd Tracking (GEANT) package offers the possibility to simulate the trajectory of particles through matter (see particle transport engine). Furthermore this package provides comprehensive detector and physics modeling capabilities, which are embedded in a flexible structure (see [CER94]). Further on, it features the capabilities for tracking, geometry description (see [Cos04]) and navigation as well as material specification and abstract interfaces to physics processes [Gea07c]. Moreover the package can manage events and handle detector response. Besides this, GEANT offers also interfaces to external frameworks as well as graphics and complete user interface systems.
Beyond that, the physics processes cover diverse interactions over an extended energy range, which reaches from optical photons and thermal neutrons to the high energy reactions at the LHC and in cosmic ray experiments. Furthermore tracked particles include leptons, photons, hadrons and ions. Moreover various implementations of physics processes are offered, which provide complementary or alternative modeling approaches. In addition to this, the package provides also interfaces to enable user interaction with their application like saving the results for instance.
Finally GEANT4 has to be mentioned apart, because it represents a completely new detector simulation package, which is written in C++ (see [Gea07a]). An overview of recent developments in diverse areas of this package, which include performance optimizations for complex setups, improvements for the propagation in fields, new options for event biasing, improvements in geometry and physics processes as well as interactive capabilities, is presented in [AAA+06].
More information on GEANT4 can be found in [Ame99], [Gea07d] and [Gea07b].

GEANE Package: The GEANE package offers the possibility for the user to compute the average motion trajectory of particles through the detector stations (see [IMN91]). In addition to this, the transport matrix and the propagated error covariance matrix are also computed, while extrapolating the particle track from detector station to station. For that purpose, a simplified track follower (see [ATFG+07]) is implemented by this package, which means that the trajectory of a charged particle is predicted in terms of mean values and errors in both forward and backward direction. Further on, three different effects are taken into account:

- The energy loss, which affects mean values and errors
- The Coulomb multiple scattering, which affects errors only
- The magnetic field, which affects mean values only

Moreover this package provides also a set of routines, which are developed by the European Muon Collaboration and integrated in the GEANT3 system. Supplementary an interface to CBMROOT exists, which requires of course modifications in the VMC classes of ROOT (see [GSI06] on page 3). Nevertheless GEANE can be easily used as good and fast implementation of the Runge-Kutta approach, which is introduced in section 2.5 on page 82.

Development With Microsoft Visual Studio: Since the interface between the event simulation and the reconstruction is well-defined, it is surely feasible to run these two tasks independent of each other. Hence this circumstance can be obviously used to work off-line from the data production by the framework. This means more precisely that ROOT based files, which are created to save the MC data by the CBMROOT framework on a GSI machine (see [GSI]), can be afterwards evidently transferred to any place for off-line working. Furthermore this functionality can be of course also used to work with these files on any local desktop computer or laptop without the need for a network connection or almost all CBMROOT framework packages. However it has to be noticed that the access to these files requires some simple object libraries replacing the complex framework interface.
Coming now to the point, the advantage of this concept is rapidly clear, because it is now possible to switch to every platform and compiler without any regard to such ones, which are supported by the CBMROOT framework including all required packages. But it has to be additionally mentioned here that the only package, which can not be omitted and must thus exist for the taken platform, is identified by ROOT.
Thence it is very easy for any user to develop an algorithm needing such an input, because there is not longer special knowledge about the complex framework, which includes packages like GEANT, necessary. So anyone can simply create a standalone program, which requires just the usage of some earlier generated standard ROOT based files.

Within this context, it should be finally noticed that I prefer to develop the Hough tracking on a machine running Microsoft Windows XP as operating

system in combination with the programming environment Microsoft Visual Studio .NET 2003 (see [Vis] and [Mic]). For this purpose, information about getting ROOT and Microsoft Visual Studio together can be found in [ROO]. Moreover the G++ compiler (see [GCC]) of Cygwin (see [Cyg] and [GNUa]) is used in conjunction with the GNU Make (see [mak] and [GNUb]) and special makefiles for cross-compiling, because the developed software should run inside the CBMROOT framework at the end.
As however this so-called framework version requires surely a few specialized source code files, the mechanism to receive all necessary files out of the Microsoft Visual Studio projects is realized by using the same makefiles (see Make) than for cross-compiling. Therefore these makefiles contain special targets, which realize this job. In this connection, it should be also mentioned here that the reason for multiple targets is determined by the functionality to choose between different build processes, which are implicitly supported by the generation of either a makefile for GNU Make, a file containing meta-tags for AutoMake or a file consisting of meta-tags for CMake.

Development With CMake And G++: At first, it has to be told that CMake (see [Theb]) in combination with G++ (see [GCC]) is the choice of the framework developers.
Further on, CMake is a cross-platform, open-source make system, which is used to control the software compilation process by using simple platform and compiler independent configuration files. Hence it is able to generate native makefiles and workspaces with regard to these files that can be used in the specific compiler environment of the user's choice (see [GSI06] page 3).
Besides this, G++ is the traditional nickname of GNU C++, which is a freely redistributable C++ compiler. Moreover this compiler is part of the GNU Compiler Collection; originally: GNU C Compiler (GCC), which signifies an integrated distribution of compilers for several major programming languages. Furthermore it is interesting that this distribution includes libraries for these languages as well as front ends.
So based on this concept, the common general startup requires the following steps:

- Install and setup all packages or login into a dedicated GSI machine
- Checkout the framework CBMROOT
  - svn co https://subversion.gsi.de/fairroot/cbmroot

- Build process
    - mkdir <build>
    - export SIMPATH=<path> (not needed at GSI machine)
    - cd <build>
    - cmake <source>
    - . ./config.sh
    - make
    - make install
- Run macros

If this startup process is then executed once, the following steps would be sufficient:

- Setup the environment
    - cd <build>
    - . ./config.sh
- Run macros

Information about the proceeding for adding a package to this common build process can be found in [Add].

Data Input

As mentioned earlier, the data, which is required for input to the Hough tracking task, is delivered by the CBMROOT framework. Therefore figure 3.7 gives an imagination about the implemented interface.

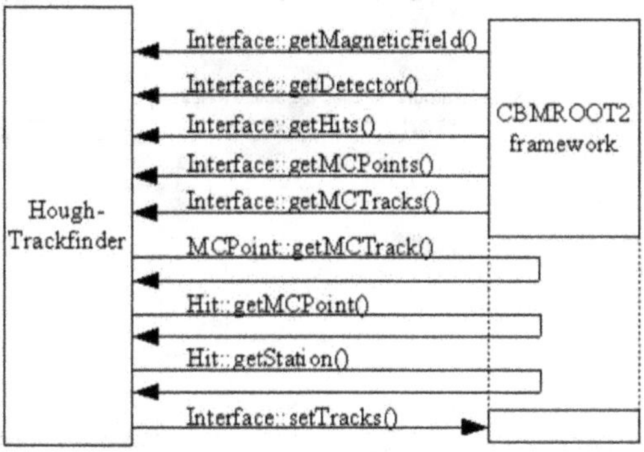

Figure 3.7: Interface between the CBMROOT framework and the Hough tracking

With regard to this figure, the major objects, which have to be accessed via this interface, are determined by the class *CbmStsHit* representing the information of the hits and the class *CbmField* featuring the magnetic field. In addition to this, specific information of the detector setup, which is actually maintained by the class *CbmStsDigiScheme*, is of course also necessary. Furthermore debugging purposes and internal quality measurements claim the entire MC information to be acquirable.
In this connection, it is important to realize that the class *CbmMCTrack* describes the tracks, which should be searched and of course also be found at the best, while the class *CbmStsPoint* poses the original detector interaction and contains therefore the correspondence from a hit to a MC Track.
As this circumstance is very important, it should be explicitly noticed here that the parameter 'RefIndex' of a hit would lead via index to the correct point, if it exists, and the parameter 'TrackID' connects further this point via index to the corresponding MC Track. However since the common index access is heavily used, but does not provide the optimal method in the C++ environment with regard to the time consumption, additional local classes, which speed up all relationships by the application of pointers, are installed and afresh setup at the beginning of each event (see event simulation).

Besides this, it has to be supplementary noticed that the index access is preferred to the pointer access by the ROOT framework developers, because all tracks, hits and points of a single event are stored in separate *TClonesArray* objects inside a ROOT based file and there is no easy possibility to save pointers in a file.
Coming now to the point, figure 3.8 depicts all necessary classes including their relations.

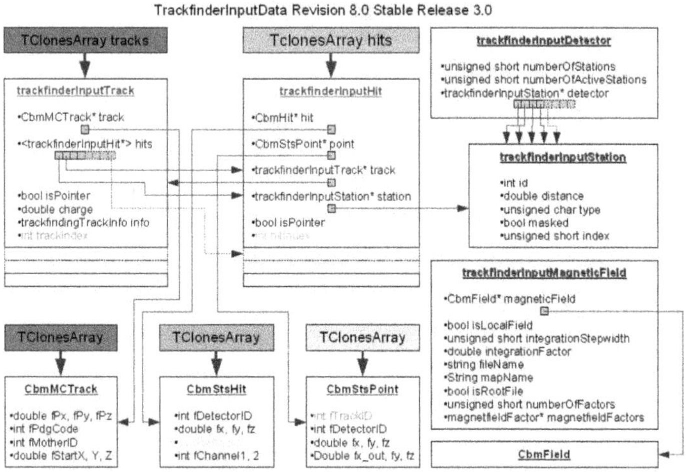

Figure 3.8: Information about all data input classes including their relationship details

So with regard to this figure, the realization of the mentioned speed up requires the class *trackfinderInputTrack* to possess for instance a pointer to each existing *trackfinderInputHit* object, which belongs to the actual track, and a variable to store the index in the original *TClonesArray* in addition to a pointer to its corresponding *CbmMCTrack* object. Collaterally there is also a variable to store the charge of the track and an object named 'info', which represents the internal quality used in an optionally applied improving algorithm after the Hough tracking.
Further on, the variable 'isPointer' simply identifies this object to be local or global. The reason for this information is quite simple, because the framework does naturally not access local data and does thus not free the respective memory by a main destructor call. So if the object is local, it would have to

be destroyed by the task itself. With regard to the class *trackfinderInputHit*, it is obvious that this concept is universally applied, because this class contains also such a variable. But in this case, that variable is obviously set for an object, which is generated by the local hit producers, and would be unset, if the object is delivered by the framework.
Moreover this class exhibits supplementary multiple other pointers. There is for instance a pointer to the original *CbmStsHit* object, a pointer to the *trackfinderInputStation* object where the hit occurs, and if the destination exists in memory, a pointer to the *CbmStsPoint* object as well as a pointer to the *trackfinderInputTrack* object. Further on, the class features finally an index for the original *TClonesArray*.

Beyond that, the representation of the detector setup is quite elementary. The main class, which is called *trackfinderInputDetector*, consists simply of two variables for the number of existing stations and the number of active stations as well as an array of pointers to each station. Within this context, it should be additionally noticed that the two initially mentioned variables offer the functionality to easily evaluate the quality of the algorithm with missing entire stations due to the disabling possibility.
Further on, the secondary class called *trackfinderInputStation* contains basically a value for each necessary information, which is determined by the id for station identification, the distance to the target, the type of the station (MAPS, hybrid pixel or micro-strip) and the mask (true, if the station is disabled and false for an active station) as well as the index for the signature in the histogram cell, if the station is active.

Finally the last but not least input to the Hough tracking is of course defined by the magnetic field, which is represented by the quite more complex class *trackfinderInputMagneticField*. Nonetheless this class owns also some objectives, which follow the general concepts, like a pointer to the *CbmField* object and the variable 'isLocalField' realizing the same feature than the earlier mentioned variable 'isPointer'.
But in contrast to this, there are additionally specific items like the variable 'integrationStepwidth', which defines the number of used steps to compute the numerical integral of the magnetic field by applying the general Simpson rule ([BSMM05] on page 923), and the constant factor 'integrationFactor', which allows the modification of this integration result by a simple multiplication.
Moreover the parameters 'fileName', 'mapName' and 'isRootFile' would be only used to simply identify the magnetic field, if it is read locally.
By the way, the possibility to represent the magnetic field by a constant factor for each detector station requires the array named 'magnetfieldFactors',

which contains these factors, in combination with parameter 'numberOfFactors', which defines the number of factors and thus the size of the array.

Before going on with the implementation of the algorithm itself, it has to be noticed here that the setup of all pointers include some cross checks about the consistency of the data, which are explicitly illustrated in section 4.2.2 on page 219.

Algorithm

The structure of the algorithm implementation is depicted in figure 3.9.

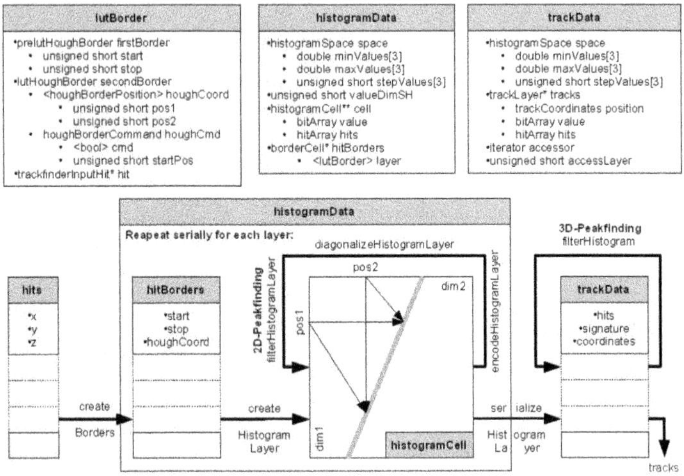

Figure 3.9: Structural software design of the Hough transform algorithm

With regard to this figure, it is obvious that the algorithm takes at first each hit and creates for each the corresponding Hough curve, which is represented by the borders. Hence a new *lutBorder* object is generated for each hit, which contains a pointer to the provoking hit and the output of both LUTs. In this connection, it has to be noticed that the output of the first LUT is determined by a *prelutHoughBorder* object, while the output of the second LUT is defined by a *lutHoughBorder* object.

So in contrast to a possible hardware implementation, both LUTs are evaluated before storing the information. Furthermore the reason for this different

circumstance is apparently ascertained by an easier debugging possibility of the source code.
Nevertheless the first LUT computes just as in hardware the entry range of the third dimension in the Hough space, which is identical to the designation of the histogram layers by its start and stop index, while the second LUT calculates the Hough curve, which is contrarily implemented for easier access by a list of histogram cell indexes or positions instead of a start position and a selection command. Moreover such a list is realized by the template *specialMem* in combination with the class *houghBorderPosition* for the content. However such a different functionality requires obviously an additional interface, which features the access of these histogram cell positions in the same way as needed for the hardware, because it is then possible to write the recently occurred values by the application of this well-defined interface into an American Standard Code for Information Interchange (ASCII) formatted file during the processing of an event (see event simulation). So this file can be thus afterwards of course used for the input to the hardware, which enables the identity comparison at the next well-defined interface to verify the correctness of the implementation. For that purpose, this interface is identified by the class *houghBorderCommand*, because such an object consists of the variable 'startPos', which represents the initial histogram layer column to define the start position, in conjunction with the variable 'cmd', which comprises a boolean value for each subsequent histogram row characterizing the selection command.
Besides this, all computed borders are then stored in a data object called *borderCell*, which is located in the class *histogramData*. At this juncture, it is important to recognize that this object stores the borders sorted with regard to the start index of the histogram layer, because this layer defines the first need for the data of the border. Of course it is needless to say that this data is possibly applied to a sequence of consecutive histogram layers, which is implicitly defined by the stop index minus the start index. So it is obvious that this object implements the functionality of the so-called HBuffer unit (see section 3.2.4 on page 136) with a small difference in the content.

Based on this context, it is clear that the first major step is successfully implemented and the second one can put into the focus. Hence the algorithm starts to fill the first histogram layer by processing all entries of the *borderCell* object for the first layer, which means apparently the read-out of each *lutBorder* object in combination with the adjustment of the corresponding bit in the signature of the affected histogram cells. For this purpose, the *histogramData* object contains a two dimensional array of the class *histogramCell*, which implements apparently the histogram layer, because each

object consists further of an object named *bitArray* to realize the signature and another object called *hitArray* to store the pointers to the originating hits.
In addition to this, it is clear that the variable 'stepValues' of the *histogramSpace* object determines the size of each histogram dimension, while the the variable 'valueDimSH' defines the width of the signature.
Further on, it has to be noticed here that each *lutBorder* object, which belongs to consecutive layers, is collaterally inserted in a temporary memory space for later anew usage. So it is self-evident that this memory has to be also operated, while filling such a histogram layer.
Subsequently the Encoding could be naturally started, if the last *lutBorder* object of the actual layer is successfully processed. For this reason, a further LUT is applied to assign a priority class to each histogram cell, which is of course directly related to the signature of the cell.
After the finished Encoding of the histogram, the Diagonalization obviously takes place, which means that the priority class of the diagonal cell one histogram row above the actual cell would be incremented to the next higher priority class, if both cells exhibit the highest available priority class.
Thereafter the 2D Peak-finding is evidently employed. This step processes the layer with a specialized window geometry, which is adapted to the layout of a peak cluster, to morph the cluster to one cell at best. Therefore special conditions and arithmetics are applied to the histogram cells, which are affected due to the previously mentioned window geometry.
Since the 2D Peak-finding determines the last step of the algorithm, which works on the histogram layer, it is clear that the next step has to combine the results of all layers and save them in the so-called LBuffer unit analog to the HBuffer unit. Thus the Serialization follows to transfer the information of the surviving track candidates from the *histogramData* object into the *trackData* object. Within this concept it is also obvious that these track candidates are sorted with regard to their layer of occurrence in the array of the template *trackLayer*.
In this connection, the necessary information of such a track candidate is of course defined by the coordinate of the histogram cell identifying the feature parameters, the signature for the successively applied 3D Peak-finding and the hits for debugging. As this data includes apparently all required information, it is clear that the other parameters are just used for support. So the variable 'accessor' and 'accessLayer' offer for instance simply the functionality to access the tracks in the sorted *trackLayer* array one after the other independent of any layer, while the *histogramSpace* object contains all information to compute the analog values of the feature parameters based on the digital histogram coordinates.

Finally the cluster problem, which occurs surely also over consecutive histogram layers, requires the 3D Peak-finding including another specialized window geometry in that Hough space dimension to be applied after the entire data is delivered to the *trackData* object. Thence the conclusive surviving track candidates of the actual event are found afterwards and the tracking for the next event (see event simulation) can be started.

Error And Warning Concept

The major interesting fact, which is related to the error and the warning concept, is that both are global to only the Hough tracking package. This means more precisely that both concepts are applied to each library in this package, but not to the CBMROOT framework itself.

Coming now to the point, an error message would occur, if the actual processing cannot be continued and an automatic solution for this problem is not available. So the developer has to catch the error, which is thrown, and can either write the error message to the standard output screen followed by an exit command or can solve the problem and try the function call causing the error again. By the way, it is clear that the user has only the chance to adjust parameters, which are available via interface, and try to execute the program again.
In contrast to this concept, a warning is not thrown, because it is just used to tell the developer or user that a problem occurred, but it can be solved automatically. Nevertheless a warning message is also printed to the standard output screen to offer information about the taken available solution.

So within this context, both concepts feature the displaying of information, but on different severity levels. Moreover it is important to realize that each possible error or warning message can be produced by just a single class, which possesses exactly this information as message string. However as it is imaginably inefficient to use a separate source code file for each of these classes, it is obvious that each library of this package contains two dedicated files, which consists respectively of all error classes or all warning classes for this library.
In addition to this, the general structure of these classes, which is enforced by the common inheritance of classes, requires evidently a single base class for the errors and another one for the warnings, which are both located in the library *MiscLIB* (see section 3.2.2 on page 130). Furthermore this library contains also the some special so-called global error and warning classes, because these ones are used in many diverse libraries. Hence a specific

identifier, which belongs to the name of the library, has to be inserted at the beginning of each message.

Besides this, it has to be finally noticed that the warning concept is additionally used to encapsulate all messages to the standard output screen. Thus no library function is able to write directly to the standard output screen and all messages of a library can be easily prevented by a simple define statement in the corresponding source code file.

3.2.2 Libraries of the Hough Tracking Package HTRACK

This section presents a short summary about the API of the Hough tracking package HTRACK in combination with all necessary libraries including their major essential classes and their inheritance concepts, which offer together the complex task of tracking with the adapted Hough transform that is further customized to the CBM experiment.

AnalysisLIB

The *AnalysisLIB* library contains all objects, which implement analyses. In this connection, the mainly affected fields are the tracking quality based on absolute or relative values, the tracking quality with regard to the momentum and the tracking quality with regard to the histogram occupancy. In addition to this, there are also many analyses, which help to set the parameter values of the algorithm.
More information about diverse analyses can be found in section 4 on page 213.

Besides this, there are two main API classes within this library named *analysis* and *histogramAnalysis*, which encapsulate all other available analyses. The only difference between them is that the *histogramAnalysis* object is specialized to each state (Histogramming, Encoding, Diagonalization, Peakfinding) of one histogram layer, while the *analysis* object realizes in contrast to this all other global examinations.
Nevertheless each analysis of both classes must be enabled, disabled, initialized or run by using the corresponding interface.

Beyond that the object oriented inheritance paradigm is heavily used in this library. For instance, figure 3.10 depicts the inheritance concept for the analysis of the magnetic field.

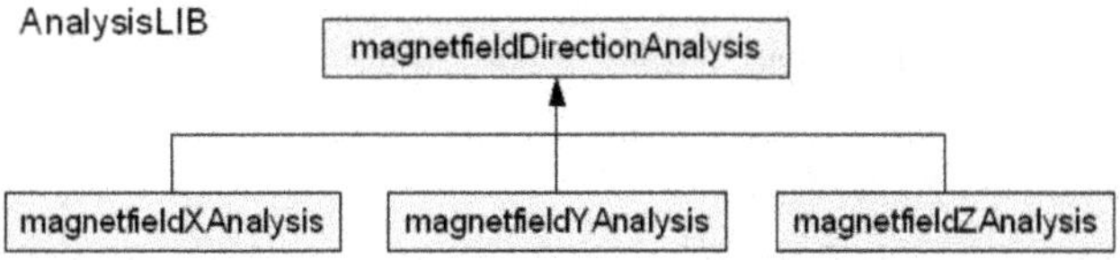

Figure 3.10: Inheritance design of the class *magnetfieldDirectionAnalysis*

With regard to this figure, it is obvious that this analysis is divided into one base class and three subclasses. Furthermore these subclasses are only required to feature the evaluation of three projections, because a three dimensional field in a three dimensional coordinate space cannot be displayed at all.

Further on, another example is determined by the quality measurement of the Hough tracking, because it is evidently important to be able to measure the quality for each event separately and accumulated for an amount of events (see event simulation). For that reason, it is clear why figure 3.11 would illustrate only two subclasses in conjunction with a base class, even if the analysis is able to give information, which is related to found or not-found MC data by the application of absolute numbers or percentages.

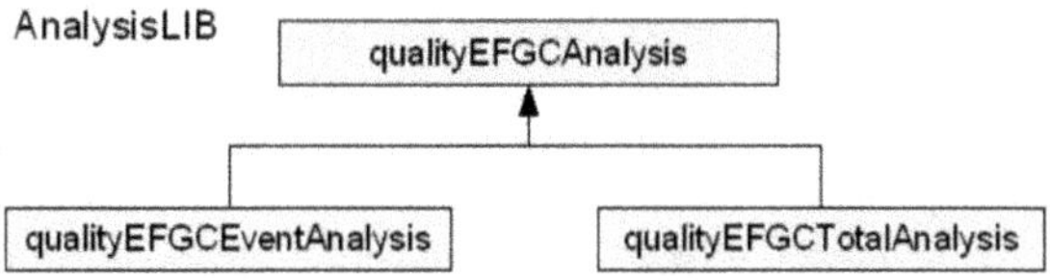

Figure 3.11: Inheritance design of the class *qualityEFGCAnalysis*

However this is obviously not the total story, because the above mentioned general quality analysis is supplemented by another quality analysis, which takes the momentum of the found or not-found MC Tracks into account. So for this purpose, another group of classes is installed, which contains a separation with regard to the momentum in addition to the event relation. Thus figure 3.12 shows four subclasses derived from a base class.

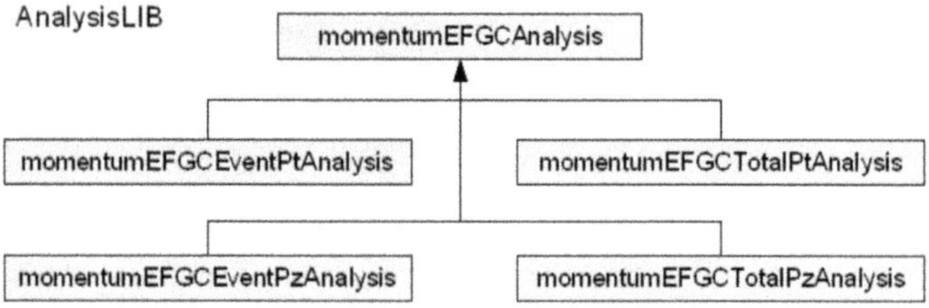

Figure 3.12: Inheritance design of the class *momentumEFGCAnalysis*

The last but not least example, which is depicted in figure 3.13, combines the quality measurement with a projection into a specified plane. As this circumstance offers apparently the most complex context, it is quite good to have a closer look to the naming convention of the analysis.
So it is clear that each analysis starts with the character of the analysis, which is here the projection. Afterwards the name exhibits all implemented measurements, which are here the tracking efficiency, the tracking fake rate, the tracking ghost rate, the tracking clones rate and the tracking not-found track rate. The following term depends then on the event relation (see event simulation). In this connection, the available options are evidently again defined by the event selection or the total selection. This means more precisely that the first option shows the results for each event separately, while the other one accumulates the result of all evaluated events. Subsequently the last term represents the direction of the character, which means here the projection planes defined by the first and second dimension, the first and third dimension as well as the third and second dimension.
Moreover the universal inheritance concept can be finally summarized by the

degrees of freedom, which are available due to the event relation (see event simulation) and the required projections of an analysis.

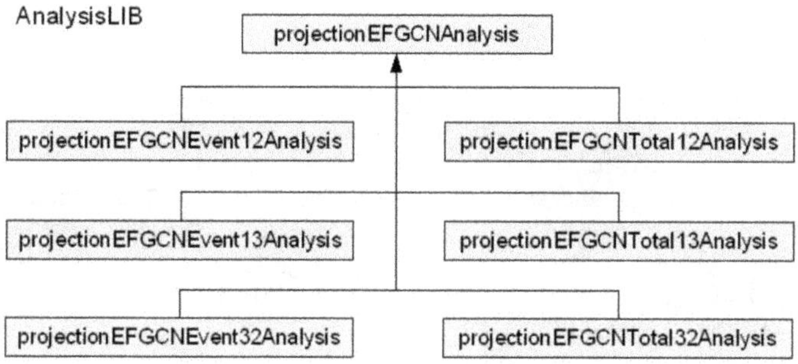

Figure 3.13: Inheritance design of the class *projectionEFGCNAnalysis*

Further on, this library contains also many unique and independent classes, which implement supporting algorithms to optimize the parameters of the Hough tracking. Some examples are determined by the quantization of the Hough space, the optimal constant magnetic field factors and the quality of both LUTs independent of each other as well as in combination.
More information about the analyses can be found in section 4 on page 213.

In addition to this, the library possesses collaterally some special classes, which implement just graphical front-ends for these analyses. Examples for such objects are given by the classes *showAnalysis*, *visualAnalysis* and *hardwareAnalysis*. Furthermore it has to be also mentioned that each graphical analysis can be automatically written into a ROOT based file.

CbmHoughStsTrackFinder

The API *CbmHoughStsTrackFinder* implements the main task for the *CBM-ROOT* framework by a class with the same name in combination with another class, which realizes a configuration file object named *inf*. Furthermore the corresponding class of this configuration file object is apparently directly derived from the class *configuration*, which is located in the library *FileioLIB* (see figure 3.14 of section 3.2.2 on page 125), because it offers the possibility to either define some standard values for the parameters or to read these parameter and value pairs from file.

DataObjectLIB

The library *DataObjectLIB* contains all classes, which are completely independent of the ROOT or the CBMROOT framework and implement special memory, file access or formulas.
Examples for memory classes are determined by the histogram cell signature classes *bitfield* and *bytefield*, the quality analysis memory classes *trackToPeak* and *peakToTrack*, the Hough space classes representing cell positions like *trackParameter* and *trackCoordinates*, the table class *table* and many more.
Referring to figure 3.14 of section 3.2.2 on page 125, examples for the file access classes are the table file class *tableFile* and the LUT access file classes *prelutAccessFile* and *lutAccessFile*.
Finally an example for a formula class is defined by the analytic formula class *analyticFormula*, which consists of all basic formulas and the equations to compute both LUTs based on the approach including a parametric description.

DataRootObjectLIB

In contrast to the library *DataObjectLIB*, this library contains all classes, which implement special memory or data objects dependent of the ROOT or the CBMROOT framework.
Examples for memory classes are determined by the histogram class *histogramData*, the histogram cell class *histogramCell*, the hit list class *inputHitStlList*, which is instantiated for each bit in the signature of histogram cell, and many more.
In addition to this, examples for data classes are the hit class *trackfinderInputHit*, the track class *trackfinderInputTrack*, the detector and station classes *trackfinderInputDetector* and *trackfinderInputStation* as well as the magnetic field class *trackfinderInputMagneticField*.
In this connection, it has to be also noticed that the class *trackfinderInputMagneticField* realizes supplementary the numerical computation of the averaged integral of the magnetic field, which is used for the factor B_y in the analytic formula, by the application of the Simpson rule (see [BSMM05] page 927).

FileioLIB

The library *FileioLIB* contains surprisingly just the three abstract classes, which are named *fileio*, *configuration* and *io*. Further on, details show that the class *fileio* determines the base class for all access to an ASCII formatted file, while both others are derived from this one. Moreover the difference between these two derived classes is simply identified by the circumstance that the class *io* offers the possibility to access a file, which contains data in addition to an information header describing related parameters, whereas the class *configuration* consists just of a such an information header. Thus a class, which is derived from the class *io*, is commonly used to realize data files featuring the serialization and deserialization of data objects, while the class *configuration* is usually applied to access parameter files.
Referring now to figure 3.14, an example for such a parameter file is defined by the class *inf*, which is used to get the parameters for the Hough tracking algorithm. Besides this, examples for data files are given by the table file class *tableFile* as well as both LUT access file classes *prelutAccessFile* and *lutAccessFile*.

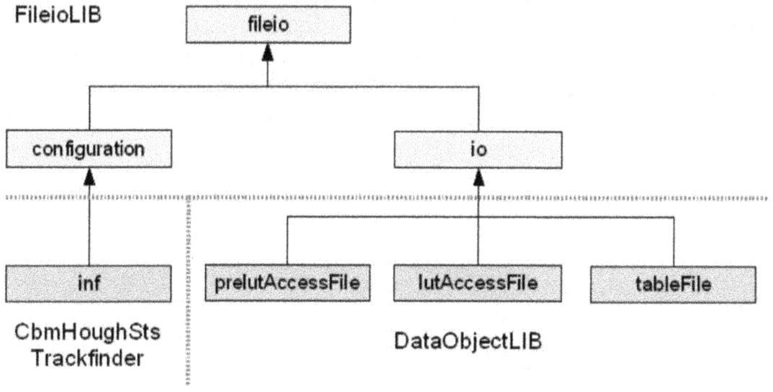

Figure 3.14: Inheritance design of the class *fileio*

So based on this concept, a data or parameter file class can be apparently implemented by deriving a class from the corresponding abstract base class including additionally some methods to deal with the information header as well as the serialization and deserialization of the data.

HistogramTransformationLIB

The library *HistogramTransformationLIB* consists of all classes, which are necessary to modify on the one hand the histogram layer by implementing the algorithm steps Encoding, Diagonalization, 2D Peak-finding and Serialization of the Hough transform as well as on the other hand the serialized track candidates by the algorithm step 3D Peak-finding.
For this purpose, the simple task of Encoding and Diagonalization require obviously just a single class, while the Serialization can be even realized internally without the need for a separate class.
In contrast to this, the complex task of Peak-finding, especially due to the logical reassembling of the 2D layers into a 3D space, necessitates multiple classes to support different conceivable strategies. Furthermore it is additionally clear that each Peak-finding geometry is decoupled from the applied arithmetic method, which processes the values inside the geometry (see section 3.2.6 on page 144), because this circumstance enables the availability of even more strategies without more classes. So it has to be explicitly noted that a complete Peak-finding strategy, except a uniquely one, has to combine a geometry class with an arithmetic class. Moreover figure 3.15 depicts the available classes, which implement diverse geometries for the 2D Peak-finding. In this case, it is important to recognize that each 'X' in the name of the base classes represents one dimension of the concerned geometry. So that means more precisely that the class *filterDimXDimX* arranges for instance the base class for a two dimensional one, which is further derived into the class *filterDim1Dim2* to realize a geometry covering dimension one and two with a privilege for dimension one.

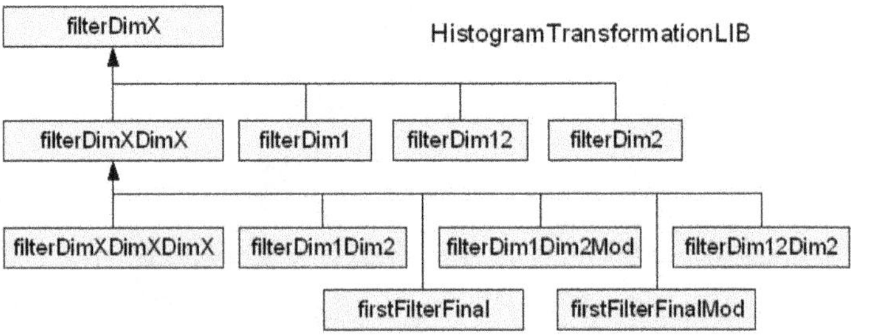

Figure 3.15: Inheritance design of the class *filterDimX*

In addition to this, figure 3.16 shows the available Peak-finding classes for the 3D Peak-finding. At this juncture, it has to be noticed that almost all of these classes take contrarily to the 2D Peak-finding all three dimensions into account. Even if this circumstance is surprising at a first glance, it would be anticipated, if the reassembling of the 3D Hough space is considered, because neighboring histogram layers have to be obviously involved in the computation.

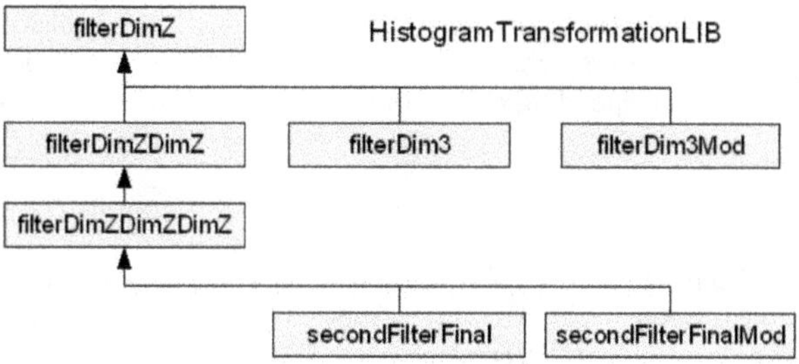

Figure 3.16: Inheritance design of the class *filterDimZ*

Besides this, the geometries, which are used for the 2D and the 3D Peak-finding, are determined in the class *maxMorphSearch*, while the definition of the applied arithmetic method is located respectively in both abstract base classes for the Peak-finding geometries, which are named *filterDimX* and *filterDimZ*. Moreover the arithmetic methods are installed in the classes *filterBasicSimple*, *filterBasicSimpleMod*, *filterBasicComplex*, *filterBasicComplexMod*, *filterBasicSpecial* and *filterBasicNeutral*. Although multiple classes exist, the challenge of these methods is evidently always identified by the evaluation of a local maximum, which is of course not that differently implemented and not that complex.
In contrast to this, the Peak-finding class *eraser* realizes as mentioned earlier a different strategy, which is more compact, but entails less performance.
Nevertheless all interfaces are encapsulated in the class *histogramTransformation*.

HoughTransformationLIB

The library *HoughTransformationLIB* consists just of a single class, which features the evaluation of the LUTs for each hit, the storing of these obtained results and further even the filling of the serially computed histogram layers. This class implements thus the algorithm step Histogramming of the Hough transform as well as the computation of all necessary prerequisites. For this purpose, the process, which evaluates the LUTs, requires obviously an interface to the corresponding classes of the library *LutGeneratorLIB.*

InputLIB

The library *inputLIB* consists of all necessary classes to implement the input interface to the outer world. For this purpose, two classes, which are named *inputAscii* and *inputRoot*, accept either an ASCII formatted file or ROOT based methods to receive the data input. Further on, these ROOT based methods are classified in two different configurations. So the class *inputFRoot* supports the special interface of the CBMROOT framework to access the data input, while the class *inputSRoot* features contrarily the standalone access to a ROOT based file, which implies explicitly access independent of the CBMROOT framework. Besides this, the following figure 3.17 depicts the relationship between all available classes.

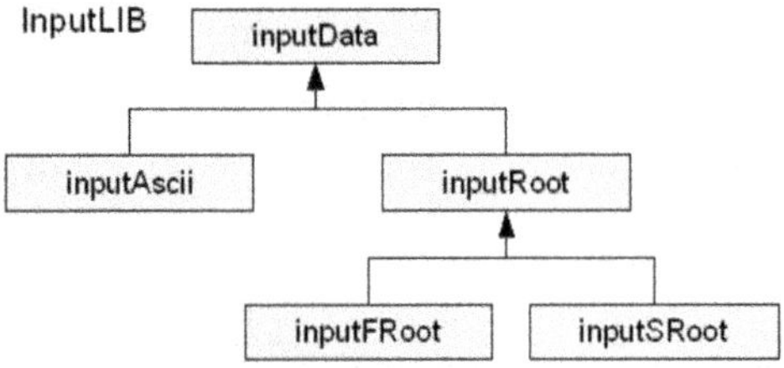

Figure 3.17: Inheritance design of the class *inputData*

Beyond that, it is important to notice that this complex structure has to be setup, because former versions of the CBMROOT framework do naturally not support the newest and most comfortable methods to access the data input. For example, the ancient version of the framework does even not involve ROOT, which requires thus the usage of a simple ASCII file. Moreover the successive framework versions offer indeed the functionality of ROOT, but do not provide an interface. Hence the possibility to read the data manually out of a standalone ROOT based file is required. In contrast to this, the newest and most convenient CBMROOT framework exhibits obviously the mentioned interface. But as this interface is up to now not setup entirely, a supplementary concept must evidently exist, which lays somewhere in between. So this circumstance defines the actual applied concept more precisely to use some input from the framework interface as well as some other input, which has to be read manually out of a ROOT based file. Thus the reason for keeping and supporting all versions to get the input is imaginably obvious.

LutGeneratorLIB

The library *LutGeneratorLIB* contains all classes, which are related to the LUTs. Examples for these classes are these ones, which realize their generation, as well as these ones, which feature the entire file access including explicitly the read and write mode. For this purpose, figure 3.18 shows that a single base class is evidently derived two times to implement all actually available versions of both LUTs.

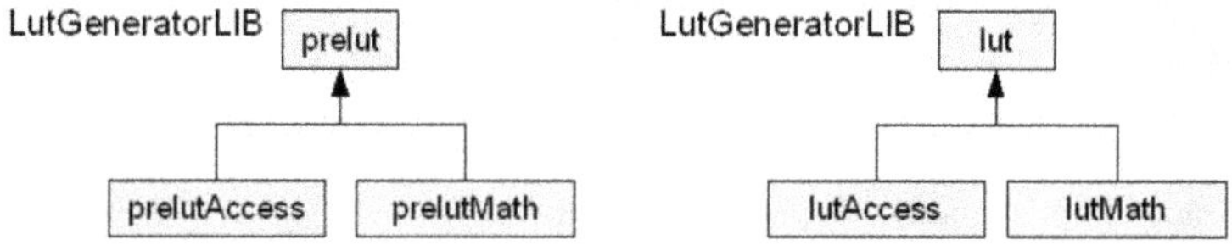

(a) *prelut* (b) *lut*

Figure 3.18: Inheritance design of the classes belonging to the LUT

A big advantage of this implementation is obviously determined by the circumstance that the 'Math' version of each class realizing respectively one of the two LUTs can be not only used to generate the LUT files, but also applied directly on-line. This fact implies further that external files including both LUTs are not required at all, because the analytic formula can be directly applied to the coordinate positions of the actual hit to compute the transformation. Hence these 'Math' classes offer the possibility to be independent from the digitization of the hits according to the detector station layout, which is apparently very useful, because the digitization is not ready for quite a long time.
Besides this, it is clear that the 'Access' version of each class realizing respectively one of the two LUTs implements the common approach, which bases

on the external files. So this means more precisely that everything, which is necessary to access the files with a digital index based on the hits, is setup there.
Finally it has to be noticed here that this library features not only the analytic formula (see section 2.4 on page 66) to compute the LUTs, but also the Runge-Kutta approach (see section 2.5 on page 82).

MiscLIB

The library *MiscLIB* contains everything, which can not be put into another dedicated library. Thus the major important objects, which are located here, are at first defined by global errors, global warnings and global definitions. So examples for such global things are given by the definition of all three different necessary coordinate systems (momentum, coordinates, Hough space) and the definition of the maximal size for the dimensions of the Hough space, if the space is allocated before runtime (see section 3.2.5 on page 138). In addition to this, the functionality related to the progress information, which is displayed on the standard output, as well as all possible conversion routines from integrated data types into strings or character arrays and vice versa are of course also located in this library.

OutputResultLIB

The library *outputResultLIB* contains all classes to implement the output interface to the outer world, which features of course the delivery of the found tracks. Furthermore these tracks, which are determined by the track parameters and a list of hits, can be either committed to a special CBMROOT framework interface or directly written into a ROOT based file.

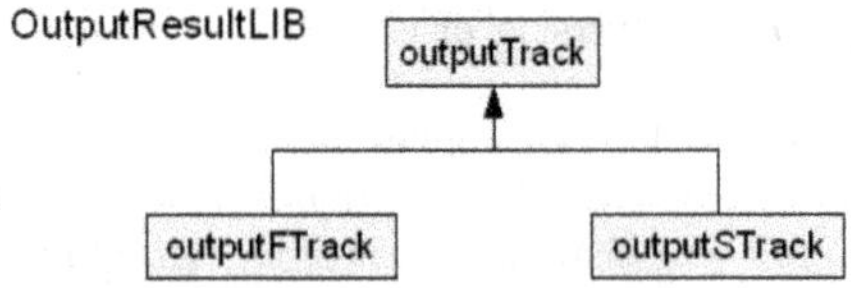

Figure 3.19: Inheritance design of the class *outputTrack*

So with regard to figure 3.19, the class *outputResult* requires analog to the class *inputRoot* (see section 3.2.2 on page 128) some small differences related to the usage of the CBMROOT framework interface, which is realized by the class *outputFResult*, and to the standalone implementation, which is achieved with the class *outputSResult*.

RootFrameworkLIB

The library *RootFrameworkLIB* consists of all classes, which are required to implement a standalone adaptation of all essential CBMROOT framework interfaces belonging naturally to the ROOT based file access. Hence this library features the tracking package to run completely independent of the CBMROOT framework.
Besides this, the major advantage, which is offered by this library, is immediately clear, because these classes can be certainly compiled on every imaginable platform independent of the CBMROOT supported ones.
However this circumstance goes obviously hand in hand with the disadvantage that the simulation chain is not working, because just the small part of the reconstruction chain, which features the tracking, is considered to work. So for instance local hit-producers based on a binning or a smearing would have to be applied to the MC Points, if the hits are not accessible via the ROOT based file.

TrackfinderLIB

The library *TrackfinderLIB* consists just of a unique class called *trackfinder*, which simply chains all necessary classes of the other libraries to implement the Hough based tracking algorithm.
So with regard to the illustrating figure 3.20, it is obvious that classes of the library *inputLIB* provides the entire data input for the tracking, which is further delivered to classes of the library *HoughTransformationLIB*. Subsequently this library fills then the histogram layers based on the data input by the usage of some classes, which are encapsulated in the library *LutGeneratorLIB*. Afterwards classes of the library *HistogramTransformationLIB* cleans the histogram layers from background noise including the clusters. And finally the library *OutputResultLIB* delivers the data output.
Nevertheless it has to be noticed here that classes of the libraries *HoughTrans-*

formationLIB and *HistogramTransformationLIB* would work commonly serial in a loop until all histogram layers are processed, even if a simple sequence is allowed in this figure. However this circumstance is just caused by the fact that the loop would not exist, if all histogram layers are instantiated in parallel.

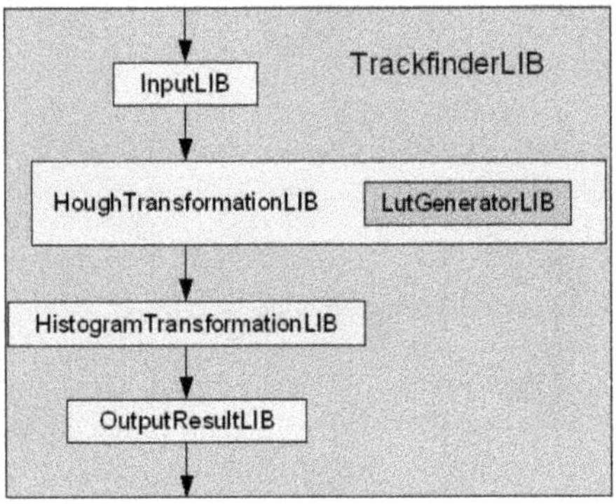

Figure 3.20: Illustration of the library *TrackfinderLIB* with regard to its internally utilized sub-libraries

3.2.3 Look-up-tables

At first, it has to be remembered that the library *lutGeneratorLIB* contains both available implementations of each Look-up-table (LUT) (see section 3.2.2 on page 129), which are defined by an on-line applicable 'Math' version and a real LUT file 'Access' version.
So based on this context, it is evident that the implementation of both file 'Access' versions are quite complex, because they need an array to hold all imaginable Hough curves on-line and a defined method to serialize and deserialize the objects of this array for the file access. However figure 3.21 depicts that the serialization and deserialization defines no crucial problem, because the two objects *prelutAccessFile* and *lutAccessFile* implement simply the common approach for the file access, which is featured by the inheritance of the class *io* located in the library *FileioLIB* (see section 3.2.2 on page 125). Nevertheless the content of these files, which is obviously equal to the content of the arrays, defines a bigger problem, because the amount of entries in each LUT is of course directly correlated by an one to one connection to the

available amount of imaginable digital hits, which are used to address these arrays. Thus it is immediately clear that a good detector resolution causes the array size to increase rapidly to many Gigabyte.

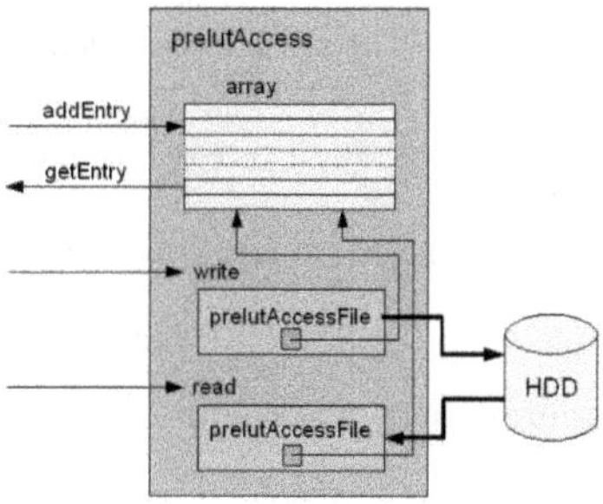

(a) Class *prelutAccess*

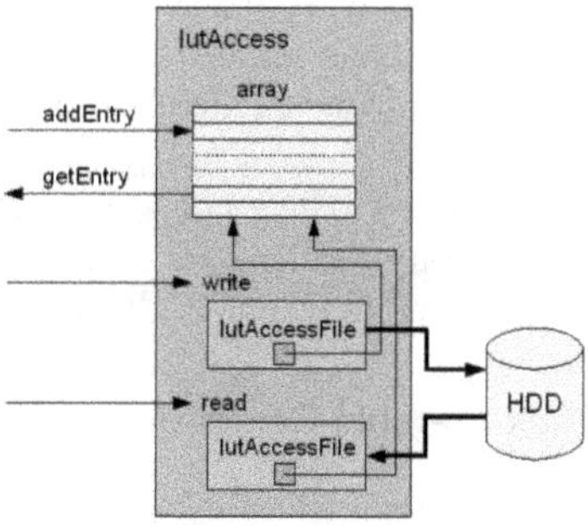

(b) Class *lutAccess*

Figure 3.21: Illustration of the classes related to the file based LUT

In contrast to this, the implementation of both 'Math' versions is quite simple, because they only make use of the equations, which are defined by the analytic formula and encapsulated in the class *analyticFormula*. Hence these on-line LUTs offer the functionality to compute the Hough curves for the three dimensional coordinates of each hit without the requirement for digital information. Collaterally this circumstance is very useful, because these implementations could be applied instead of the file based LUTs, if the digital information for the hits is unknown. However the generation of the LUT files based on the analytic formula require apparently both versions.

Look-up-table Generation by the Analytic Formula

The algorithm, which is able to generate both Look-up-tables (LUTs) based on the analytic formula, is very simple. Nevertheless an important prerequisite, which is already mentioned earlier, is determined by the circumstance that the digital hit information must be accessible in addition to the corresponding three dimensional coordinate, because this information defines the index in the LUTs for the recent computation of the equations requiring the coordinate.

So if this prerequisite is satisfied, the entire content of each LUT could be simply computed by the evaluation of equation 2.13 of section 2.4.1 on page

73 and equation 2.18 of section 2.4.2 on page 82 for all digital hit information, which can imaginably occur in the detector stations. In this connection, the reason for the existence of the prerequisite is immediately clear, because each unique calculation in the loop over all digital hit information requires apparently the three dimensional coordinate for input to the equations, whose result determines respectively the entry in the LUTs at the address defined by the corresponding digital information.

Besides this, it is obvious that the primarily involved classes are determined by *prelutMath*, *prelutAccess*, *lutMath* and *lutAccess*, which are all members of the library *LutGeneratorLIB*. Hence more information about these classes can be found in section 3.2.2 on page 129.

Look-up-table Generation by the Runge-Kutta Approach

The algorithm, which offers the generation of both Look-up-tables (LUTs) based on the Runge-Kutta approach, is evidently not as simple as the one for the analytic formula, because this approach applies just a single indivisible model for the entire motion trajectory of a charged particle in an inhomogeneous magnetic field. Hence the resulting exclusive LUT has to be collaterally decomposed into the two searched LUTs.
In addition to this, it is also clear that the prerequisite, which is introduced in section 3.2.3 on page 133 for the generation of both LUTs based on the analytic formula, counts here as well.

So if this prerequisite is again satisfied, the algorithm could start to compute the exclusive LUT by the application of a loop over all available track parameters, which correspond to a histogram cell in the Hough space. Furthermore the 4th order Runge-Kutta method is naturally used inside this loop to evaluate the result for equation 2.23 of section 2.5.1 on page 91 according to each track parameter set. Hence it is clear that equation 2.25 of section 2.5.1 on page 94 is taken into account to compute a list of hit coordinates, which correspond to one track defined by the actual parameters. But due to the subsequent need for the digital hit information, it is by now useful to store this information in the list instead of the three dimensional hit coordinate. As this concept is further on not that easy, the following example illustrates a single entry of such a LUT.

- track parameter 42 $\mapsto$ hit 1, hit 2, hit 9, hit 17, hit 23, hit 56, hit 123

Moreover this exemplary entry shows immediately that such a LUT possesses the converse direction with regard to the needs. For that reason, the successive step, which is defined by an inversion, is obviously expectable and the next list, whose entries look like the subsequent example, can be directly obtained.

- hit 1 $\mapsto$ track parameter 42, track parameter 53, track parameter 64

Although this list exhibits now the correct direction, it has to be noticed that the generation implicates apparently two consistency problems. This means more precisely that there is obviously neither the guaranty for the inverted list to contain an entry for each imaginable hit nor the guaranty that a single entry consists of a list of track parameters, which form the required Hough curve without holes. Hence the subsequent step is naturally just required to interpolate this hit indexed list including its entries in the way that it will fulfill afterwards both requirements. Finally the last step is formed as mentioned earlier by the decomposition of this single list into the two needed lists, which form the required LUTs.

3.2.4 HBuffer

The design of the HBuffer unit in software is quite simple, because it applies just an array of Standard Template Library (STL) *std::vector* templates, which contain further objects of the class *lutBorder* including objects of the classes *prelutHoughBorder* and *lutHoughBorder* as well as a pointer to the causative hit. Moreover it has to be noticed here that the array is of course selected due to the circumstance that the number of elements is fixed by the number of histogram layers, while the STL *std::vector* template is naturally chosen for the next level because of an unknown number of elements, which depend obviously on the data input. So for that reasons, it is clear that this implementation can be easily assumed to realize the optimum for the storage holding the LUT results for all hits in a sorted order belonging to the start index of the histogram layer. Further on, figure 3.22 features a deeper insight about all involved classes, their relation and available information.

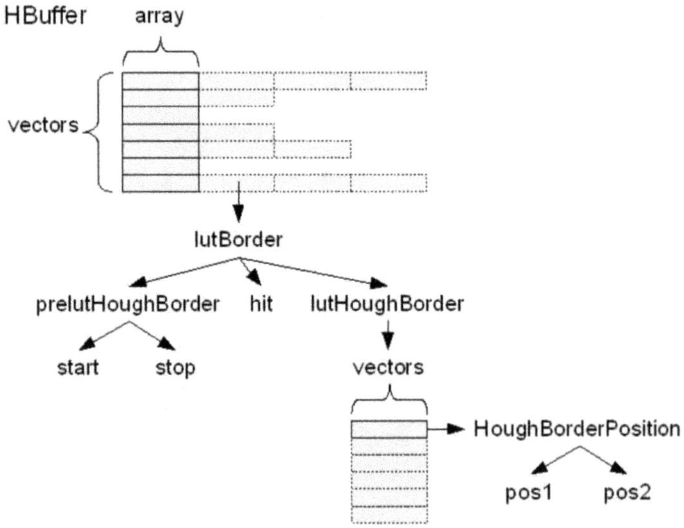

Figure 3.22: Illustration of the classes related to the HBuffer unit

Coming now to a short summary about the usage of this HBuffer unit in the software tracking algorithm, it is at first clear that this implementation has to store a single *lutBorder* object for each occurring hit ordered in the array with

regard to the value of the parameter 'start' in the object *prelutHoughBorder*. Furthermore it has to be mentioned within this context that these objects are apparently unsorted in the corresponding STL *std::vector* template, because there is no need to sort. Further on, the corresponding value of the parameter 'stop' in the object *prelutHoughBorder* defines of course the last histogram layer, which is affected by the actual *lutBorder* object.
Besides this, the required information for the affected bit in the signature of a histogram cell causes evidently an additional link to the hit including this information, which is of course implemented by a pointer. Although this information can be realized by a simple parameter as well, the link to the causative hit is used, because there is then additional information available, which can be supplementary used for debugging.
Beyond that, the object *lutHoughBorder* contains a STL *std::vector* template of *houghBorderPosition* objects, which establish a curve in the Hough space, because each of these objects represent the coordinate of a histogram cell in the two dimensions of a histogram layer.

So based on this information concept, the algorithm uses this HBuffer unit to store the results of both evaluated LUTs for each hit in the data input sorted to the histogram layer of the first occurrence. As furthermore the write access to this unit is straight forward and nothing has to be regarded, the read-out, which is caused by the Histogramming, can be obviously set into the focus.
But before starting, it has to be mentioned that the implemented nondestructive read-out of the HBuffer unit enables in combination with the serial processing of the histogram layers the requirement of an additional temporary memory, which is realized by a *std::list* object from the STL. In this connection, it is not surprising that a single list is sufficient, because it comprises every *lutBorder* object, which is needed for the computation of the recent histogram layer except those ones, which are already saved in the HBuffer unit according to this layer.
Further on, the nondestructive read-out, which means explicitly that the content of the HBuffer unit is never changed until the end of the corresponding event is reached, enables this information to be taken into account for analysis during all steps of processing.

Coming now to the point, all *lutBorder* objects of the STL *std::vector* template at the array position with index zero are accessed at startup one after the other to compute the first histogram layer.
As however the process of Histogramming defines no secret, the conditional insertion of each *lutBorder* object into the temporary memory, which depends apparently on the circumstance whether this object is needed in consecutive

histogram layers, is of course not less essential for a successful tracking. Furthermore it has to be noticed that the evaluation of this decision is quite easy, because it is determined by the result of the comparison between the value of the parameter 'stop' in the *prelutHoughBorder* object and the index of the actual processed histogram layer. So at this juncture, it is needless to say that the *lutBorder* object would be only inserted, if the value of the parameter 'stop' is bigger than the index of the actual processed histogram layer. Otherwise this object can be discarded, because it has no effect on the following layers.
By the way, it is now immediately clear that this temporary memory has to be processed before accessing the *lutBorder* objects of the STL *std::vector* template at the array position with index one, because otherwise an object can be possibly inserted in the accumulator twice, once for the STL *std::vector* template and a second time for the temporary memory. Beyond that, it is also obvious that the objects in the temporary memory can only remain there until their exigency expires. Hence the earlier introduced decision has to be made certainly as well, but with the difference that the object would be removed from the memory, if the value of the parameter 'stop' is equal to the index of the actual processed layer, and kept otherwise.

3.2.5 Histogram

Although the three dimensional histogram is decomposed into serially computed two dimensional layers, the allocation of the huge amount of memory, which is however needed to implement a single histogram layer, has to be taken into account, because it defines the central part of the algorithm and affects thus evidently the major aim of a fast simulation. For this purpose, the two generally available methods for memory allocation, which are named static and dynamic, have to be considered.

A closer look to the so-called static memory allocation shows immediately its characteristic, which is identified by the fact that the memory is allocated at compile time before the associated program is executed. Thus no time is consumed for the allocation during the program run, which obviously enables this method to be the fastest imaginable one. Nevertheless the main disadvantage is surely determined by the reorganization of the memory allocation and the ascertainment of the needed size, because the size has to be fixed to an unknown amount before the program run. For that reason, a common approach is simply defined by the allocation of the maximal amount of memory ever needed during the runtime, which realizes apparently an inefficient memory usage.

In contrast to this, the so-called dynamic memory allocation is characterized by the circumstance that the memory is allocated during the runtime of the program, which offers certainly the advantage that the amount of memory is at all times precisely known before the allocation takes place. However the biggest disadvantage is unfortunately antidromic to the major aim, because the search for a block of unused memory of sufficient size, which includes evidently problems like internal and external fragmentation as well as the size inflation due to the allocator's meta data, impacts naturally the performance significantly. For instance, [DDZ94] shows the lowest average instruction path length, which is required to allocate a single memory slot, to be 52. Hence the usage of this method results in a slower simulation with regard to the static memory allocation.

So based on these two methods, it is clear that the optimal strategy with regard to the major aim of a fast simulation is located somewhere in between, which means precisely that a minimal amount of memory is allocated statically at compile time, while the adjustment to the effective size is realized dynamically during runtime.
This means in detail that the cells of the histogram and the signature objects, which are both apparently always needed, are allocated statically with a fixed predefined amount, while the rest is allocated and destroyed during runtime. In this connection, it is important to realize that an error would be thrown during runtime, if the sizes for the dimensions of the histogram are set by the user via the configuration file parameters 'trackfinderThetaStep' (see table A.40 of appendix A.2 on page A-21), 'trackfinderLutRadiusStep' (see table A.43 of appendix A.2 on page A-22) and 'trackfinderGammaStep' (see table A.37 of appendix A.2 on page A-20) to values exceeding the static range. Otherwise the simulation will deliver results, while not using the entire allocated memory. Furthermore this maximum static sizes can be set via the define statements 'maxDimTheta', 'maxDimRadius', and 'maxDimGamma' in the source code file *memoryDef.h* of the library *MiscLIB* (see section 3.2.2 on page 130 and appendix A.3 on page A-79).

Moreover figure 3.23 depicts a deeper insight about all involved classes, their relation and available information as well as the allocation method, which is marked with dashed object lines to show a dynamic allocation and a continuous line to highlight a static allocation.

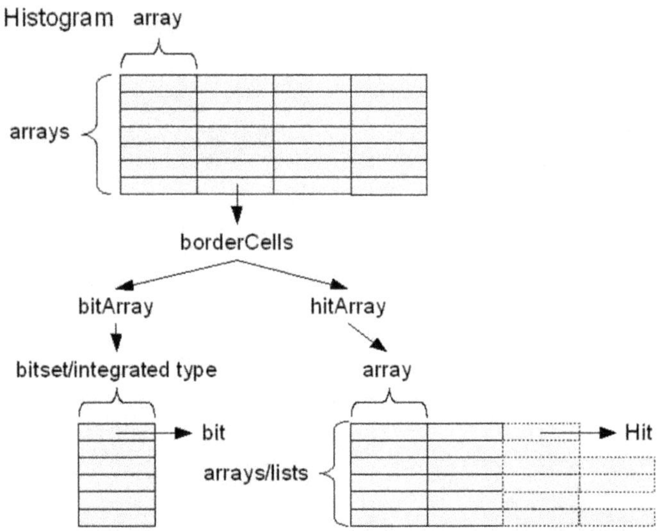

Figure 3.23: Illustration of the classes related to the histogram

Coming now to a short summary about the usage of this histogram unit in the software tracking algorithm, it is at first clear that this implementation realizes a cell by the class *borderCell*, which contains further a *bitArray* object to store the signature and optional a *hitArray* object to save pointers to the causative hits related to the signature. So within this context, it is immediately palpable that the class *bitArray* consists of a fixed amount of bits, which is equal to the fixed number of detector stations, while the class *hitArray* includes this size to realize an also fixed array of dynamic memory slots for the pointers to the hits, which can be either implemented by simple lists or a dynamic augmented static array. Furthermore these memory slots have to feature additional dynamic memory, because the amount of hits, which set the corresponding bit in the signature, is obviously diverse, varying and dependent on the data input.

So based on this information concept, it is clear that the algorithm applies this histogram unit to implement the accumulator. Since each modification step of the algorithm requires thus object access to the histogram and especially to the signatures inside the cells, the implementation of the class *bitArray* is surely of major importance and interest with regard to the major aim of a fast simulation. Therefore this aim necessitates obviously the used data structure to optimally support on the one hand the access to single bits for the Histogramming step, and on the other hand the global access for the arithmetic operations, which are necessary for the Encoding and the Peakfinding step. As such a data structure is however apparently not realizable that easy, two different implementations are developed under the essential constraint of needing as less memory as possible. But before presenting now to the concrete classes, the following table 3.1 summarizes at first the operators, which have to be supported.

<table>
<tr><th>Operation</th><th>Symbols</th></tr>
<tr><td>Assignement and Indexing</td><td>=, [ ]</td></tr>
<tr><td>Logic</td><td>~, !, ∧, &, &&, |, ||,
∧=, &=, |=,
<<, <<=, >>, >>=</td></tr>
<tr><td>Comparison</td><td><, <=, >, >=, !=, ==</td></tr>
<tr><td>Stream</td><td><<, >></td></tr>
<tr><td>Arithmetic (not on bit-level)</td><td>+, -, ·, ÷, %,
+=, -=, ·=, ÷=, %=
++(), ()++, --(), ()--</td></tr>
</table>

Table 3.1: Overview of the required histogram signature operations

Besides this, a more detailed look to the source code file bitArray.h in the library *MiscLIB* shows that this file does not contain the description of a class, but contrarily a simple define statement, which realizes an easy switch between the two available implementations located both in the library *DataObjectLIB* and defined by the classes *bitfield* and *bytefield*. Moreover the class *bitfield* bases on the STL template *std::bitset*, while the class *bytefield* adopts an integrated data type like *unsigned char*. So based on these different concepts, table 3.2 exhibits a direct comparison chart between the main features.

Objective	Class *bitfield*	Class *bytefield*
Indexing	Basic operation (fast)	Bit-shift and type conversion (slow)
Arithmetic	Conversion to integrated type or simple hardware algorithms (slow)	Basic operation (fast)
Memory (compiler dependent)	Stepwidth: 4 Byte req: 6 bit ⇒ res: 8 Byte req: 32 bit ⇒ res: 8 Byte	Stepwidth: 4 Byte req: 6 bit ⇒ res: 8 Byte req: 32 bit ⇒ res: 8 Byte
Restriction	Theoretically none	Biggest integrated type

Table 3.2: Comparison table for the classes implementing the signatures

With regard to this table, it is obvious that the optimal implementation can be only selected with regard to the most frequented operations of the algorithm in combination with their exact time consumption. But as these results are apparently really hard to evaluate for all required operations, it is of course preferred to simply measure the time of a sample event and take this implementation for standard, whose result is faster. In this connection, it has to be finally noticed that the actual result for such a sample event causes the selection of the class *bitfield*.

Beyond that, the implementation of the class *hitArray* is also not completely negligible, because a possible slow down of the algorithm could be avoided, if this optional class is also realized optimal in the case of usage. But before having a closer look to the implementation, it is certainly important to describe the functionality.

So with regard to the quality of the Hough tracking algorithm, whose corresponding analysis is presented in section 4.7 on page 258, it is clear that the signature of a finally surviving track candidate is of major importance and must be recoverable, because it is destroyed during the Encoding. Even if this information can be surely stored during the Encoding in an additional memory, the information about the contributing hits for each bit in the signature allows even more complex analysis, which might use the exact determination of the hits causing a final track candidate. For instance, the signature of such a detailed track candidate can be nonetheless reconstructed under the constraint that some special hits are excluded for the reconstruction, or something else. For that reason, it has to be explicitly noticed here that each histogram cell requires a *hitArray* object just to be able to offer these kinds of special analysis and not for the algorithm itself. Thence this optional object would be unused, if no quality analysis is evaluated. Furthermore the common tracking run naturally discards the evaluation of this extra information for the signatures to avoid the time consumption for such dispensable information.

Coming now to the precise implementation, it is not surprising that the concept of the *hitArray* realizes the same switch as the *bitArray*. This means in detail that the source code file hitArray.h in the library *MiscLIB* does not contain the description of a class, but a simple define statement, which realizes an easy switch between the two available implementations located both in the library *DataRootObjectLIB* and defined by the classes *inputHitStlMem* and *inputHitSpecialMem*. Moreover the class *inputHitStlMem* bases on the STL template *std::list*, while the class *inputHitSpecialMem* adopts a statically reserved and if required an additional dynamically allocated array.
Regardless of that, both implementations instantiate at first an array of the size equal to the number of bits in the signatures, which is of course further equal to the number of detector stations, consisting either of *inputHitStlList* objects or of *inputHitSpecialArray* objects located both again in the library *DataRootObjectLIB*. Further on, the class *inputHitStlList* contains now the mentioned STL template *std::list*, whereas the class *inputHitSpecialArray* possesses two different arrays. In this connection, it is important to recognize that the size of the statically reserved array can be controlled by the parameter 'minimalArraySize' in the source code file inputHitSpecialArray.h. So the functionality is easily characterized by the general usage of the fast accessible static array, which is only augmented for the histogram cells, whose static array size is not sufficient, by the application of the dynamic array. Hence the earlier mentioned define statement implements obviously the trade off between memory usage and fast processing time.
Besides this, the content of this additional memory in each histogram cell for each signature bit is remembered to be identified by a pointer to the hit, which has set the corresponding bit in the signature. With regard to this, a hint is given for the determination of the value for the define statement, because the average of the number of hits, which set a single bit in a signature is statistically computable. However this circumstance advices to form regions inside the histogram, which feature diverse values. But due to the complexity and the variability of the histogram dimensions, such an approach is actually not implemented.

Beyond that, it is as obvious as for the *bitArray* that the optimal implementation can be only selected with regard to the frequented memory operations *push(...)* and *pop(...)* in combination with their exact time consumption. But as these results are also really hard to evaluate, it is again preferred to simply measure the time of a sample event and take this implementation for standard, whose result is faster. In this connection, it has to be finally realized that the actual result for such a sample event causes the selection of the class *inputHitSpecialMem*. Nevertheless it has to be additionally noticed that this result depends evidently on the size of the recently used statically reserved array.

3.2.6 Filtering

While remembering the diverse steps of the Hough tracking algorithm, the entire process of reducing the background noise in the histogram is evidently separated into different steps, which are called Encoding, Diagonalization and Peak-finding including the Serialization. In addition to this, it has to be also reminded that the Hough space decomposition has caused the Peak-finding to be collaterally divided into two separate steps, which are named 2D Peak-finding and 3D Peak-finding.
Although this situation seems at a first glance confusing, figure 3.24 depicts a simple illustration of the ordered algorithm steps in combination with the major involved classes.

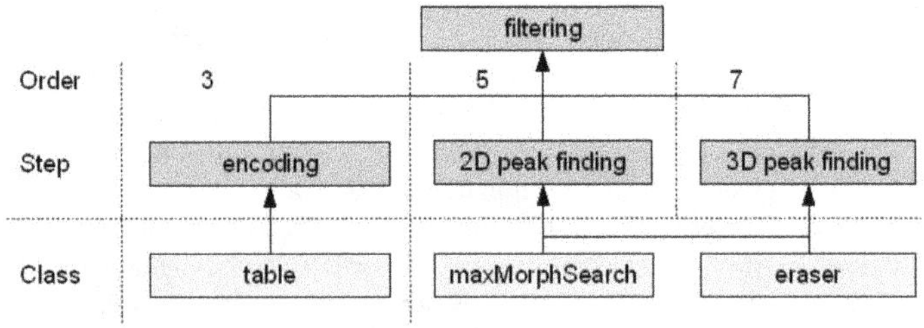

Figure 3.24: Illustration of the classes related to the filtering process including the corresponding ordered algorithm steps

Encoding

The Encoding of the histogram is naturally realized by the application of this one of the three used tables, which is called 'codingTable'. Furthermore this table contains obviously priority classes, which represent the encoded signatures. Therefore the evaluation of the corresponding priority class is basically performed by indexing this table with the signature of the histogram cell. As such a signature of a cell is moreover of course in place exchanged with the evaluated priority class, it is clear that they are both represented by a *bitfield* or *bytefield* data object dependent on the recently used implementation (see section 3.2.5 on page 138). So this context presumes that no additional memory is needed to store the priority class in a histogram cell. Further on, this circumstance leads then certainly to a very simple algorithm, which examines each histogram cell to replace the actual signature according to the connected priority class.

Besides this, it has to be supplementary noticed that the implementation and generation of the three tables including the table, which is used for the Encoding of the histogram, is contrarily quite complex. For this purpose, it is at first important to realize that a single object, which is identified by the class *tables*, holds all three tables and controls the access.
In addition to that, this class offers also the possibility to generate the content of each table automatically by algorithms, which range from very simple to very complex ones. However it is obvious within this context that the efficiency of a simple algorithm is not as good as the efficiency of a complex one. Nevertheless the best automatic algorithm to generate the table for the Encoding seems to be this one, which takes the Hough transform into account, because then the contours of real peaks are known.
So even if there is an automatic way to compute such a table, there would be supplementary a need to make a good logical cut by hand. For that reason, the best way to get an applicable table is to use the analysis for the LUT benignity (see section 4.5 on page 229) and to insert the first maybe three signatures with the highest frequency of occurrence into the table.

Coming now to the point, figure 3.25a depicts that each table object is apparently realized by the class *table*, which is illustrated in figure 3.25b in detail.

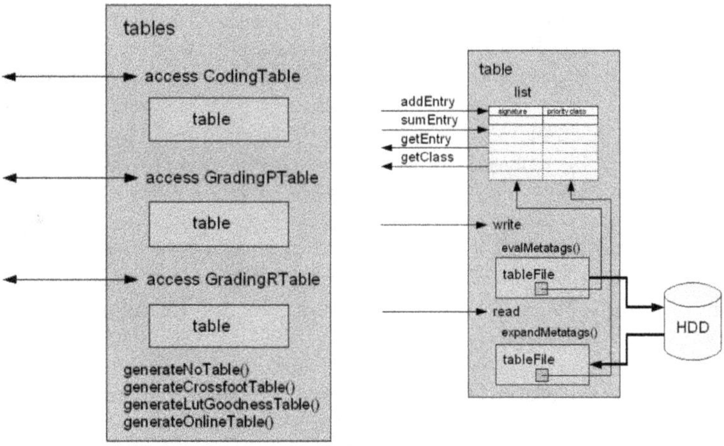

(a) Class *tables* (b) Class *table*

Figure 3.25: Illustration of the classes related to the Encoding

So with regard to this figure, it is immediately clear that the key element of the class *table* is determined by a list, which has to store the signatures in combination with the corresponding priority classes. In this connection, it is furthermore obvious that this list is realized by the template *specialList*, because the main advantage of this template is determined by the circumstance that its entries are sorted and can be thus searched very fast. However a prerequisite for the usage of this template is defined by the encapsulation of each signature and priority class pair into a *tableEntry* object, because the class *specialList* requires some supplementary member functions to unfold the entire functionality. Within this context, it has to be addionally noticed that there are also extra member variables, which store for instance the minimum and the maximum priority class.

Moreover the file access is evidently realized by the class *tableFile*, which is again derived from the class *io* located in the library *FileioLIB* (see section 3.2.2 on page 125). However it is important to mention an additional functionality at this juncture, which belongs to a special form of data com-

pression in the file. This means more precisely that the signatures would be compressed by using the Quine-McCluskey method during the write access, if the corresponding priority classes are equal. So for this purpose, Don't-cares are explicitly allowed characters in the signatures inside of a file. Furthermore it is of course clear that this compressed data is then in turn expanded to the origin during the read access.
Besides that, it has to be realized that this functionality is not implemented to save disk space, but to offer the possibility to immediately identify detector stations, which are useless in specific signatures. Further on, this possibility would appear evidently in another light due to the automatic table generation, even if it seems to be not that interesting at a first glance, because unprofitable detector stations are automatically identified.
Regardless of that, the major advantage of this concept is apparently determined by the circumstance that entire detector stations can be simply disabled by setting a Don't-care in the corresponding position of all signatures.

Beyond that, the signatures and the priority classes can be surely written into the file featuring different radixes, which are merely put in front of the number. Furthermore the advantage of this concept is immediately clear, because the signatures can be then written in binary mode, while the priority classes occur in decimal mode. Nevertheless it should be explicitly mentioned that any radix, which is compatible to appendix B on page B-1, is allowed.

Diagonalization

The Diagonalization of the histogram is as easy as it seems to be, because since the signatures in the histogram cells are encoded into priority classes, the Diagonalization process defines just a modification of the actual cell's priority class based on the diagonal cell, which is located below left. Furthermore this modification would be attained by moving the actual priority class of a cell to the next higher priority class, which is commonly realized by incrementing its value by one, if the diagonal value is equal and not in the lowest priority class. In this concept, the exception for the lowest priority class is really essential, because this class represents no priority and should be thus not incremented in any case.
Moreover it has to be noticed that the intention of this modification originates certainly in a simplified detection of the peaks for the Peak-finding due to the diagonal orientation of the clusters.

Besides this, it is clear that the modification process has a general hidden problem, because there will be obviously a priority class overflow, which is caused by two diagonal cells possessing the highest available priority class. Although this overflow can not happen in software due to the implementation of the class *bitArray*, which offers always enough resources to prevent this situation, this circumstance has to be mentioned here to be able to keep it in mind with regard to other implementations and especially the hardware ones.

Peak-finding and Serialization

Before going into detail, section 3.2.2 on page 126 has to be remembered, because the introduced decoupling of the geometry and the arithmetic operation in the Peak-finding process allows evidently the completely independent examination of both as well as the combination of each geometry with each arithmetic. In this connection, it has to be noticed that this circumstance can be easily accomplished by delivering a simple array, which contains only pointers to the correct elements of the actual geometry, and a mark for the examined element to the arithmetic operation, because the input to the arithmetic is then independent of the geometry except for the number of elements.

While having now a closer look exclusively to the geometry at first, it is obvious that the optimal one has to be able to cope perfectly with the cluster problem, which is introduced in section 2.1 on page 45. For that reason, it is unsurprising that this circumstance implies the best match of the geometry to almost all surrounding histogram cells of each peaks, which possess still a priority class before the Peak-finding starts. However it is immediately clear that these peak contours vary certainly from one to the other. Hence the optimal geometry is apparently determined by statistics about the contours of the occurring peaks. But as the evaluation of such an optimal geometry requires evidently the inspection of each peak and it is thus very complex to get satisfying statistics, many diverse static geometries are supplementary supported.

Moreover the decoupling aspect in combination with the fact that multiple different arithmetic operations are supported as well leads to a quite huge amount of available Peak-finding strategies. Furthermore this amount would even rise, if the differentiation between the 2D and 3D Peak-finding is taken into account, because each 2D strategy can be obviously combined with each 3D strategy.
So based on this concept, it is now clear that the entire process of Peak-finding combines the initial application of the 2D Peak-finding geometry with its arithmetic operation to each histogram cell with the subsequent SERIALIZATION of the surviving track candidates as well as the final utilization of

the 3D Peak-finding geometry with its arithmetic operation to the surviving track candidates of the parallel histogram layers.
Beyond that, it has to be realized that the 3D Peak-finding has to be implemented completely different to the 2D Peak-finding, because the regular structure of a histogram layer can not be used anymore. Hence the 3D Peak-finding has to identify a neighborhood relation by an additionally saved attribute of the track candidates, which represents the original coordinate of the corresponding histogram cell.

Figure 3.26a depicts now the actually used static 2D Peak-finding geometry, while figure 3.26b shows the projection of the three dimensional upgrade in the plane spanned by the third and second coordinate dimension.

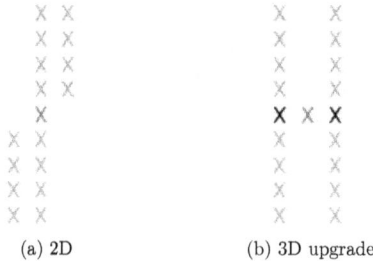

(a) 2D (b) 3D upgrade

Figure 3.26: Visualization of the actual Peak-finding geometry

With regard to both figures, the red 'X' identifies obviously the coordinate of the value, which is actually examined. Further on, the neighborhood histogram layers are evidently processed in the 3D version with the 2D geometry, while the actual layer includes naturally just the main cell. Furthermore it is clear that the neighborhood histogram layers in the three dimensional geometry implies the two dimensional geometry. By the way, this circumstance can be of course not seen here due to the projection.

In addition to this, the simplest form of arithmetic operations, which is actually also applied for standard, is defined by the following lines of source code.

```
pos = positionOfElementToEvaluate;
returnValue = array[pos];
for (unsigned int i = 0; i < length; i++) {
        if ((array[i] >= array[pos]) && (i != pos))
                returnValue = bitArray(0);
}
```

With regard to these lines, it is obvious that the value of the element, which is actually examined, would only remain, if all other elements in the considered geometry exhibits a smaller value. Otherwise it is clear that the value of the actually examined element is reset to the initial priority class.

In contrast to this complex process of Peak-finding, the Serialization of the track candidates, which survive the 2D Peak-finding, is quite simple, because this process must just copy the track candidates including some supplementary information like the originating histogram cell coordinate into the LBuffer unit to store them completely sorted.

3.2.7 LBuffer

The design of the LBuffer unit is unsurprisingly analog to the HBuffer unit. However there are two major differences, which are obviously characterized by the stored information and the used STL template. So the array is again chosen due to the circumstance that the number of elements is fixed by the number of histogram layers, while the STL *std::list* template is selected for the next level because of an unknown number of elements, which depend obviously on the surviving track candidates of the 2D Peak-finding. In contrast to the HBuffer unit, the LBuffer unit prefers in this connection the STL *std::list* template to the STL *std::vector* template, because the efficiency of element deletion, which characterizes the main operation here, is simply better. For that reason, this implementation can be easily assumed to realize the optimum for the storage holding the needed information in a sorted order. Moreover figure 3.27 depicts a deeper insight about all involved classes, their relation and available information.

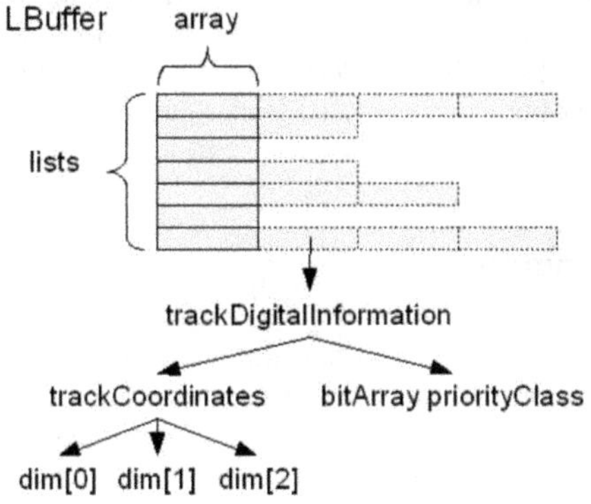

Figure 3.27: Illustration of the classes related to the LBuffer unit

Coming now to a short summary about the application of this LBuffer unit in the software tracking algorithm, it is immediately clear that this implementation has to store a single *trackDigitalInformation* object for each track candidate, which survives the 2D Peak-finding, ordered in the array according to the index of the histogram layer. Furthermore it has to be mentioned within this context that these objects are apparently supplementary sorted in the corresponding STL *std::list* template due to the order, which is implemented by the Serialization process.
Besides this, it is needless to say that such a *trackDigitalInformation* object has to contain generally all information of a track candidate, which is required for the reverse transformation of the feature parameters and by the 3D Peak-finding. For that reasons, it is obvious that this information includes the *bitArray* object, which represents the priority class needed by the 3D Peak-finding, in combination with the three dimensional coordinate of the histogram cell defining the feature parameters.

So based on this concept, the algorithm uses this LBuffer unit apparently to cope with the Hough space decomposition in relation to the 2D and 3D Peak-finding. Hence the write access to fill the memory happens in blocks, which are separated due to each serialized histogram layer, while the final read-out delivers the results of the tracking algorithm. However it is clear that the 3D Peak-finding, which lays in between, has to operate on the data stored in this unit. As furthermore the sort order is at first explicitly defined by the third dimension and subsequently implicitly by the second followed by the first dimension, the 3D Peak-finding process uses this information to navigate fast through the data to find the corresponding elements for the actual Peak-finding geometry. Further on, this process would apparently decide, if the actual element is deleted from the memory or kept. Hence a successful 3D Peak-finding causes the LBuffer unit to contain only the finally found tracks, which are identified by their feature parameters.

3.3 Hardware

This chapter presents two different hardware implementations of the adapted and customized general Hough transform. Moreover the only difference between them is apparently determined by the circumstance that the first one bases directly on FPGA chips, while the second one uses a Cell BE based embedded system. So the following sections describe the functionality and developed concepts of the main processing units for the FPGA implementation before their mapping to the special characteristics of the Cell BE based system is portrayed.

3.3.1 FPGA Implementation

At a first glance, one tends certainly to implement the algorithm on a single FPGA chip. But due to the flexibility of the algorithm and the restricted hardware resources of such a chip, the possibility to distribute it among a multi-chip system with an interconnection network between the chips should be not forgotten. For that reason, an abstract representation for the concept of each main functional unit is developed separately in the following sections, while their precise implementation, which concerns sometimes multiple realizations requiring diverse hardware resources of a FPGA chip, is shifted into section 5.5 on page 327 exhibiting directly the results as well. Thence the concrete implementation of the entire algorithm can be assembled of functional components with regard to their corresponding tangible resource consumption. In this connection, it is then immediately clear that the resulting resources depend evidently on the selected realizations for each unit, which leads then further apparently to an amount of FPGAs required to fulfill the needs for the CBM experiment.
But before starting to illustrate the abstract concepts of the main functional units, a short introduction into the FPGA technology is given, which is afterwards followed by the characterization of the assumed data input and the adjusted structure of the algorithm.

Field Programmable Gate Array

Although FPGAs are usually slower than their Application Specific Integrated Circuit (ASIC) (see [App]) counterparts, cannot handle as complex designs and draw more power, their advantages include a shorter time to market, lower non-recurring engineering costs and the ability to reprogram in the field to fix bugs (see [Fie]).

The historical roots of the FPGAs are found in the 1980s in the Complex Programmable Logic Devices (CPLDs) (see [Com]), which can be still an alternative for simpler designs. Further on, the usage of a FPGA requires as well as the CPLD a design specification with a Hardware Description Language (HDL) (see [Har]) at first. Subsequently the HDL source files are fed to a software suite from the FPGA or CPLD vendor. This software produces then through different steps a file, which is finally transferred to the device via a Joint Test Action Group (JTAG) (see [Joi]) interface or to an external memory device like an Electrically Erasable Programmable Read-Only Memory (EEPROM) (see [EEP]).
In addition to this, these HDL source files are commonly developed and managed by the application of additional programs like the Integrated Development Environment (IDE) HDL Designer Version 2007.1 (Build 19) with ModuleWare dynamic library version 1.9 (see [HDL]) and the HDL simulator ModelSim SE 6.3d Revision 2007.11 (see [Mod]) from the company Mentor Graphics (see [Men]) in combination with the synthesis tool Synplify Pro Version 9.0.1 (see [Synb]) from the company Synplicity (see [Syna]) and the vendor specific software suite ISE 8.2.03i Application Version I.34 (see [ISE]) from the company Xilinx (see [Xila]).
In this connection, it is obvious that these tools define my preferred ones and are thus used to produce the results, which are presented in this thesis. More detailed information about the concrete generation of this final file can be found in appendix C on page C-1.

Coming now to more hardware specific details, FPGA (see [Fie]) names generally a semiconductor device, which is composed of an array of Configurable Logic Blocks (CLBs) and programmable interconnects. Furthermore these CLBs consist in most FPGAs of a LUT with four up to six inputs in combination with memory elements, which may be simple flip-flops or more complete blocks of memory. Moreover this LUT is generally able to realize any combinatorial function like AND, NAND and XOR based on the corresponding inputs. Nevertheless combinatorial functions, which require a high fan-in because of the need for more inputs than one single LUT can provide, can be built by the direct connection of multiple LUTs to each other. For this purpose, a hierarchy of the programmable interconnects allows logic components to be interconnected as needed by the system designer after the FPGA is manufactured. This circumstance enables then obviously the realization of any logical function. In addition to this, special-purpose dedicated routing networks, which are separately managed, exist in commercial FPGAs to handle clock signals and often other high-fanout signals.

So until the common FPGA technology is now briefly introduced, figure 3.28 depicts an example, which shows a simplified Xilinx Virtex 4 slice including two CLBs. As this figure is of course rudimentary, more information about this particular chip can be found in [Virc].

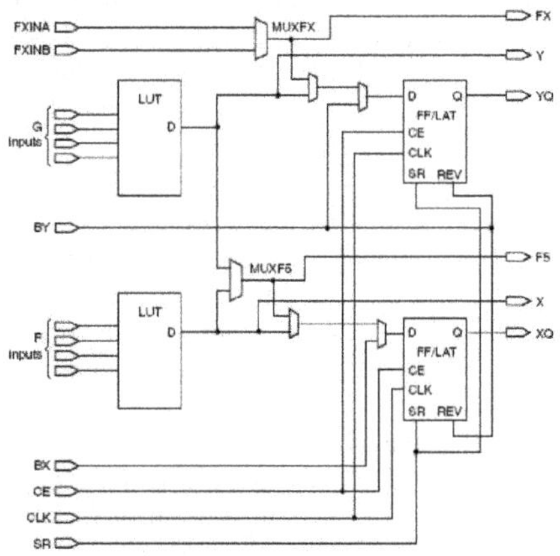

Figure 3.28: Simplified block diagram of a Virtex 4 slice (Source: Virtex 4 User Guide on page 204)

Besides this, modern FPGA families expand upon the above capabilities to include higher level functionality fixed into the silicon like multipliers, generic Digital Signal Processing (DSP) blocks, Phase Locked Loop (PLL) components, high speed IO logic and embedded memories for instance. So having these common functions embedded into the silicon reduces evidently the required area and gives those functions increased speed compared to building them from primitives. Moreover modern developments take the coarse-grained architectural approach even a step further by combining the logic blocks and interconnects of traditional FPGAs with embedded microprocessors and related peripherals to form a complete system on a programmable chip.

Examples of such hybrid technologies can be found in the Xilinx Virtex II PRO (see [Virg]) and Virtex 4 device family (see [Virb]), which include both one or more PowerPC (PPC) processors embedded within the FPGAs logic fabric. In contrast to this, an alternative approach to utilize processors is defined by soft processor cores, that are implemented within the FPGA logic, like the implementation of a Intel Pentium processor on a Xilinx Virtex 4 FPGA (see [LYK+07]) for instance.
Beyond that, many modern FPGAs have also the ability to be reprogrammed at run time, which leads then apparently to the idea of reconfigurable computing (see [Reca]) or reconfigurable systems. This means more precisely that the FPGAs or parts of them can be reconfigured to suit a special task (see [Para]) as presented in [FHC] for example.

Finally it is also interesting to know that the biggest manufacturer of FPGAs are Xilinx (see [Xilb]), Lattice Semiconductor Corporation (see [Lat]), Altera (see [Alta]) and Actel (see [Act]). Besides this, the fields of application for FPGAs include digital signal processing (see [SC05]), software-defined radio (see [Sie05]), aerospace (see [Xil05]) and defense systems (see [SE05]), robotic control systems (see [Har06]), ASIC prototyping (see [ASIb] and [ASIa]), medical imaging (see [Mul05]), computer vision (see [GSTB05]), speech recognition (see [MQR02] and [VFJ01]), cryptography (see [WGP04] and [HLSH00]), bioinformatics (see [KRM+06]), computer hardware emulation (see [RF05]) and a growing range of other areas. Moreover FPGAs find especially applications in any area or algorithm, that can make use of the massive parallelism offered by their architecture.

Data Input

In the experimental setup, the data input of the trigger is in principle produced by the FEE of the detector stations and transported via the DAQ system (see [Dat]) to the trigger system. So with regard to this structure, it is obvious that the preliminary determination of both, the FEE and the DAQ, does not allow to fix the data input of the trigger.
Hence it can be actually only proposed to apply an optical network system, which bases on the RocketIO Multi-Gigabit Tranceiver (MGT) technology, which is introduced in the Xilinx Virtex 4 FPGA device family. Although these devices are advertised with up to 24 MGTs realizing a speed of up to 6.5 $\frac{\text{Gbit}}{\text{s}}$ each (see [Virb]), colleagues of mine have reached just 5 $\frac{\text{Gbit}}{\text{s}}$ per MGT in the first test implementations. But in addition to this circumstance, it is really interesting that the consecutive device family, which is called Xilinx Virtex V, supports only a decreased number of 16 MGTs realizing a reduced maximum speed of up to 3.75 $\frac{\text{Gbit}}{\text{s}}$ (see [Vird]).
So within this context, it is clear that the essential facts can be summarized by the usage of 3.75 $\frac{\text{Gbit}}{\text{s}}$ serial optical network channels transferring 32 bit for a single digitized hit of an event. Furthermore a single digitized hit of an event is recently assumed to consume 32 bit, which represent the identifier for the detector station in the STS with three bit, each station dimension coordinate with 13 bit and some reserved special information with the last three bit.

Algorithm

The adjusted structure of the Hough tracking algorithm is very simple.
In the first step, all hits of one event are naturally delivered by the network, partly transformed by the first LUT and then stored in the HBuffer unit. As the major characteristic of this unit is further on defined by sorting the input data implicitly based on the results of the first LUT, the second step has to wait obviously for the successful delivery of all data from the network.
Afterwards the sorted and partly transformed hits are apparently read out of the HBuffer unit in blocks dependent of the instantiated histogram layers and completely processed. Therefor this process surely includes the second LUT transformation, the Histogramming, the Encoding, the Diagonalization, the 2D Peak-finding and the Serialization. In this connection, it has to be additionally noticed that the specific realization of the LBuffer unit and the 3D Peak-finding unit causes the Serialization unit to require a supplemental small amount of memory, which stores the results for such a single processed

block until the IBM PPC 405 RISC processor core and especially the Processor Local Bus (PLB) is ready to fetch it.
Hence the final step three has to fetch the results of the processed histogram layers, store it in the LBuffer unit, which is realized by memory of the PPC, and apply the 3D Peak-finding. Although this 3D Peak-finding process is apparently also realizable in hardware, it seems to be less effort to use the PPC, which would be anyway inside the FPGA device, if used or not.

So based on this concept, the general structure of the algorithm including the partitioning steps, which are already introduced in section 3.1 on page 96, is depicted in figure 3.29.

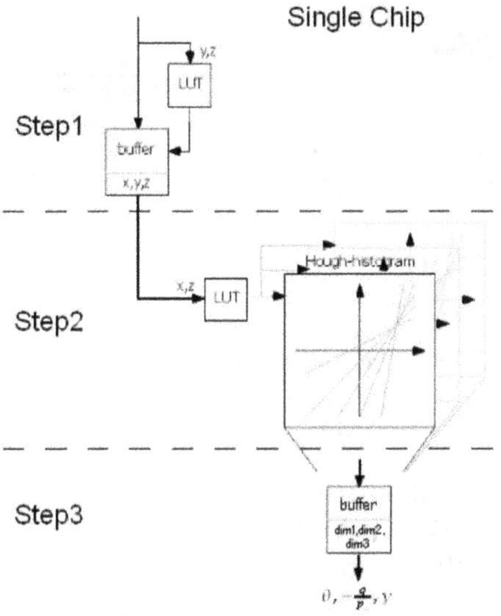

Figure 3.29: Sketch of a pipelined low-level hardware design for the Hough transform algorithm

With regard to this figure, it is obvious that each step is encapsulated and decoupled from the surrounding steps by the application of a memory buffer, which certainly stores the result of the previous step and delivers this data for input to the next step. Furthermore it has to be noticed here that this arrangement offers the possibility to pipeline the steps or parallelize the functional units inside each step without changing the interface memories.

Besides this, figure 3.30 shows the concrete connection chain of the functional units, which implement the entire Hough TRACKING algorithm. In this connection, it has to be additionally mentioned that the dotted units are not realized up to now. However this circumstance is negligible, because these ones are evidently uncritical.

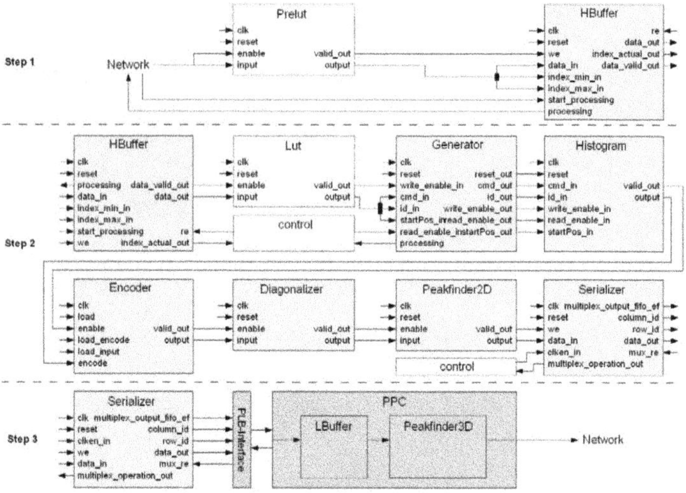

Figure 3.30: Structural FPGA design of the Hough transform algorithm

Since the possible parallelization of the functional units is exemplarily presented in section 3.3.2 on page 191 for the Cell BE based implementation and the pipelining strategy is obvious, the focus of the following sections is respectively set to the precise description of each functional unit, which is not dotted in the figure. Further on, it has to be also mentioned at this juncture that the LBuffer unit and the three dimensional Peak-finding unit define exceptions, because both units are planned to be realized by the IBM PPC 405 RISC processor core, which is included in a modern FPGA like a Xilinx Virtex 4 device for example (see [Virb]).

Look-up-tables

Remembering now the different tasks of the Look-up-tables (LUTs) at first, the first one has to provide the precise area boundaries of the histogram

layers, while the second one has to supply the concrete Hough curve.
Further on, the information gained by the first LUT, whose location in the data path can be seen in figure 3.29 of section 3.3.1 on page 157, is collaterally used in the HBuffer unit to sort the hits with regard to their first histogram layer of occurrence, which is obviously equal to the minimal or starting area boundary (see section 3.3.1 on page 160).
Besides this, the information acquired from the second LUT, whose location in the data path can be also seen in figure 3.29 of section 3.3.1 on page 157, is afterwards used in the generator unit of the histogram to fill the actual layer.

Coming now to the concrete implementation, it is clear that the unfixed data input, which is stated in section 3.3.1 on page 156, defines a really crucial problem, because it forms naturally the address range, determines thus the number of entries and influences further of course the size of the LUTs.
Furthermore it has to be noticed here that the size of the LUTs is apparently absolutely essential for the implementation in hardware, because it configures the usable types of memory due to their characteristics like maximal amount, speed and accessibility. For instance, a FPGA offers commonly the possibility to implement memory with internal distributed Random Access Memory (RAM) modules, internal block RAM modules or external RAM like Static Random Access Memory (SRAM) or Dynamic Random Access Memory (DRAM), which feature all surely different characteristics limiting their applicability. Hence it is not expedient to implement the LUTs until the data input is well-known.
However it has to be finally mentioned here that the algorithm itself would

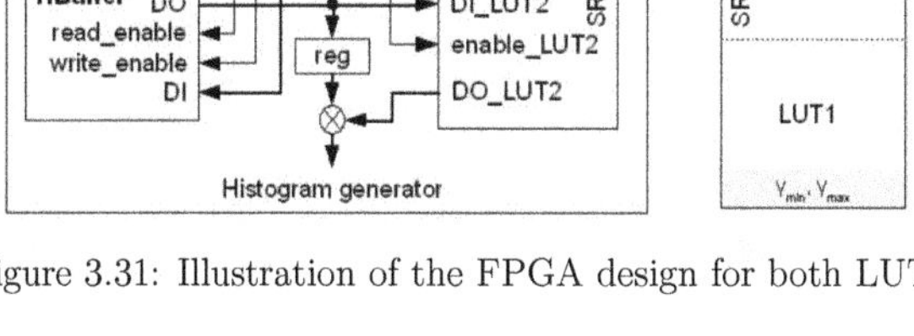

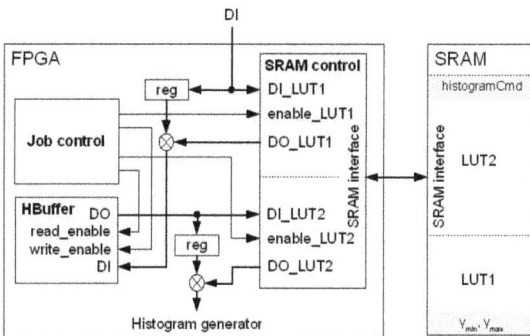

Figure 3.31: Illustration of the FPGA design for both LUTs

cause additionally an optimal memory to deliver a random access data in a single clock cycle to avoid a slowdown, even if the required size tends to a huge memory, which implies usually a lower speed. Nevertheless figure 3.31 depicts a theoretical and up to now the most reasonable exemplary implementation of the LUTs by the application of external SRAM, which sufficiently exhibits the required speed and hopefully the needed size.

HBuffer

In contrast to the software implementation, it has to be primarily noticed here that the FPGA implementation of the HBuffer unit does explicitly not store the results of both LUTs in combination with the corresponding hit. This unit has rather to store the histogram layer stop index, which is part of the result of the first LUT, in combination with a fragment of the digitized hit, which represents the address for the second LUT and the index of the detector station used to set the corresponding bit in the signatures. Furthermore figure 3.31 of section 3.3.1 on page 159 has already illustrated this circumstance. Moreover it has to be noted that the histogram layer start index, which is of course also part of the result of the first LUT, is implicitly saved, because the entry is stored in the order, which is identified by this value. So based on these facts, figure 3.32b depicts now the design of the HBuffer unit.

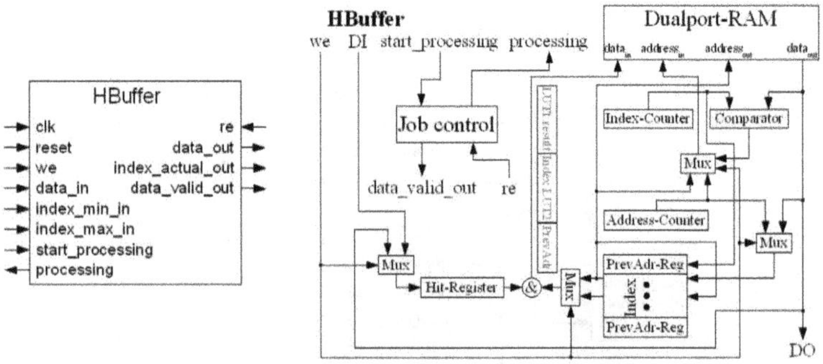

(a) Interface (b) Blockdiagram

Figure 3.32: Illustration of the FPGA design for the HBuffer unit

With regard to this figure, the major elements of this unit are apparently identified by a Dual-Ported Random Access Memory (DPRAM) and a register set including one register for each histogram layer.
Furthermore it is clear due to the functionality of this unit that the DPRAM accommodates the above defined data in combination with a link in linked lists with each consisting of the entries according to a single histogram layer. For that reason, it is now obvious that the essential initial link to access these lists is realized by the register set, which contains simply the DPRAM address of each last list entry. In addition to this, it is then clear that the link of an entry in the DPRAM is determined by the address of the previous entry in the list.
So the insertion of a new entry into a list, which is obviously only possible at the end, combines just the modification of the corresponding register with the write access to the DPRAM. However two facts have to be particularly highlighted in this connection. The first and more hidden one belongs to the identification of the link address for a predecessor element in a list required for the actual entry in the DPRAM, because this link is of course simply defined by the address, which is actually overwritten in the corresponding register. The second and almost evident fact is determined by the circumstance that a Single-Ported Random Access Memory (SPRAM) is sufficient to realize such an insertion.
But in contrast to this, the common processing of these lists necessitates the DPRAM due to the required update of the corresponding link address in an entry, which is caused by the movement of this entry to the consecutive list. The reason for this circumstance is quite simple, because the necessary write access has to be realized in parallel to the continuing read-out of the actual list to implement the fastest possible data delivery of one entry per clock cycle. In addition to this update, it is clear that the entries in the DPRAM are read-out list after list, while the corresponding register provides always the correct address due to the fact that it is overwritten with the link address of the actual read-out, which defines naturally either the link address to the previous entry of the list or an identifier for an empty list. So while processing, one port of the DPRAM is used for read access and the other one for write access.
Besides this, it has to be mentioned here that the identifier representing an empty list is actually just set to the address zero, because this circumstance simplifies the implementation, but accepts the side effect that this DPRAM address is forbidden to contain a correct entry.
Moreover it is also self-evident that the same entry would be prevented to occur more than once in the DPRAM, even if it is used in more than one list, which coincides with more than one histogram layer.

The advantages of this approach are immediately clear, because this concept entails naturally no limitation for the lists in the memory, while the moving of an entry to a consecutive list, which is certainly required by the common processing, is realized in place by modifying only the link address. So even if this concept is not that easy, it would perfectly fit the requirements, because the memory is used optimally, a list overflow is implicitly avoided and an assumption for a typical amount of list entries is needless. Hence this concept defines a key element of the Hough transform implementation for a FPGA and the following small example should be used to illustrate the really complex processing.

For this purpose, the HBuffer unit is assumed to receive the following data sets in the subsequent order at startup:

1. $\gamma_{min} = 4, \gamma_{max} = 4, indexLUT2 = A$

2. $\gamma_{min} = 3, \gamma_{max} = 4, indexLUT2 = B$

These data sets lead then to the next lists, which have to be processed:

1. NULL

2. NULL

3. NULL $\leftarrow \gamma_{max} = 4, indexLUT2 = B$

4. NULL $\leftarrow \gamma_{max} = 4, indexLUT2 = A \leftarrow \gamma_{max} = 4, indexLUT2 = B$

So based on this example, figure 3.33 depicts the intermediate states of the hardware implementation.

Operation	Dual-Port RAM in-port				Dual-Port RAM out-port				Register	
	Act. Adr.	V_{max}	idxLUT2	Prev. Adr.	Act. Adr.	V_{max}	idxLUT2	Prev. Adr.	List 3	List 4
Initialization	1	-	-	-	-	x	x	x	0	0
Write 10 sets behind list 4	...	...	...	...	-	x	x	x	0	0
Write to list 4	11	4	A	0	-	x	x	x	0	11
Write to list 3	12	4	B	0	-	x	x	x	12	11
Read from list 3	12	4	B	11	12	4	B	0	0	12
Read from list 4	-	-	-	-	12	4	B	11	0	11
Read from list 4	-	-	-	-	11	4	A	0	0	0
Proceeding	...	...	...	...	...	...	...	...	...	...

Figure 3.33: Visualization of a HBuffer example processing

With regard to this figure, it is apparently clear that the initialization phase is used to set the actual address of the DPRAM write access input port or simply in-port to one, because the address zero marks as introduced earlier an empty list and represents thus not a valid link address. After this initialization phase, the example assumes further that ten values are written to some lists, which correspond to histogram layers out of interest, to show a common utilization.
Subsequently the first entry of the HBuffer unit, which is of interest in the example, exhibits the address eleven at the DPRAM in-port in combination with the link address zero, because it defines the first item of this list. Additionally the register, which corresponds to list four, has to be updated because of the new entry in the DPRAM at address eleven. Furthermore it is evident that this procedure is repeatedly executed for the next entry. As these two entries are now assumed to form the last two ones, the HBuffer unit is successfully filled and the entries can be processed.

But in this context, the read-out of list one and two is skipped, because they are both empty. Hence it is continued with list three, which starts by applying the value of the corresponding initial link register three to the address input of the DPRAM read access output port or simply out-port. Furthermore this address, which is apparently determined by the value twelve, causes then the

DPRAM to deliver the data stored at that location. Moreover this data is immediately analyzed with regard to the histogram layer stop index, which is four in this case, because this value defines whether the actual entry must be appended to the consecutively following list or not. So as the actual processed list or histogram layer is three, the recent data is required to be added to the end of list four. Therefore the register representing list four has to be obviously overwritten with the value twelve, because this is the recent address of the actually processed data in the DPRAM, while the predecessor value eleven is used to adjust the link address of this data. Further on, it is clear that there is the need to rewrite this entry including the modified link address to the address twelve in the DPRAM. In addition to this, the register of list three has to be of course updated with the link address of the actually processed entry as well. But as the recently processed entry establishes the only one in list three, the value is set to zero to mark this list to be successfully processed and thus empty. So list four can be now set into the focus of processing to evaluate histogram layer four.

For this purpose, register four, which contains now the value twelve, is apparently applied to the address input of the DPRAM out-port to receive the data at that location, because this data defines of course the last entry of this list. Further on, the analysis of the histogram layer stop index causes no additional movement, because the value is equal to the actually processed list. Thus an update of the corresponding register with the new link address eleven is sufficient and the list can be proceeded.
However it has to be noticed in this connection that the free in-port of the DPRAM can be surely used to reset the memory to any initial value like zero for example. But it has to be supplementary noted that defined initial values of the memory are not required for a successful processing, because the address links define valid data.

Besides this, a last common problem has to be mentioned, because a read access in the DPRAM to an address, whose value is modified by a write access one clock cycle before, might generally result not in the delivery of the theoretical correct value because of the amount of clock cycles, which are required to setup the new value. Although this situation can easily occur due to the movement of an entry from one list to the end of the consecutive list combined with a directly following list switch, this problem is practically insignificant, because the systolic processing of the histogram layer causes the HBuffer unit to stall the data delivery for a certain amount of clock cycles, which is at least big enough to propagate the last entry through all affected histogram cells and to realize the entire read-out of the histogram layer.

Histogram

Figure 3.34 illustrates at first the connection between the second LUT and the histogram via the generator unit.

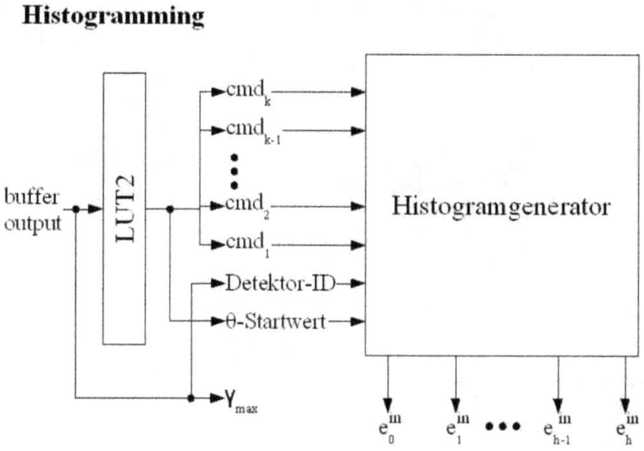

Figure 3.34: Demonstration of the histogram interface including the connection to the second LUT

At a first glance, the limited resources of a FPGA would cause the plan to process each histogram layer separately, even if the algorithm has the ability to compute any number of layers in parallel. Nevertheless this possibility opens the question for the number, which represents the best trade-off between necessary hardware resources and event time consumption. But for this purpose, I simply assume that the optimal trade-off between resources and parallelism is determined by using as many histogram layers in parallel as the biggest number of list or layer entries for a single hit is, because then it is obvious that each hit has to be read just twice in the worst case. However this interesting circumstance counts not until multiple FPGAs are used to increase the hardware resources.
So the actual and initial plan is explicitly defined by the instantiation of just a single histogram layer, which is processed serially. Regardless of that, it has to be additionally noticed here that an optimal abstract implementation, which suits best for each available FPGA, is naturally not conceivable, because each FPGA device features for instance different wiring capacity, different location and number of registers or RAM. Hence the variability of the

histogram with regard to its implementation requires two different general approaches to be presented in the next paragraphs.

Realization With Registers The standard histogram is realized with simple cells, which are connected to the neighborhood by a regular connection structure to form a systolic array. For this purpose, figure 3.35 shows the basic units implementing such a histogram cell.

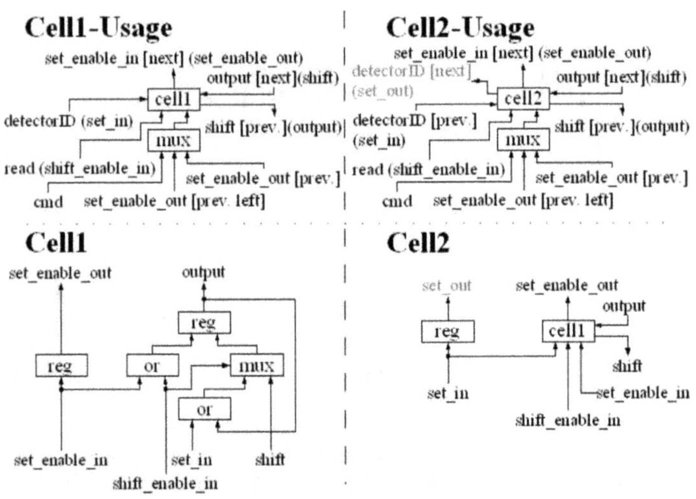

Figure 3.35: Possible FPGA designs of a histogram cell

With regard to this figure, the implementation of the standard cell type 'Cell1' is quite simple. It consists just of a multiplexer, two logical OR-gates and two registers, which store respectively the signature and the enable-flag. In contrast to this, a small modification of the functionality, which is marked red in the figure, causes certainly also a modified cell type named 'Cell2', which requires additionally a register to store the signature for the actual processing detector station index.
Moreover the principle of operation is quite simple for both types, because the enable-flag register and the detector-id-register are only used to delay the enable signal and the detector station index signature for the next line of the systolic array, while the OR-gates and the multiplexer are used to set the two modes of the signature register. Further on, it is clear that the first

mode is defined by the write access, which belongs to the setting of a single bit in the signature register, and the second mode by the read access, which corresponds to the setting of all bits in the signature register in parallel. So combining now the functionality of the read access with the concept of systolic processing, it is obvious that the output of each signature register is connected line by line, which enables them to act like simple pipelined shift registers.
An illustration of the resulting regular structure for such a histogram layer is depicted in figure 3.36b.

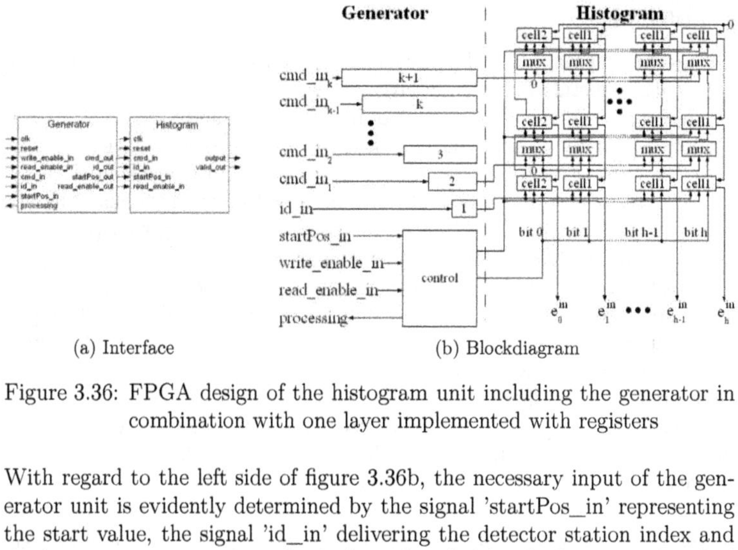

(a) Interface (b) Blockdiagram

Figure 3.36: FPGA design of the histogram unit including the generator in combination with one layer implemented with registers

With regard to the left side of figure 3.36b, the necessary input of the generator unit is evidently determined by the signal 'startPos_in' representing the start value, the signal 'id_in' delivering the detector station index and the histogram command word, which is identified by the bit vector 'cmd' of length k equal to the second dimension of the histogram, as well as the two supplementary input signals 'write_enable_in' and 'read_enable_in' to set the operating mode of the histogram. Further on, this unit requires only some shift registers for the input and a control unit to implement the entire functionality. These shift registers define furthermore just delay lines, which are needed for the systolic processing. Based on this circumstance, it is now clear why the delay length of the histogram command word increases by one for each line.
In addition to this input, the output is apparently defined by the signal 'processing', which informs about an unready unit due to the required stall by

the last processed hit insertion for instance, and the histogram data output, which is tagged by the bit vector 'e' also of length k.

Having further the possibility to instantiate a certain number of parallel histogram layers in mind, it is clear that the dashed line in figure 3.36b determines the separator, which isolates the parallel scaling part on the right side from the independent non-scaling part on the left side. So within this context, it has to be explicitly told that the left part is only required once independent of the parallelization, while the right part can be evidently instantiated as often as wanted. The reason for this fact originates of course in the circumstance that just the same entry can be inserted in all layers at a specified time.

Besides this, the right side of figure 3.36b exhibits also the implementation of the regular structured histogram layer. So with regard to this figure as well as the corresponding cell usage part of figure 3.35, an additional multiplexer is applied in front of each cell to essentially feature slopes bigger than 45 degrees for the inserted curves. Obviously this multiplexer is controlled by one bit of the histogram command word, because the entire command builds the slope of the curve. Moreover there has to be a strategy to support the two modes of operation, which are introduced earlier.
For that reason, the internal control signal is used to allow the compare units, which are used to find the correct column for the start value, to start the bottom-up systolic processing during the write mode. Thus all higher lines in the systolic array can work normally.
In contrast to this, the earlier mentioned connection of the signature registers leads to a top-down systolic processing during the read mode, which ends in the delivery of the output values at the bottom line in parallel and delayed by the order of the line. Furthermore it is obvious that this read-out process can be concurrently used to reset the histogram by simply inserting an initial signature for each signature register at the top of the systolic array, because these ones are apparently also shifted down the layer. However it is important to note here that the read-out can not only be exclusively realized top-down, but also from left to right, because there is no restriction. The only difference in this left-right approach belongs evidently to the consecutively utilized functional units, because the neighborhood relation and thus their dimensioning is rotated. More precise information about this circumstance is following later in section 3.3.1 on page 174.
Returning now to the concrete structure of the histogram layer, which is depicted in figure 3.36b, it is eye-catching that the first column of the histogram layer consists of cells of type 'Cell2', while the rest applies the type 'Cell1'. However the reason for this circumstance is quite simple, because the cell

in each row of the type 'Cell2' is able to provide the entire rest in this row with the corresponding detector index signature. So within this context, it is immediately clear that an alternative structure is defined by the sole usage of type 'Cell2' cells. Moreover the disadvantage of this structure is then of course defined by the more resources, which are obviously needed, because each cell requires an additional register. But the advantage of this structure is in contrast to this determined by the fact that the register of the cell type 'Cell2' does not imply such a huge fan-out and thus such a huge routing requirement.

Finally it has to be mentioned here that some special attributes, which are set inside the HDL source files and processed by the synthesis software, are able to influence the result. More information about this topic can be found in section 5.5.3 on page 334.

But as this is still not the total story, figure 3.37 depicts a third possible implementation.

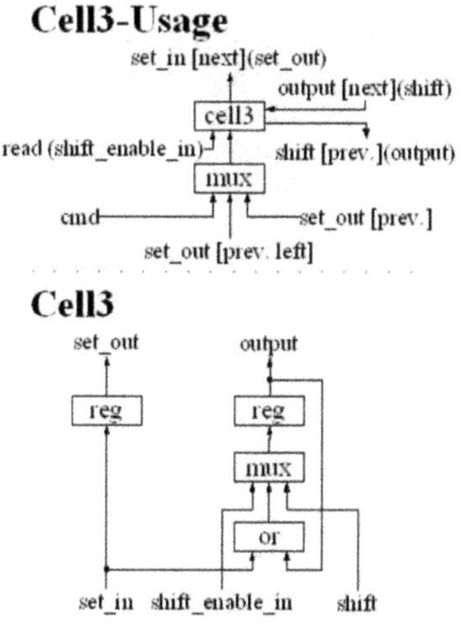

Figure 3.37: Feasible FPGA design of a modified histogram cell

With regard to this figure, the basis for this unit is apparently cell type 'Cell2' including all advantages and disadvantages compared to cell type 'Cell1'. However the difference to type 'Cell2' is determined by the enable-flag signal, which is now combined with the detector station index signature. So

based on this concept, it is further on important to realize that this modified signal can be generated in the histogram layer itself at the bottom and must be not delivered from other units outside. Hence the only requirement for such a realization is obviously defined by the usage of a single additional multiplexer in each column, which would insert either the detector station index, if the comparison is true, or the zero index otherwise.
Finally the advantage, which is identified by needing one control signal and one OR-gate less in each cell, has to be evidently noticed in comparison to the disadvantage, which is characterized by the missing possibility to stall cells for maybe occurring incorrect input.

Beyond that, additional modified implementations can be of course easily created by a different encoding of the signal for the detector station index as well, because all already introduced structures assume the index to be a one-hot encoded signature. Even if this one-hot encoding is naturally practical due to the combination of this value with another signature, the advantage of a missing conversion requirement would have to be compared to the disadvantage, which is characterized by the supplementarily needed resources like bits in registers or even wiring capacity. For that reason, these additional implementations apply simply the binary encoding for the detector station index to prevent this circumstance. However the thus smaller registers and less needed routing resources entail a conversion unit for each histogram cell, which converts the binary value into one-hot encoding for the signature. But as all these differences care only about possible implementations on the data path level, it is clear that the multiple available hardware realizations for the functional descriptions expand this circumstance further in section 5.5 on page 327.

Realization With Random Access Memory In contrast to the realization with registers, this more complex implementation of the histogram layer is determined by the application of internal DPRAM modules, which are commonly available in any FPGA device, but with varying amount. Furthermore figure 3.38a and figure 3.38b depicts two different possibilities to realize a single line of a histogram layer with RAM, which can be then used to build the regular structure.

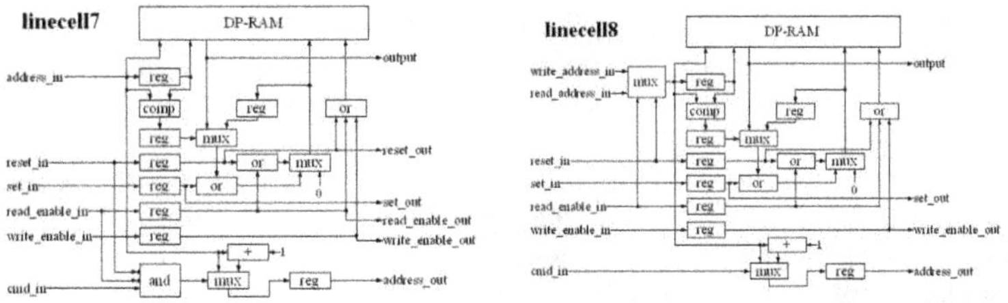

(a) One address input for read and write (b) Different read and write address inputs

Figure 3.38: Potential FPGA designs of a single histogram line implemented with Block RAM

With regard to figure 3.38a, the major components are eight delay registers, three multiplexers, three OR-gates, one AND-gate, one adder and one comparator. Further on, the only difference of figure 3.38b to this one belongs apparently to the 'address_in' input and the generation of the 'address_out' output, because the 'address_in' is split into an address input for the write and a separate one for the read mode, while an additional multiplexer selects the correct address for the DPRAM ports and the output address generation due to the actual mode. Moreover it is clear that this additional multiplexer offers simply the removal of the AND-gate at the 'cmd_in' input, because the 'adress_out' output is already correctly selected by this multiplexer. Thence the number of functional units remains the same, although the two implementations are different.

Besides this, the functionality of both units is as simple as the ones, which are realized with registers. Nevertheless a difference compared to them is defined by another form of systolic processing via the lines of the histogram layer and especially the read-out, which has to be apparently realized from left to right instead of the top-down approach.
So the write mode causes at first the unmodified start value to be used at the first line to read the corresponding histogram cell signature of the DPRAM

at this address. Subsequently this address is then of course also applied to write the modified signature in the next clock cycle back into the DPRAM. In addition to this insertion process, it is obvious that the address for the next line of the systolic processed histogram is certainly calculated in parallel to the modification of the signature. Furthermore this computation is very simple, because the accepted curves, which are characterized by a slope bigger than 45 degrees, lead to the circumstance that this address has to be either equal or simply incremented by one. This selection depends naturally analog to the realization with registers on the corresponding bit of the histogram command word. By the way, it has to be noticed here that the DPRAM is surely only required to avoid the stall of the line by the signature rewrite, because this rewrite causes evidently a concurrent write and read access in a single clock cycle.
Moreover the read mode can be doubtless easily implemented by the application of a decreasing counter, which generates one read address after the other to read the DPRAM out. However such a counter is not sufficient for the unit depicted in figure 3.38a, because the systolic processed histogram requires the address to be passed from line to line with a delay of respectively one clock cycle independent of the read or write mode, which leads obviously to the circumstance that the output of the histogram is delivered with a clock cycle delay equal to the ordering number of the line. So this means more precisely that the applied address in the read mode delivers the corresponding output of the first line one clock cycle later, of the second line two clock cycles later, and so on. But as the consecutive units should not complain about this fact, a solution is obviously defined analog to the histogram generator unit, which has to cope with the systolic processing for the write mode, by applying some simple delaying shift registers, which are in the size of course determined by the corresponding line of the histogram. Moreover some bits of these delaying shift registers can be naturally saved by moving them behind the Encoding unit, because this unit changes also the representation of the value from one-hot for the signatures to binary for the priority classes, which necessitate commonly less bits. For instance, this movement will actually avoid the spending of five additional bits for each applied shift register, because the bit width of a signature is set actually to seven due to the used number of detector stations, while a priority class needs recently only two bits to cover the values one, two and three. Further on, it has to be explicitly said that this movement is possible, because the Encoding depends just on the signature of each histogram cell respectively and does not involve any relationship to neighboring cells. In this connection, the major advantage of the unit, which is depicted in figure 3.38b, compared to the other one has to be apparently highlighted, because the split address in-

put features the application of the same address for all lines of the histogram at the same time, which avoids surely the requirement of the delaying shift registers in total. Further on, it has to be collaterally noticed that the top-down read-out, which is usually used in the version of the histogram layer implemented with registers, is evidently impossible to realize. Hence such a realization supports the bottom-up approach for the Histogramming, but only the left-right approach for the read-out.

Besides this, figure 3.39b shows now a good illustration for the structure of one histogram layer. Furthermore it is obvious that the dashed line indicates the same circumstance as already evinced in figure 3.36b, which means more precisely that the instantiation of a certain number of parallel histogram layers requires the right side to be parallelized as often, while the left side is regardless needed just once.

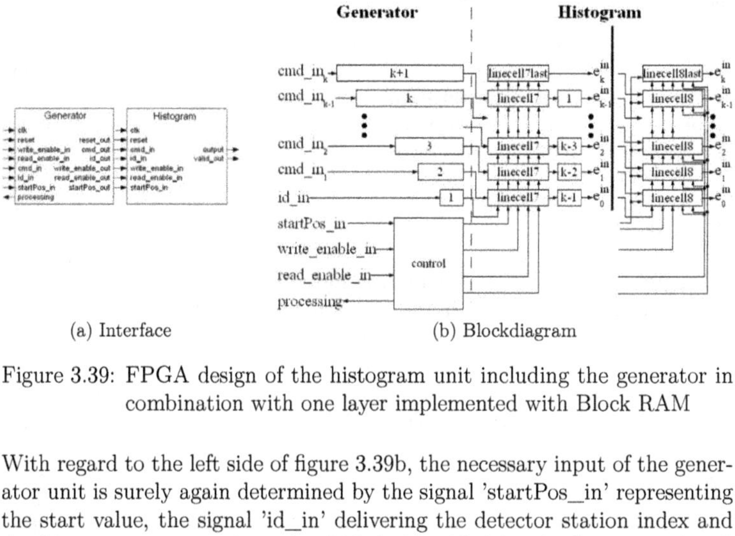

(a) Interface (b) Blockdiagram

Figure 3.39: FPGA design of the histogram unit including the generator in combination with one layer implemented with Block RAM

With regard to the left side of figure 3.39b, the necessary input of the generator unit is surely again determined by the signal 'startPos_in' representing the start value, the signal 'id_in' delivering the detector station index and the histogram command word, which is identified by the bit vector 'cmd' of length k equal to the second dimension of the histogram, as well as the two supplementary input signals 'write_enable_in' and 'read_enable_in' to set the operating mode of the histogram. Further on, this unit requires also some shift registers for the input and a control unit to implement the entire functionality. These shift registers define obviously again delay lines, which are in analogousness needed for the systolic processing.

In addition to this input, the output is naturally just as well defined by the signal 'processing', which informs about an unready unit, and the histogram data output, which is tagged by the bit vector 'e' also of length k. In this connection, it has to be supplementarily noted that an unready state of this unit is not only caused by the identical reasons as for the version of the histogram layer implemented with registers, but also by the manually initiated reset of the histogram, because the reset process needs due to the impossible parallel access of each signature evidently as many clock cycles as signatures or more precisely as addresses of the DPRAM are valid. Nevertheless the usual reset is of course again implied in the read-out, because the functionality of the DPRAM features certainly the read-out on one port, while the other one is used to write an initial signature to the same address.

Moreover the right side of figure 3.39b exhibits the implementation of the regular structured histogram layer by the usage of the basic units for each line.

Besides this, the two modes of operation are obviously supported by the internal control signals, which differ between the bottom-up systolic processing for the write mode and the left-right systolic processing for the read mode. Nevertheless the functionality is apart from that apparently equal to the histogram layer implemented with registers.

Filtering

Before presenting the precise hardware implementation, which is applied to reduce the background noise in the histogram, it has to be noticed here that the entire process is just as separated as in software. Hence there are analog to section 3.2.6 on page 144 detached sections for the Encoding and the Diagonalization as well as a single combined section for the Peak-finding and Serialization, which includes also the differentiation between the 2D and 3D Peak-finding.

Encoding The Encoding of the histogram layer is in hardware obviously as easy as in software (see section 3.2.6 on page 145), because there is just a need for a unit, which transforms each signature of a cell into a priority class without the usage of neighborhood information. However the major difference of this implementation compared to the software is apparently determined by the circumstance that the entire process of Encoding is at the best parallelized and pipelined with regard to the read-out of the histogram, the Diagonalization and the two dimensional Peak-finding.

Further on, it is additionally essential to notice that the number of supported priority classes is determined by the number of used bits, which means more precisely that the available amount can be computed to $nopc = 2^{bits-1}$. In this connection, the minus one modifies the common calculation of available values for a certain amount of used bits, because the smallest value zero is reserved to represent no priority class and the highest value $2^{bits} - 1$ is required to avoid the priority class overflow, which can otherwise happen in the Diagonalization process. So based on this circumstance, the two recent necessary different priority classes utilize two bits under the constraint that the highest priority class value three remains reserved for the Diagonalization, which is exhibited in section 3.3.1 on page 176.
Since multiple implementations are apparently imaginable to realize such a simple transformation, figure 3.40 depicts a unit, which applies just some simple logic gates.

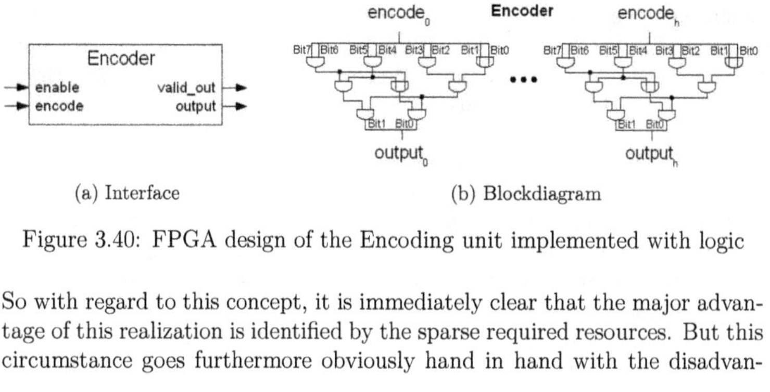

(a) Interface (b) Blockdiagram

Figure 3.40: FPGA design of the Encoding unit implemented with logic

So with regard to this concept, it is immediately clear that the major advantage of this realization is identified by the sparse required resources. But this circumstance goes furthermore obviously hand in hand with the disadvan-

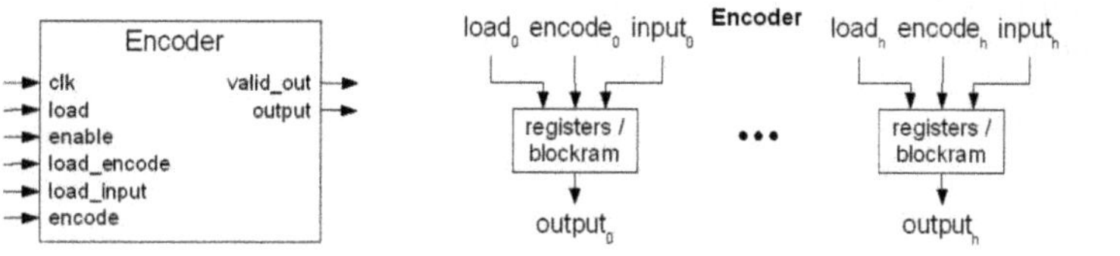

(a) Interface (b) Blockdiagram

Figure 3.41: FPGA designs of the Encoding unit implemented either with registers or with Block RAM

tage defined by the fixed transformation, which means more precisely that a modification of the priority class evaluation for the signatures implies the development of another functional description and thus implementation.
Besides this, the realization by the application of a LUT, which is depicted in figure 3.41, determines evidently a simple concept swapping these two facts.

So based on these two concepts, the most important selection criterion for one of them is apparently characterized by the circumstance that more required hardware resources might be acceptable to feature a more flexible and adjustable transformation, because a LUT can be evidently reloaded and updated at any time.

Diagonalization The Diagonalization of the histogram is in hardware apparently also just as easy as in software (see section 3.2.6 on page 147), because there is again just the requirement for a unit, which would simply increment the encoded priority class of each cell, if the priority class of the diagonal cell located below left is equal and represents not the smallest value zero representing no priority class. Moreover it is additionally not surprising that the major difference of this implementation compared to the software is as well determined by the circumstance that the entire process of Diagonalization is at the best parallelized and pipelined with regard to the read-out of the histogram, the Encoding and the two dimensional Peak-finding.
Besides this, it is important to notice that the two supported read-out directions of the histogram, which are both introduced in section 3.3.1 on page 165, necessitate respectively a specialized implementation of the Diagonalization unit, because some kind of neighborhood relationship is in contrast to the Encoding unit taken into account. Nevertheless a single block diagram, which is depicted in figure 3.42b, is sufficient to illustrate both implementations.

With regard to this figure, the structure of this unit is obviously regular and quite simple, because it consists just of some delaying shift registers in combination with a routing network to some arithmetic units. Furthermore it is clear that this routing network conjoins only the necessary diagonal located histogram cells with each corresponding arithmetic unit. Moreover the delaying shift registers are apparently connected in the same direction as the histogram read-out, which means either top-down or left-right. So within this context, it is self-evident that the arithmetic units, which are located at the bottom, correspond to the top-down histogram read-out, while the arithmetic units, which are located at right, belong to the left-right histogram read-out. Thence just one these blocks, which is marked either red with continuous lines or blue with dotted lines, is needed for a specialized implementation.

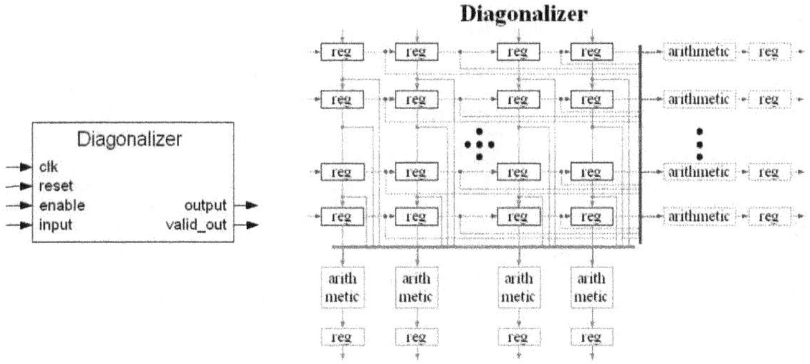

(a) Interface (b) Blockdiagram

Figure 3.42: FPGA design of the Diagonalization unit

However it has to be explicitly highlighted in this connection that even if the structures are equal, the corresponding resource consumption would be different, because the number of parallel units scale certainly either with the histogram dimension one or two. In contrast to this, the expected identical amount of main shift registers for both implementations causes the routing resources to be equal as well, but differently arranged.
As the functionality is now clarified, figure 3.43 depicts a small example with regard to both directions.

Further on, a closer look to this figure shows a filled 3×3 histogram layer in the upper left corner in combination with zeros, which are subsequently following according to the appropriate read-out direction. This means more precisely that the red colored zeros are used for a row-wise read-out at the bottom, while the blue colored zeros are applied for for a column-wise read-out at right. As the involved cells in an arithmetic operation, which are simply defined by the cell to evaluate and the diagonal one located one row below and one column left, are furthermore obviously connected with regard to the corresponding read-out direction of the histogram layer, the yellow colored cells mark the resulting output cells of the Diagonalization respectively. In this connection, it has to be noted that the colored zeros outside of the cells are required just for the arithmetic operations at the borderline of the parallelized units, because the histogram layer is of course not

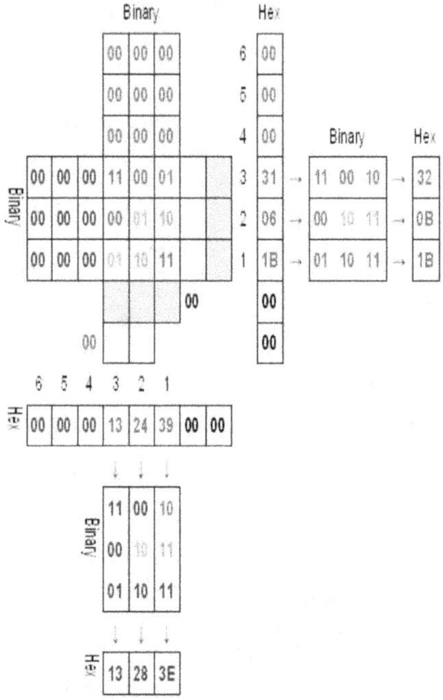

Figure 3.43: Visualization of a Diagonalization example processing

able to deliver values at these outside positions. Moreover the differently colored values except the magenta ones inside the filled 3×3 histogram layer highlights interesting situations for the Diagonalization, whose result is surely also equally colored and can be viewed either on the right for the row-wise read-out at the bottom or at the bottom for the column-wise read-out at right. In addition to that, it has to be also noticed that the standard color for unmodified values is changed to the standard color of the read-out direction to give a better imagination about the togetherness.

Having now a closer look to the highlighted interesting situations, the major important one is expectedly identified by the green values, because the Diagonalization results apparently in the highest available priority class because of the incrementation. Furthermore this situation is evidently set to be the most important one, because a deletion would occur due to an overflow to the priority class zero representing no priority, if the highest priority class is not reserved for the Diagonalization process.

Peak-finding and Serialization Before going into details, a closer look has to be spent on figure 3.44, because this figure exhibits the connection structure of the algorithm steps 2D Peak-finding, Serialization and 3D Peak-finding.

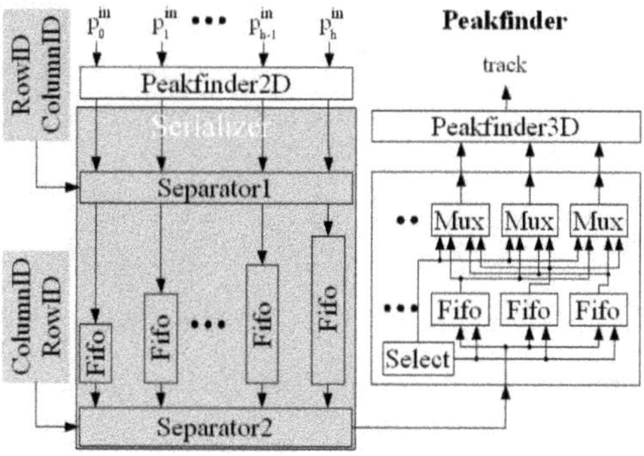

Figure 3.44: Abstract FPGA design of the Peak-finding units in combination with the Serialization unit

So with regard to this figure, the 2D Peak-finding defines evidently the first step directly after the Diagonalization. For that reason, it is furthermore clear that the general parallel layout of the 2D Peak-finding unit has to be quite similar to the one of the Diagonalization unit.
Further on, the result of this step is subsequently feed into the Serialization unit, which has to apparently serialize the track candidates and add an identifier for the corresponding histogram layer row and column to be able to recover the position in the layer. In this connection, it is important to mention that an identifier for the histogram layer itself is dispensable, because the afterwards applied memory has to store this data sorted according to the index of the corresponding layer.
Moreover it is now obvious that the unit, which is finally applied, is determined by the 3D Peak-finding.

Coming now to the concrete realization of the 2D Peak-finding, it is self-evident that the decoupling of the geometry and the arithmetic operation, which is introduced in section 3.2.2 on page 126 and already utilized in section

3.2.6 on page 148 and similarly in section 3.3.1 on page 176, is certainly also applied for this implementation. So based on the re-use of this concept, the figure 3.45b, which depicts the block diagram of this unit, looks as expected.

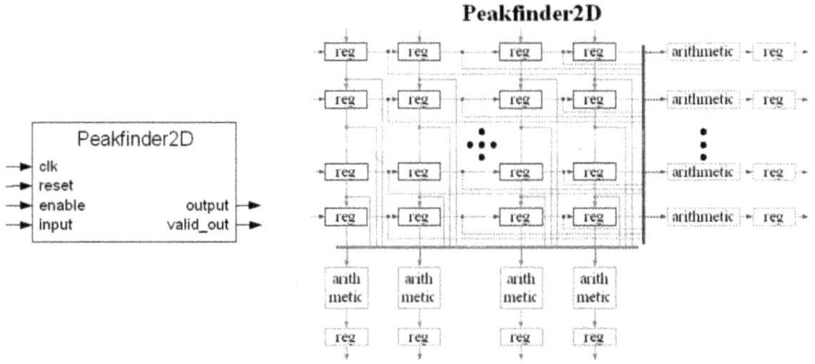

(a) Interface (b) Blockdiagram

Figure 3.45: FPGA design of the 2D Peak-finding unit

With regard to this figure, the earlier made allusion to the analogousness of the general structure for the 2D Peak-finding unit and the Diagonalization unit, which includes naturally the adjustment according to the read-out direction of the histogram as well, is apparently proven. Nevertheless there are surely differences due to the required connections of the cells to the arithmetic units and to the applied arithmetic operations in these units.
Furthermore precise information about the geometry, which is implemented by the connection of the cells to the corresponding arithmetic units, and the arithmetic operations can be found in section 3.2.6 on page 148.
As the functionality is currently clarified, figure 3.46 depicts a small example with regard to both directions.

A more detailed look to this figure illustrates now an identical setup compared to the Diagonalization example of section 3.3.1 on page 176, which shows a filled 3×3 histogram layer in the upper left corner in combination with zeros subsequently following according to the appropriate read-out direction. So while remembering this section, it is obvious that the red colored zeros are again used for a row-wise read-out at the bottom, while the blue colored zeros are afresh applied for a column-wise read-out at right. But in contrast to the Diagonalization example, the involved cells in an arithmetic operation are not illustrated here due to the complexity. Nevertheless it is clear that

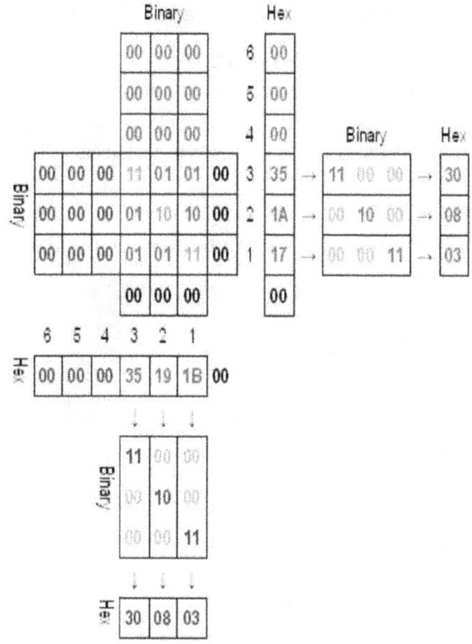

Figure 3.46: Visualization of a 2D Peak-finding example processing

they are also connected with regard to the corresponding read-out direction of the histogram layer, which is indicated by the leading zeros in each direction. Besides this, the green colored values inside the filled 3×3 histogram layer highlights the the 2D Peak-finding surviving track candidates from the magenta colored background. Moreover the result of this processing step is surely also colored and can be viewed either on the right for the row-wise read-out at the bottom or at the bottom for the column-wise read-out at right. Further on, the standard color for unmodified values, which represent thus surviving track candidates, is again changed to the standard color of the read-out direction to give a better imagination about the togetherness, while the color of the deleted track candidates and the background is modified to cyan.

Having now a closer look to the surviving track candidates clarifies immediately due to the functionality of the 2D Peak-finding that even smaller values, which are located directly next to a higher one, could survive this process, if the geometry does not include the corresponding neighborhood relationship.

Hence this example emphasizes the importance of an appropriate geometry.

In contrast to this complex process of Peak-finding, the Serialization of the track candidates, which survive the 2D Peak-finding, is quite simple, because the responsible unit has just to add a position identifier for the histogram layer row and column to each track candidate as already illustrated in figure 3.44 and to correspondingly delay each of them to become a sorted serial data stream. Thence the Serialization unit, which is depicted in figure 3.47b, prepares the actual amount of track candidates for the 3D Peak-finding.

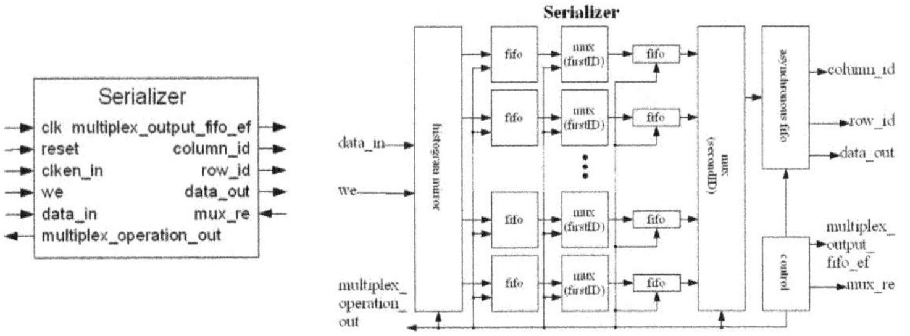

(a) Interface (b) Blockdiagram

Figure 3.47: FPGA design of the Serialization unit

With regard to this figure, it is obvious that both position identifiers are added in two separate steps, which are respectively realized in a multiplexer stage. As the assignment of these identifier to the two stages is furthermore definitely important, it is not surprising that this mapping depends directly on the the read-out direction of the histogram layer. Moreover it is additionally clear that this circumstance belongs also to the number of parallel required input ports, which is obviously determined by either histogram layer dimension one or two.
Besides this, it is evident that the correct delay of each track candidate can be simply realized by a hierarchy of First In First Out (FIFO) memories, because all track candidates are already sorted with regard to the histogram layer dimension, which belong to the read-out direction. That means more precisely that the read-out at the bottom implies the sorting to the dimension two, while the read-out at right implies the sorting to the dimension one. For that reason, it is self-evident that the first part of the identifier to add for a read-out at the bottom is fixed by the row and the second one by the column, whereas the order is vice versa for the read-out at right.

Finally it has to be mentioned that the simulation of the Serialization unit with the HDL simulator ModelSim SE 6.3d Revision 2007.11 (see [MOD]) from the company Mentor Graphics (see [MEN]) requires the argument 'SERIALIZER_LIB.glbl' to be used.

LBuffer

As the LBuffer unit consumes obviously a lot of memory and the 3D Peak-finding defines apparently a process with heavy memory access, it is planned to implement both on a PPC inside the FPGA. Thus it is clear that the LBuffer unit has to be located in the memory structure of this PPC architecture, which extends the functionality of this unit to three features.
In this connection, it is self-evident that the first one in order is of course determined by merging and storing the sorted surviving track candidates of all processed histogram layers in blocks analog to the HBuffer unit. For this purpose, it is additionally clear that the data transfer is initiated by the PPC via a network connection from the Register Transfer Level (RTL) logic to the LBuffer unit.
Moreover the successive feature is of course identified by holding these track candidate during the final 3D Peak-finding process.
And finally the last one is naturally characterized by saving the final result of the Hough transform until the data has to be delivered to the next following unit via an external network.

3.3.2 Cell BE Microprocessor Implementation

The complexity and required resources of the tracking algorithm based on a general Hough transform suggests itself that a multi-chip solution is more suitable for a hardware realization. So at a first glance, a multi-chip FPGA implementation, whose proposed design is depicted in figure 3.48, is the choice.

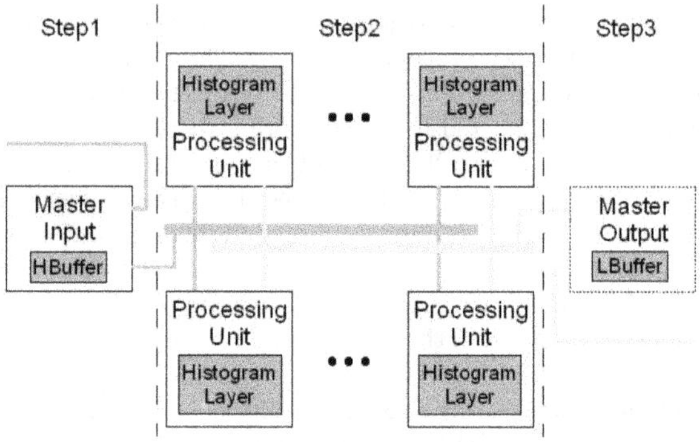

Figure 3.48: Proposed multi-chip configuration for the Hough tracking algorithm

With regard to this figure, it is obvious that a single master processes the data input with regard to the transformation. As the following part is naturally decoupled in the well-known algorithm by the HBuffer unit, it is clear that this unit forms the interface. For this purpose, it is evident that the data is already stored fragmented into jobs due to the computable amount of histogram layers on a single processing unit. Afterwards these jobs are then transported to the available processing units via an internal network. Moreover the result of each processing unit is then certainly merged in the LBuffer unit, which can be either located on another master or on the same one used recently for the transformation.
Besides this, the advantage of the additional master is immediately defined by a pipelined processing, while the other approach necessitates less hardware and in particular just a single internal network.

Regardless of that, the necessary development costs for such a system are apparently immense, because it identifies surely a custom circuit and the available FPGAs are quite expensive (e.g. XC5VLX110-1FFG1760C costs $\$\,1.6 \cdot 10^3$, see [Vire] and [Avn]). For that reason, it is preferred to use a rapid prototyping system, which offers the possibility to prove the concepts, before the precise development should be started. Thus the next step is obviously characterized by the search for such an applicable system, which might be perhaps able to support the required processing speed as well to exhibit a serious cheaper competitor. For this purpose, the common Sony Playstation III (e.g. $\$ 0.4 \cdot 10^3$, see [Ama]) is accounted, because the combination of the Cell BE microprocessor, which is developed by the STI, with an explicitly supported Linux operating system seems to meet the requirements for rapid prototyping perfectly.

Cell BE Microprocessor in a Sony Playstation III System

Before presenting now details about the entire Sony Playstation III system, a brief summary about the capabilities of the Cell BE microprocessor is given to emphasize its applicability.

Cell BE Microprocessor The Cell BE is a microprocessor architecture jointly developed by an alliance of the Sony Computer Entertainment Incorporated (SCEI) (see [Sonb] and [Sona]), the Toshiba corporation (see [Lap] and [Tos]) and the International Business Machines corporation (IBM) (see [IBMb] and [IBMa]), which is known as Sony-Toshiba-IBM joint venture (STI).

The first-generation Cell BE microprocessor, which is depicted in figure 3.49,

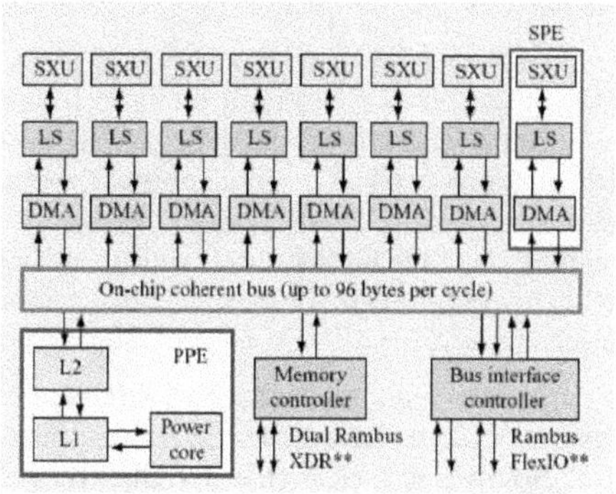

Figure 3.49: Block diagram of the Cell BE microprocessor (Source:[KDH+05])

is a multi-core chip comprised of a 64 bit Power Architecture processor core (Power Processing Element (PPE)) and eight synergistic processor cores (Synergistic Processing Elements (SPEs)). Furthermore these cores are capable of massive floating point processing as well as optimized for compute-intensive workloads and broadband rich media applications. Moreover a high-speed memory controller and high-bandwidth bus interface (Memory Interface Controller (MIC)) are unlike the Element Interconnect Bus (EIB) integrated on-chip.

Further on, figure 3.50 gives a good imagination about the required chip resources and the location of the different units on the die.

Figure 3.50: Chip die design of the Cell BE microprocessor (Source:[KDH+05])

Beyond all that, the PPE and SPE are both Reduced Instruction Set Computer (RISC) architectures with a fixed-width 32 bit instruction format. Furthermore the PPE contains a 64 bit general purpose register set, a 64 bit floating point register set and a 128 bit AltiVec register set. In contrast to this, the SPE contains only 128 bit registers, which can be used for scalar data types from 8 bit to 128 bit in size or for Single Instruction Multiple Data (SIMD) computations on a variety of integer and floating point formats.

Besides this, system memory addresses for the PPE and SPE are respectively expressed as 64 bit values for a theoretic address range of 2^{64} Byte ($\approx 16.8 \cdot 10^6$ Terabyte). However it is clear that not all of these bits are in practice implemented in hardware. Moreover local store addresses, which are apparently internal to the SPE processor, are expressed as a 32 bit word. In this connection, it has to be also noted that the documentation of the Cell

BE determines a word to represent always 32 bit, a double word to be 64 bit, and a quadword to mean 128 bit.

Regardless of that, the MIC is separated from the dual channel next-generation Rambus XIO macro, which interfaces the Rambus Extreme Data Rate Dynamic Random Access Memory (XDRDRAM) memory. Furthermore the XIO-XDR link runs at 3.2 $\frac{\text{Gbit}}{\text{s}}$ per pin, while two 32 bit channels can provide a theoretical maximum of 25.6 $\frac{\text{GB}}{\text{s}}$. Moreover the system interface, which is used in the Cell BE, is known as FlexIO and is organized in twelve lanes with each lane being a unidirectional 8 bit wide point-to-point path. In addition to this, four inbound plus four outbound lanes are supporting memory coherency. However five 8 bit wide point-to-point paths are inbound lanes to the Cell BE, while the remaining seven are outbound. This circumstance provides a theoretical peak bandwidth of 62.4 $\frac{\text{GB}}{\text{s}}$ (36.4 $\frac{\text{GB}}{\text{s}}$ outbound, 26 $\frac{\text{GB}}{\text{s}}$ inbound) at 2.6 GHz. Further on, the FlexIO interface can be clocked independently, but is typically set to 3.2 GHz.

Aside from this, the EIB is presently implemented as a circular ring comprised of four 16B-wide unidirectional channels, which counter-rotate in pairs. Further on, when traffic patterns permit, each channel can convey up to three transactions concurrently. As furthermore the EIB runs at half the system clock rate, the effective channel rate is given by 16 Byte every two system clocks. So at maximum concurrency with three active transactions on each of the four rings, the peak instantaneous EIB bandwidth is 96 $\frac{\text{Byte}}{\text{clock}}$ (12 concurrent transactions $\cdot \frac{16\,\text{Byte wide}}{2\,\text{system clocks per transfer}}$).

So by summarizing now all the mentioned facts into a conclusion, the breakthrough multi-core architecture and ultra high-speed communication capabilities of a Cell BE deliver vastly improved real-time response, which is in many cases ten times the performance of the latest personal computer processors. Moreover the Cell BE architecture is operating system neutral and supports multiple operating systems simultaneously, while applications may range from next generation game systems with dramatically enhanced realism, via systems that form the hub for digital media and streaming content at home, or systems used to develop and distribute digital content, to systems, which accelerate visualization or supercomputing applications.
By the way, more exhaustive information about the Cell BE microprocessor can be found in [Cela], [Celb], [Celc], [KDH+05] and [Celd].

Sony Playstation III Sony Playstation III names the third home video game console, which is produced by SCEI, and defines the successor to the Sony Playstation II as part of the Sony Playstation series. This console system (see [Plad]), which is depicted in figure 3.51, competes apparently with Microsoft's Xbox 360 (see [Xbo]) and Nintendo's Wii (see [Us.]) as part of the seventh generation of video game systems. Moreover the key characteristics of its hardware are determined by the 3.2 GHz Cell BE CPU and the 550 MHz nVIDIA/SCEI Reality Synthesizer Graphics Processing Unit (GPU) (see [Son05]). Further on, this system includes additionally a Blu-ray drive, a 2.5" SATA hard drive with different possible memory sizes, up to seven Sixaxis/DualShock 3 controllers and other connectivity accessories (see [Plaa] and [Plab]).

Figure 3.51: Photo of the Sony Playstation III system (Source: [Plaa])

A closer look to the CPU shows that the Cell BE microprocessor is made up of one 3.2 GHz PPC-based PPE and six accessible SPEs. Furthermore the seventh runs in a special mode and is dedicated to aspects of the operating system and security, while the eighth is disabled (see [Lin07] and [Proa]). Moreover, the GPU, according to nVIDIA, is based on the nVIDIA G70, which was previously known as NV47, architecture (see [Coo06]) with 256 MB Graphics Double Data Rate 3 (GDDR3) RAM clocked at 700 MHz (see [Son05]).
Further on, the main memory is defined by 256 MB of XDRDRAM clocked at CPU die speed (see [Son05]). In this connection, it has to be additionally noticed that the firmware update 2.0101478 and consecutive ones reserve 32 MB of the XDRDRAM memory for the XrossMediaBar (XMB) user interface of the Sony Playstation III (see [Plaa]).

Besides this, the floating point performance of the whole system, which is commonly measured in Floating Point Operations Per Second (FLOPS), is reported for the CPU and GPU together to be 2 TFLOPS (see [Son05]). Furthermore the Cell BE CPU achieves 204 GFLOPS single precision and 15 GFLOPS double precision (see [Don06]).
Beyond that, the console weights approximately 4.4 kg and exhibits the dimensions of $325 \times 98 \times 274$ mm (see [Son07]). The power consumption of the initial Sony Playstation III unit ranges from 180 – 200 Watt (see [Plac]), despite having a 280 Watt power supply (see [Son07]). However the power consumption of newer Sony Playstation III (65 nm process) units range from 120 – 140 Watt (see [Plac]). Additionally the cooler noise is also decreased from 1.3 Sone to 0.8 – 1.0 Sone (see [Plac]).

But regardless of all that, the most impressive fact of the Sony Playstation III is that Sony announced the console to be able to natively run a Linux operating system (see [Lina] and [Linc]) including X11 (see [Linb]). So by combining this circumstance with the availability of an assembler description and a GCC compiler, which understands almost c/c++ with some extensions for the PPE and SPE stuff (see [Proa], [Prob] and [PPU]), it is clear that the software development of third parties is explicitly supported. Hence the reasons for selecting the Sony Playstation III system as rapid prototyping is now obvious. By the way, a detailed summary about the used system can be found in appendix D on page D-1.

Data Input

While remembering the facts of section 3.3.1 on page 156, it is clear that the data input of the trigger implementation using the Sony Playstation III is already unfixed as well. Since this implementation is furthermore also not planned to be applied to the real experiment, there is obviously only the need for the specification of the data input with regard to the simulation of the concepts, which is apparently independent of the available 1000BASE-T ethernet network connection (see [Son05]) featuring a theoretical maximal transfer rate of 1 $\frac{\text{Gbit}}{\text{s}}$ or 125 $\frac{\text{MByte}}{\text{s}}$.

For that reason, the circumstance that the CBMROOT framework does actually not support the Sony Playstation III platform requires evidently an indirection, which involves data files containing framework produced data. This means more precisely that the CBMROOT framework, which runs on a common personal computer to simulate a well-defined physics event related to particle collisions, produces this data and creates the corresponding files containing all necessary input information. Furthermore it is clear that these files are afterwards simply transferred to the Sony Playstation III system via the available network connection and can be then used by the algorithm as input data to prove the developed concepts.

Finally it has to be noted here that three different files are necessitated, which consist of either the hit information or both LUTs respectively. In addition to this, it is also really essential to understand that the unavailable digitization of the hit causes the LUT files to contain only the entries, which occur in the CBMROOT framework simulation, instead of the entire general content, while the hit file comprises just numbers indexing these special LUT files. The reason for this setup, which is obviously more complex than necessary, is surely defined by the circumstance that it will be someday easy to replace this one by the correct setup including the correct general LUTs.

Algorithm

Compared to the modulated algorithm, which suits for an implementation based on the utilization of a single FPGA (see figure 3.29 of section 3.3.1 on page 157), the major adjustment for an implementation based on a Cell BE microprocessor belongs apparently to the different parallelism, because there is a dedicated amount of processing units featuring fixed resources available in such an embedded system. Furthermore this system offers two different levels of parallelism, which are defined by the number of parallel processing units and the number of parallel histogram layers evaluated by such a single vector-capable unit. So while paying attention to this fact, figure 3.52 depicts now the general structure of the algorithm including the partitioning steps, which are already introduced in section 3.1 on page 96.

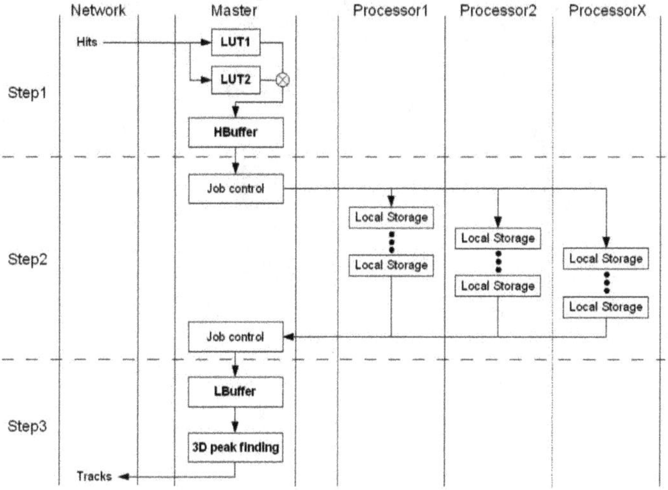

Figure 3.52: Sketch of the Cell BE utilization for the Hough transform algorithm concerning insufficient resources to instantiate the entire histogram in parallel layers

With regard to this figure, multiple slave processing units or slave processors are used to evaluate the second step of the Hough transform algorithm, while the first and the third step is realized by a unique master processing unit or master processor.

So based on this concept, the arriving hits are transformed by applying both

LUTs consecutively, before the HBuffer unit stores all results sorted with regard to the start index of the histogram layer. Furthermore this sorted data is then fragmented into job packages or jobs due to some limitations and afterwards delivered to these slave processors. Besides this, the slave processors are able to evaluate the results for their histogram layers, which are installed in their own local storage, independent of all others. This means more precisely that each slave processor has to realize the algorithm steps Histogramming, Encoding, Diagonalization, 2D Peak-finding and Serialization. Moreover it is clear that the termination of the last step activates the transport of the results back to the master processing unit, which is then able to reassemble the jobs in the LBuffer unit and further to apply finally the 3D Peak-finding.

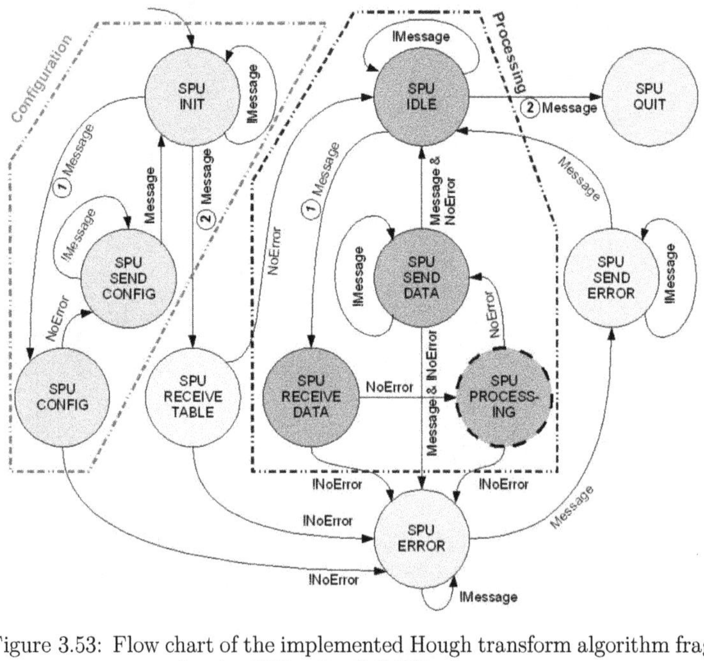

Figure 3.53: Flow chart of the implemented Hough transform algorithm fragments for the SPEs of a Cell BE microprocessor system

Beyond that, figure 3.54 depicts now the flow chart of the implemented algorithm including the communication for the master processor, which is surely defined by the PPE, and figure 3.53 for the slave processors, which are certainly determined by the SPEs.

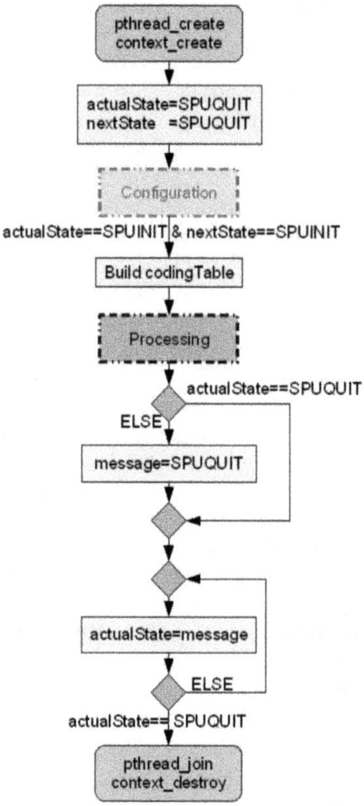

Figure 3.54: Flow chart of the implemented Hough transform algorithm fragments for the PPE of a Cell BE microprocessor system

With regard to these figures, it is immediately clear why the PPE implements the so-called master and why the SPEs realize the slaves. Nevertheless it should be told here that this circumstance is simply defined by the fact that the PPE controls all slave SPEs at any time by sending command messages and receiving only acknowledgment messages via the mailbox system of the Cell BE. Further on, it is explicitly highlighted within this context that the common handshaking protocol is used used in combination with watchdog timers to transfer commands.

Besides this, the necessary hit input data and track output data is apparently transported by the Direct Memory Access (DMA) engine of the Cell BE with a connection, which is established by the appropriate SPE, because the mailbox system is certainly not designed for a huge amount of data.
As this data is furthermore packaged into jobs as mentioned earlier, the focus is now set to the very interesting topic of job creation and especially to the major important limitation. Further on, this limitation is obviously determined by the available memory in the local storage of the SPEs, because the job data and the number of histogram layers require space as well as the program code or even dynamic data like the program heap and stack, whose sizes are surely really hard to evaluate. So for that reason, there is evidently a need for an additional algorithm step in front of the data processing, which helps to find a practical solution. Since this extra step, which is called configuration, is furthermore naturally characterized by the clarification of the memory status, the master simply requests all slaves to send their spendable amount of memory in the local storage. Based on these hopefully uniform sizes, the master is then of course able to determine the maximal amount of data for the job packages, which does not exceed the minimal limitation. However it has to be mentioned here that this determination is not as easy as it seems to be, because the transformed hit does not respect the borders, which are defined by the fixed arrangement of the number of parallel implemented histogram layers on a processing unit, and can thus occur in more than one job package. Beyond that, this concept obviously enables each job package to be certainly generated and delivered to each ready slave processor independent of any ordering information, because there is no restriction for and between any job package. Figure 3.55 shows now the corresponding flow chart of the additional configuration step for the master, while the one for the slave is already depicted in 3.53.
Moreover further information about other limitations of the job creation like the maximal number of parallel processable histogram layers on a single slave, the load balanced number of histogram layers and the load balanced maximal number of hits, which play of course a not less significant role for a fast implementation, can be found in section 3.3.2 on page 203 and in [Str09].

Returning now to the actual focus, it is further on obvious that multiple strategies exist, which feature the generation of these job packages.

So according to the intention, it is immediately clear that the first and easiest one is characterized by a loop over all hits, which generates simply each job by assigning the hits to the corresponding jobs before any job is delivered or

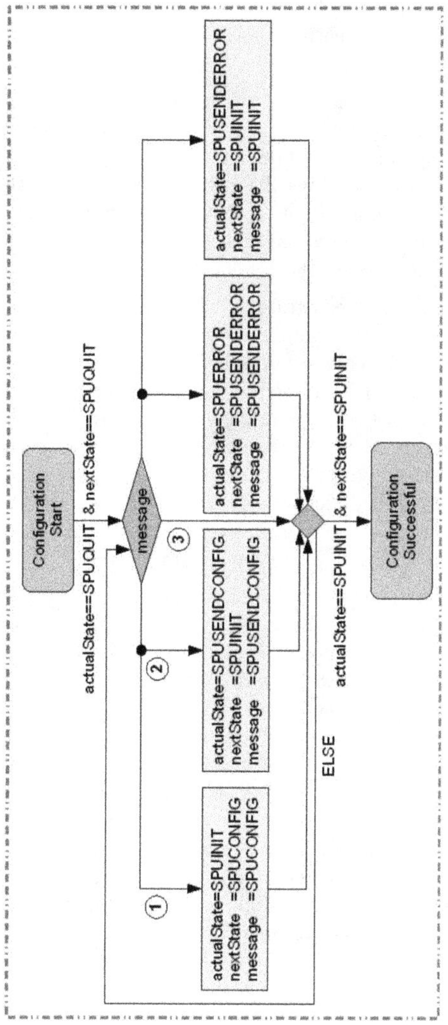

Figure 3.55: Flow chart of the implemented configuration algorithm fragment for the PPE of a Cell BE microprocessor system

processed. The advantage of this strategy is of course directly connected to the elapsed time, because each hit has to be accessed just once independent of the number of jobs.

But as the aim of the implementation is a PPE, which generates all jobs in parallel to the processing SPEs, this loop must be obviously restructured. Thus the resulting loop, which fulfills the prerequisite for the parallelization and enables further a benchmark for the accessibility of the hits, creates job after job by searching for the corresponding hits. However it has to be noticed here that this job creation strategy generates still all jobs before any job is delivered or processed.

In contrast to this, the next strategy, which defines apparently the goal, is simply characterized by the parallelization of the restructured loop with the processing SPEs. This means more precisely that the PPE generates job by job, while each created job can be immediately delivered to each ready SPE. So the needed time for the job creation is then successfully hidden behind the time, which is consumed for the Hough transform processing by the SPEs. A very good illustration about these three different job creation strategies offering supplementary their direct comparison can be found in figure 3.56, which depicts the flow chart of the implemented processing algorithm fragment for the PPE of a Cell BE microprocessor system.

Besides this, it has to be noticed here that the memory limit for the job packages of all introduced strategies is set to the minimum available memory of all SPEs, which is evaluated in the configuration fragment of the algorithm, because the jobs are required to be independently processable by each SPE. But as this fragment evaluates the memory limit of each SPE separately, there is apparently also the possibility to build specialized jobs, which are SPE assigned. In addition to this, it has to be highlighted that such a job creation strategy can be surely parallelized to the Hough transform processing as well, but the SPE must be naturally ready before the PPE can even create the corresponding specified job with regard to the specialized memory limit, because this job has to be directly delivered to this assigned SPE. Thus such a strategy offers certainly the possibility to cope with heterogeneous systems, which are for instance characterized by a dramatically varying memory. Nevertheless this one can be easily skipped here, because each SPE exhibits the same amount of memory and runs identical program code.

Finally it should be mentioned that each of these four implemented job creation strategies are called code versions in the following and can be selected in the source code file PpeCode/include/config.h by the parameter 'CODEVERSION'.

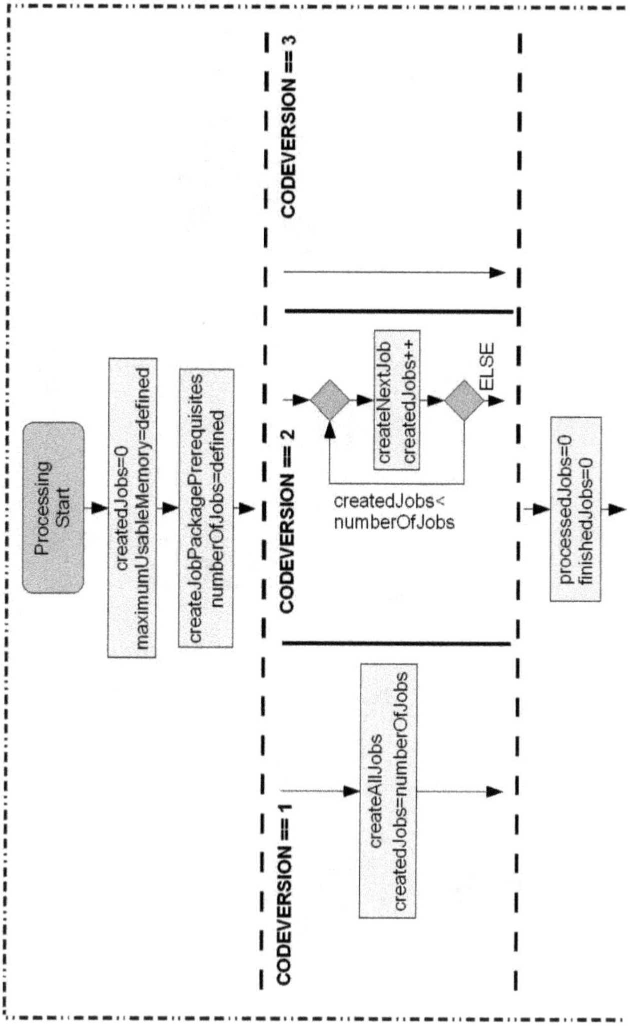

(a) Part one

Figure 3.56: Flow chart of the implemented processing algorithm fragment for the PPE of a Cell BE microprocessor system

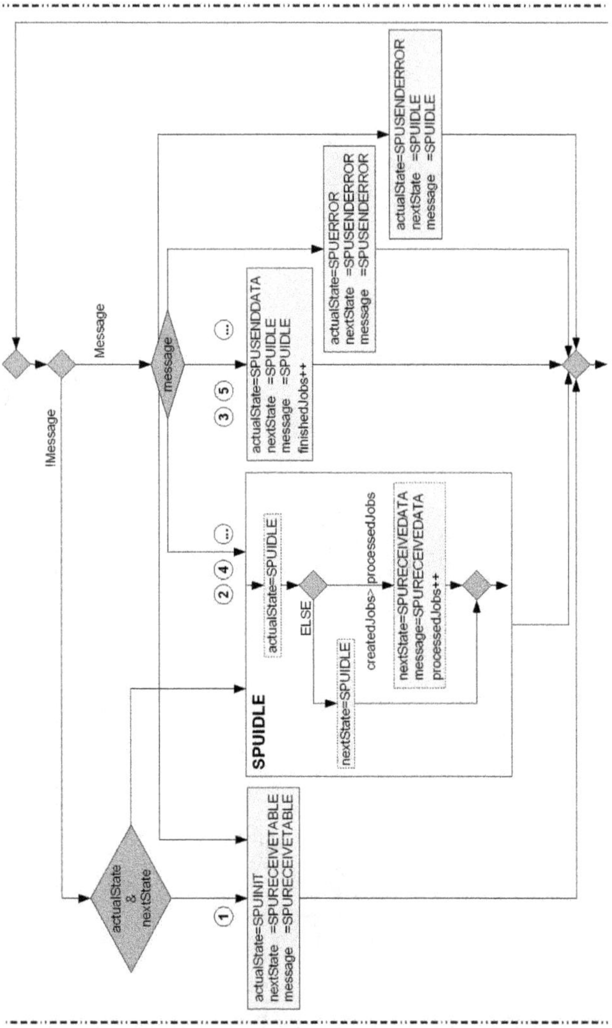

(b) Part two

Figure 3.56: Flow chart of the implemented processing algorithm fragment for the PPE of a Cell BE microprocessor system

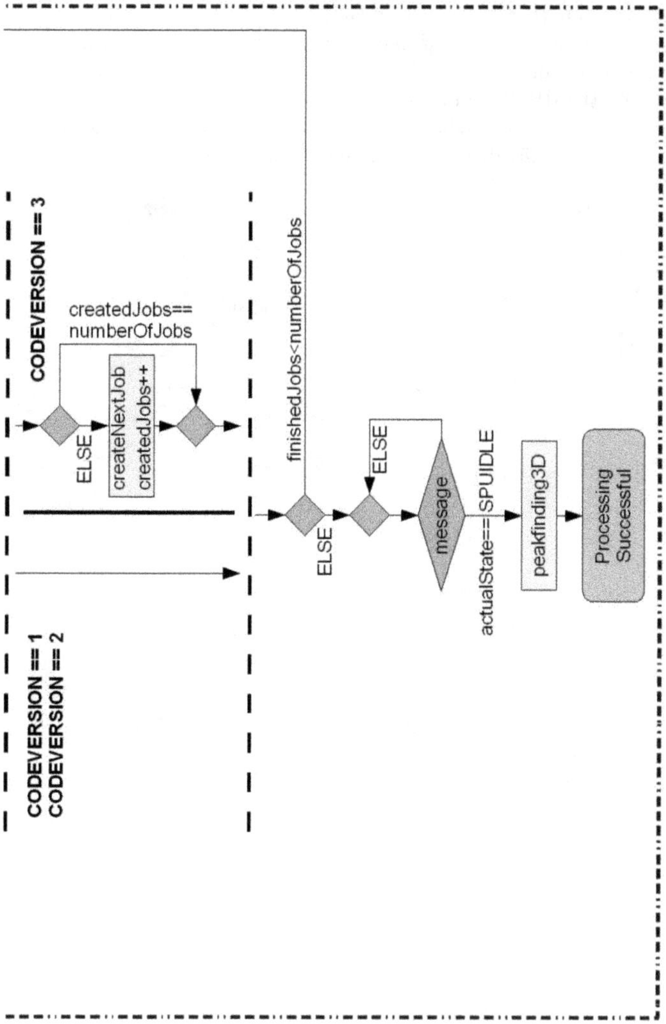

(c) Part three

Figure 3.56: Flow chart of the implemented processing algorithm fragment for the PPE of a Cell BE microprocessor system

Besides this, another special case related to the job packages has to be additionally noted here, because if enough processors are considered to guarantee the parallel instantiation of all histogram layers, there would be apparently no need for the HBuffer unit due to the circumstance that all transformed hit information can be immediately delivered to the corresponding processors. Therefore figure 3.57 depicts the modified implementation.

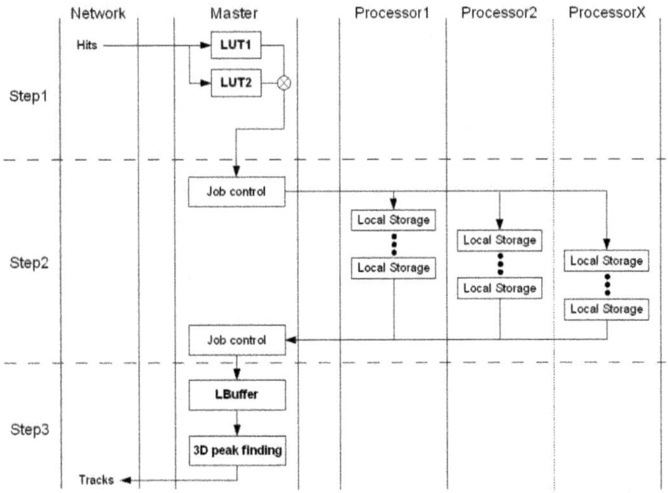

Figure 3.57: Sketch of the Cell BE utilization for the Hough transform algorithm concerning sufficient resources to instantiate the entire histogram in parallel layers

Look-up-tables

Due to the fixed Cell BE architecture in the Sony Playstation III system, it is obvious that only limited possibilities exist to implement the Look-up-tables (LUTs). So with regard to them, a fragment of the XDRDRAM memory seems to be the best solution for this purpose. Moreover figure 3.58 depicts now the corresponding implementation in combination with the first step of the algorithm.

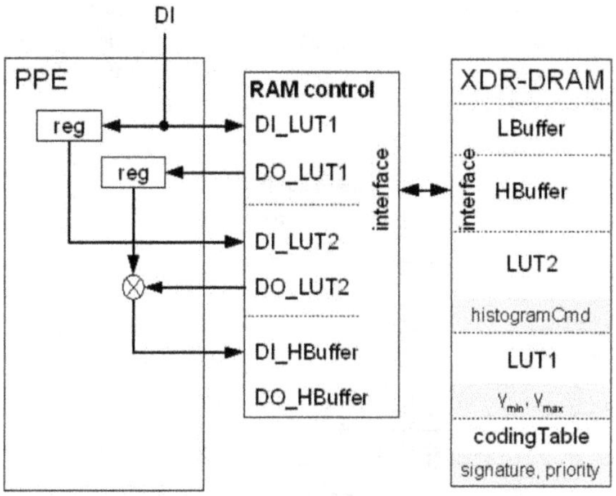

Figure 3.58: Illustration of both LUT implementations

With regard to this figure, it is clear that the arriving hit data is applied to evaluate the corresponding value of the first LUT by accessing the XDR-DRAM memory via a control interface of the PPE. Furthermore this data is supplementary delayed to be also usable for indexing the second LUT via the same interface. As the result of the first LUT is then naturally delayed as well, both results can be afterwards stored in the HBuffer unit, which is likewise located in the XDRDRAM memory.

HBuffer

The HBuffer unit, which serves in turn for the decoupling of the LUT evaluation on the PPE and the Histogramming on the SPEs, is apparently also located in the XDRDRAM because of two reasonable advantages. While the first one is of course simply identified by the availability of the required amount of space and even much more, the second advantage is determined by the circumstance that this memory supports DMA requests, which can be originated from the SPEs themselves. Hence the PPE has explicitly not to deliver the job packages, but only to offer the address of the corresponding data in the XDRDRAM to the SPEs, which are then able to fetch it. Moreover figure 3.59 depicts now the accordant implementation of the HBuffer unit in combination with the transport network.

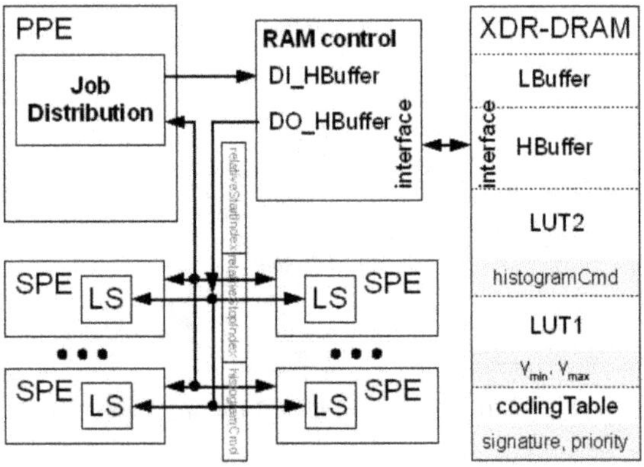

Figure 3.59: Illustration of the HBuffer implementation

With regard to this figure, it is clear that each SPE reads as many transformed hit information out of the HBuffer unit as necessary to process all locally parallel implemented histogram layers. Furthermore this corresponding job package is certainly transported directly via the network from the XDRDRAM into the local storage of the SPE by using the DMA engine.

Histogram

By remembering section 3.3.2 on page 191, it is clear that the histogram is distributed in layers via the local storage of each processing SPE, because the steps of the Hough transform algorithm, which belong to the modification of these layers, are already introduced to be implemented there. Thence the Histogramming, which is schematically illustrated in figure 3.60, can be simply characterized by an execution loop, which generates the affected histogram layers based on the corresponding job package.

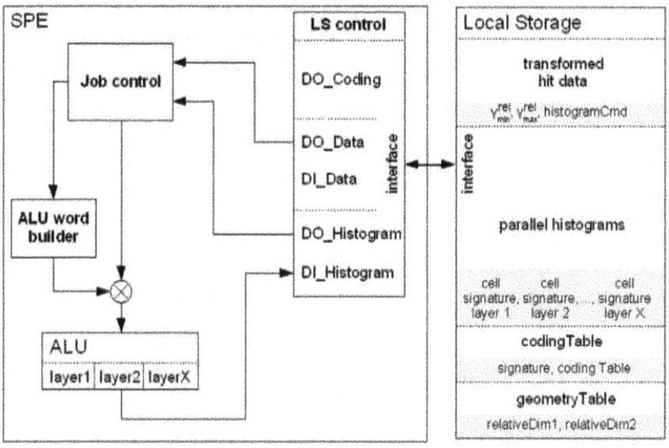

Figure 3.60: Illustration of the the histogram implementation

With regard to this figure, it is just a simple job control unit required, which accesses each transformed hit information of the job in combination with its affected histogram cells respectively, transport both information to the Arithmetic Logic Unit (ALU) and stores further the modified value of each histogram cell back into the local storage at the original address.

Moreover it is also exhibited in this figure that the whole necessary data for the recent job, which contains the histogram layers, all transformed hits of the job, the table used for the Encoding and the Peak-finding geometry, has to be located in the local storage for a high-performance processing. In this connection, it has to be supplementary mentioned that the found track candidates, whose amount is certainly less compared to the number of hits in the job, do not need a reservation in the local storage, because the released memory space due to the destructive read-out of the transformed hit information is surely sufficient to absorb the resulting data.

However these facts tell obviously not the total story about the local storage, because the architecture of the SPE is a typical von Neumann, which causes generally the program code, the heap and the stack to be additionally located there (see [Proa]). Since this situation is now quite complex, figure 3.61 gives a good imagination.

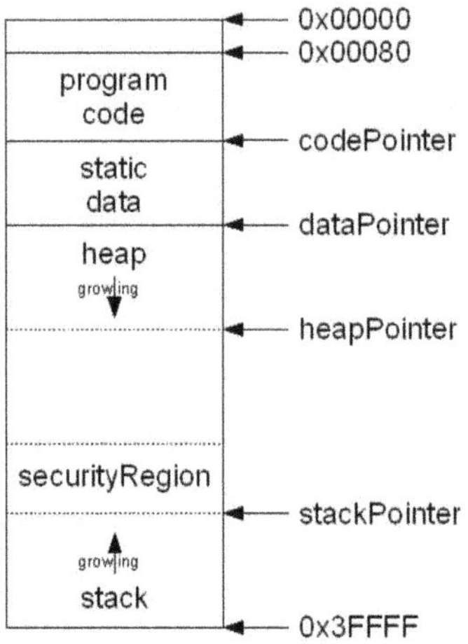

Figure 3.61: Composition of a local storage with regard to its content

With regard to this figure, the reason for the configuration fragment of the algorithm, which is supplementary added to the common Hough transform as introduced in section 3.3.2 on page 191, is immediately obvious, because it has to evaluate naturally the available amount of memory in the local storage of each SPE, which can be applied for the user data of the algorithm and determines thus the limitation for the job creation. So based on this context, it is clear that the relation of the job size and the memory limit is defined by the sum of the required almost variable memory for the histogram layers, the required variable memory for the hits corresponding to these layers and the constant memory for the Encoding table and the Peak-finding geometry. Hence the key elements of a job are apparently characterized by the number of histogram layers in combination with the number of corresponding hits.

Besides this, it is of course additionally important to think about the workload balancing for the multiple SPEs because of the implementation goal speed. So with regard to these key elements, it is now evident that only two possibilities exist to control the balance. But as both are apparently directly related to each other, because a limit for the number of hits will surely restrict the amount of corresponding histogram layers and vice versa, there is just a single possibility in the strict sense. For that reason, the selected key element is naturally determined by the number of histogram layers, which can maximally correspond to a job, because it is easier to implement such a restriction due to the circumstance that the total amount of layers and the number of SPEs are both fixed. Thence the job restriction based on the number of histogram layers is certainly defined by the total amount of 191 layers divided by 6 SPEs, which leads then to a limit of 32 histogram layers per job. Even if this situation seems to be quite easy, more information about the workload balancing can be found in [Str09].

Beyond that, one of the two featured parallelism levels requires additionally another considerable job restriction, which is apparently closer to the hardware than the general workload balancing, because this one belongs to the SIMD processing technique of the SPEs as stated in section 3.3.2 on page 185. So with regard to this technique, it is immediately clear that the processing time would decrease enormously compared to the usual implementation applying the Single Instruction Single Data (SISD) technique, if the algorithm is capable to use the SIMD processing efficiently. For this purpose, multiple analyses have shown that the best applicable parallelization level belongs to the insertion of the transformed hit plane, which concerns the same histogram cells in multiple consecutive layers.
So the combination of this concept with the histogram cell signatures for a setup of five usable detector stations, which is introduced in section 1.3.1 on page 12, and the 128 bit ALU of a SPE in a Cell BE environment leads immediately to the capability of processing theoretically $\frac{128}{5} = 25$ histogram layers in parallel. But as the local storage and the ALU offer both a minimal accessibility of eight bit, it is clear that it is more feasible to reserve three bits in the signature of each histogram cell, because bit manipulations will increase the required processing time dramatically. Hence the parallelization level decreases to only $\frac{128}{8} = 16$ histogram layers.
By considering further the strategy of memory usage, the first suggested approach, which is characterized by randomly access the memory of the affected histogram cells for a single SIMD computation, leads to the requirement of integrating the vectors in front of a ALU calculation and disintegrating them afterwards. Thence 16 different random memory accesses are necessary for each vector, which is involved in the recent ALU operation, in conjunction with 16 ones for the result. As however this circumstance constitutes obviously a really bad situation due to the consumed time for the construction and destruction of the vectors, the strategy or better memory layout, which

is depicted in figure 3.62, is developed, because this one avoids the heavily addressed memory access by a clever arrangement of the histogram layers in the local storage.

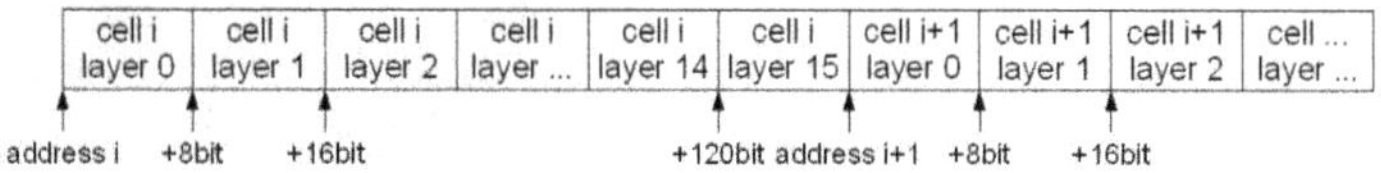

Figure 3.62: Strategy demonstration for the histogram layer memory layout

So with regard to this figure, the mentioned clever arrangement is simply characterized by the location of the affected cells next to each other, because then the random access of 16 single eight bit cells is removed by the parallel access of the entire 128 bit vector of cells. Further on, it has to be noticed here that this parallel access is only efficient due to the circumstance that the affected histogram cells in a single ALU operation are always located in a plane, which is spread via consecutive histogram layers.
Nevertheless a big disadvantage of this approach is obviously determined by the circumstance that all 16 histogram layers would be always implemented in the local storage, even if many of them might be unused, because a job package is evidently able to contain less data due to of another job restriction. In addition to this, the efficiency of such an implementation is also questionable, because there are naturally many transformed hits, which are spread into layers via job borders or should not be inserted in all 16 parallel layers. Thence the result of this approach depends directly on the distribution quality of the input data and thus on the transformation quality.
Since these two introduced approaches implement apparently the absolute opposite strategy due to the memory usage and vector requirement, it is expected that a third strategy can be realized, which extenuates each respective disadvantage by accepting the softened advantages. Hence the new approach combines the memory access of the common data types for 8, 16, 32 and 64 bit with their corresponding vector load mechanisms, which leads then certainly directly to the circumstance that the number of parallel instantiated histogram layers can be varied in steps according to the size of these data types. So based on this concept, it is immediately clear that the corresponding vector can be surely either integrated or disintegrated with n = 1, 2, 4 or 8 histogram cells by a single access instruction or referenced itself with n = 16 histogram cells. For that purpose, it is self-evident that the selected data type has to be at least big enough to include the required parallelism level of the job. Moreover the correlation between the parallelism level of the job and the data type size, which corresponds to the number of

instantiated histogram layers, is shown in table 3.3.

Parallelism level	1	2	3	4	5	6	7	8
Size of data type	1	2	4	4	8	8	8	8
Parallelism level	9	10	11	12	13	14	15	16
Size of data type	16	16	16	16	16	16	16	16

Table 3.3: Compendium of data type size correlations for the possible parallelism levels of a job

A closer look to this table shows now immediately the advantage of this approach, because at the utmost $\lceil \frac{n}{2} \rceil - 1$ unused histogram layers are implemented. Finally it should be additionally mentioned that each of these three implemented approaches are called memory access types or memory versions in the following and can be selected in the source code file PpeCode/include/-config.h by the parameter 'MEMORYVERSION'.

Filtering

Before presenting now the specific implementation, which is applied to reduce the background noise in the histogram, it has to be noticed here that the entire process is again just as separated as in software. Thus detached sections exist in analogousness to section 3.2.6 on page 144 for the Encoding and the Diagonalization as well as a single combined section for the Peak-finding and Serialization, which includes also the differentiation between the 2D and 3D Peak-finding.

Encoding The Encoding of the histogram layers is nearly as easy as in software (see section 3.2.6 on page 145), because there is just a need for a loop over all cells, which transforms each signature without the usage of neighborhood information into a priority class. But in contrast to the software implementation, the parallel processing, which is featured by the available SPEs, requires the used table to be initially transferred into the local storage of each SPE to consume as less time as possible for the Encoding. So based on this concept, such a process can be apparently implemented by using the common execution loop to modify the histogram layers, which is depicted in figure 3.63.

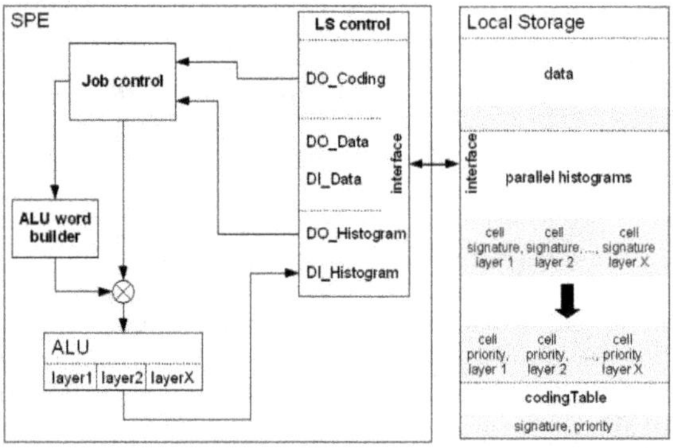

Figure 3.63: Illustration of the Encoding implementation

However this implementation is in detail not really that easy, because an ordinary histogram access contains not just one signature value due to the clever arrangement of the layers in the memory, which is introduced insection 3.3.2 on page 203, but a number of histogram signatures in parallel. Thence the operations inside the execution loop are determined by an initial read operation of a vector containing a certain amount of histogram cell signatures, which is obviously followed by an access of the Encoding table for each signature respectively and finally completed by a write operation of the resulting vector to the same memory address. Moreover it is clear that this process could be simply improved by the parallelization of the Encoding table so that a single entry features the result for the parallel histogram signatures, if the local storage exhibits enough space, because the multiple table access is then of course replaced by just a single access. Nevertheless it has to be noted here that this parallelization requires a rapidly and dramatically increasing memory space.

Diagonalization The Diagonalization of the histogram layers is obviously just as easy as in software (see section 3.2.6 on page 147), because there is in turn only the requirement for a loop over all cells, which would increment the encoded priority class of each histogram cell, if the priority class of the diagonal cell located below left is equal and represents not the smallest value zero representing no priority class. So based on this very easy concept, such a process can be apparently implemented by using again the common execution loop to modify the histogram layers, which is depicted in figure 3.64.

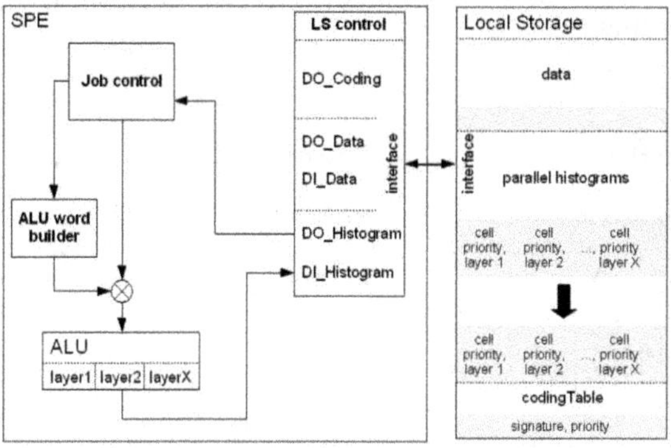

Figure 3.64: Illustration of the Diagonalization implementation

With regard to this figure, it is evident that the same facts as already presented in section 3.3.2 on page 207 count here as well. This circumstance means more precisely that the clever arrangement of the histogram layers requires the operations inside the execution loop to combine an initial read operation of two diagonal related vectors containing a certain amount of histogram cell signatures with some consecutively following arithmetic operations for the corresponding signatures respectively and finally a write operation of the resulting vector to the accordant memory address.

Peak-finding and Serialization Before going into details, the general sequence of the algorithm steps 2D Peak-finding, Serialization and 3D Peak-finding, which is already shown in figure 3.44 of section 3.3.1 on page 179, has to be remembered, because the actual implementation combines the 2D Peak-finding and the Serialization in a single process on the SPEs, while the 3D Peak-finding follows on the PPE after the final DMA transfer delivers the last data set. Further on, this combined 2D Peak-finding and Serialization process can be apparently implemented by using again the common execution loop, which is depicted in figure 3.65, because the process is quite similar to the Diagonalization.

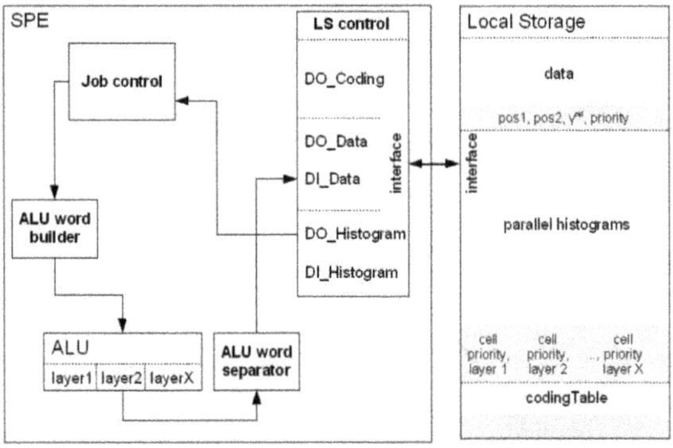

Figure 3.65: Illustration of the 2D Peak-finding and the Serialization implementation

With regard to this figure, the combined 2D Peak-finding and Serialization process is not that surprising, because the usual loop over all cells simply evaluates a result for each cell due to the geometry and the applied arithmetics, which is afterwards not written back into the histogram cell, but into the memory region of the local storage containing earlier the destructively read transformed hit information. So within this context, it is clear that the Serialization order belongs directly to the direction of the loop through the histogram layers. In this connection, the vector arrangement of the layers in the memory has to be reminded, but it defines evidently no significant problem.

LBuffer

In analogousness to the HBuffer unit, the LBuffer unit serves in turn for the reassembling of the track candidates, which survive the 2D Peak-finding on the SPEs, to be ready for the final 3D Peak-finding realized on the PPE. For that reason, this unit is apparently also located in the XDRDRAM due to the same advantages, which count for the HBuffer unit. In this connection, the two major advantages, which are identified by the supported DMA requests from the SPEs and the availability of the required amount of space and even much more, have to be especially highlighted again. So based on this concept, it is once again evident that the PPE has explicitly not to fetch the results, but only to offer the address of the corresponding space to place the data in the XDRDRAM to the SPEs, which are then able to deliver their results. Furthermore figure 3.66 depicts now the accordant implementation of the LBuffer unit in combination with the transport network.

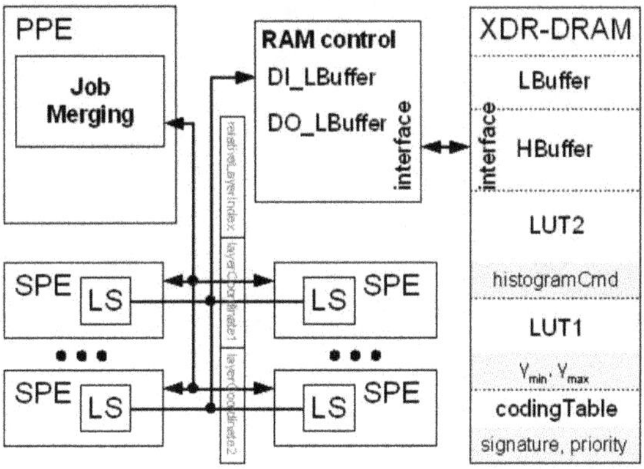

Figure 3.66: Illustration of the LBuffer implementation

With regard to this figure, it is obvious that the corresponding final track candidates are certainly transported directly via the network from the local storage of the SPE into the XDRDRAM by using the DMA engine. Thence the 3D Peak-finding, which is realized by reading the track candidates according to the Peak-finding geometry out of the LBuffer unit, applying the arithmetics and writing the results back, can finally not start until the last set of track candidates is delivered.

Chapter 4

Analyses

As the general Hough transform is a very abstract algorithm, which can be applied to many diverse problems in many varying fields (see section 2 on page 45), it is self-evident that each special implementation has to be customized to the given problem. So without loss of generality, such a customization process is commonly realized by simply adjusting the parameters of the Hough transform algorithm to the needs for a well-conditioned environment, because then the performance can be obviously assumed to be optimal.

For that reason, this chapter contains some basic analyses, which offer support by either adjusting these parameters automatically well-conditioned or by giving some hints, which ease the manual adjustment. So having now a more detailed look to the content of this chapter, it can be directly seen that the first two sections care about the input to the algorithm, the section afterwards about the required resources, the subsequent three sections about algorithm specific key elements, the next section about the quality of the algorithm and the final ones care about the automatically evaluated parameters of the algorithm.

Besides this, it should be additionally noticed here that all of these in the following presented analyses are implemented in the software framework CBM-ROOT, which is introduced in section 3.2.1 on page 102, and can be usually switched on or off by setting just some simple enable/disable parameters almost in the configuration file, which is precisely explained in appendix A.2 on page A-4.

4.1 Input Summary

This section presents analyses, which summarize the most abstract key indicators of the Hough tracking for the actual reconstruction run. So in this connection, it is clear that these indicators, which are determined by the values of all configuration parameters, the necessary information of the detector setup, some information of the recent event and some additional information related to the priority classes occurring in the actual event, are of major importance, because they can be used to get a first imagination about the quality of the data input to the algorithm and thus the complexity of the Hough transform.

4.1.1 Configuration Parameters

As often mentioned in diverse parts of this thesis, the Hough transform algorithm has many configuration parameters, which have to be defined in the configuration file (see appendix A.2 on page A-4). So with regard to this amount of parameters, a common problem for a potentially not that good evaluation performance is identified by an error during the initialization phase, which might be for instance a totally unset or wrong set parameter. Caused by this circumstance, the analysis presented in this section is able to print a summary to the standard output screen, which displays the total available information of all Hough tracking configuration parameters. Furthermore it is clear that this information includes the parameter name and value pair as well as the detail, if the standard value is taken, or whether the value is customized by the user. Moreover a typical exemplary output of such an analysis, which is enabled in the configuration file (see appendix A.2 on page A-4) by the parameter 'analysisInitConfiguration', is depicted in figure 4.1.

```
Setup for the Hough tracking:
----------------------------------------
CommandID:   1111111100001111110000111111111100101011101...

[PARAMETERS FOR THE INPUT OBJECT]
inputFileName : ../../Data-Source/auau.25gev.centr.sts4pix3strips.mc.root
...
```

Figure 4.1: Input summary about the configuration of the algorithm

By taking now a closer look to this figure, it is immediately clear that these few lines show just the constitution of the output and list not the whole configuration. By the way, the most important line contains obviously the value for the parameter 'CommandID', because this value contains a single bit for each configuration parameter, which identifies the corresponding value to be the standard one by an unset bit or to be customized via reading the configuration file by a set bit. Further on, it should be also noticed here that this circumstance implies apparently that each parameter of the tracking is definitely set to a configured value at any time. But as it is nevertheless important to know which bit position is linked to which parameter, this fact can be easily discovered by the order of the printout on the standard output screen or by the order in the standard configuration file, which can be automatically produced by the software package (see appendix A.2 on page A-4). Finally it should be additionally mentioned in this connection that the constructor of the main class features the overwriting of dedicated configuration parameters, whose values must be then of course checked manually.

4.1.2 Detector Stations

Since it is evidently good to have an imagination about the used detector system setup, the analysis, which is presented in this section, offers the possibility to print a summary of the necessary details to the standard output screen. So a typical exemplary output of such an analysis, which is enabled in the configuration file by the parameter 'analysisInitDetector' (see appendix A.2 on page A-4), is depicted in figure 4.2.

```
Number of detector stations:      7
...
Station number 3:
        Id: 17
        Distance: 20
        isNo: 1
        isMaps: 0
        isHybrid: 0
        isStrip: 0
        Mask: 0
        Index: 2
Station number 4:
...
```

Figure 4.2: Input summary about the detector setup

By having now a more detailed look to this figure, it is in analogousness to the previous analysis obvious that these few lines show also just the constitution of the output and list not the whole setup. However the general displayed information is composed of the number of detector stations in the setup in combination with details for each station. Furthermore these details consist of a station number for identification, the GEANT id, the distance from the target in *cm*, the type of the station, the mask for the computation and the index position of the histogram cell signature. Further on, the most surprising fact in this connection is apparently determined by the least self-explanatory properties, which are the mask and the index, because both are at the same time the most important ones for the Hough tracking algorithm. So coming now to a short explanation, the mask represents a boolean value, which enables or disables the corresponding detector station during the computation by the algorithm, while the index typifies an unsigned decimal number, which defines the unique connection between the detector station and the corresponding bit in the hit signature of the peaks (see section 3.1 on page 96). Moreover it is immediately clear within this context that a masked detector station is not able to contribute anything to the tracking, because it is not linked to a representing bit in the signatures. Therefore it has to be finally noticed that such a functionality offers evidently the possibility to evaluate the tracking performance with missing detector stations by just setting such a mask in the configuration file (see appendix A.2 on page A-4).

4.1.3 Event Information

A very important summary for the actually processed event can be printed to the standard output screen by enabling the analysis, which is presented in this section, with the parameter 'analysisInitEvent' in the configuration file (see appendix A.2 on page A-4). As furthermore all the displayed information is self-explanatory, a typical exemplary output of such an analysis is depicted in figure 4.3.

```
Number of event:                                          0
Number of tracks:                                         921
Number of hits:                                           15658
Number of tracks (class priority >= 2x0000001 and P > 1):  346
```

Figure 4.3: Input summary about event characteristics

A closer look to this figure shows now that the information consists of the ordering number of the event, the number of simulated MC Tracks, the number of delivered hits and the number of findable tracks, which is determined by applying the prerequisites of a momentum higher than 1 $\frac{\text{GeV}}{\text{c}}$ and a minimum priority class of actually one to the total number of existing MC Tracks. Besides this, it has to be finally noticed here that a comparison of the number of existing MC Tracks to the total amount of findable MC Tracks distinguishes the maximal available data reduction, while the comparison of the number of delivered hits to the number of hits, which survive the computation by the algorithm (see section 4.7.1 on page 258), constitutes the real achieved data reduction of the Hough tracking.

4.1.4 Priority class Information

The analysis presented in this section simply extends the event information analysis, which is introduced in the previous section 4.1.3 on page 216, by adding some extra information about the priority classes arising for the MC data of the actual event. This accessory information shows then details about the number of MC Tracks, which correspond to the available priority classes based on the actual used table files (see appendix A.4 on page A-101). So a typical exemplary output of such an analysis, which is enabled in the configuration file by the parameter 'analysisInitClassPriority' (see appendix A.2 on page A-4), is depicted in figure 4.4.

```
Number of event:                                           0
Number of tracks:                                          921
Number of hits:                                            15658

ANALYZED SOURCE DATA
Actual number of tracks with class priority 3 and P > 1:        308
Actual number of tracks with class priority 2 and P > 1:        10
Actual number of tracks with class priority 1 and P > 1:        28
Number of tracks (class priority >= 2x0000001 and P > 1):   346
```

Figure 4.4: Input summary about the priority classes of an event

Finally it has to be noticed here that the standard analysis for the event must be obviously collaterally enabled to receive this extra information.

4.2 Input Details Summary

In contrast to to all other sections, it has to be at first noticed here that almost all analyses, which are presented in this section, can be only enabled or disabled by define statements in the source code (see appendix A.3 on page A-79). Furthermore the reason for this circumstance is obviously not that surprising, because each analysis summarizes precise information about the data input to the Hough tracking algorithm, which is mainly used for debugging purpose.

4.2.1 Object Information

The analysis presented in this section enables a printout to the standard output screen, which displays the total available information for each detector station, for some MC Tracks and for some hits. So in this connection, it is again obvious that such a printout contains just an excerpt of the MC Tracks and hits, because otherwise the huge amount of data will overfill the buffer of the output console. Further on, the define statements 'DETECTORINFO', 'TRACKINFO' and 'HITINFO', which enable this analysis, can be found in the source code file InputLIB/include/inputData.h. Coming now to the point, a typical exemplary output of such an analysis is depicted in figure 4.5.

```
numberOfStations: 7
...
Station number 2:
    Id: 17
    Distance: 20
    isMaps: 0
    isHybrid: 0
    isStrip: 0

Station number 3:
...
```

(a) Available information for the detector stations

```
numberOfTracks: 921
...
trackId: 16
    Hits: 7
    Points: 7
    PdgCode: 2212
    Charge: 1
    momentumX: 0.19
    momentumY: -0.62
    momentumZ: 10.80
...
```

(b) Available information for the MC Tracks

```
NumberOfHits: 15658
...
HitID: 6
    PointID: 3181
    TrackID: 605
    DetectorID: 15
    PosX: -1.99
    PosY: 0.85
    PosZ: 5
    FieldX: -0.02
    FieldY: -6.38
    FieldZ: 0.01
...
```

(c) Available information for the hits

Figure 4.5: Excerpt of the diverse algorithm input data for an event

4.2.2 Data Consistency Information

The group of analyses, which is presented in this section, cares universally about the consistency of the input data. So for this purpose, all of these analyses automatically check the input data before the evaluation of the Hough transform is started and would produce further a warning message, if something seems to be strange. Furthermore figure 4.6 depicts now some of the most appearing warning messages.

DataObjectLIB WARNING: There are 2 hits with a wrong trackId!!!

DataObjectLIB WARNING: There are 13 tracks which have no hits!!!

DataObjectLIB WARNING: There are 11045 hits which have no point!!!

DataObjectLIB WARNING: There are 11047 hits which have no track!!!

Figure 4.6: Demonstration of the most frequently occurring data consistency warning messages for a typical event

Coming now to the detailed information, the first possibly occurring warning message of such a consistency report is caused by MC Tracks, which do not possess any MC Point in the detector setup, because these tracks are obviously strange and thus recognizable.
Further on, the second interesting warning message is produced by MC Points, which refer to an id of a MC Track that does not exist in the input data, because this circumstance is explicitly not correct.
Moreover it is of course evidently good to know the number of MC Tracks, which exhibit no corresponding hits for the existing MC Points, because this information potentially features the detection of wrong simulated hit data. However it has to be noticed here that a small number of such MC Tracks typically occurs due to the modeled detector station inefficiencies.

Besides this, there are also two further warning messages in the report, which care about hits that do not correspond to MC Points and not to a MC Track. But considering now the recent detector setup in this connection, these two warning messages are less significant due to the usage of micro-strip detector stations, because such stations produce many so-called fake hits, which are characterized exactly by this attribute (see section 1.3.1 on page 19).

Beyond that, it has to be finally mentioned, that all of these analyses can be enabled by define statements in the source code file DataRootObjectLIB/include/trackfinderInputData.h, while more detailed analyses, which can be enabled in this file as well, feature a deeper insight into the affected data section.

4.2.3 Data Propagation Information

In addition to section 4.2.2 on page 219, this section presents an analysis, which takes the physical context of the input data into account. So based on this circumstance, it is clear that the data could be finally only approved to be valid, if the consistency and the physical quality is checked.
However as the input data is mainly produced by hit producers preceded by the GEANT package (see section 3.2.1 on page 109), which propagates collided particles (see section 1.2 on page 7) from a fixed target through a detector system (see section 1.3 on page 10) in an inhomogeneous magnetic field, the huge complexity does not allow a fast afresh computation for verification. But fortunately such an afresh computation by the GEANT package is not necessary, because the GEANE package (see section 3.2.1 on page 109) offers the possibility to do almost the same with quite less computation effort.
Hence the analysis is implemented pretty simple, because the GEANE package is applied to propagate the existing MC Tracks again, while the result for the subsequent comparison of each computed detector station interaction to the corresponding GEANT supplied MC Points or hit producer supplied hits is then inserted into a distance distribution for each detector station.
So based on this concept, it is now obvious that the quality of the input data is measured by the distances of the twice computed sampling points for each MC Track. Moreover it is self-evident that the quality gets better, the smaller the distances are.
However it has to additionally noticed here that a big lack of this analysis is determined by the fact that it can not be enabled at the present time, because the GEANE package, which is a feature of the CBMROOT framework (see section 3.2.1 on page 102), is not ready yet. Nevertheless, if it will be ready in the future, the parameters 'analysisInitTrackPropagationEventPoint', 'analysisInitTrackPropagationEventHit', 'analysisInitTrackPropagationTotalPoint', 'analysisInitTrackPropagationTotalHit', 'analysisInitTrackPropagationDisplay', 'analysisInitTrackPropagationToRoot', 'analysisTrackPropagationPointDisplayMask' and 'analysisTrackPropagationHitDisplayMask', which are located in the configuration file (see appendix A.2 on page A-4), would enable the analysis.

4.3 Resources Summary

At first, it has to be noticed in this section that all analyses, which are presented here, are specialized to the resource consumption of the Hough transform algorithm. So in contrast to all other analyses, these ones do not help to optimize the algorithm in any way. However such analyses are important, because they offer the possibility to compare the required resources of the general software implementation with the resource consumption of other implementations, which are specialized to some hardware like the Sony Playstation III or the FPGA.
Further on, the concerned resources, which are wanted to be compared, are as usual the memory and the time. However it has to be mentioned here that a direct comparison is lacking, because the huge overhead information in the software implementation, which is used only to debug the algorithm, needs additional memory and time. Nevertheless the order of magnitude is interesting to get an imagination about the complexity of the Hough tracking.

4.3.1 Memory

The first analysis introduced in this section prints a detailed summary about the memory consumption of the Hough tracking, which contains information about the statically reserved memory, the allocated memory during runtime and the exact used memory, to the standard output screen. In this connection, it has to be additionally noted that such an analysis implies evidently a hidden feature, because since it is well-known that the dynamic memory allocation during runtime slows down algorithms compared to the static memory, this analysis is able to support the determination of the amount of memory, which should be statically reserved, and that one, which should be allocated during runtime. Hence the trade-off between these two types of memory allocation strategies can be certainly evaluated close to the optimum, which is characterized by a fast processing time without much unused but reserved memory. However since this optimization can be obviously not realized automatically, there are some parameters to define the amount of statically reserved memory (see appendix A.3 on page A-79).
Further on, the printout of the summary, which is exemplary depicted in figure 4.7, can be enabled by the parameter 'analysisInitMemory' in the configuration file (see appendix A.2 on page A-4).

```
Size of used memory for
            the MC data:                    5.0 MB
            the MC data addon:              380 kB
      the input data:                            5.4 MB
            the Hbuffer::Prelut:            61.2 kB
            the Hbuffer::Lut:               7.6 MB
            the Hbuffer::Hits:              61.2 kB
      the Hbuffer unit:                          7.7 MB
            the signature:                  41.6 kB
            the hits:                       41.6 MB
      the Histogram:                             41.6 MB
      the Lbuffer unit:                          367.7 kB
      the output data:                           351.0 kB
                                                 ----------------
      the tracking:                              55.4 MB
```

Figure 4.7: Resource summary about the memory consumption of the algorithm

A closer look to this figure shows now that just a single item, which is called ' the MC data addon', is not self-explanatory, because it is special to this implementation. However the idea of this item is not that mysterious, because it simply represents the amount of memory, which is used for pointers that speed up and simplify the index access of related input data objects. Finally it has to be noticed in this connection that the data consistency analyses, which are introduced in section 4.2.2 on page 219, can be made during the initialization of these pointers with a negligible timing effect.

4.3.2 Time

The second analysis introduced in this section prints a detailed summary about the time consumption of the Hough tracking, which contains information about the CPU-time and real-time for each algorithm step separately, to the standard output screen. Further on, the printout of this summary, which is exemplary depicted in figure 4.8, can be enabled by the parameter 'analysisInitTime' in the configuration file (see appendix A.2 on page A-4).

Needed realtime for:

creating the borders:	2.7 s
creating the histogram layers:	3.0 s
encoding the histogram layers:	3.6 s
diagonalizing the histogram layers:	1.0 s
peak finding in the histogram layers:	8.0 s
finalizing the histogram layers:	1.1 s
resetting the histogram layers:	3.4 s
peak finding in the track candidates:	0.02 s
	----------
the tracking:	22.8 s
the event:	28.7 s

Figure 4.8: Resource summary about the time consumption of the algorithm

By having now a more detailed look to this figure, it is interesting to notice that the 2D Peak-finding consumes about 35% of the overall real-time. Thus the 2D Peak-finding embodies a key issue with regard to the time consumption of the software implementation.

4.4 Magnetic Field

This section presents all analyses, which are related to the magnetic field. So with regard to the analytic formula, which can be used to form the computational kernel of the Hough transform algorithm (see section 2.4 on page 66), the main purpose is evidently determined by the effect of the inhomogeneous magnetic field to the quality of the formula adaptation (see section 2.4.2 on page 79).
Therefore the related fields of analysis are obviously given by the visualization of the magnetic field itself, the evaluation of the optimal constant factors, which represent the optimal field fixed at each detector station (see section 2.4.2 on page 79), and the visualization of the mathematical computed averaged integrated magnetic field (see section 2.4.2 on page 80). So based on this concept, it is clear that the major interesting information about the quality, which is presented by the last analysis of this section, is defined by the comparison of the optimal constant factors and the mathematical computed averaged integrated magnetic field.

4.4.1 Visualization of the Magnetic Field

As stated in the introduction, it is at first useful to have a look at the visualized magnetic field. However the analysis presented in this section has to solve a complexity problem before, because the coordinate space and the magnetic field components are both three dimensional. Thus figure 4.9 depicts nine different graphs, which display each magnetic field component in dependency of each coordinate direction.

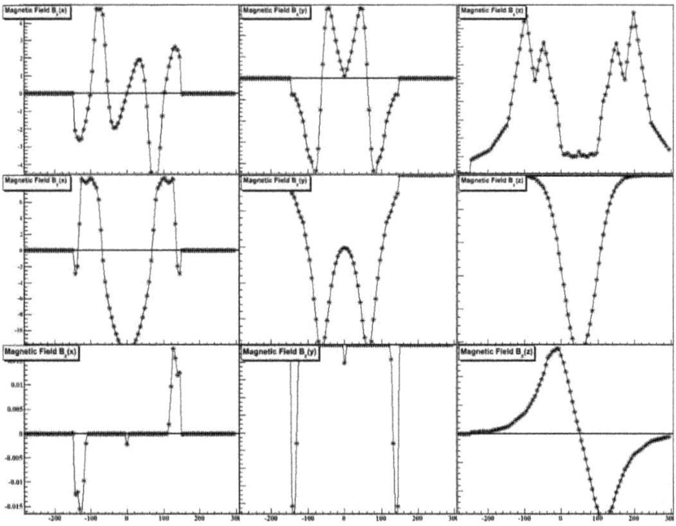

Figure 4.9: Visualization of the magnetic field components dependent on different coordinate dimensions

A closer look to this figure shows now immediately that the graphs are ordered in the way that the columns contain the magnetic field components in dependency of the same coordinate direction, while the rows consists of the same magnetic field components in dependency of each discriminative coordinate direction.
Besides this, it is important to notice that the magnetic field can be apparently always shown only dependent of a single coordinate direction due to the complexity. Hence it is self-evident that the coordinate directions of each graph, which are not displayed, are constant during the evaluation. More-

over it is really essential to realize that the result could vary dramatically, if the constant is moved to another coordinate in the allowed range. For this purpose, these constants can be obviously set in the source code via an initialization function call, which looks like:

```
initMagnetfieldXAnalysis(bool enable = true,
    int nBins = 100, double min = -300,
    double max = 300, double constValDim1 = 20,
    double constValDim2 = 50)
```

With regard to this example, the chosen concept is self-explanatory, because the initialized graph simply evaluates 100 supporting points in the coordinate range $[-300; +300]$ for the x direction at the constant coordinates $y = 20\,\text{cm}$ and $z = 50\,\text{cm}$.
Finally it should be also mentioned that all initialization function call declarations can be found in the source code file AnalysisLIB/include/analysis.h, while the parameters, which enable the entire visualization, can be found in the configuration file (see appendix A.2 on page A-4). Furthermore the names of these parameters are 'analysisInitMagnetfieldX', 'analysisInitMagnetfieldY', 'analysisInitMagnetfieldZ', 'analysisInitMagnetfieldDisplay' and 'analysisInitMagnetfieldToRoot'.

4.4.2 Constant Factors

The analysis presented in this section is able to compute and visualize the constant factors, which can be used in equation 2.17 of section 2.4.1 on page 78 to generate the second LUT.
For this purpose, the algorithm, which is introduced in section 2.4.2 on page 79, is used to evaluate the optimal ones bearing the common problems of the accumulator, which are mentioned in section 2.1 on page 45, and especially the fact that not all transformed hits encounter the correct histogram cell in mind. Thus the decision criteria for the selection of a factor for the corresponding station is obviously determined by the computation of the sum of the shortest distances from the transformed hit entries to the correct histogram cell. So it is self-evident that the best summand, which is apparently zero, represents the case that all transformed hits encounter the correct histogram cell, while the bigger the value of the summand is, the worse is the actual factor.
Further on, figure 4.10 depicts now a typical distribution of these added

distances in dependency of the correspondingly used constant factor.

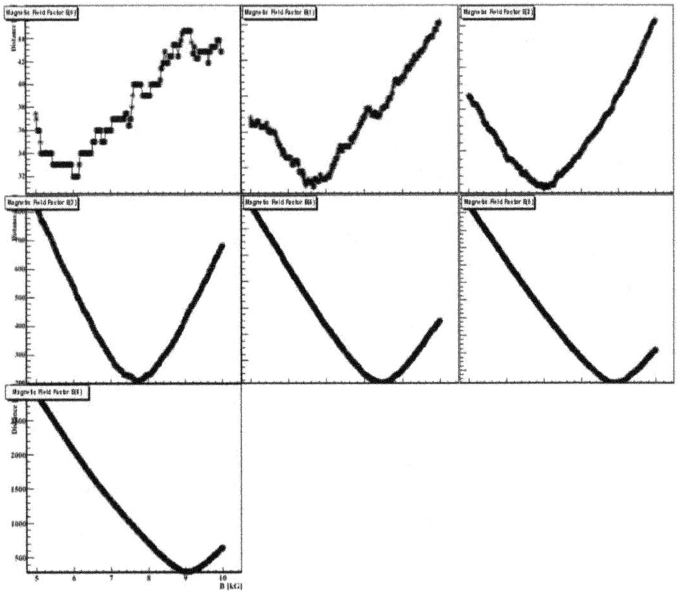

Figure 4.10: Visualization of the constant factor distributions representing the averaged integral of the magnetic field

Coming now to an implementation specific issue, it has to be particularly pointed out that the huge amount of computing time for this algorithm is practically avoided by restricting the analyzed values to an interesting range, which has to be defined by the user via the define statements 'NUMBEROFFACTORS', 'FACTORMIN' and 'FACTORMAX' in the source code file AnalysisLIB/include/analysis.h. Furthermore it has to be additionally mentioned in this connection that this range has to be set really carefully, because the factors are used in the denominator of equation 2.17 of section 2.4.1 on page 78, which implies consequentially that the constant factor zero is emphatically not allowed. In contrast to this, negative factors are explicitly allowed, even if the actually chosen ones are commonly positive. Moreover a typical user range for the actual experimental setup is defined by $[5; 10]$ in combination with 501 factors, which leads apparently to an evaluation exactness in the order of 10^{-2}.

Besides this, the special interest in the accurate value of each selected factor can be satisfied by an additional printout to the standard output screen, which is simply produced by warning messages. Therefore the corresponding bit, which is labeled with 'MAGNETFIELDFACTORINFORMATION' (see appendix A.3 on page A-79), has to be set in the parameter 'analysisInit-AnalysisResultWarnings' in the configuration file (see appendix A.2 on page A-4). Furthermore such a typical printout of all factors is shown in figure 4.11.

```
AnalysisLIB WARNING: The magnetic field factor of the station with index 0 is set to 5.98!!!
AnalysisLIB WARNING: The magnetic field factor of the station with index 1 is set to 6.66!!!
AnalysisLIB WARNING: The magnetic field factor of the station with index 2 is set to 7.01!!!
AnalysisLIB WARNING: The magnetic field factor of the station with index 3 is set to 7.70!!!
AnalysisLIB WARNING: The magnetic field factor of the station with index 4 is set to 8.37!!!
AnalysisLIB WARNING: The magnetic field factor of the station with index 5 is set to 8.89!!!
AnalysisLIB WARNING: The magnetic field factor of the station with index 6 is set to 9.04!!!
```

Figure 4.11: Exhibition of the used precise value for the constant factor representing the averaged integral of the magnetic field at each detector station

Finally it should be additionally mentioned that the parameters 'analysisInit-MagnetfieldConstantForEachEvent', 'analysisInitWeightedMagnetfieldConstant', 'analysisInitMagnetfieldConstantDisplay', 'analysisMagnetfieldConstantDisplayMask' and 'analysisInitMagnetfieldConstantToRoot', which are also located in the configuration file (see appendix A.2 on page A-4), enable this analysis.

4.4.3 Averaged Magnetic Field Integration vs. Constant Factors

As the optimal constant factors (see section 2.4.2 on page 79) and the averaged integrated magnetic field (see section 2.4.2 on page 80) can be both applied in the generation process of the second LUT, it is obvious that a comparison between their quality is of major interest. Therefore this section presents a visualization of the magnetic field itself in combination with the averaged integrated one detailed for the important coordinate range. So within this context, it is clear that the earlier addressed comparison of the recent visualizations to the constant factors, which are exhibited in section 4.4.2 on page 225, can be realized by enabling this analysis with the parameter 'analysisInitMagnetfieldVSConstants' in the configuration file (see appendix A.2 on page A-4).
Moreover typical results of such an analysis are depicted in figure 4.12, which

shows the original inhomogeneous magnetic field, and in figure 4.13, which displays the averaged integration. In this connection, it is obviously as important as mentioned in section 4.4.1 on page 224 that the coordinate in the x and the y dimension must be fix to evaluate all values in dependency of the z direction.

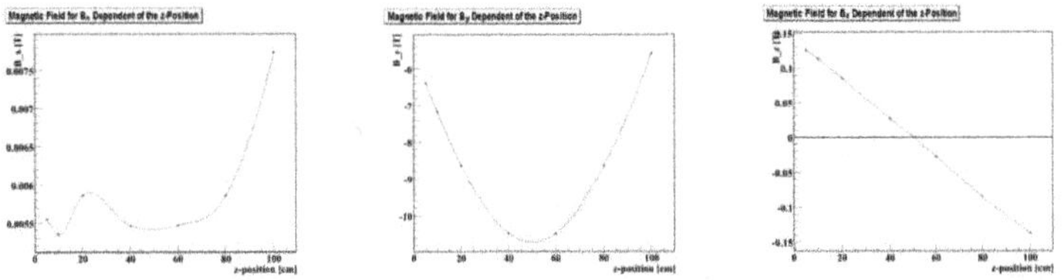

(a) Component in x direction

(b) Component in y direction

(c) Component in z direction

Figure 4.12: Visualization of the magnetic field components in dependency of the z coordinate

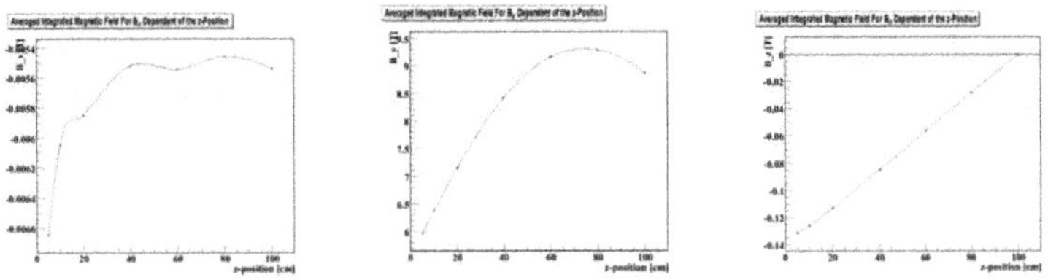

(a) Component in x direction

(b) Component in y direction

(c) Component in z direction

Figure 4.13: Visualization of the averaged integrated magnetic field components in dependency of the z coordinate

Although the comparison of the two different approaches is practicable now, a better imagination can be surely achieved by visualizing both possibilities in the same manner. Thus figure 4.14 depicts the optimal constant factors in dependency of the z direction.

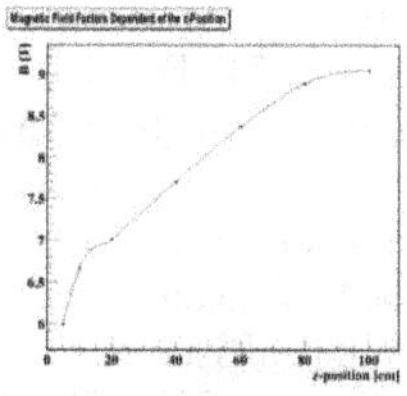

Figure 4.14: Graph of the the automatically evaluated optimal constant magnetic field factors for the detector stations

So coming now to the point, the similarity of the figures 4.13b and 4.14 advices in the end that the averaged integrated magnetic field is applicable in the generation process of the second LUT to avoid the dependency of the MC information, which is a prerequisite for the evaluation of the optimal constant factors. Furthermore it is self-evident that these two figures can not be identical, because the averaged integral is, in contrast to the optimal constant factors, also dependent of the coordinates in x and y.

Finally it should be also noticed here that the constant factors, which are used in this comparison, can be possibly defined by choosing one of two options. In this connection, it is evident that the first one is apparently determined by the automatic generation via the enabled analysis, which is presented in section 4.4.2 on page 225. If this analysis is however disabled, the second option, which is characterized by define statements in the source code file DataRootObjectLIB/src/trackfinderInputMagneticField.cxx, would be naturally applied.

4.5 Look-up-tables

This section presents dedicated analyses, which evaluate the quality of the Look-up-tables (LUTs). For this purpose, the first two analyses care about each LUT separately, while the final one takes the combination of both into account. Moreover it has to be noticed here that the reason for this differential analysis concept is determined by the possibility to track everything directly to its exact origin.

4.5.1 First Look-up-table

The analysis, which is presented in this section, features detailed information about the quality of the first LUT. For that objective, the evaluation of this LUT has to be evidently decoupled from the second one. Even if this intention seems to be complicated, such a process can be surprisingly easily implemented by assuming the second LUT to be perfect, which means that all transformed hits are assumed to encounter the correct histogram cell. So errors in the occurring histogram peaks are then directly related to the corresponding transformed hits by the first LUT. But before having a closer look to the first LUT, the focus has to be initially set to the following prerequisite, which is assumed in the analytic formula of section 2.4 on page 66.

$$\frac{p_y}{p_z} = \frac{y}{z} = constant \tag{4.1}$$

As there is further on evidently the need to check this prerequisite for a statistical amount of data, all hits of the findable MC Tracks are analyzed with regard to this relation by applying a small security region for the equality comparison. Furthermore this circumstance results obviously in a concept, which offers the computation of the number of hits satisfying this prerequisite for each findable MC Track. So keeping this feature in mind, figure 4.15 depicts a typical distribution of the relative number of findable MC Tracks, which possess a certain percentage of hits with the correct slope.

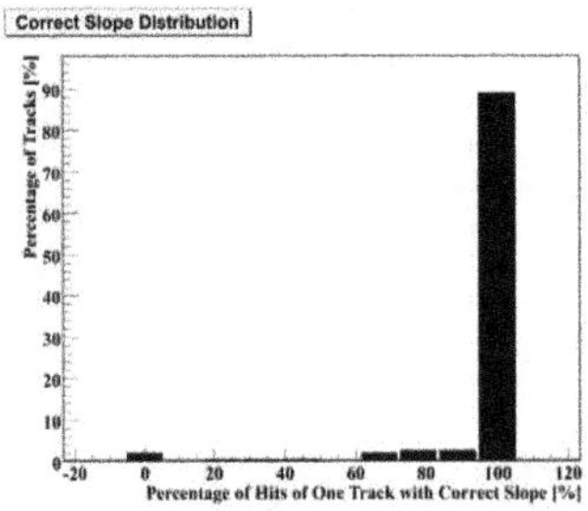

(a) Graphical front-end

```
AnalysisLIB WARNING: Distribution
of the tracks which have the percentage
of hits with the correct slope:
Hits:  0% - <  10% => #Tracks:  2% (  7)
Hits: 20% - <  30% => #Tracks:  1% (  2)
Hits: 30% - <  40% => #Tracks:  1% (  2)
Hits: 40% - <  50% => #Tracks:  1% (  2)
Hits: 60% - <  70% => #Tracks:  2% (  7)
Hits: 70% - <  80% => #Tracks:  3% (  9)
Hits: 80% - <  90% => #Tracks:  3% (  9)
Hits: 90% - <100% => #Tracks: 89% (308)
```

(b) Screen output

Figure 4.15: Resulting distribution for the correct slope analysis of the MC Tracks

With regard to this figure, it is now proved that the assumed prerequisite is applicable for almost all MC Tracks. But as few of them violate equation

4.1, it is necessary to get an imagination about these ones. Therefore figure 4.16 offers the possibility to display the projected hits of a single MC Track in the y-z coordinate plane.

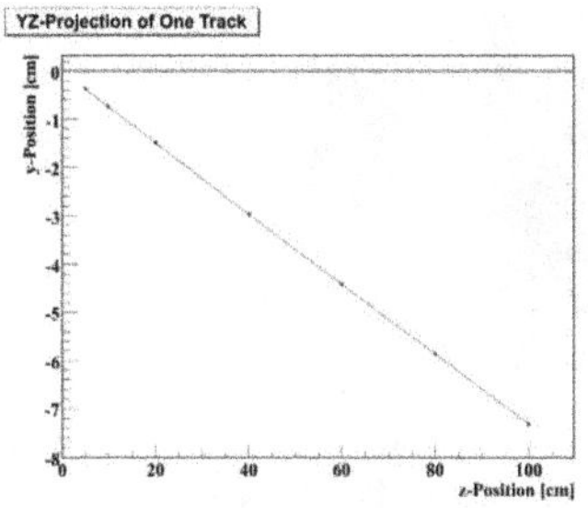

Figure 4.16: Projection of a MC Tracks's motion trajectory into the y-z-plane

Since this prerequisite is exhaustively analyzed up to here, it is feasible to change the focus of the analysis to the quality of the transformation process. For this purpose, the two initial key elements have to be noted, which are defined by the facts that each findable MC Track defines with its momentum the corresponding correct histogram layer, and that each hit of a findable MC Track can be used to evaluate the histogram layer containing the found track. So within this context, it is clear that both results have to be identical at the best, but the consistency is obviously disturbed by the transformation inexactness.
Furthermore figure 4.17 depicts the distribution of the relative number of findable MC Tracks, which exhibit a certain percentage of hits in the correct histogram layer.

Besides this, it is self-evident that the MC Tracks are as more likely to be found as more hits encounter the correct histogram layer. But with regard to the figure 4.17, it is clear that the inexactness of the transformation can lead to a result, which is quite bad. Although the LUT is tuned for the hit transformation into the correct histogram layer, it is imaginable that this is not a prerequisite of a peak, because it would be enough, if the hits encounter the same layer, which must be not the correct one. Thence the only effect of identifying such a peak will be an inexact momentum approximation for the correct found MC Track (see figure 4.27 of section 4.5.3 on page 239).

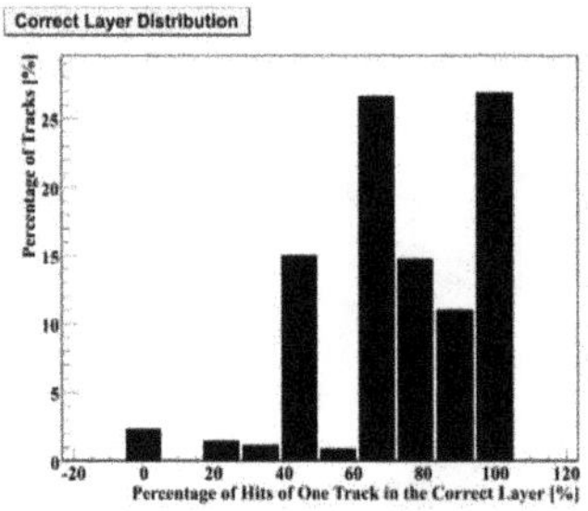

(a) Graphical front-end

```
AnalysisLIB WARNING: Distribution
of the tracks which have the percentage
of hits in the correct layer:
Hits:  0% - <  10% => #Tracks:  1% (  5)
Hits: 20% - <  30% => #Tracks:  2% (  6)
Hits: 30% - <  40% => #Tracks:  1% (  2)
Hits: 40% - <  50% => #Tracks:  6% ( 22)
Hits: 50% - <  60% => #Tracks:  1% (  4)
Hits: 60% - <  70% => #Tracks: 29% (100)
Hits: 70% - <  80% => #Tracks: 26% ( 89)
Hits: 80% - <  90% => #Tracks: 14% ( 50)
Hits: 90% - <100% => #Tracks: 20% ( 68)
```

(b) Screen output

Figure 4.17: Resulting distribution for the analysis of the correct histogram layer with regard to the first LUT

Moreover figure 4.18 shows now the distribution, which is evaluated by identifying the peak with the histogram layer containing the maximal available number of hits for the corresponding MC Track.

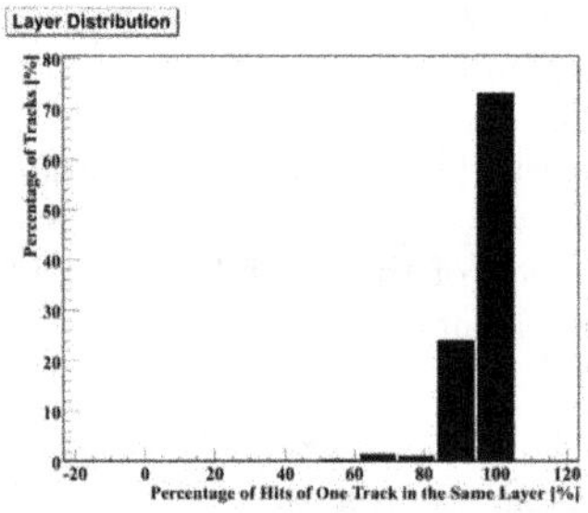

(a) Graphical front-end

```
AnalysisLIB WARNING: Distribution
of the tracks which have the percentage
of hits in one layer:
Hits: 50% - <  60% => #Tracks:  0% (  1)
Hits: 60% - <  70% => #Tracks:  1% (  5)
Hits: 70% - <  80% => #Tracks:  1% (  4)
Hits: 80% - <  90% => #Tracks: 24% ( 83)
Hits: 90% - <100% => #Tracks: 73% (253)
```

(b) Screen output

Figure 4.18: Resulting distribution for the analysis of the identified histogram layer with regard to the first LUT

So with regard to this figure, such a concept leads obviously to a result, which is much better and thus more applicable. Besides this, it should be additionally mentioned here that this analysis can be apparently also used to compare the quality of both options, which are able to generate the first LUT (see section 3.2.3 on page 133 and section 3.2.3 on page 134).

As the previous introduced concept offers the possibility to evaluate the distribution of the relative number of findable MC Tracks, which own a certain percentage of hits in the same histogram layer, the distance between the correct layer and the identification layer might be of additional interest. Hence figure 4.19 shows the computed result for a typical distribution.

```
AnalysisLIB WARNING:
The average distance of the closest layer, which contain the maximum number of
hits, to the correct layer of the MC Track is: 0.84!!!
The average distance of the farest layer, which contain the maximum number of
hits, to the correct layer of the MC Track is: 0.97!!!
The average distance of the closest layer, which contain the maximum number of
hits, to the correct layer of the MC Track is: 0.90!!!
```

Figure 4.19: Resulting output for the average distance between the correct histogram layer and the identified histogram layer of the MC Tracks

Further on, it is clear that these values have to be as small as possible, because the magnitude is directly involved in the computation of the momentum error for the found track with regard to the corresponding MC Track.

Finally it should be mentioned here that the parameter 'analysisInitPrelutGoodness', which is located in the configuration file (see appendix A.2 on page A-4), offers the possibility to enable or disable the related analysis. Corresponding to the others, the bits in the parameters 'analysisInitAnalysisResultWarnings' and 'analysisInitAnalysisResultDisplays', which are labeled with 'TRACKWITHHITWITHCORRECTSLOPEDISTRIBUTION', 'TRACKWITHYZPROJECTION', 'TRACKWITHHITINCORRECTLAYERDISTRIBUTION', 'TRACKWITHHITINBESTLAYERDISTRIBUTION' and 'DISTANCEOFBESTANDCORRECTLAYER' (see appendix A.3 on page A-79), control the printout to the standard output screen and the display containing the graphical front-ends.

4.5.2 Second Look-up-table

The analysis, which is presented in this section, reveals detailed information about the quality of the second LUT. For that purpose, the evaluation of this LUT has to be evidently decoupled from the first one analog to the previous section. Thus the first LUT is simply assumed to be perfect here, which means that all transformed hits are assumed to encounter the correct histogram layer. So errors in the occurring histogram peaks are then obviously directly related to the corresponding transformed hits by the second LUT.

Further on, the two initial key elements, which have to be noticed in analogousness to the previous section, are apparently determined by the two facts that each findable MC Track defines with its momentum the corresponding correct histogram cell, and that each hit of a findable MC Track can be used to evaluate the histogram cell containing the found track. So within this context, it is again clear that both results have to be also identical at the best, but the consistency is as well evidently disturbed by the transformation inexactness.
Furthermore figure 4.20 depicts the distribution of the relative number of findable MC Tracks with regard to the hit signatures, which occur in the correct histogram cell defined by the momentum of each MC Track.

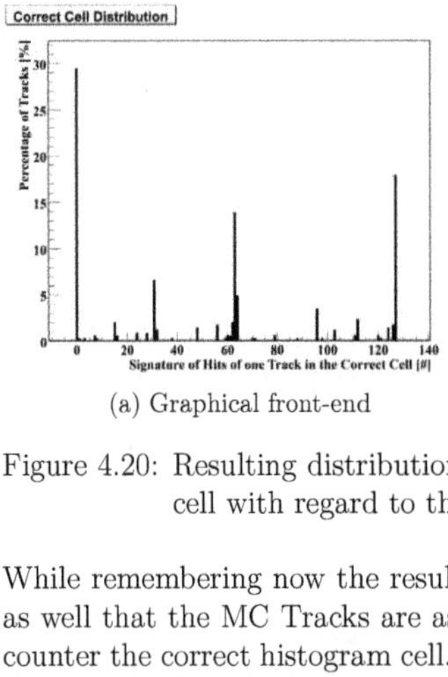

(a) Graphical front-end

```
AnalysisLIB WARNING: Distribution
of the tracks which have the signature
of hits in the correct cell:
...
Signature: 2x0000000 => #Tracks: 27% ( 95)
Signature: 2x0000001 => #Tracks:  3% ( 12)
Signature: 2x0001111 => #Tracks:  3% ( 11)
Signature: 2x0011000 => #Tracks:  3% ( 11)
Signature: 2x0011111 => #Tracks: 16% ( 55)
Signature: 2x0100000 => #Tracks:  7% ( 25)
Signature: 2x0111111 => #Tracks: 11% ( 38)
Signature: 2x1111111 => #Tracks:  3% (  9)
```

(b) Screen output

Figure 4.20: Resulting distribution for the analysis of the correct histogram cell with regard to the second LUT

While remembering now the results for the first LUT, it is manifested here as well that the MC Tracks are as more likely to be found as more hits encounter the correct histogram cell. So analog to the results there, figure 4.20 shows again that the inexactness of the transformation can lead to a result, which is quite terrible. Although this LUT is apparently also tuned for the hit transformation into the correct histogram cell, it is in turn imaginable that this is not a prerequisite of a peak, because it would be enough, if the hits encounter the same cell, which must be not the correct one. Thence the only effect of identifying such a peak will be once again an inexact momentum approximation for the correct found MC Track (see figure 4.27 of section 4.5.3 on page 239).
Moreover figure 4.21 shows now the distribution, which is evaluated by identifying the peak with the histogram cell containing the maximal available number of hits for the corresponding MC Track.

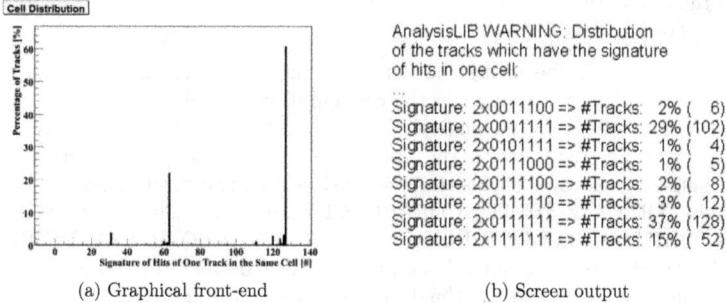

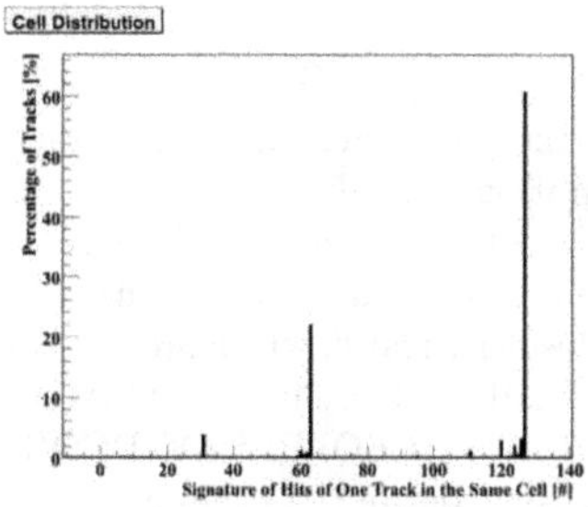

(a) Graphical front-end

```
AnalysisLIB WARNING: Distribution
of the tracks which have the signature
of hits in one cell:
...
Signature: 2x0011100 => #Tracks:  2% (  6)
Signature: 2x0011111 => #Tracks: 29% (102)
Signature: 2x0101111 => #Tracks:  1% (  4)
Signature: 2x0111000 => #Tracks:  1% (  5)
Signature: 2x0111100 => #Tracks:  2% (  8)
Signature: 2x0111110 => #Tracks:  3% ( 12)
Signature: 2x0111111 => #Tracks: 37% (128)
Signature: 2x1111111 => #Tracks: 15% ( 52)
```

(b) Screen output

Figure 4.21: Resulting distribution for the analysis of the identified histogram cell with regard to the second LUT

So with regard to that figure, this concept leads evidently in turn to a result, which is much better and thus more applicable. Besides this, it should be additionally mentioned here that this analysis can be obviously also used to compare the quality of both options, which are able to generate the second LUT (see section 3.2.3 on page 133 and section 3.2.3 on page 134).

As the previous introduced concept offers the possibility to evaluate the distribution of the relative number of findable MC Tracks with regard to the maximal available number of hits in the same histogram cell, the average distance between the correct cell and the identification cell might be of additional interest. Thus figure 4.22 shows the computed result for a typical distribution.

```
AnalysisLIB WARNING:
The average distance of the closest cell, which contain the maximum number of
hits, to the correct cell of the MC Track is: 1.70!!!
The average distance of the farest cell, which contain the maximum number of
hits, to the correct cell of the MC Track is: 2.88!!!
The average distance of the closest cell, which contain the maximum number of
hits, to the correct cell of the MC Track is: 2.18!!!
```

Figure 4.22: Resulting output for the average distance between the correct histogram cell and the identified histogram cell of the MC Tracks

Further on, it is in turn self-evident that these values have to be as small as possible, because the magnitude is also directly involved in the computation of the momentum error for the found track with regard to the corresponding MC Track.

Finally it should be mentioned here that the parameter 'analysisInitLut-Goodness', which is located in the configuration file (see appendix A.2 on page A-4), offers the possibility to enable or disable the related analysis. Corresponding to the others, the bits in the parameters 'analysisInitAnalysis-ResultWarnings' and 'analysisInitAnalysisResultDisplays', which are labeled with 'TRACKWITHHITINCORRECTCELLDISTRIBUTION', 'TRACK-WITHHITINCELLDISTRIBUTION' and 'DISTANCEOFBESTANDCOR-RECTCELL' (see appendix A.3 on page A-79), control the printout to the standard output screen and the display containing the graphical front-ends.

4.5.3 Transformation

As mentioned in the introduction, the final analysis presents now a quality measurement, which takes both LUTs into account. But before having a closer look to the result, it is important to analyze the hit signature of the MC Tracks to get an imagination about the best result, which is achievable (see section 3.1 on page 96). For this purpose, the idea is analog to the previous sections, which contain the analysis for the first LUT and the second LUT, except the difference that the hit signatures are independent of any transformation characteristic.
Hence figure 4.23 depicts the hit signature distribution of the findable MC Tracks.

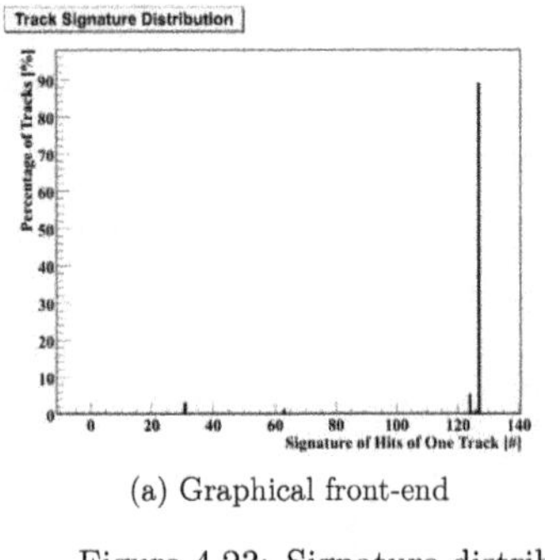

(a) Graphical front-end

```
AnalysisLIB WARNING: Distribution
of the tracks which have the signature
of hits:
...
Signature: 2x0011111 => #Tracks:  3% ( 10)
Signature: 2x0111111 => #Tracks:  1% (  5)
Signature: 2x1011111 => #Tracks:  0% (  1)
Signature: 2x1111100 => #Tracks:  5% ( 18)
Signature: 2x1111101 => #Tracks:  0% (  1)
Signature: 2x1111110 => #Tracks:  1% (  3)
Signature: 2x1111111 => #Tracks: 89% (308)
```

(b) Screen output

Figure 4.23: Signature distribution of the MC Tracks for an event

So at a first glance, it is clear that the distribution is very good, because almost all MC Tracks possess a hit in each detector station. Nevertheless each hit of a findable MC Track has to be evidently transformed to evaluate the computed hit signature in the corresponding correct histogram cell analog to the previous sections. However the transformation inexactness is here contrarily not the only analysis focus, because the comparison of the result to figure 4.23 allows a more reasonable interpretation of the quality aspect. For that reason, figure 4.24 depicts now the distribution of the relative number of findable MC Tracks with regard to the hit signatures, which occur in the correct histogram cell defined by the momentum of each MC Track.

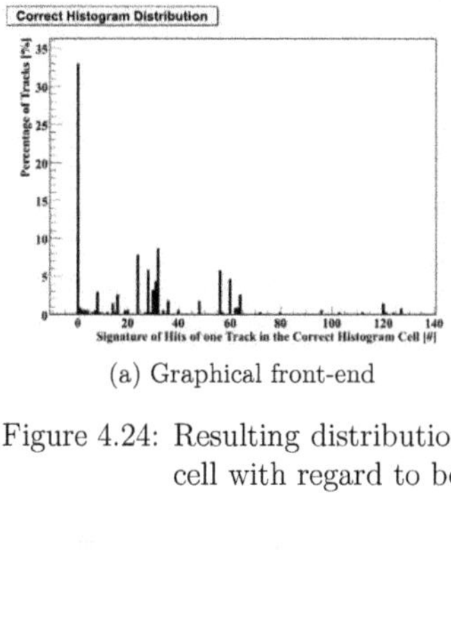

(a) Graphical front-end

```
AnalysisLIB WARNING: Distribution
of the tracks which have the signature
of hits in the correct histogram cell:
...
Signature: 2x0000000 => #Tracks: 33% (114)
Signature: 2x0001000 => #Tracks:  3% ( 10)
Signature: 2x0011000 => #Tracks:  8% ( 27)
Signature: 2x0011100 => #Tracks:  6% ( 20)
Signature: 2x0011111 => #Tracks:  4% ( 15)
Signature: 2x0100000 => #Tracks:  9% ( 30)
Signature: 2x0111000 => #Tracks:  6% ( 20)
Signature: 2x0111100 => #Tracks:  5% ( 16)
```

(b) Screen output

Figure 4.24: Resulting distribution for the analysis of the correct histogram cell with regard to both LUTs

Remembering further the results of both uniquely analyzed LUTs, it is again assured that the MC Tracks are as more likely to be found as more hits encounter the correct histogram cell. However since this result is getting surely worse, it is also in turn obvious that the hits must not encounter the correct cell, but rather sufficiently the same cell. Furthermore it has to be notified once again that the only effect of identifying such a peak will be an inexact momentum approximation for the correct found MC Track (see figure 4.27 of section 4.5.3 on page 239).
Therefore figure 4.25 shows the distribution, which is evaluated by identifying

the peak with the histogram cell containing the maximal available number of hits for the corresponding MC Track.

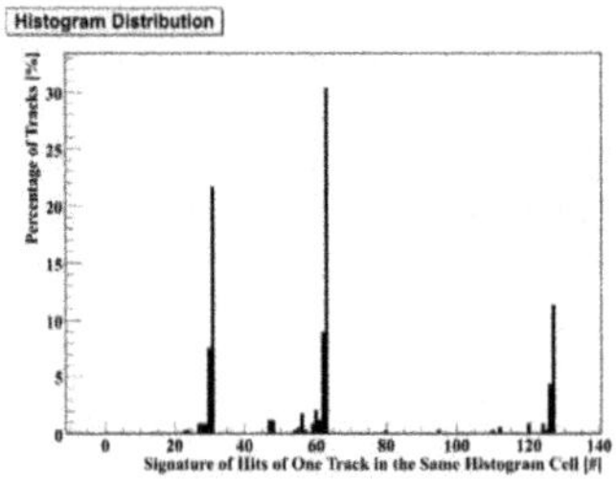

(a) Graphical front-end

```
AnalysisLIB WARNING: Distribution
of the tracks which have the signature
of hits in one histogram cell:
...
Signature: 2x0011110 => #Tracks:  8% ( 26)
Signature: 2x0011111 => #Tracks: 22% ( 75)
Signature: 2x0111000 => #Tracks:  2% (  6)
Signature: 2x0111100 => #Tracks:  2% (  7)
Signature: 2x0111110 => #Tracks:  9% ( 31)
Signature: 2x0111111 => #Tracks: 30% (105)
Signature: 2x1111110 => #Tracks:  4% ( 15)
Signature: 2x1111111 => #Tracks: 11% ( 39)
```

(b) Screen output

Figure 4.25: Resulting distribution for the analysis of the identified histogram cell with regard to both LUTs

So with regard to that figure, this concept leads over and over again to a result, which is much better and thus more applicable. Besides this, it should be additionally noticed here that the presented analysis can be also adducted to automatically generate table files by applying an algorithm, which is chosen by the parameter value 'ONLINETABLE' (see appendix A.4 on page A-101).

As the previous introduced concept offers the possibility to evaluate the distribution of the relative number of findable MC Tracks with regard to the maximal available number of hits in the same histogram cell, the average distance between the correct cell and the identification cell might be of additional interest. Thus figure 4.26 shows the computed result for a typical distribution.

```
AnalysisLIB WARNING:
The average distance of the closest histogram cell, which contain the maximum
number of hits, to the correct histogram cell of the MC Track is: 2.10!!!
The average distance of the farest histogram cell, which contain the maximum
number of hits, to the correct histogram cell of the MC Track is: 2.93!!!
The average distance of the histogram cell, which contain the maximum
number of hits, to the correct histogram cell of the MC Track is: 2.45!!!
```

Figure 4.26: Resulting output for the average distance between the correct histogram cell in the correct histogram layer and the identified histogram cell of the MC Tracks

Beyond that, it is once again self-evident that these values have to be as small as possible, because the magnitude is directly involved in the computation of the momentum error for the found track with regard to the corresponding MC Track. But nevertheless the most impressive aspect is apparently determined by the distribution, which displays the momentum errors, because the general nature of quality is given by the comparison of the major characteristic for each reconstructed track to its corresponding MC Track. Hence a good imagination about such a distribution is given in figure 4.27, because it depicts a typical one.

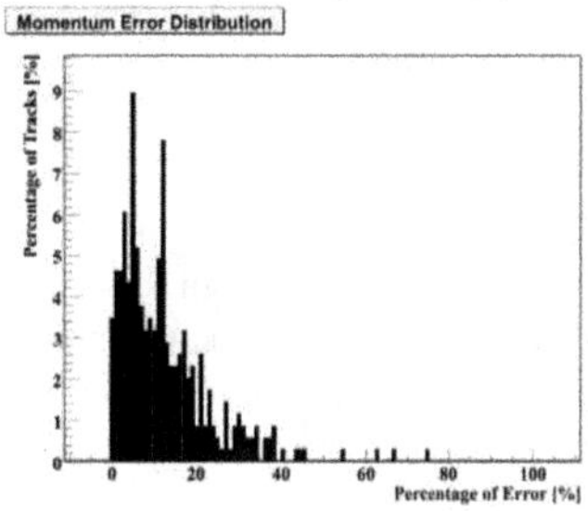

(a) Graphical front-end

```
AnalysisLIB WARNING: Distribution
of the tracks which have the percentage
of error in the momentum:
Error:  0% - <  1% => #Tracks:  9% ( 31)
Error:  1% - <  2% => #Tracks: 19% ( 67)
Error:  2% - <  3% => #Tracks: 12% ( 41)
Error:  3% - <  4% => #Tracks: 14% ( 48)
Error:  4% - <  5% => #Tracks: 10% ( 35)
Error:  5% - <  6% => #Tracks:  7% ( 24)
Error:  6% - <  7% => #Tracks:  7% ( 25)
Error:  7% - <  8% => #Tracks:  4% ( 15)
Error:  8% - <  9% => #Tracks:  3% ( 10)
...
```

(b) Screen output

Figure 4.27: Relative momentum error distribution for an event evaluated by the momentum comparison of each found track with the corresponding MC Track

Finally it should be also mentioned here that the parameter 'analysisInitHoughTransformGoodness', which is located in the configuration file (see appendix A.2 on page A-4), offers the possibility to enable or disable the related analysis. Corresponding to the others, the bits in the parameters 'analysisInitAnalysisResultWarnings' and 'analysisInitAnalysisResultDisplays', which are labeled with 'TRACKWITHHITINCORRECTHISTOGRAMDISTRIBUTION', 'TRACKWITHHITINHISTOGRAMDISTRIBUTION', 'DISTANCEOFBESTANDCORRECTHISTOGRAM', 'TRACKWITHSIGNATUREDISTRIBUTION' and 'TRACKWITHMOMENTUMERRORDISTRIBUTION' (see appendix A.3 on page A-79), control the printout to the standard output screen and the display containing the graphical front-ends. In addition to this, the parameter 'analysisInitTotalAnalysis', which is located also in the configuration file (see appendix A.2 on page A-4), features the accumulation of all distances outlined in figure 4.26 for multiple events.

4.6 Histogram

As the central part of the Hough transform, which is introduced in section 2.1 on page 45, is identified by the histogram, it is obviously essential to get an imagination about each aspect of the implementation. Thus the next sections present different analyses, which summarize information about the quantization, the corresponding peak distances, the resulting occupancy and the layers itself. Referring further to other sections, it is important to notice that the define statement 'DEBUGJUSTONEGOODTRACK' in the source code file MiscLIB/include/defs.h features the evaluation of all analysis for just a single well-defined findable MC Track.

4.6.1 Quantization

Before presenting the analysis for the quantization of the Hough space, it has to be at first noticed that this aspect identifies the primary key element of the algorithm, because it is directly connected to the required amount of resources, the peak separation ability and thus the tracking quality (see section 4.7 on page 258).

So while concerning now the imaginable consequences of this aspect, it is a common approach to simply quantize the Hough space with the same resolution than the original space. But even if the original coordinate space of the experimental setup will be fixed in the future (see section 3.3.1 on page 156), such an approach would be not suitable here, because the resulting quantization of 29 bit would require $2^{(13+13)} \rightarrow 64\,\mathrm{M} \approx 67 \cdot 10^6$ histogram cells (x and y coordinate) containing each $2^3 \rightarrow 8\,\mathrm{bit}$ signatures (z coordinate). So in addition to this unrealizable amount of $512\,\mathrm{M} \approx 537 \cdot 10^6$ flip-flops or 64 MB memory, another problem is determined by the actually not quantized momentum space.
Hence it is necessary to quantize the histogram independent of the original space, but dependent of some other requirements.

For this purpose, there are two different objectives presented, which pay attention to the peak separation ability in the Hough space and the resolution of the reconstructed momentum. In contrast to this, the required amount of resources for the histogram are of minor importance, because it has to be at least big enough to fulfill the tracking task.
Setting the focus now to the peak separation ability related to the MC Tracks of a sample event, figure 4.28 depicts a typical result, which is projected into each dimension of the histogram. As usual, this figure is produced by the pa-

rameter 'analysisInitQuantizationGoodness' in the configuration file (see appendix A.2 on page A-4), while the output is controlled by setting the corresponding bit, which is labeled with 'QUANTIZEDHOUGHSPACEDIMDISTRIBUTION', in the parameters 'analysisInitAnalysisResultWarnings' and 'analysisInitAnalysisResultDisplays' (see appendix A.3 on page A-79). Furthermore it is self-evident that the first parameter enables the results to be printed to the standard output screen while the second one displays graphical front-ends.

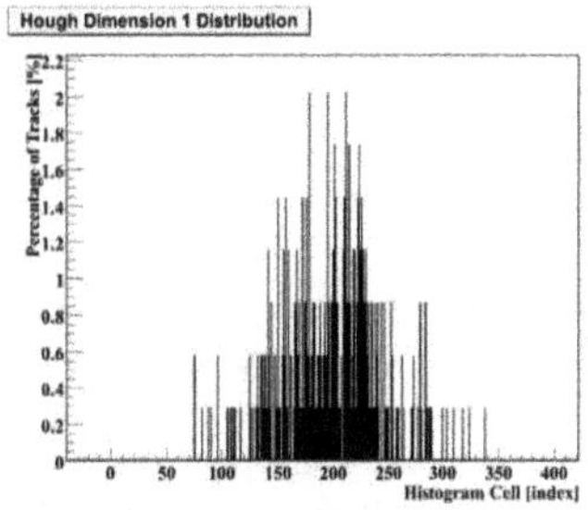

(a) Dim1 graphical front-end

```
AnalysisLIB WARNING:
Distribution of the tracks which
have the Hough space dim1:
...
Index: 176 => #Tracks:  1% (  5)
Index: 177 => #Tracks:  1% (  2)
Index: 178 => #Tracks:  2% (  7)
Index: 179 => #Tracks:  0% (  1)
Index: 180 => #Tracks:  1% (  2)
...
```

(b) Dim1 standard output screen

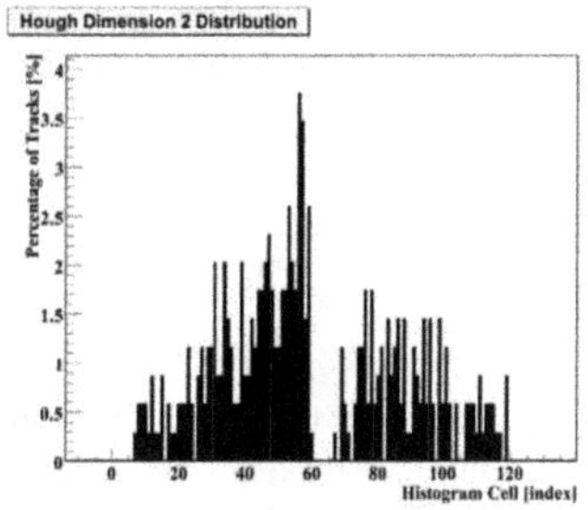

(c) Dim2 graphical front-end

```
AnalysisLIB WARNING:
Distribution of the tracks which
have the Hough space dim2:
...
Index:  55 => #Tracks:  1% (  5)
Index:  56 => #Tracks:  3% ( 11)
Index:  57 => #Tracks:  3% ( 10)
Index:  58 => #Tracks:  1% (  4)
Index:  59 => #Tracks:  2% (  7)
...
```

(d) Dim2 standard output screen

Figure 4.28: Resulting distributions for the quantization analysis of the projected Hough space dimensions

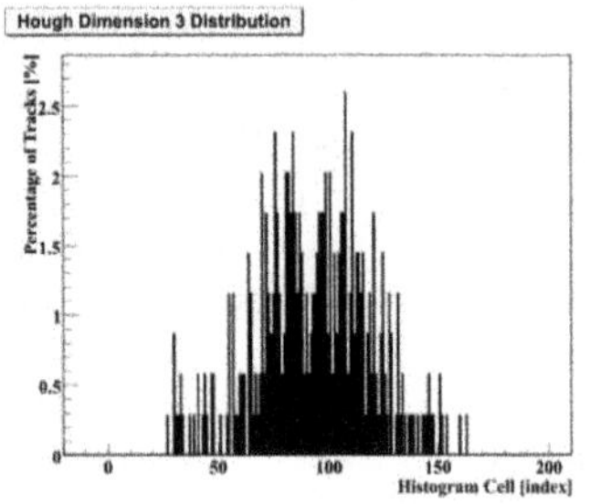

(e) Dim3 graphical front-end

```
AnalysisLIB WARNING:
Distribution of the tracks which
have the Hough space dim3:
...
Index: 107 => #Tracks:  2% (  6)
Index: 108 => #Tracks:  3% (  9)
Index: 109 => #Tracks:  1% (  2)
Index: 110 => #Tracks:  1% (  4)
Index: 111 => #Tracks:  2% (  8)
...
```

(f) Dim3 standard output screen

Figure 4.28: Resulting distributions for the quantization analysis of the projected Hough space dimensions

Having now a closer look to figure 4.28, it is clear that multiple MC Tracks could possibly encounter the same histogram cell, if the three dimensional histogram is projected into a single dimension. Although such a result is not very well, because these MC tracks can be evidently only identified just once, this circumstance is at a first glance negligible, because it is not sure that the MC Tracks would encounter also the same histogram cell, if the unprojected histogram is studied. However such an investigation is not needless, because an improvement here will directly lead to an improvement in the unprojected histogram.
Besides this, it has to be additionally mentioned here that another related analysis, which exhibits more information about the quantization dependent geometry of a peak, can be found in section 4.8.4 on page 279.

Beyond that, figure 4.29 shows now a typical result for the reconstructed momentum, which base on the same quantization than used for the evaluation of figure 4.28. Even if such a result is still worse, it is obvious that the same rules than before count here. Accordingly the production of this figure can be enabled also by the parameter 'analysisInitQuantizationGoodness' in the configuration file (see appendix A.2 on page A-4) and the bit, which is labeled with 'QUANTIZEDMOMENTADIMDISTRIBUTION', in the parameters 'analysisInitAnalysisResultWarnings' and 'analysisInitAnalysisResultDisplays' (see appendix A.3 on page A-79).

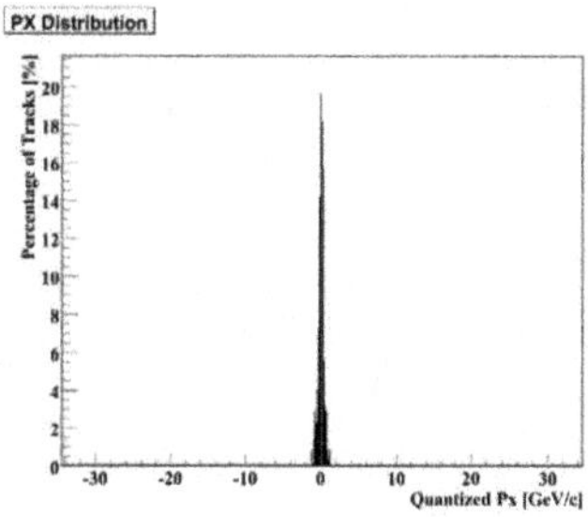

(a) P_x graphical front-end

```
AnalysisLIB WARNING:
Distribution of the tracks which
have the momenta [GeV/c] in X:
...
Momentum: -0.2998 => #Tracks:  7% ( 25)
Momentum: -0.1499 => #Tracks: 14% ( 49)
Momentum: -0.0000 => #Tracks: 20% ( 68)
Momentum:  0.1499 => #Tracks: 18% ( 63)
Momentum: -0.2998 => #Tracks: 16% ( 54)
...
```

(b) P_x standard output screen

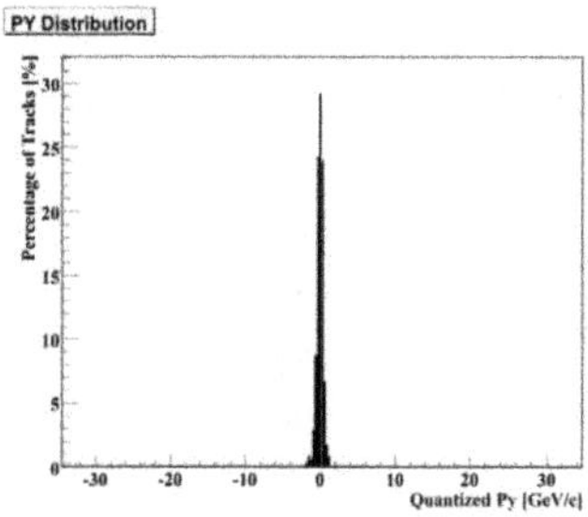

(c) P_y graphical front-end

```
AnalysisLIB WARNING:
Distribution of the tracks which
have the momenta [GeV/c] in Y:
...
Momentum: -0.6028 => #Tracks:  9% ( 30)
Momentum: -0.3014 => #Tracks: 24% ( 84)
Momentum: -0.0000 => #Tracks: 29% (101)
Momentum:  0.3014 => #Tracks: 24% ( 83)
Momentum: -0.6028 => #Tracks:  7% ( 23)
...
```

(d) P_y standard output screen

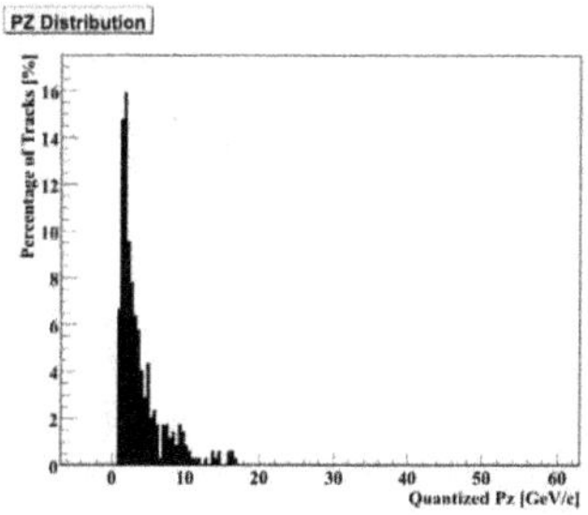

(e) P_z graphical front-end

```
AnalysisLIB WARNING:
Distribution of the tracks which
have the momenta [GeV/c] in Z:
Momentum:  0.9000 => #Tracks:  7% ( 23)
Momentum:  1.3428 => #Tracks: 15% ( 51)
Momentum:  1.7857 => #Tracks: 16% ( 55)
Momentum:  2.2285 => #Tracks: 10% ( 33)
Momentum:  2.6714 => #Tracks:  8% ( 27)
Momentum:  3.1142 => #Tracks:  6% ( 22)
...
```

(f) P_z standard output screen

Figure 4.29: Resulting distributions for the projected momentum component analysis of the MC Tracks

As the details are now satisfyingly analyzed, the focus is changed to the unprojected results. Therefore all figures have to be combined into a three dimensional representation. But because of the limited amount of display dimensions, the visualization is flattened again by simply concatenating the dimensions. Thus figure 4.30 depicts this distribution for the histogram cells, which implies the peak separation ability, and the distribution for the reconstructed momentum, which implies the resulting resolution, with regard to the MC data of a common event. Within this concept, it is immediately clear that this output is also controlled by the usage of the parameter 'analysisInitQuantizationGoodness' in the configuration file (see appendix A.2 on page A-4) and the setting of the corresponding bits, which are labeled with 'QUANTIZEDHOUGHSPACEDISTRIBUTION' and 'QUANTIZED-

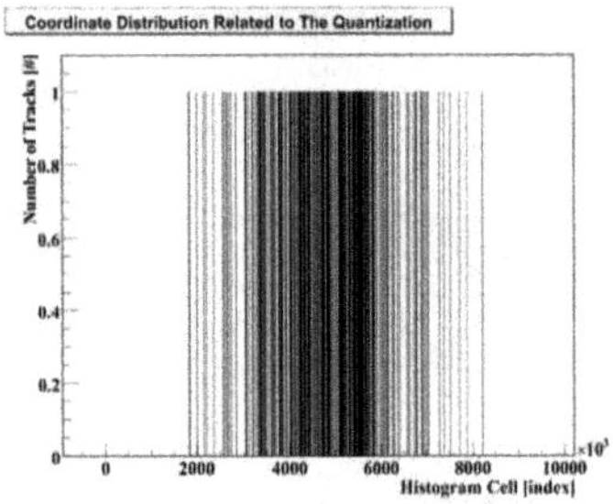

(a) Hough space graphical front-end

```
AnalysisLIB WARNING: Distribution
of the tracks which have the Hough space:
...
Index: 4620014 => #Tracks:   1
Index: 4638182 => #Tracks:   1
Index: 4639656 => #Tracks:   1
Index: 4667588 => #Tracks:   1
Index: 4668153 => #Tracks:   1
...
```

(b) Hough space standard output screen

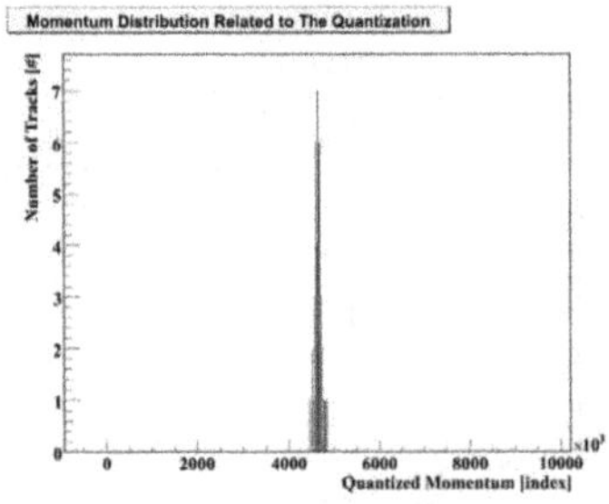

(c) $\vec{P}$ graphical front-end

```
AnalysisLIB WARNING: Distribution
of the tracks which have the momenta:
...
Index: 4645279 => #Tracks:   1
Index: 4645280 => #Tracks:   7
Index: 4645281 => #Tracks:   7
Index: 4645282 => #Tracks:   3
Index: 4645283 => #Tracks:   1
...
```

(d) $\vec{P}$ standard output screen

Figure 4.30: Resulting distributions for the Hough space quantization and momentum vector analysis with regard to the MC Tracks of an event

MOMENTADISTRIBUTION', in the parameters 'analysisInitAnalysisResultWarnings' and 'analysisInitAnalysisResultDisplays' (see appendix A.3 on page A-79). Furthermore it is in turn self-evident that the first parameter enables the results to be printed to the standard output screen while the second one displays graphical front-ends.

With regard to this figure, the major important fact, which can be seen there, is evidently that the used quantization is not to small, because multiple MC Tracks never encounter the same histogram cell. However it has to be kept in mind that possibly a smaller quantization might be also able to offer this characteristic and another momentum distribution of the MC Tracks might affect it.

Beyond this, the relation of the complexity of the produced figures to the only really necessary information leads naturally to the usage of two simple summarizing warning messages, which are shown in figure 4.31. These warning messages can be further on enabled by setting the corresponding bits, which are labeled with 'SAMEHOUGHSPACECELLDISTRIBUTION' and 'NUMBEROFTRACKSWHICHCANNOTBEFOUND', in the parameter 'analysisInitAnalysisResultWarnings' (see appendix A.3 on page A-79) in addition to the usage of the parameter 'analysisInitQuantizationGoodness' in the configuration file (see appendix A.2 on page A-4).

AnalysisLIB WARNING: Distribution of the tracks which have the same histogram cell:

No histogram cells consisting of more than one track found!!!

(a) Distribution of the histogram cells containing more than a single MC Track

AnalysisLIB WARNING: There are 0 tracks which cannot be found, because they are in the same histogram cell as another track!!!

(b) Number of MC Tracks occurring in an already frequented histogram cell

Figure 4.31: Resulting output for the quantization analysis with regard to the not findable MC Tracks

Since the optimal quantization is obviously determined by the best trade-off between the required histogram resources and the surveyed distribution of the histogram cells containing more than a single MC Tracks for multiple events, a general optimal quantization is evidently not realizable, but statistics help naturally to define a quite good one.
For this purpose, the parameter 'analysisInitTotalAnalysis' in the configuration file (see appendix A.2 on page A-4), which has to be used in addition to the parameter 'analysisInitQuantizationGoodness' and the bit 'NUMBEROFTRACKSWHICHCANNOTBEFOUND' in the parameter 'analy-

sisInitAnalysisResultWarnings' (see appendix A.3 on page A-79), enables the production of a warning message presenting the combined result of the not findable MC Tracks for multiple events. In this connection, it is apparently clear that a not findable MC Track is defined by the characteristic that it encounters the same histogram cell than another MC Track.
Besides this, such a message is not depicted here, because it is quite similar to the one shown in figure 4.31b with the only difference in the presented number. Nevertheless figure 4.32 depicts the distribution of this warning message with regard to different identifiers, which represent a predefined quantization configuration of the histogram dimensions.

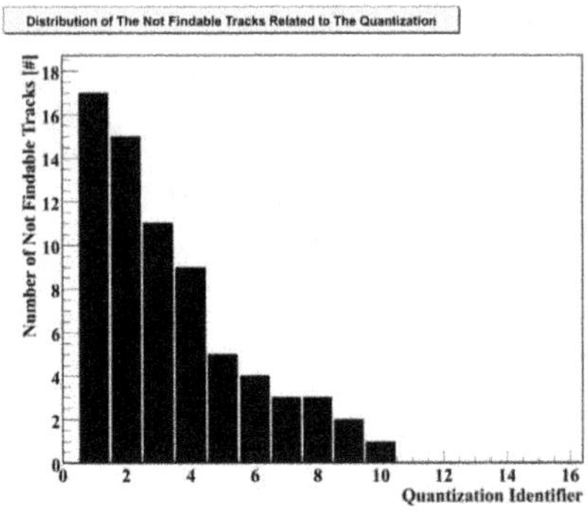

Figure 4.32: Resulting distribution for the not findable MC Tracks with regard to different quantizations

Further on, the production of this figure can be again simply enabled by the parameter 'analysisInitTotalAnalysis' in the configuration file (see appendix A.2 on page A-4). To this, the graphical front-end is controlled by setting the corresponding bit, which is labeled with 'NUMBEROFTRACKSWHICHCANNOTBEFOUND', in the parameter 'analysisInitAnalysisResultDisplays' (see appendix A.3 on page A-79). Finally it should be also mentioned here that the usage of this bit in the parameter 'analysisInitAnalysisResultWarnings' enables completely different features.
Beyond that, the content of this figure can be evidently not produced automatically. Thus it depicts just a visualization of user evaluated statistics for each quantization, which are computed by the previous introduced warning message and setup in the source code file AnalysisLIB/include/totalAnalysisDef.h.

4.6.2 Peak Distance

Although the analysis, which is presented in section 4.6.1 on page 240, offers the possibility to evaluate the theoretical minimal histogram quantization limit with regard to the peak separation ability, it is clear that this is not the last word on this subject, because it is quite hard to separate peaks, which are located next to each other so that they comprise overlapping clusters. Hence the distance of the peaks caused by the MC Tracks define obviously a further measurable quality aspect, which leads to a more practicable peak separation ability. For this purpose, the analysis, which is presented in this section, is able to give detailed information about diverse aspects of the peak distance.

So at the beginning, figure 4.33 depicts at first a typical peak distance distribution, whose number of entries would be apparently computed to $\sum_{k=0}^{n} k = \frac{n\cdot(n+1)}{2}$ (see [BSMM05] on page 19), if n represents the number of findable MC Tracks. Further on, this figure is naturally likewise produced and shown by the usage of the parameter 'analysisInitPeakDistanceGoodness' in the configuration file (see appendix A.2 on page A-4) and the setting of the corresponding bit, which is labeled with 'PEAKDISTANCEDISTRIBUTION', in the parameters 'analysisInitAnalysisResultWarnings' and 'analysisInitAnalysisResultDisplays' (see appendix A.3 on page A-79). Furthermore it is self-evident that the first parameter enables the results to be printed to the standard output screen while the second one displays graphical front-ends.

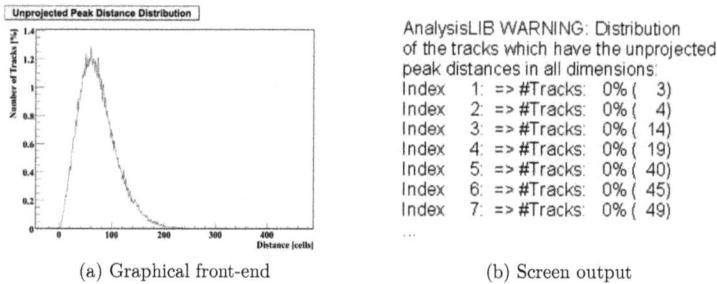

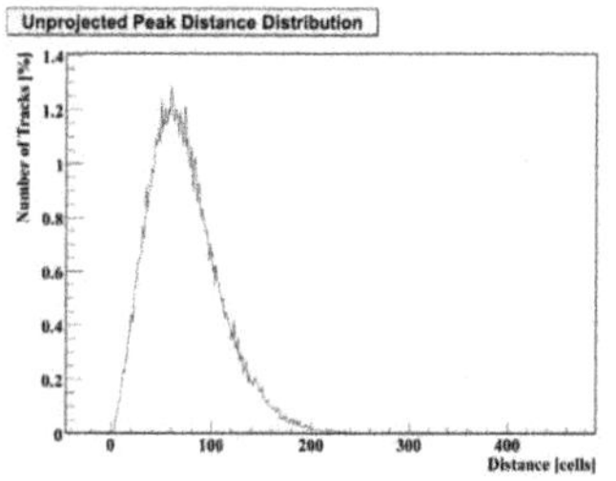

(a) Graphical front-end

```
AnalysisLIB WARNING: Distribution
of the tracks which have the unprojected
peak distances in all dimensions:
Index   1: => #Tracks:  0% (  3)
Index   2: => #Tracks:  0% (  4)
Index   3: => #Tracks:  0% ( 14)
Index   4: => #Tracks:  0% ( 19)
Index   5: => #Tracks:  0% ( 40)
Index   6: => #Tracks:  0% ( 45)
Index   7: => #Tracks:  0% ( 49)
...
```

(b) Screen output

Figure 4.33: Resulting distribution for the unprojected peak distances

With regard to this figure, the essence is apparently determined by the circumstance that the more distant the peaks are from each other, the more easy is the Peak-finding process, because it is more likely that peak clusters do not overlap. So at a first glance, it seems to be good to take the highest realizable quantization. But a closer look to higher quantizations illustrates also the side effect of degrading peak altitudes, which leads immediately to a worsened Peak-finding geometry (see section 4.8.4 on page 279) and a downgraded Peak-finding result. Therefore it is self-evident that these antidromic characteristics have to be taken into account for a well-configured quantization.

Returning now to figure 4.33, and especially to the warning message shown in figure 4.33b, it is proven that the chosen quantization leads to a peak separation, which is quite good, because the small distances occur less frequent. Nevertheless a more detailed impression can be obtained by each peak distance distribution, which is separately projected into a single dimension. Thus the figures 4.34a, 4.34b, 4.34c, 4.34d, 4.34e and 4.34f, which depict the corresponding results, are also important. Referring further to other sections, these figures are produced by the parameter 'analysisInitPeakDistanceGoodness' in the configuration file (see appendix A.2 on page A-4), while the output is controlled by setting the corresponding bit, which is labeled with 'PEAKDISTANCEDIMDISTRIBUTION', in the parameters 'analysisInitAnalysisResultWarnings' and 'analysisInitAnalysisResultDisplays' (see appendix A.3 on page A-79). Furthermore it is self-evident that the first parameter enables the results to be printed to the standard output screen while the second one displays graphical front-ends.

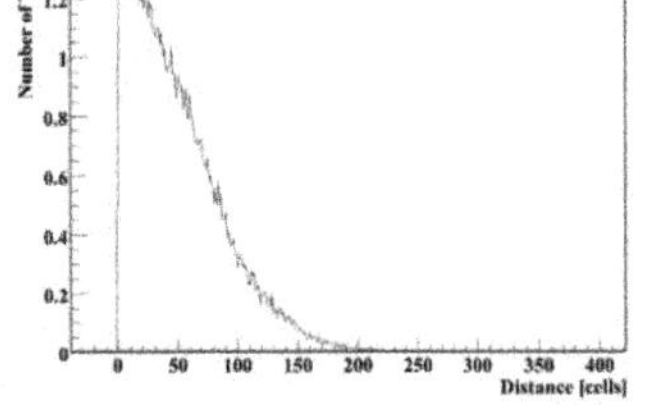

(a) Dim1 graphical front-end

```
AnalysisLIB WARNING: Distribution
of the tracks which have the peak distances in dim1:
Index  0: => #Tracks:  0% (385)
Index  1: => #Tracks:  1% (806)
Index  2: => #Tracks:  1% (804)
Index  3: => #Tracks:  1% (776)
Index  4: => #Tracks:  1% (783)
Index  5: => #Tracks:  1% (760)
Index  6: => #Tracks:  1% (771)
Index  7: => #Tracks:  1% (775)
...
```

(b) Dim1 standard output screen

Figure 4.34: Resulting distribution for the projected peak distances

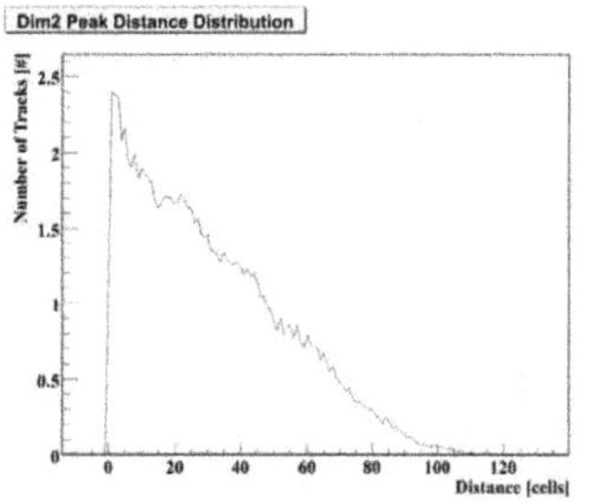

(c) Dim2 graphical front-end

```
AnalysisLIB WARNING: Distribution
of the tracks which have the peak distances in dim2:
Index   0:  => #Tracks:  1% ( 713)
Index   1:  => #Tracks:  2% (1435)
Index   2:  => #Tracks:  2% (1421)
Index   3:  => #Tracks:  2% (1404)
Index   4:  => #Tracks:  2% (1241)
Index   5:  => #Tracks:  2% (1289)
Index   6:  => #Tracks:  1% (1158)
Index   7:  => #Tracks:  1% (1136)
...
```

(d) Dim2 standard output screen

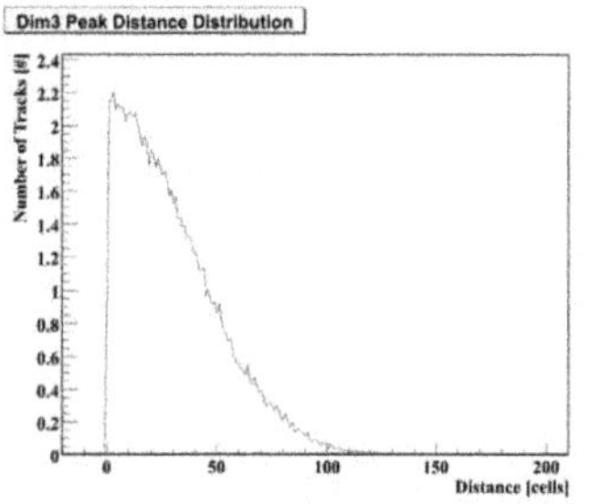

(e) Dim3 graphical front-end

```
AnalysisLIB WARNING: Distribution
of the tracks which have the peak distances in dim3:
Index   0:  => #Tracks:  1% ( 622)
Index   1:  => #Tracks:  2% (1282)
Index   2:  => #Tracks:  2% (1283)
Index   3:  => #Tracks:  2% (1317)
Index   4:  => #Tracks:  2% (1252)
Index   5:  => #Tracks:  2% (1271)
Index   6:  => #Tracks:  2% (1256)
Index   7:  => #Tracks:  2% (1263)
...
```

(f) Dim3 standard output screen

Figure 4.34: Resulting distribution for the projected peak distances

Although these figures show obviously worsened distributions compared to the unprojected one, this circumstance is again negligible, because just the unprojected distribution counts. But if one can improve such a projected distribution, this would lead apparently directly to an improvement in the unprojected one.
Beyond that, it is clear that there is in turn the requirement for a measurable aspect, which is related to the quantization and can be evaluated for multiple events. Hence the usage of the parameter 'analysisInitPeakDistanceGoodness' in the configuration file (see appendix A.2 on page A-4) and the bit, which is labeled with 'AVERAGEPEAKDISTANCE', in the parameter 'analysisInitAnalysisResultWarnings' (see appendix A.3 on page A-79) enable the production of the warning messages, which are depicted in figure 4.35.

Since this figure informs about the average peak distance for the unprojected histogram in combination with each available projection, such simple

AnalysisLIB WARNING: The average peak distance in dim1 is
50 cells, which is about 13% of the total available amount!!!

AnalysisLIB WARNING: The average peak distance in dim2 is
31 cells, which is about 25% of the total available amount!!!

AnalysisLIB WARNING: The average peak distance in dim3 is
30 cells, which is about 16% of the total available amount!!!

AnalysisLIB WARNING: The average unprojected peak distance is
74 cells, which is about 17% of the total available amount!!!

Figure 4.35: Exhibition of the average peak distance with regard to all imaginable histogram dimension projections

values can be obviously averaged for multiple events. Thus the parameter 'analysisInitTotalAnalysis' in the configuration file (see appendix A.2 on page A-4), which has to be used in addition to the parameter 'analysisInitPeakDistanceGoodness' and the bit 'AVERAGEPEAKDISTANCE' in the parameter 'analysisInitAnalysisResultWarnings' (see appendix A.3 on page A-79), prepares another warning message, which offers the feature of averaging the average peak distance for the unprojected histogram over multiple events. However such a message is not depicted here, because it is quite similar to the last one shown in figure 4.35.

Nevertheless figure 4.36 depicts the distribution of this warning message with

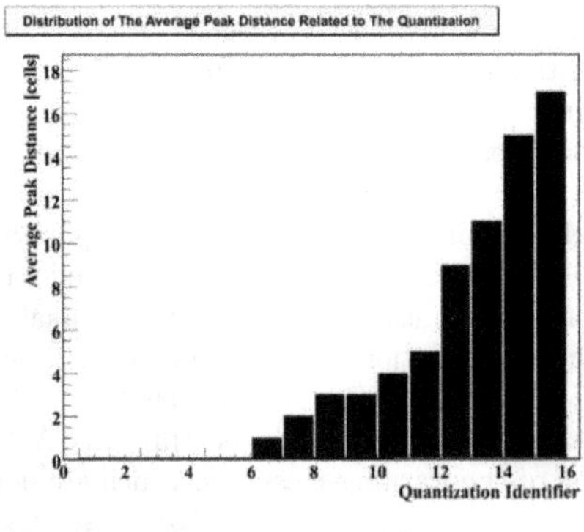

Figure 4.36: Resulting distribution for the average peak distance with regard to different quantizations

regard to different identifiers, which represent a predefined quantization configuration of the histogram dimensions.

Further on, the production of this figure can be certainly enabled by the parameter 'analysisInitTotalAnalysis' in the configuration file (see appendix A.2 on page A-4). To this, the graphical front-end is controlled by setting the corresponding bit, which is labeled with 'AVERAGEPEAKDISTANCE', in the parameter 'analysisInitAnalysisResultDisplays' (see appendix A.3 on page A-79). Finally it should be also mentioned here that the usage of this bit in the parameter 'analysisInitAnalysisResultWarnings' enables completely different features.
Beyond that, the content of this figure can be evidently not produced automatically. Thus it depicts just a visualization of user evaluated statistics for each quantization, which are computed by the previous introduced warning message and setup in the source code file AnalysisLIB/include/totalAnalysisDef.h.

4.6.3 Quality Occupancy

As the previous section 4.6.2 on page 247 analyzes the distribution of the peak distances, the peak occupancy of the three dimensional histogram is definitely of further interest, because it might be possible to classify and remove peaks based on their location or to use a non-equidistant quantization instead of the actual used equidistant one. Therefore the analysis, which is presented in this section, is able to display the projection of the histogram into each possible plane while separating the different track candidate classifications. For this purpose, the figure 4.37 depicts the projections for a common event into the planes, which are spanned by the first and second, the first and third and finally the third and second dimension. In addition to this, a more detailed look to each figure shows further on the existence of a separate display for all available track candidate classifications, which are defined by the MC Track correspondents, the fake tracks, the ghost tracks, the clone tracks and the not-found tracks (see section 4.7 on page 258).
Moreover the parameters, which enable this analysis, can be found in the configuration file (see appendix A.2 on page A-4). Their names are 'analysisInitProjectionEFGCNEvent12EFGCN', 'analysisInitProjectionEFGCNEvent13EFGCN', 'analysisInitProjectionEFGCNEvent32EFGCN', 'analysisInitProjectionEFGCNTotal12EFGCN', 'analysisInitProjectionEFGCNTotal13EFGCN', 'analysisInitProjectionEFGCNTotal32EFGCN', 'analysisInitProjectionEvent', 'analysisInitProjectionTotal', 'analysisInitProjectionDisplay' and 'analysisInitProjectionToRoot'.

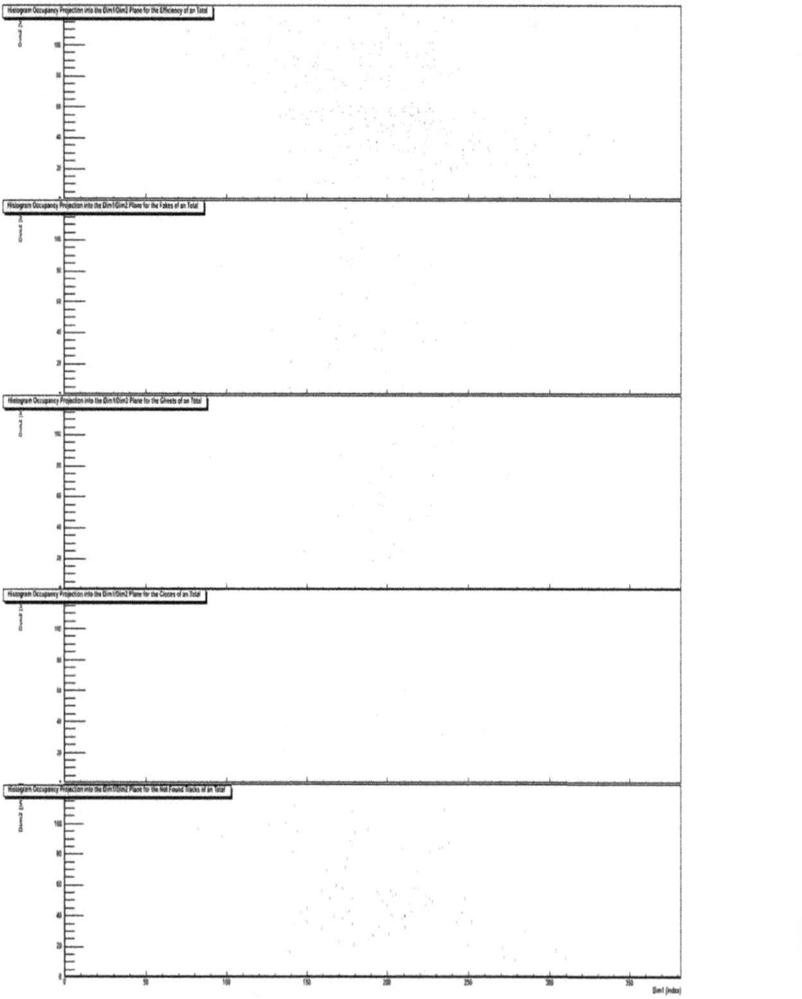

(a) Dim1/Dim2

Figure 4.37: Histogram occupancy projection of the classified track candidates into the planes spanned by predefined dimensions (top-down: found MC Tracks, fake tracks, ghost tracks, clone tracks and not-found tracks)

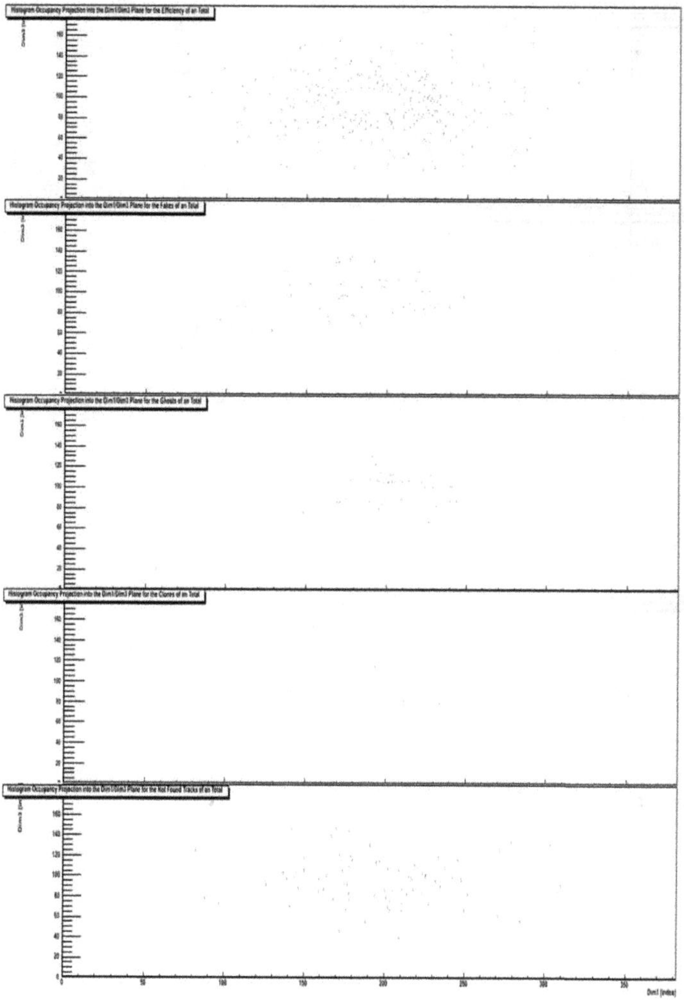

(b) Dim1/Dim3

Figure 4.37: Histogram occupancy projection of the classified track candidates into the planes spanned by predefined dimensions (top-down: found MC Tracks, fake tracks, ghost tracks, clone tracks and not-found tracks)

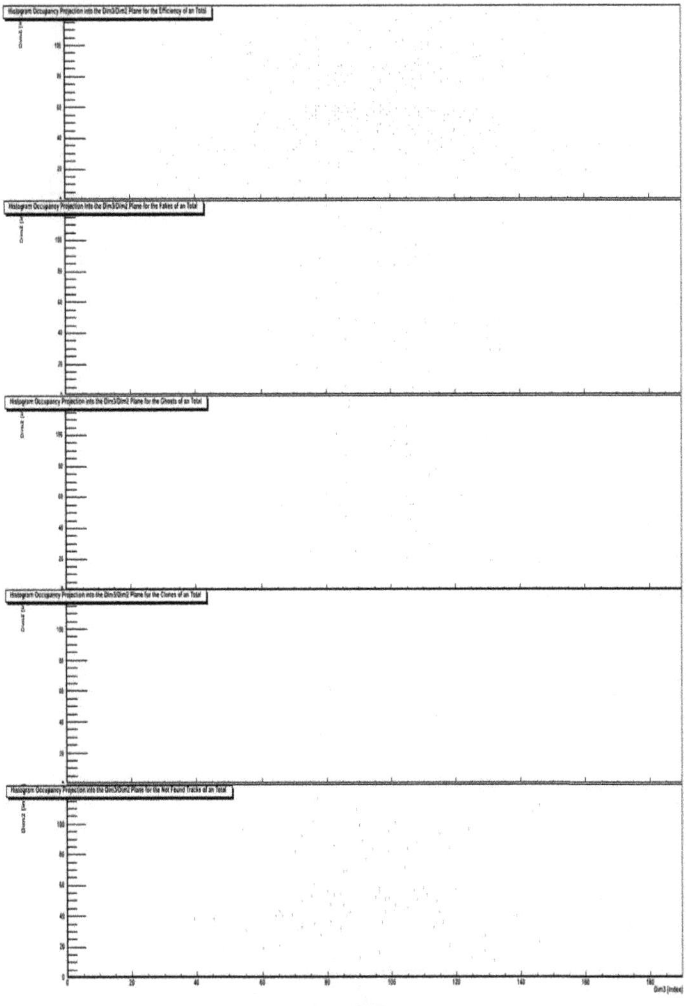

(c) Dim3/Dim2

Figure 4.37: Histogram occupancy projection of the classified track candidates into the planes spanned by predefined dimensions (top-down: found MC Tracks, fake tracks, ghost tracks, clone tracks and not-found tracks)

Referring to these figures, it is now clear that the frequency of occurrence can not be used to classify the track candidates, because their distributions are not locally concentrated, but totally spread into each other.
However the usage of a non-equidistant quantization for the histogram seems to be applicable, because the peaks are mainly aggregated in the central part. So resources could be saved with regard to the histogram, if the outer regions are quantized with bigger steps than in the inner area.

4.6.4 Layer Occupancy

As multiple static histogram analyses are introduced in the previous sections, the analysis, which is presented here, offers the possibility to visualize each histogram layer at fix configured positions in the algorithm. So a closer look to the differences among two visualizations feature a deeper insight of the algorithm, because the realized changes can be directly attached to the specific steps in between. Within this context, figure 4.38, 4.39 and 4.40 depict for instance the visualizations of the identical layer after the three major algorithm steps, which are called Histogramming, Encoding and 2D Peak-finding, by using the graphical front-end.

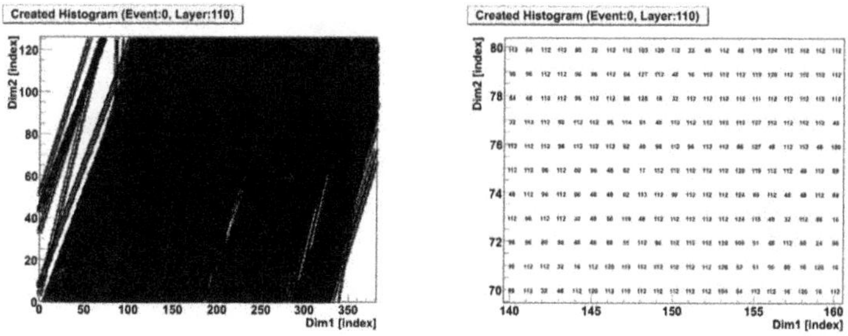

(a) Total scale of the layer

(b) Zoomed scale of the layer to the region of interest shows two overlapping peak clouds

Figure 4.38: Graphical front-end visualizing a single histogram layer after the Histogramming

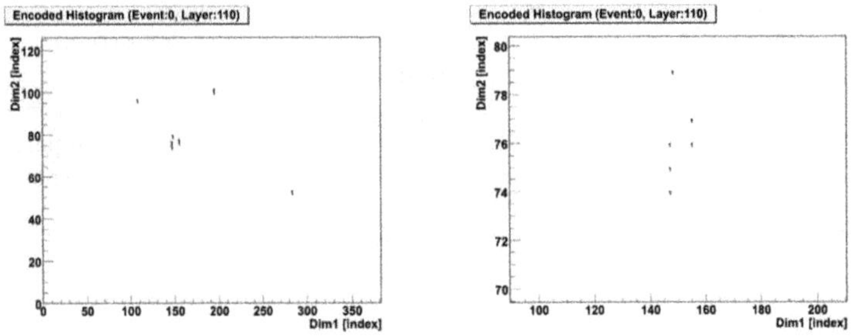

(a) Total scale of the layer

(b) Zoomed scale of the layer to the region of interest shows two close peak clusters

Figure 4.39: Graphical front-end visualizing a single histogram layer after the Encoding

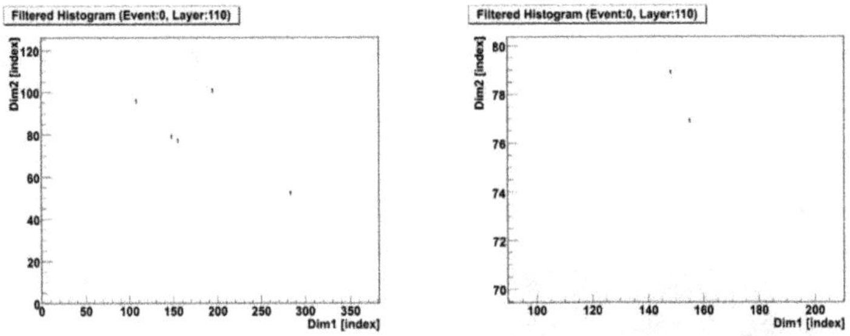

(a) Total scale of the layer

(b) Zoomed scale of the layer to the region of interest shows two separated peaks located in a single cell respectively

Figure 4.40: Graphical front-end visualizing a single histogram layer after the 2D Peak-finding

With regard to the first figure 4.38, it is obviously really hard to visually find the finally surviving peaks in the histogram, because the layer is massively filled during the Histogramming phase. Further on, the result of the afterwards applied Encoding step, which is shown in figure 4.39, is very impressive, because this process is mainly characterized by the huge reduction of background noise and thus the remaining of just some simple cluster structures around the peaks. Moreover the last figure 4.40 gives a good imagination about the result of the 2D Peak-finding and therefore the corresponding finally surviving track candidates for the shown layer.

Besides this, it should be noticed here that the parameters 'analysisInitCreatedHistogramToRoot', 'analysisInitEncodedHistogramToRoot', 'analysisInitFilteredHistogramToRoot', 'analysisInitJustOneCreatedHistogramToRoot', 'analysisInitJustOneEncodedHistogramToRoot', 'analysisInitJustOneFilteredHistogramToRoot', 'analysisInitCreatedHistogramToShow', 'analysisInitEncodedHistogramToShow', 'analysisInitFilteredHistogramToShow' and 'analysisInitHistogramLayer', which enable the analysis, can be found in the configuration file (see appendix A.2 on page A-4).
To this, some define statements in the source code files HoughTransformationLIB/src/houghTransformation.cxx and HistogramTransformationLIB/src/histogramTransformation.cxx would additionally offer the generation of human readable data files in the ASCII format, which contain all histogram layers in detail before and after each algorithm step, if the graphical visualization is not sufficient. As these files are furthermore created automatically, it is not surprising that each filename is simply built by a basic identifier for each algorithm position, which is defined also in the source code, concatenated by the index of the actual histogram layer. So within this context, it is clear that a separate file is constructed for each position and each layer.
In this connection, it has to be also noticed that the created histogram borders for the hits can be written into a separate file as well.
So based on this concept, it is finally clear that all available data, which occurs during the evaluation of the Hough transform algorithm, can be visualized and analyzed.

4.7 Tracking Quality

This section presents different analyses, which are able to evaluate the quality of the Hough tracking algorithm. For this purpose, they expand the computation of the momentum error distribution, which is introduced in section 4.5.3 on page 236, with more details belonging to the Hough transform itself. Furthermore the efficiency of the tracking is summarized with absolute or relative numbers for all relevant characteristics of the algorithm in combination with the reconstructed momentum distribution for the tracks.

Besides this, it should be noticed here that even if these analyses give a statistical benchmark for the Hough tracking, the define statement 'DEBUGJUSTONEGOODTRACK' in the source code file MiscLIB/include/defs.h would restrict the input data to consist just of a single special findable MC Track. Thence this feature offers the possibility to analyze just a single MC Track processed by the algorithm.

4.7.1 Absolute Quality

The analysis presented in this section gives detailed information about the tracking quality by summarizing each relevant characteristic of the Hough transform algorithm with absolute numbers. For this purpose, the parameters 'analysisInitQualityEFGCEventAbsolute' and 'analysisInitQualityEFGCTotalAbsolute' produce such a summary, which typically looks like figure 4.41. In this connection, it is clear that the first parameter features an event based result, while the second one offers the result for an accumulation of all events. Further on, these parameters are evidently located in the configuration file (see appendix A.2 on page A-4) analog to other analysis.

With regard to this figure, the summary is evidently divided into two major sections, which show on the one hand details about the found peaks, which represent the final track candidates, and on the other hand details about the correspondingly analyzed MC Tracks.

Moreover the peak section is divided into two additional sections, which connect either the peaks to the MC Tracks or the peaks to the available classifications. In this connection, it has to be additionally noticed here that these classifications are given by 'track', 'clone', 'no track', 'ghost' and 'fake'. So while having now a closer look to these classifications, it is obvious that the easiest one is the 'track' definition, because such a peak identifies directly a real occurring findable MC Track.
Further on, a 'clone' or clone track simply indicates a copy of a peak, which

Absolute Analysis:	
Peak Summary	
Total number of peaks corresponding to no track:	76
Total number of peaks corresponding to one track:	321
Total number of peaks corresponding to more tracks:	1
Total number of peaks:	398
Total number of peaks classified as track:	271
Total number of peaks classified as clone:	3
Total number of peaks classified as fake:	42
Total number of peaks classified as ghost:	34
Total number of peaks classified as no track:	48
Total number of peaks:	398
Track Summary	
Total number of tracks corresponding to no peak:	74
Total number of tracks corresponding to one peak	269
Total number of tracks corresponding to more peaks:	3
Total number of tracks:	346
Total number of well identified findable tracks:	173
Total number of not well identified findable tracks:	99
Total number of identified not findable tracks:	47
Total number of identified tracks:	319
Total number of identified findable tracks:	272
Total number of findable tracks:	346

Figure 4.41: Tracking quality represented by absolute numbers

is determined to be a 'track'.
In addition to this, the 'no track' definition is also self-explanatory, because it belongs just to a misinterpreted peak, which is caused by the problems of the accumulator (see section 2.1 on page 45).
Aside from this, a 'ghost' or ghost track identifies a peak, which exists in contrast to the real 'track' due to hit contributions from different MC Tracks. And finally a 'fake' or fake track denotes a peak, which is commonly identified as 'ghost', but would not exist, if the fake hits are removed from the input data.

Besides this, the MC Track section is apparently divided into three sections. Further on, the first two ones feature evidently the analog information as the peak section, because the first one connects the MC Tracks to the peaks, while the second one illustrates the classification. However the classification is obviously different, because the key characteristic is determined by the MC

Track attribute 'findable', which is more precisely specified by the three decreasing values 'well findable', 'not well findable' and 'not findable'. Moreover a more intensive look to the the final third section shows immediately that this one is only required to evince the absolute numbers for the two possible base qualifiers with regard to the relative efficiency, which is introduced in the following section 4.7.2 on page 260.

4.7.2 Relative Quality

The analysis, which is introduced in this section, shows information about the tracking quality by applying relative numbers. For this purpose, figure 4.42 depicts a typical output, which can be enabled by the parameters 'analysisInitQualityEFGCEventRelative' and 'analysisInitQualityEFGCTotalRelative' in the configuration file (see appendix A.2 on page A-4). In analogousness to the previous section 4.7.1 on page 258, the first parameter features an event based result, while the second one offers the result for an accumulation of all events.

Relative Analysis:	
Total efficiency [e] (aim 100%):	79%
Total fake rate [f] (aim 0%):	11%
Total ghost rate [g] (aim 0%):	9%
Total clones rate [c] (aim 0%):	1%
Total identification rate [i] (aim 100%):	68%
Total data reduction rate [r] (aim near 85%):	78%

Figure 4.42: Tracking quality represented by relative numbers

With regard to this output, it is clear that the definition of the utilized terms is necessary to understand the displayed information.
So the term 'efficiency', which represents obviously the tracking efficiency, is commonly defined by the quotient of the number of peaks or final track candidates, which can be identified to correspond to a real occurring findable MC Track, and the number of all findable MC Tracks.
Remembering further the definitions of the previous section 4.7.1 on page 258, it is self-evident that the term 'clones rate', which typifies the tracking clones rate, is defined by the quotient of the number of peaks or final track candidates, which correspond to an already found MC Track, and the number of existing track candidates. Thence it is implicitly clear that a multiple found MC Track contributes only once to the tracking efficiency and otherwise to the tracking clones rate. Thusly the tracking clones rate is able to inform about the multiplicity of the track candidates.

Besides this, the term 'ghost rate', which embodies the tracking ghost rate, is defined by the quotient of the number of peaks or final track candidates, which exist just because of hit contributions from different MC Tracks, and the number of existing track candidates. Furthermore these peaks or track candidates does obviously not belong to a real occurring findable MC Track. Moreover the term 'fake rate', which covers the tracking fake rate, is defined by the quotient of the number of peaks or final track candidates, which exist just because of the fake hit contributions to the common hit contributions from different MC Tracks, and the number of existing track candidates. So in contrast to the tracking ghost rate, the tracking fake rate contains just peaks or final track candidates, which would not exist, if the fake hits are removed from the input data to the tracking. Thus the tracking fake rate is a specialty of the tracking ghost rate.
Beyond that, another important term is apparently determined by the 'identification rate', which represents the tracking identification rate, because its definition is set by the quotient of the number of peaks or final track candidates, which can be identified to correspond to a real occurring findable MC Track, and the number of all track candidates. Furthermore it has to be explicitly highlighted in this connection that the clone tracks are naturally filtered out as well as for the tracking efficiency, which implies that peaks or final track candidates corresponding to the same MC Track are just counted once.
Last but not least, the term 'reduction rate', which typifies the tracking reduction rate, is defined by one minus the quotient of the idealized number of hits for the peaks or final track candidates and the number of hits in the input data. Thence this quotient is able to approximate the real occurring data reduction, because the amount of output data is directly correlated to the amount of input data. However this character is obviously not perfectly fitting, because the number of hits for the peaks or final track candidates is idealized, while a single data object of the input and the output can be additionally assumed to be different in bit size due to the constituting information or just the encoding for instance.

Since almost all rates are up to now specified to be relative to the number of findable MC Tracks, it is clear that the definition of this term is of major interest. So within this context, the number of findable MC Tracks is simply determined by the number of MC Tracks, which own the characteristics of a predefined momentum cut (commonly higher than $1\,\frac{\mathrm{GeV}}{\mathrm{c}}$) and particular predefined hit signatures (see appendix A.4 on page A-101).

Having finally a closer look to each of these definitions, it is self-evident that a single one is obviously not able to establish a criterion to measure the quality of the Hough tracking, because each term has to be rated in relation to the others. Thence the best quality, which can be achieved theoretically, is defined by a tracking efficiency and a tracking identification rate of 100 % in combination with a tracking fake rate, a tracking ghost rate and a tracking clones rate of 0 %, because this result will then directly imply that each MC Track corresponds to just one finally found track candidate. But as such an aim is imaginably hard to achieve, the result would be certainly optimal, if these values are obtained as good as possible. However this circumstance entails obviously another problem, because these five degrees of freedom complicate naturally the comparison of two different results. Nevertheless such a comparison can be easily realized by adducting an aim based weight for each degree.

4.7.3 Relative Quality Related to the Momentum

At a first glance the two analyses, which are introduced in section 4.7.1 on page 258 and section 4.7.2 on page 260, seem to be enough to evaluate the quality of the Hough tracking algorithm. But as a findable MC Track is distinguished by two characteristics, which are determined by a dedicated hit signature and a minimal momentum, it is surely interesting to gain an imagination about the quality with regard to each characteristic independent of the other.

So since two analyses for the hit signatures are already introduced in section 4.5 on page 229 and section 4.8.3 on page 272, the actual section features apparently an already earlier awaited analysis related to the momentum. For this purpose, the parameters 'analysisInitMomentumEFGCEventPzEFGC', 'analysisInitMomentumEFGCEventPtEFGC', 'analysisInitMomentumEFGCTotalPzEFGC', 'analysisInitMomentumEFGCTotalPtEFGC', 'analysisInitMomentumEvent', 'analysisInitMomentumTotal', 'analysisInitMomentumDisplay' and 'analysisInitMomentumToRoot', which are all located in the configuration file (see appendix A.2 on page A-4), offer the possibility to display the momentum distribution for each available classification of a peak or final track candidate with regard to an event base or accumulated for all events.

Accordingly figure 4.43 and figure 4.44 depict typical results for an additionally segmented momentum in either the beam axis p_z or the transversal p_t.

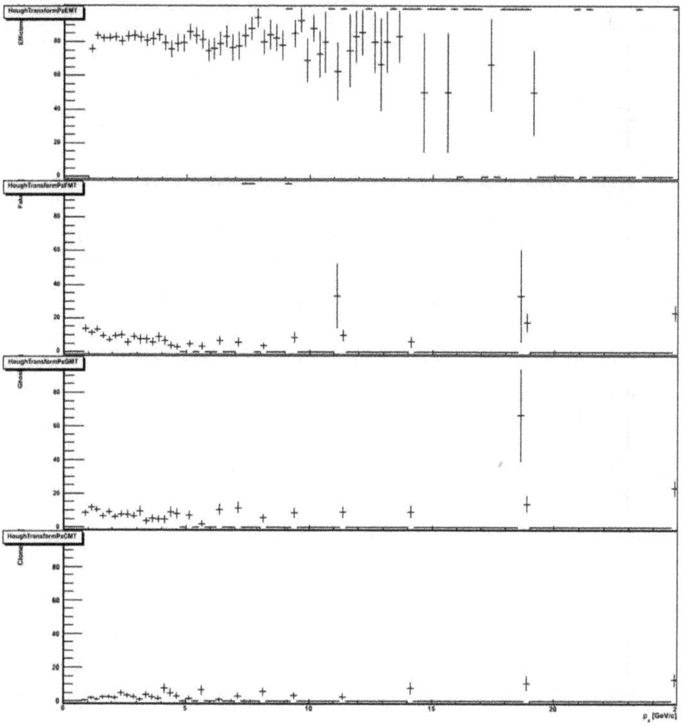

Figure 4.43: Tracking quality with regard to p_z (top-down: found MC Tracks, fake tracks, ghost tracks, clone tracks)

With regard to these figures, a detailed look to the tracking efficiency shows now that the quality is almost constant over the whole momentum distribution just with a small degradation at the minimum cut of $1\,\frac{\mathrm{GeV}}{\mathrm{c}}$.
In addition to that, the same characteristic counts obviously also for the clone tracks.
In contrast to this, the distributions of the ghost tracks and the fake tracks are a quite different, because they include a degraded percentage, which occurs for the medium momenta. As the high percentage of the low momenta can be explained by the highest track multiplicity, which leads thus to more likely combinatorial coincidences, and the increase for the high momenta is

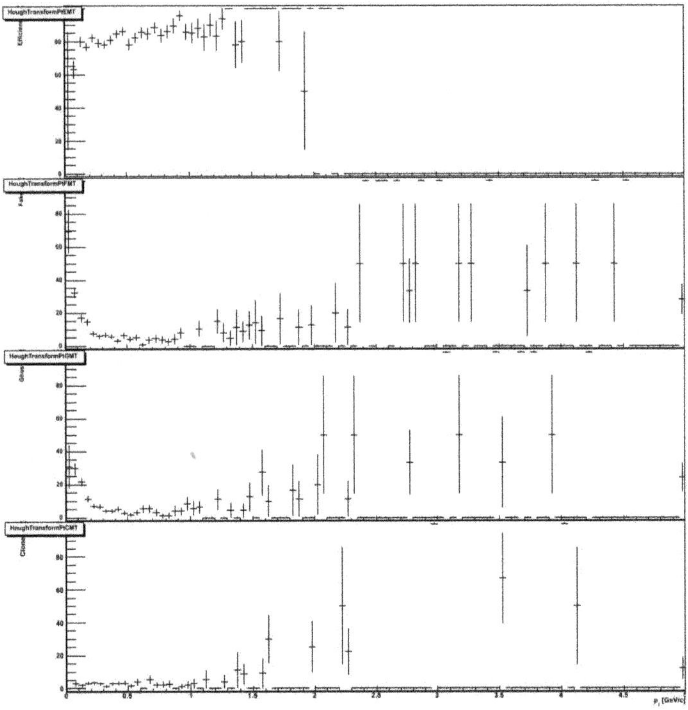

Figure 4.44: Tracking quality with regard to p_t (top-down: found MC Tracks, fake tracks, ghost tracks, clone tracks)

obviously caused by the less numerous tracks, which correspond to almost straight lines near the beam pipe with high track densities, it is anticipated that the medium momenta quality is less, because both facts can not be addicted there.

4.8 Visualization of the Results for the Automatic Optimizations

This section presents analyses, which belong to the results of internally evaluated algorithms featuring automatic optimizations. Thus it is clear that different independent aspects of the Hough tracking count here.
So at first, there is an analysis, which offers the possibility to display the results for the effect of the slope enforcement on the Hough curves by applying different restrictive rules. This one is then followed by an analysis, which offers a detailed visualization of the results causing the selection of the overlapping range m in equation 2.13 of section 2.4.1 on page 73. Afterwards the subsequent analysis is able to display the signatures, which are delivered inside the table files. And finally the last two analyses provide the functionality to show the Peak-finding geometry and to compare the motion trajectories of the MC Tracks to the ones of the found reconstructed tracks.

4.8.1 Analytic Formula Correction Belonging to Hardware Requirements

Remembering now the general idea of the Hough transform implementation, which combines a systolic processable histogram with Hough curves restricted to a slope only between 45 and 90 degrees (see section 3.1 on page 96), it is suggested that these conditions have to be just considered in the hardware realization. So the software implementation, which is introduced in section 3.2 on page 102, contains only a simulation model of such a histogram without the requirement for any restriction to the Hough curves, which should be inserted. Hence there is obviously the need for the possibility to enforce an allowed slope by introducing an error to the Hough curves, while producing the LUTs with a software algorithm. However it is self-evident in this connection that this error depends also on the actual chosen histogram quantization and has a not negligible effect on the transformation quality. As it is thus apparently essential to analyze the effects of this enforcement, the source code file LutGeneratorLIB/include/lut.h contains five different levels, which feature the evaluation of raising restrictive strategies summarized in table 4.1.

With regard to this table, it is immediately clear that the higher the level gets, the more aggressive the strategy becomes. So the first one, which realizes thus the least one, avoids obviously just Hough curves with negative slopes. Subsequently the second one, which is closer to the systolic processing

Level	Rules	Message
1	$pos1_{i+1} \geq pos1_i$ $pos2_{i+1} \geq pos2_i$	no
2	$pos1_{i+1} \geq pos1_i$ $pos2_{i+1} = pos2_i + 1$	no
3	$pos1_{i+1} = pos1_i + \begin{cases} 0 \\ 1 \end{cases}$ $pos2_{i+1} \geq pos2_i$	no
4	$pos1_{i+1} = pos1_i + \begin{cases} 0 \\ 1 \end{cases}$ $pos2_{i+1} \geq pos2_i$	if rule break once
5	$pos1_{i+1} = pos1_i + \begin{cases} 0 \\ 1 \end{cases}$ $pos2_{i+1} \geq pos2_i$	always

Table 4.1: Overview of the available strategies enforcing different restrictive rules for the slope of all Hough curves

requirement, prevents also Hough curves with a zero degree slope, but allows further slopes smaller than 45 degrees. Finally all higher levels implement evidently the total restrictions for the general hardware requirements and differ only in the way of message production on the standard output screen. Coming now to the point, such a typical message is depicted in figure 4.45.

```
ANALYSISLIB WARNING: There are 11565 corrections which are done while lut
calculation. 129 different transformations of a hit (0.82%) are affected. So in
the middle there are 90 positions corrected per transformation of a hit.
```

Figure 4.45: Resulting output for the number of Hough curve corrections with regard to a given histogram layer dimension ratio

With regard to this figure, it is immediately clear that such a message is evidently produced for each event and related to the corresponding enforcement level respectively. But as such statistics has to be apparently generated for multiple accumulated events, the parameter 'analysisInitTotalAnalysis', which is located in the configuration file (see appendix A.2 on page A-4), offers exactly this functionality. However it is important to realize in this connection that the printout of the result to the standard output screen is controlled by setting the corresponding bit, which is labeled with 'LUT-CORRECTIONCOUNTER', in the parameter 'analysisInitAnalysisMoreRe-

sultWarnings' (see appendix A.3 on page A-79) in the same configuration file.
Moreover the result for multiple accumulated events exhibits naturally not just yet enough information, because the introduced error depends apparently on the quantization of the histogram. So for that reason, the evaluation of these statistics with regard to different ratios between the first and the second histogram dimension is of additional interest. For this purpose, figure 4.46 depicts a typical distribution for the percentage of corrected transformations with regard to different ratios.

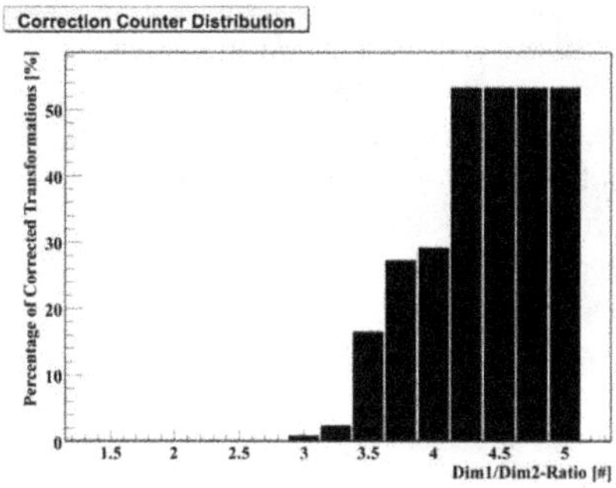

Figure 4.46: Percentage distribution of corrected Hough curve transformations related to different given histogram layer dimension ratios

With regard to this figure, it is clear that a smaller percentage of corrections is better, because more transformations can be handled without this implementation issues.
Further on, the production of this figure can be naturally enabled by the parameter 'analysisInitTotalAnalysis' in the configuration file (see appendix A.2 on page A-4), while the graphical front-end is controlled by setting the corresponding bit, which is labeled with 'LUTCORRECTIONCOUNTER', in the parameter 'analysisInitAnalysisMoreResultDisplays' (see appendix A.3 on page A-79). However it should be additionally mentioned in this connection that the output, which is generated by setting this bit in the parameter 'analysisInitAnalysisMoreResultWarnings', contains completely different content.
Moreover it is self-evident that the content of this figure can not be produced automatically. Thus it is just a visualization of user evaluated statistics for each ratio, which are computed by the previous introduced analysis feature and setup in the source code file AnalysisLIB/include/totalAnalysisDef.h.

Besides this, it should be finally noticed here that such a distribution is very useful for the decision about the quantization ratio of the histogram dimensions, which is also a key element in the analysis presented in section 4.6 on page 240.

4.8.2 Analytic Formula Prelut Cut

As the overlapping range m, which is used to cope with side effects in the y-z projection of the analytic formula introduced in section 2.4.1 on page 69, has a huge impact on the generation of the first LUT, it is suggested that an automatic computation is required to optimize the tracking performance. By remembering the mentioned section more precisely, it is furthermore clear that such an optimization process has to take care about the evaluation of a range, which enables the transformation to connect each hit to the minimal number of required histogram layers including at least the one containing the accurate peak, because the otherwise resulting situation is of course not optimal and leads thus to the behavior described in the following.
On the one hand, a wide range will lead to more involved consecutive histogram layers for each hit and therefore to a multiplied amount of occurring peaks. In combination with the hit signature definitions, which are used for the Encoding of the histogram cells, this situation can then easily result in a higher tracking ghost rate, an increased tracking fake rate and an exploding tracking clones rate with an decreasing tracking identification rate. Although the used hit signatures can prevent such a performance degradation, this is at any time not longer possible, because the signatures of the multiplied peaks turn equal to the accurate peaks.
On the other hand, a very small range will lead contrarily to very few and at least one involved histogram layers for each hit and thus to downgraded hit signatures. Even if it is possible to cope with this situation by the adaption of the selected Encoding signatures, the resulting performance effect would remain the same.
So within this concept, it has to be noted here that section 4.9.1 on page 284 offers a statistical analysis about the distribution of the hits with regard to their belonging number of histogram layers.

Coming now to the point, the analysis, which is presented in this section, seems to be very sensitive and complex at a first glance. But as the optimization comprises only the minimization of the number of histogram layers, which belong to a hit, in conjunction with the side condition of including at least the accurate layer, there are only two facts to analyze. Nevertheless section 4.5.1 on page 230 suggests at first that it is important to define the accurate layer, because there are two possibilities. So the first one in this connection is obviously identified by the correct layer, which is defined by the momentum of the MC Track, while the second and apparently more promising one is determined by the identification layer, which is characterized by holding the maximal available number of hits for the corresponding MC Track. Moreover the additional employment of the MC data to approximate the required range with regard to the hit signatures leads to a further simplification.

So based on this concept, the algorithm calculates now merely a distribution for each condition based on all MC Tracks with regard to some sample ranges, which are influenced by the approximation. Further on, typical results are depicted in figure 4.47 and figure 4.48.

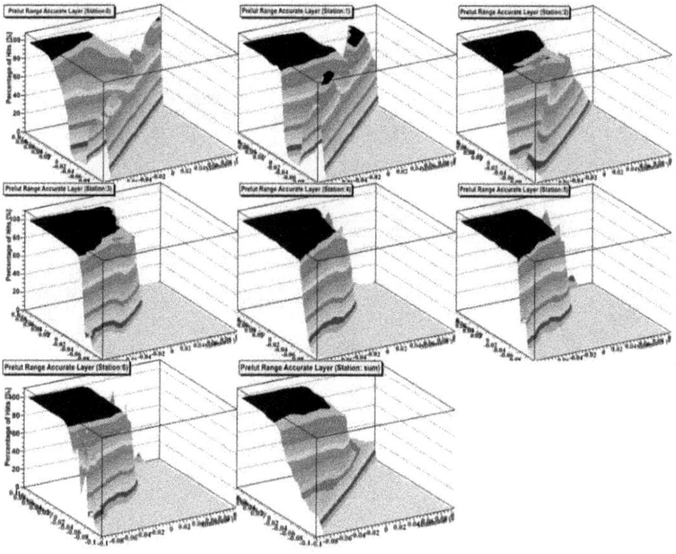

Figure 4.47: Hits percentage distribution for the accurate histogram layer with regard to sample ranges

Within this context, the first figure 4.47 shows generally the percentage of hits, which encounter the accurate histogram layer for a MC Track, with regard to the sample ranges. Furthermore this figure contains a certain amount of distributions, because the overall result of the last one, which is labeled

with 'Station: sum', is additionally subdivided into distributions for each available detector station.
In addition to this, the second figure 4.48 displays the accumulated number of histogram layers, which are affected by the insertion process of each transformed hit, with regard to the sample ranges. In this connection, the general distribution is obviously also subdivided into distributions for each available detector station.

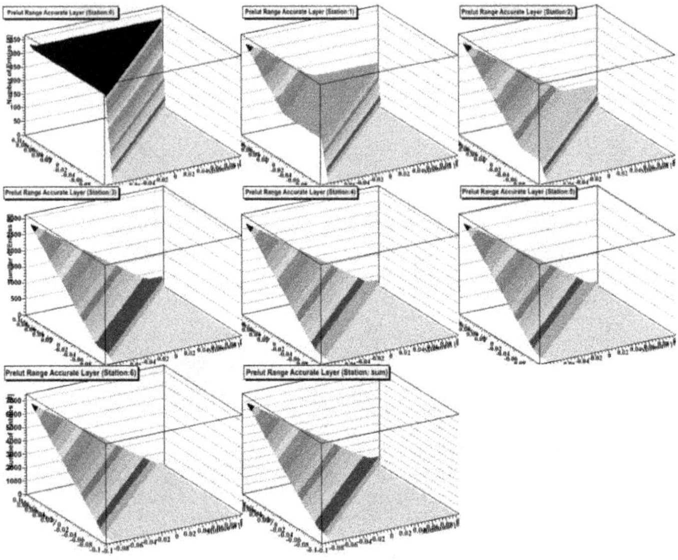

Figure 4.48: Hits distribution for the accumulated number of affected histogram layers with regard to sample ranges

As two distributions, which represent the result for each condition, are now available, it is clear that the optimum can be only found by favoring one distribution, while optimizing the other with regard to this one. For that reason, the side condition, that at least the accurate histogram layer must be included in the transformation, is certainly selected to be favored. But as such a condition determines obviously a harsh environment, it is softened by applying an accuracy percentage for the hits, which must encounter the accurate histogram layer. This process leads then directly to the circumstance that the required amount of degraded hit signatures, which are necessary for

the Encoding of the histogram cells, can be controlled.
So based on this concept, it is self-evident that the whole optimization process is actually simplified to a search for the sample range, which is characterized by at least the fixed accuracy percentage for hits in the accurate histogram layer in the distribution 4.47 in conjunction with the minimal accumulated number of affected histogram layers in the distribution 4.48.
However it is further on apparent that the fixation of this accuracy percentage is not trivial. Thence another analysis is required, which gives information about the possible region of interest in both distributions. For this purpose, figure 4.49 depicts the maximal reachable accuracy percentage in combination with the corresponding sample range and the average number of affected histogram layers per hit. So it is now possible to fix this percentage with regard to the maximum, which is reachable.

```
AnalysisLIB WARNING: The maximum hit percentage in the prelut is 99%!!!
AnalysisLIB WARNING: The prelut range causes an average of 8.8 entries per hit!!!
AnalysisLIB WARNING: The prelut range is actually set to [-0.1, 0.9]!!!
```

Figure 4.49: Exhibition of the maximal reachable hits accuracy percentage for the accurate histogram layer

But as the knowledge about this maximum provides of course also not yet enough information, another analysis has to be obviously taken into account, which is able to explore the characteristics of the region of interest defined by the actual used accuracy percentage. So a closer look to the produced figure 4.50 shows now that this summary should evidently present a high number of ranges, which fulfill the requirement of at least the accurate percentage, in combination with a sparse average of affected histogram layers per hit.

```
AnalysisLIB WARNING: There are 56 prelut ranges detected which fullfills the requirements!!!
AnalysisLIB WARNING: The prelut range causes an average of 2.9 entries per hit!!!
AnalysisLIB WARNING: The prelut range is actually set to [-0.01, 0.05]!!!
```

Figure 4.50: Exposition of the region of interest based on the actual used accuracy percentage

Besides this, it has to be reminded that the only control of the result is determined by the definition of the accuracy percentage for the hits, which must encounter the accurate histogram layer.
Nevertheless the major problem of this optimization process is apparently

determined by the saturation region of both distributions, which is commonly accomplished by a too low assumed accuracy percentage, because no useful information related to a good range can be gained then from the distributions. Furthermore it is then even possible that a wrong single-layer transformation of a hit is preferred to a multi-layer transformation, which contains the accurate histogram layer. Thence it is really important to notice here that such a saturated region can be commonly identified by the fact that each hit affects just a single histogram layer.

Finally it should be mentioned that the automatic part of the analysis is enabled with the parameters 'analysisInitPrelutRangeForEachEvent' and 'analysisInitWeightedPrelutRange' in the configuration file (see appendix A.2 on page A-4). In addition to this, the parameter 'analysisInitPercentageOfHitsForPrelutRange' defines the minimum accuracy percentage, while the parameters 'analysisChooseAccuratePrelutRange' and 'analysisChooseConstraintPrelutRange' offer the possibility to choose between the optimization goals. Besides this, the graphical front-ends in combination with the common file storage are enabled with the parameters 'analysisInitPrelutRangeDisplay', 'analysisInitPrelutRangeDisplayMode', 'analysisPrelutRangeStationDisplayMask', 'analysisPrelutRangeStationSumDisplayMask' ,'analysisPrelutRangeConstraintDisplayMask', 'analysisPrelutRangeConstraintSumDisplayMask', 'analysisPrelutRangeRelativeDisplayMask' and 'analysisInitPrelutRangeToRoot'. Accordingly the printout to the standard output screen can be achieved by setting the corresponding bit, which is labeled with 'PRELUTRANGEINFORMATION', in the parameter 'analysisInitAnalysisResultWarnings' (see appendix A.3 on page A-79). And last but not least, the predefined search region and sample rate for the range are set by the parameters 'analysisPrelutRangeMinStart', 'analysisPrelutRangeMinStop', 'analysisPrelutRangeMinSteps', 'analysisPrelutRangeMaxStart', 'analysisPrelutRangeMaxStop' and 'analysisPrelutRangeMaxSteps'.

4.8.3 Hit Signatures Delivered With Table Files

As the general concept of hit signatures serves a central role in the Hough tracking and the quality measurement, it is obvious that there is a need for analyses, which help to create the signatures in the tables. For this purpose, the appendix A.4 on page A-101 contains many different possibilities to produce such table files more or less optimal and automatic. But as all of them are very sensitive, they have to be adapted to the requirements carefully. Therefore the analysis, which is presented here, visualizes information about these table files to offer an impression about the quality and complexity. But before starting, it has to be at first noticed that the base parameters to enable the displays or messages are apparently located in the configuration file (see appendix A.2 on page A-4) and named 'analysisInitPrelutGoodness', 'analysisInitLutGoodness' and 'analysisInitTotalAnalysis'. In addition to this, the output of the analysis is controlled by setting the corresponding bits, which are labeled with 'GOODSIGNATURES', 'GOODSIGNATURETABLES' and 'USEDSIGNATURETABLES', in the parameters 'analysisInitAnalysisMoreResultWarnings' and 'analysisInitAnalysisMoreResultDisplays' (see appendix A.3 on page A-79). By the way, it is clear that the first parameter enables the results to be printed to the standard output screen while the second one displays graphical front-ends.

Coming now to the point, it is self-evident that the first important impression is determined by the hit signatures of the MC Tracks, because they form the groundwork for the generation of the table files. Therefore figure 4.51 depicts the number of accepted MC Tracks according to both LUTs, while figure 4.52 shows the number of hit signatures, which are necessary to accept these amount of MC Tracks.

ANALYSISLIB WARNING: There are 340 MC Tracks, which are accepted according to the signatures of the prelut!!!

ANALYSISLIB WARNING: There are 319 MC Tracks, which are accepted according to the signatures of the lut!!!

Figure 4.51: Resulting output for the number of accepted MC Tracks according to both LUTs independently

ANALYSISLIB WARNING: There are 2 signatures related to the prelut, which are required to accept the amount of MC Tracks!!!

ANALYSISLIB WARNING: There are 3 signatures related to the lut, which are required to accept the amount of MC Tracks!!!

Figure 4.52: Resulting output for the number of required signatures to accept a certain amount of MC Tracks

Since the results, which are shown in these two figures, are obviously strongly connected to each other, it is supposed that the control parameters are also strongly connected. So the first one, which is named 'analysisInitPercentageOfHitsInSignature' and located in the configuration file (see appendix A.2 on page A-4), defines the percentage of hits, which must be minimally contained in a hit signature to be accepted, while the second one, which is called 'analysisInitPercentageOfTracksForSignature' and also located there, determines the percentage of MC Tracks, which are minimally required for the actual amount of acceptable hit signatures. Keeping these meanings in mind, it is furthermore clear that the connection of both is appointed by the generation of more and more hit signatures as long as one of these minimums are achieved. In addition to this, the given information is evidently also related to the quality of the table files, because the number of accepted MC Tracks influences of course the maximal amount of findable track candidates, while the more hit signatures are required, the more identified track candidates might not correspond to a real existing MC Track.

So based on this context, a qualitative good table file maximizes apparently the number of accepted MC Tracks while minimizing the number of required signatures. However it is clear that these aims are almost antidromic and their optimization depends directly on the quantization of the histogram as well as the quality of both LUTs. But as such a dependency offers obviously the possibility to show the variation of these values with regard to the histogram dimension ratio, the resulting distributions, which are depicted in figure 4.53 for the first LUT and figure 4.54 for the second LUT, are very useful to measure an aspect of the quality related to such a ratio. Moreover it has to be mentioned in this connection that this aspect is also a key element in the analysis, which is presented in section 4.6 on page 240.

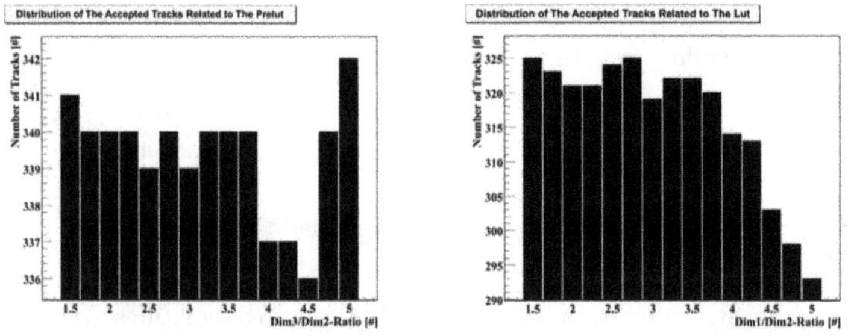

(a) First LUT

(b) Second LUT

Figure 4.53: Resulting distribution for the number of accepted MC Tracks with regard to the histogram dimensions

Further on, it should be noticed here that even if all introduced figures contain different information, the single parameter 'analysisInitTotalAnalysis'

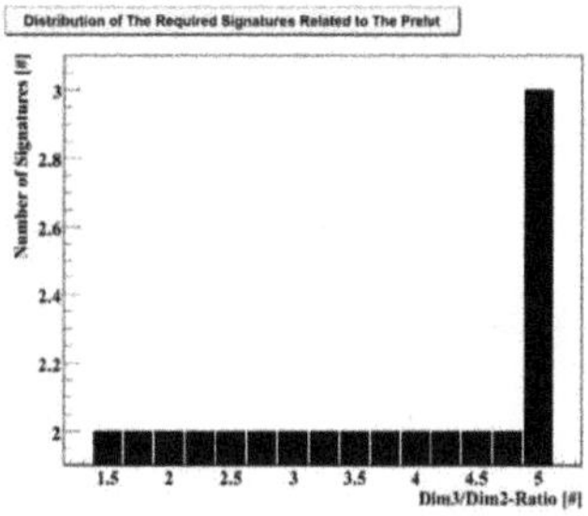

(a) First LUT

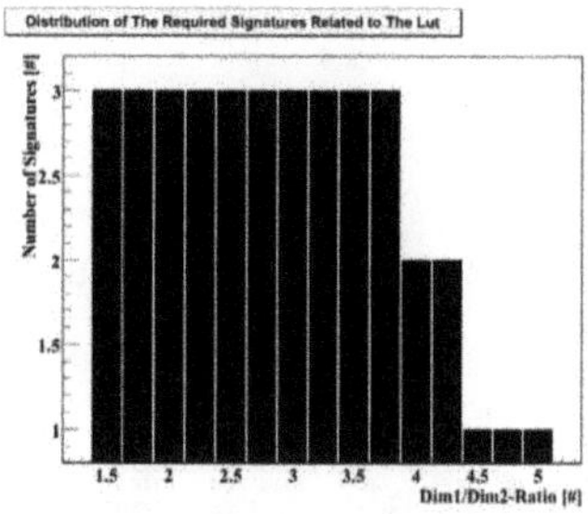

(b) Second LUT

Figure 4.54: Resulting distribution for the number of required signatures with regard to the histogram dimensions

could offer their generation. Just the visualization differs by setting the corresponding bit, which is labeled with 'GOODSIGNATURES', in the parameters 'analysisInitAnalysisMoreResultWarnings' or 'analysisInitAnalysisMoreResultDisplays' (see appendix A.3 on page A-79).
Besides this, it is self-evident that the content of the figures 4.53 and 4.54 can be not produced automatically. Thus it is simply a visualization of user evaluated statistics for each ratio, which are computed by the previous introduced analysis feature and setup in the source code file AnalysisLIB/include/totalAnalysisDef.h.

As the major important facts are clarified up to here, the focus can be set to an additional topic, which is not less interesting. So the consecutive step is characterized by the study of each hit signature including their frequency of occurrence. Therefor the parameters 'analysisInitPrelutGoodness', 'analysisInitLutGoodness' and 'analysisInitTotalAnalysis', which are located in the configuration file (see appendix A.2 on page A-4), enable the generation and visualization of figure 4.55 and figure 4.56 by the corresponding bit, which is labeled with 'GOODSIGNATURETABLES', in the parameters 'analysisInitAnalysisMoreResultWarnings' and 'analysisInitAnalysisMoreResultDisplays' (see appendix A.3 on page A-79). Moreover it should be additionally mentioned here that this functionality can be also used to get detailed information about the automatic table production algorithm, which can be enabled by the parameter value 'LUTGOODNESSTABLE' (see appendix A.4 on page A-101).

Since these figures offer now the possibility to analyze the hit signature degra-

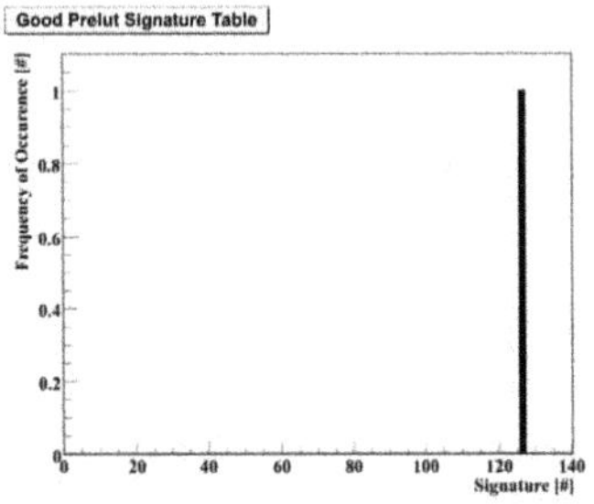

(a) Graphical front-end

AnalysisLIB WARNING: The used signatures belonging to the prelut are:
Signature: 2x1111110, Classification: 2x0000001
Signature: 2x1111111, Classification: 2x0000001

(b) Screen output

Figure 4.55: Signature distribution with regard to the first LUT

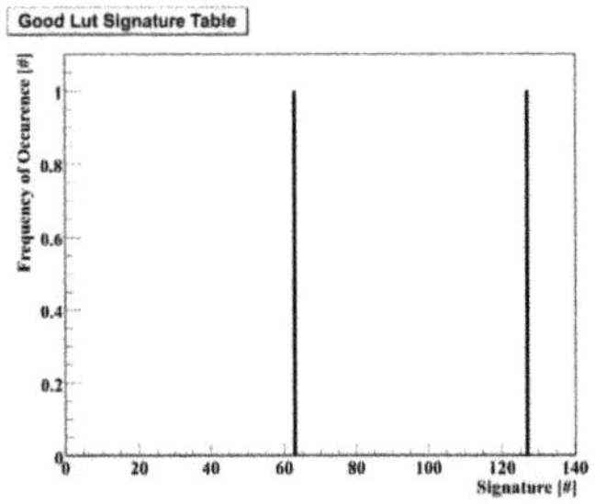

(a) Graphical front-end

AnalysisLIB WARNING: The used signatures belonging to the lut are:
Signature: 2x0111111, Classification: 2x0000001
Signature: 2x1111111, Classification: 2x0000001

(b) Screen output

Figure 4.56: Signature distribution with regard to the second LUT

dation for each LUT independent of the other one, it is obvious that the optimization potential can be directly derived. But before taking a closer look to these figures, it has to be noticed that only six detector stations would be used to produce the results, even if the hit signatures exhibit seven bit. Consequentially two signatures, which differ only in the leading bit position, have to be combined for the analysis of the recent figures, because the correct adaption of the bit width for the hit signatures will of course result in just a single signature for that case.
So with regard to figure 4.55, it is self-evident that almost no optimization is possible, because just a wide single signature remains.
In contrast to this, figure 4.56 shows a very small but higher potential anyway, because the two remaining signatures differ at the leading position.

Hence an optimization, which can be perhaps realized by adapting either the quantization of the histogram or the second LUT in the way that this corresponding hit encounters the accurate cell as well, might possibly result in a combined single signature.
In addition to that, it is also important to discover that the y-axis in both figures shows the corresponding frequency of occurrence for each hit signature standardized to an event, because this value represents then the number of events containing this hit signature instead of the total amount, which entails obviously the identification and removal of bad events.

Although the hit signatures can be analyzed in multiple fashions up to here, it is convenient to offer finally a last feature, which simply visualizes the used table files. For this purpose, the parameter 'analysisInitTotalAnalysis' in the configuration file (see appendix A.2 on page A-4) offers such a visualization in combination with the corresponding bit, which is labeled with 'USEDSIGNATURETABLES', in the parameters 'analysisInitAnalysisMoreResultWarnings' and 'analysisInitAnalysisMoreResultDisplays' (see appendix A.3 on page A-79). As usual, the first parameter enables the results to be printed to the standard output screen, while the second one displays graphical front-ends. Further on, typical results, which represent the used 'codingTable' object, the used 'gradingPTable' object and the used 'gradingTable' object, are respectively depicted in the following figures 4.57, 4.58 and 4.59.

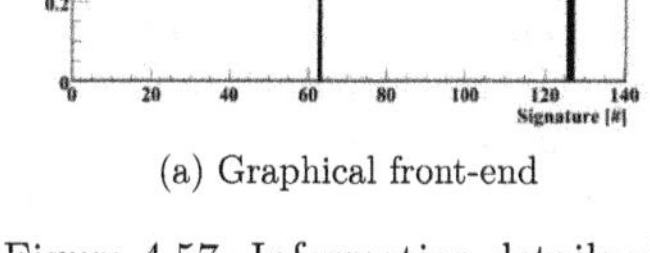

(a) Graphical front-end

```
AnalysisLIB WARNING: The used tables are:
CodingTable:
Signature: 2x0111111, Classification: 2x0000001
Signature: 2x1111110, Classification: 2x0000001
Signature: 2x1111111, Classification: 2x0000001
```

(b) Screen output

Figure 4.57: Information details about the signature distribution of the codingTable object

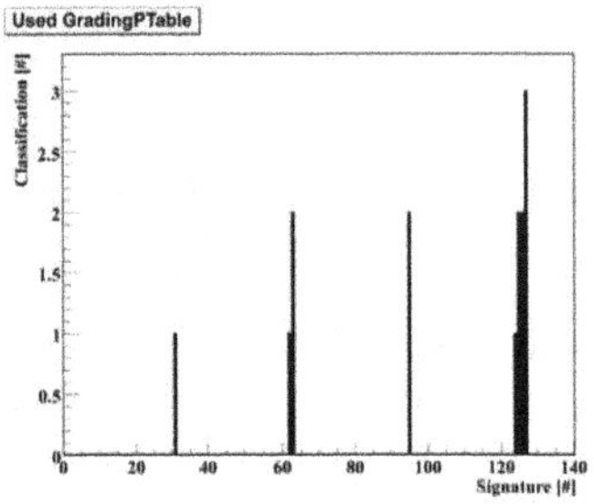

(a) Graphical front-end

```
AnalysisLIB WARNING: The used tables are:
GradingPTable:
Signature: 2x0011111, Classification: 2x0000001
Signature: 2x0111110, Classification: 2x0000001
Signature: 2x0111111, Classification: 2x0000010
Signature: 2x1011111, Classification: 2x0000010
Signature: 2x1111100, Classification: 2x0000001
Signature: 2x1111101, Classification: 2x0000010
Signature: 2x1111110, Classification: 2x0000010
Signature: 2x1111111, Classification: 2x0000011
```

(b) Screen output

Figure 4.58: Information details about the signature distribution of the gradingPTable object

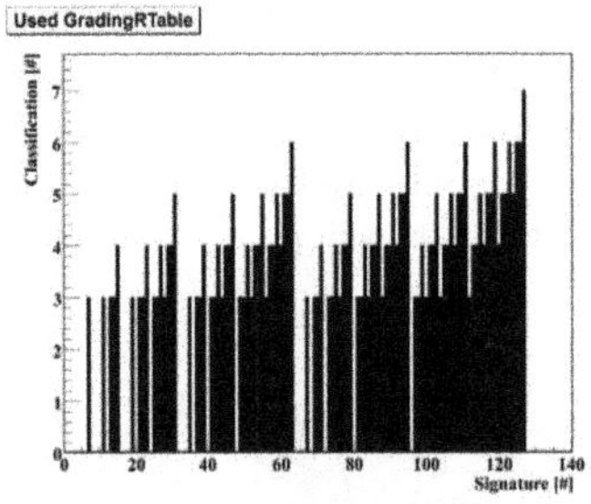

(a) Graphical front-end

```
AnalysisLIB WARNING: The used tables are:
GradingRTable:
...
Signature: 2x1110011, Classification: 2x0000101
Signature: 2x1110100, Classification: 2x0000100
Signature: 2x1110101, Classification: 2x0000101
Signature: 2x1110110, Classification: 2x0000101
Signature: 2x1110111, Classification: 2x0000110
Signature: 2x1111000, Classification: 2x0000100
Signature: 2x1111001, Classification: 2x0000101
Signature: 2x1111010, Classification: 2x0000101
Signature: 2x1111011, Classification: 2x0000110
Signature: 2x1111100, Classification: 2x0000101
Signature: 2x1111101, Classification: 2x0000110
Signature: 2x1111110, Classification: 2x0000110
Signature: 2x1111111, Classification: 2x0000111
```

(b) Screen output

Figure 4.59: Information details about the signature distribution of the gradingRTable object

With regard to these figures, a closer look reveals a regularity, which suggests the production of the table files by the application of an automatic algorithm introduced in appendix A.4 on page A-101. Furthermore the inspection of this suggestion verifies that the 'codingTable' is generated by the parameter value 'LUTGOODNESSTABLE', the 'gradingPTable' is created by the parameter value 'ONLINETABLE' and finally the 'gradingRTable' is constructed by the parameter value 'CROSSFOOTTABLE'. So based on these facts, the 'codingTable' is obviously connected to the transformation process, the 'gradingPTable' respects the detector stations and the 'gradingRTable' belongs to the sum of digits of the signatures in combination with a minimum classification.

4.8.4 Peak-finding Geometry

While remembering the cluster problem of the Hough transform applying an accumulator, which is introduced in section 2.1 on page 45, it is self-evident that an analysis must be taken into account to evaluate a more or less optimal geometry for the 2D and 3D Peak-finding process. Therefore this section presents an algorithm, which offers such a functionality.

By searching the field of digital image processing, an applicable theory is apparently identified by the connected component algorithms (see [RP66], [YK07], [PLC] and [DST92]), because it is possible to interpret the two dimensional histogram as an image, which contains connected components formed by the cluster of each peak. Thence the output of such an algorithm is defined by multiple labeled connected component clusters, which lead then directly to multiple special geometries. Moreover the accumulation of all cluster geometries finally results in statistical information about the frequency of occurrence for each geometry cell.
However a not negligible problem of such an algorithm is evidently determined by the circumstance that detailed information about the peaks, clusters and even the histogram is required, because otherwise two overlapping clusters might be identified for example as a single cluster or the correct accumulation of geometries might be impossible for instance. Nevertheless a simple modification can enable such an algorithm to cope with these problems.
So for that reason, the analysis of the once filled histogram, which contains the hits of all MC Tracks, is exchanged with multiple analysis loops separated for the hits of each MC Track, because then it is guaranteed that all data in the histogram corresponds to the same cluster and overlapping is not possible any more.
Moreover this concept solves also the problem for the correct accumulation of cluster geometries, because it is simply assumed that the connection between clusters is defined by the histogram cell in each cluster, which contains the corresponding MC Track. Consequentially the accumulation can be easily realized by accumulating each geometry cell with the corresponding cell in another geometry, which exhibits the identical relative coordinate to the cell containing the MC Track. In this connection, it has to be additionally noticed that this concept obviously causes the cluster representation to be changed at this point of the algorithm from absolute histogram coordinates

to relative coordinates with respect to a geometry origin.

Coming now to the implemented algorithm, it is clear that a loop across all MC Tracks is required, which performs for each MC Track separately the common Hough transform with a modified Peak-finding step characterized by the accumulation of cluster geometries. Further on, this algorithm will then finally result in an optimal Peak-finding geometry, which includes also statistical information about the frequency of occurrence for each geometry cell. Although this algorithm needs of cause a lot of computing time, this circumstance is not essential, because the time has to be spent just once to evaluate a typical standard geometry like the one, which is exemplary depicted in figure 4.60.

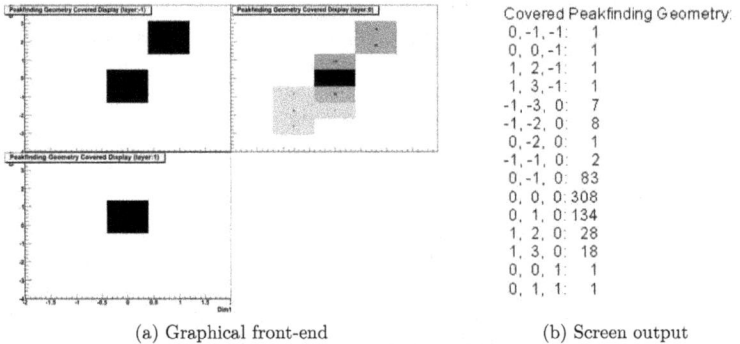

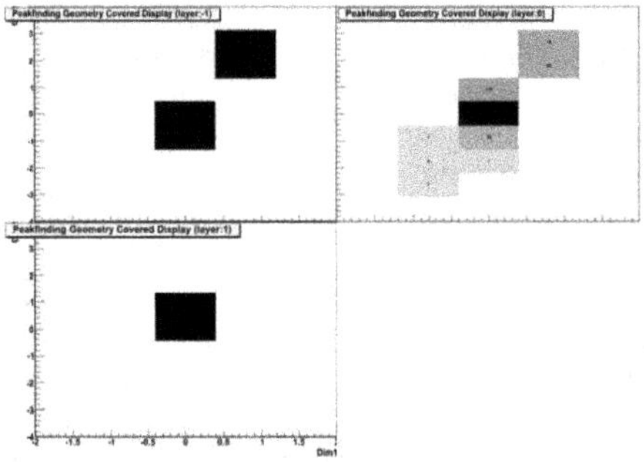

(a) Graphical front-end

```
Covered Peakfinding Geometry:
 0,-1,-1:   1
 0, 0,-1:   1
 1, 2,-1:   1
 1, 3,-1:   1
-1,-3, 0:   7
-1,-2, 0:   8
 0,-2, 0:   1
-1,-1, 0:   2
 0,-1, 0:  83
 0, 0, 0: 308
 0, 1, 0: 134
 1, 2, 0:  28
 1, 3, 0:  18
 0, 0, 1:   1
 0, 1, 1:   1
```

(b) Screen output

Figure 4.60: Resulting specification details for the automatically generated Peak-finding geometry

A closer look to this figure shows now the impressive fact that the frequency of occurrence for the geometry cells is almost completely concentrated around the origin. Hence the removal of the outer geometry cells reduces the complexity, while the effect on the performance is possibly negligible, because their frequency of occurrence is much less compared to the frequency of the inner cells.

Finally it should be mentioned that the automatic production of the geometry is turned on by setting the value of the parameter 'trackfinderFilterType' in the configuration file to either 'AUTOMATICFIRSTEVENTFILTER', 'AUTOMATICEACHEVENTFILTER' or 'AUTOMATICUPDATEEVENTFILTER' (see appendix A.2 on page A-4). In this context, it is furthermore obvious that the first value utilizes only the first event for the evaluation, the second one employs all events separately and the third one simply accumulates the geometries across the events. In addition to this, the graphical front-end is apparently enabled by setting the corresponding bits, which are labeled with 'PEAKFINDINGGEOMETRYVISUALIZATION', 'PROJECTEDPEAKFINDINGGEOMETRYVISUALIZATION', 'COVEREDPEAKFINDINGGEOMETRYVISUALIZATION' and 'COVEREDPROJECTEDPEAKFINDINGGEOMETRYVISUALIZATION', in the parameter 'analysisInitAnalysisMoreResultDisplays' (see appendix A.3 on page A-79). Accordingly the printout to the standard output screen can be achieved by setting the same bits in the parameter 'analysisInitAnalysisMoreResultWarnings' (see appendix A.3 on page A-79).

4.8.5 Track Information

The analysis, which is presented in this section, is able to give an imagination about the complexity and quality of the Hough tracking by offering the possibility to visualize the findable MC Tracks and the found tracks with regard to their motion trajectory through the actual STS detector setup. For this purpose, the production of figure 4.61a and figure 4.61b can be enabled by setting the corresponding bits, which are labeled with 'MCTRACKVISUALIZATION' and 'FOUNDTRACKVISUALIZATION' (see appendix

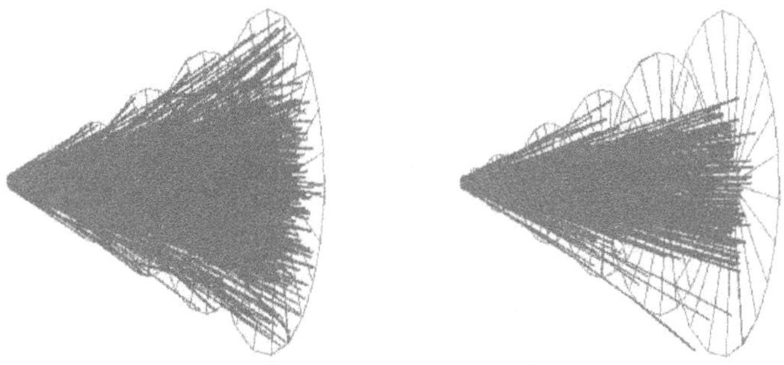

(a) Findable MC Tracks (b) Found tracks

Figure 4.61: Visualization of the MC Track input data and the found track output data of the algorithm for a single event

A.3 on page A-79), in the parameters 'analysisInitAnalysisMoreResultWarnings' and 'nalysisInitAnalysisMoreResultDisplays' in the configuration file (see appendix A.2 on page A-4). As obviously each bit is accessible in each parameter, the first one enables the results to be printed to the standard output screen, while the second one displays graphical front-ends.

In addition to this, the parameter 'DEBUGJUSTONEGOODTRACK', which is located in the source code file MiscLIB/include/defs.h, launches another big advantage of this analysis, which is commonly used for debugging purpose, because it features the visualization of a single findable MC Track in conjunction with its corresponding found track candidate. So the following figure 4.62 depicts a typical example of such a visualization.

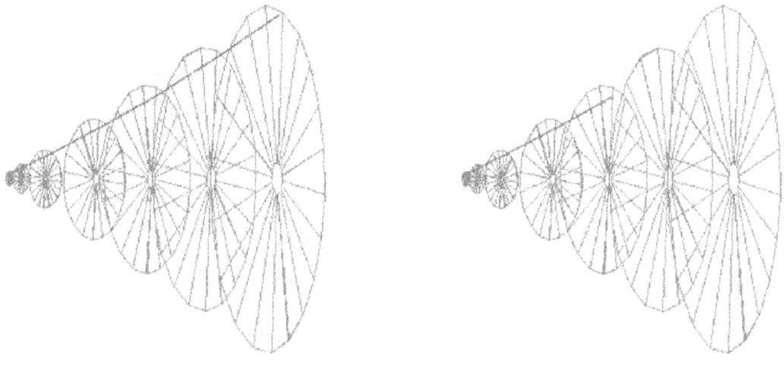

(a) Findable MC Track

(b) Found track

Figure 4.62: Visualization of a single MC Track input and the correspondingly found track output of the algorithm

Within regard to this figure, it is immediately clear that this concept features a detailed look on each motion trajectory of a found track in combination with its corresponding MC Track through the actual STS detector setup. But as this visualization contains obviously sometimes too few details, it is additionally possible to enable a summarizing printout to the standard output screen, which contains the total available and thus very detailed information. Further on, such a printout looks then typically like the following figure 4.63.

```
MCTrack Visualization:
px = -0.514, py = 0.401, pz = 1.760
  hitId =    31, trackId = 956, x = -  1.473 y =  1.143, z =   5, time = 0.177
  hitId =   713, trackId = 956, x = -  2.998 y =  2.288, z =  10, time = 0.356
  hitId = 1394, trackId = 956, x = -  6.088 y =  4.608, z =  20, time = 0.715
  hitId = 2108, trackId = 956, x = -12.858 y =  9.273, z =  40, time = 1.438
  hitId = 1508, trackId = 956, x = -20.458 y = 14.019, z =  60, time = 2.172
  hitId = 4832, trackId = 956, x = -28.889 y = 18.915, z =  80, time = 2.916
  hitId = 9248, trackId = 956, x = -37.976 y = 24.062, z = 100, time = 3.671

Found Track Visualization:
px = -0.513, py = 0.414, pz = 1.756, dim1 = 75, dim2 = 94, dim3 = 142
  hitId =    31, trackId = 956, x = -  1.473 y =  1.143, z =   5, time = 0.177
  hitId =   713, trackId = 956, x = -  2.998 y =  2.288, z =  10, time = 0.356
  hitId = 1394, trackId = 956, x = -  6.088 y =  4.608, z =  20, time = 0.715
  hitId = 2108, trackId = 956, x = -12.858 y =  9.273, z =  40, time = 1.438
  hitId = 1508, trackId = 956, x = -20.458 y = 14.019, z =  60, time = 2.172
```

Figure 4.63: Output summary of a single MC Track input and the correspondingly found track output of the algorithm

With regard to this figure, it is now obvious that the total available information for a MC Track consists of the momentum and additional information, which belong to each attached hit. Furthermore this hit specific information contains collaterally the id of the hit, the id of the corresponding MC Track, the coordinates and the time of flight. Although the id of the MC Track offers no supplementary information here due to the circumstance that it is always equal, it might be of course a different story for the found tracks.
Besides this, the total available information for a found track consists evidently of the reconstructed momentum, which is directly connected to the histogram coordinates, and additional information for each hit contributing to this track. So within this context, this additional information for each hit is evidently equal to the information mentioned earlier. But in contrast, it is immediately clear that hits, which belong originally to another MC Track, can appear here as well. Hence the information about the id of the MC Track is really essential.

4.9 Hardware Related Analyses

In the following, it is at first important to realize that this section consists only of analyses, which are directly related to statistics required for some estimations in the hardware implementation. So this means more precisely that the parameters of the Hough transform algorithm are explicitly not any longer in the center of interest, but quite contrary to this, the main focus is changed in almost all cases to the computation of the size of data, which occurs on a special interface in the algorithm. Furthermore it is clear that this information can be then of course used to determine for instance the amount of memory, which is needed for that interface to decouple algorithm parts.

So with regard to this purpose, the obviously expected statistics exhibit information about the hit distribution via the histogram layers, the number of track candidates per histogram layer column and row as well as the number of track candidates per histogram layer itself. Although this information seems to be at a first glance dispensable, a closer look to section 5 on page 295 and each subsequent section shows that it is really essential to approximate for example the number of hit read-outs for the HBuffer unit or the required memory size in the Serialization unit as well as the required memory size and thus strategy for the LBuffer unit.

4.9.1 Statistic of the Hit Distribution via the Histogram Layers

This section presents an analysis, which is of major importance for the prediction of the timing behavior related to the hardware implementation, because it helps to estimate the time critical number of transformed hit read-outs from the HBuffer memory. However it has to be noticed in this connection that such an estimation is obviously directly related to the analysis, which is presented in section 4.8.2 on page 268, because a wider range will lead immediately to more histogram entries for each hit and thus more HBuffer read-outs.
Nevertheless the following figure 4.64 is able to illustrate the number of hits, which correspond to a certain number of HBuffer memory read-outs. Besides

this, it is also clear within this context that such a distribution gets better, the closer the number of read-outs moves to one, because each hit is then only read once, which defines of course the optimum.

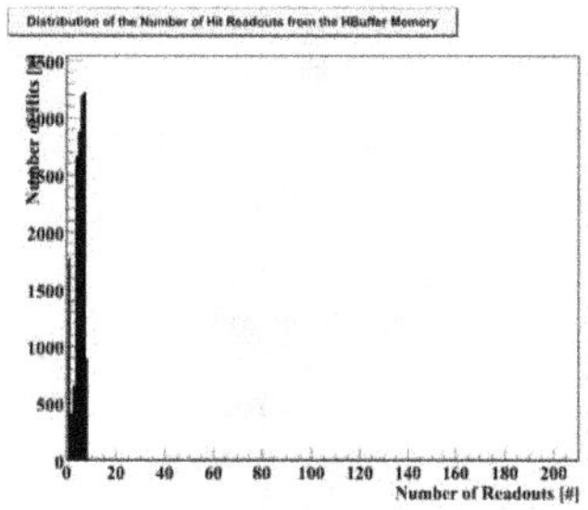

(a) Graphical front-end

AnalysisLIB WARNING: Distribution of the hit readout/processing to build all histogram layers:
Number of readouts: 1 => #Hits: 1767
Number of readouts: 2 => #Hits: 404
Number of readouts: 3 => #Hits: 650
Number of readouts: 4 => #Hits: 2659
Number of readouts: 5 => #Hits: 2873
Number of readouts: 6 => #Hits: 3193
Number of readouts: 7 => #Hits: 3222
Number of readouts: 8 => #Hits: 890

(b) Screen output

Figure 4.64: Typical distribution of the hit to histogram layer affiliation ratio for a single event

With regard to this figure, it is rapidly clear that such a distribution can be now used to compute the total number of transformed hit read-outs from the HBuffer unit by simply adding the results of the multiplication between the number of hits and their corresponding number of read-outs.
However it has to be additionally noted in this connection that the implicit made assumption of instantiating just a single histogram layer is a prerequisite for this computation, because if multiple layers are considered, a transformed hit, which corresponds to multiple histogram layers, could be apparently read-out out just once anyway. But even if such a situation is at a first glance astonishing, a more detailed look could immediately show that another circumstance, which is characterized by the spreading of the transformed hits via the serial processing borders of the parallel instantiated histogram layers, has naturally also to be taken into account.
For this purpose, figure 4.65 is produced to illustrate the distribution of the average number of read-outs per hit with regard to the parallel instantiated histogram layers.

So with regard to this figure, the spreading of the transformed hits leads apparently to the anticipated distribution, because as more histogram layers are instantiated as less transformed hits has to be read-out repeatedly, which decreases obviously the average. But having figure 4.64 in mind, it is clear that this process will lead quite early to a saturation region, because

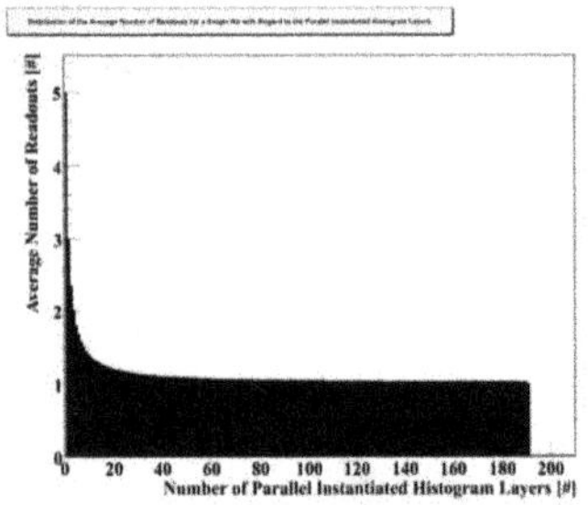

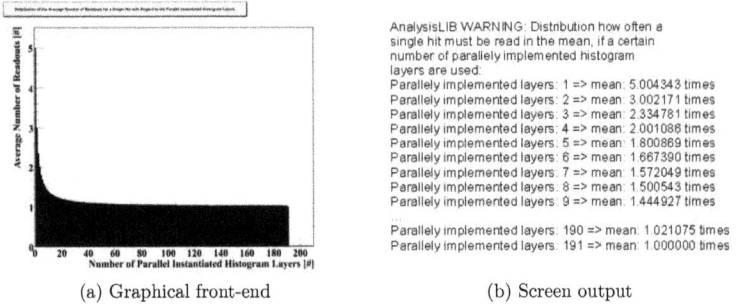

(a) Graphical front-end

```
AnalysisLIB WARNING: Distribution how often a
single hit must be read in the mean, if a certain
number of parallely implemented histogram
layers are used:
Parallely implemented layers: 1 => mean: 5.004343 times
Parallely implemented layers: 2 => mean: 3.002171 times
Parallely implemented layers: 3 => mean: 2.334781 times
Parallely implemented layers: 4 => mean: 2.001086 times
Parallely implemented layers: 5 => mean: 1.800869 times
Parallely implemented layers: 6 => mean: 1.667390 times
Parallely implemented layers: 7 => mean: 1.572049 times
Parallely implemented layers: 8 => mean: 1.500543 times
Parallely implemented layers: 9 => mean: 1.444927 times
...
Parallely implemented layers: 190 => mean: 1.021075 times
Parallely implemented layers: 191 => mean: 1.000000 times
```

(b) Screen output

Figure 4.65: Distribution of the average number of read-outs for a single hit with regard to the parallel instantiated histogram layers

the maximal number of read-outs for a single hit is actually determined by eight times. So the instantiation of more histogram layers has evidently a negligible improving effect, because a transformed hit has to be read out at the utmost once. However such an improving effect occurs anyhow, because the correct statistic for the spreading of the transformed hits is approximated by assuming a unique probability distribution for each hit to be spread.

Furthermore a detailed analysis of this assumed probability shows that a hit, which contributes only to a single histogram layer, must be committed with the spreading probability zero, because it is impossible to be spread. Thence the probability of all other hits can be computed to $\frac{number\ of\ affected\ layers-1}{number\ of\ instantiated\ layers}$. So if three instantiated histogram layers are for instance considered in combination with a hit, which contributes to two layers, the corresponding probability would be evaluated to $\frac{1}{3}$. Although this approach is at a first glance surprising, figure 4.66 depicts a good illustration for this example, which features the interpretation that just a single possible spreading exists with

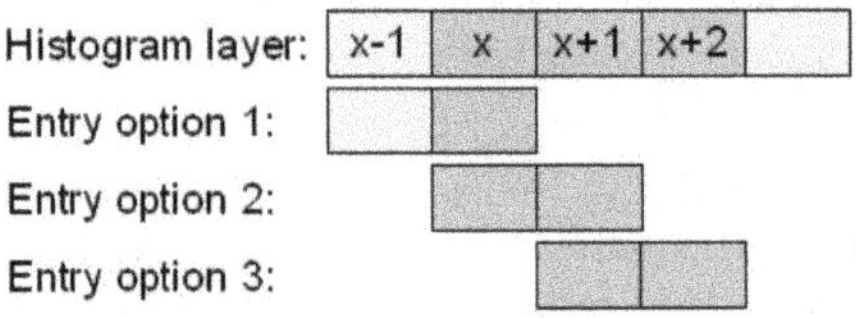

Figure 4.66: Illustration for the hit spreading probability

regard to three possible histogram layer insertion options for that hit. Thus it is immediately clear that such an approach defines a quite good approximation.

Coming now to the end, the final figure, which is introduced by this analysis, offers a prediction for the approximated number of clock cycles, which is needed for the entire histogram processing of a common FPGA implementation with regard to the number of instantiated parallel histogram layers. Thence the covered algorithm steps include the complete Histogramming as well as the whole histogram read-out.
Moreover it should be additionally mentioned here that a pipeline stall, which can obviously occur during the histogram read-out, is not considered. Further on, the corresponding choice due to the circumstance that the histogram can be read-out row-wise at the bottom or column-wise at right must be set by the parameter 'analysisInitReadoutColumnsInParallel' in the configuration file (see appendix A.2 on page A-4).
So based on these facts, figure 4.67 depicts a typical result of such an analysis for the column-wise read-out at right, while the result for the other read-out mode is skipped, because the only difference is located inside the calculation and the graphical front-ends look thus comparable.

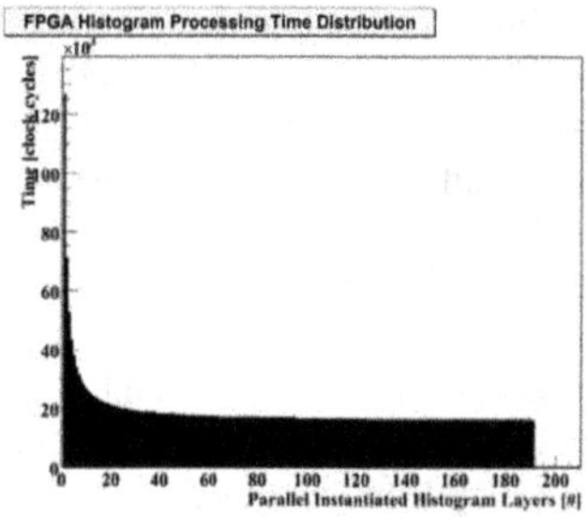

(a) Graphical front-end

```
AnalysisLIB WARNING: Distribution of the time needed for
processing the histogram, if a certain number of parallely
implemented layers are used:
Parallely implemented layers: 1 => time: 102615 clock cycles
Parallely implemented layers: 2 => time: 59200 clock cycles
Parallely implemented layers: 3 => time: 44686 clock cycles
Parallely implemented layers: 4 => time: 37429 clock cycles
Parallely implemented layers: 5 => time: 33152 clock cycles
Parallely implemented layers: 6 => time: 30172 clock cycles
Parallely implemented layers: 7 => time: 28172 clock cycles
Parallely implemented layers: 8 => time: 26544 clock cycles
Parallely implemented layers: 9 => time: 25419 clock cycles
Parallely implemented layers: 10 => time: 24468 clock cycles
Parallely implemented layers: 11 => time: 23644 clock cycles
Parallely implemented layers: 12 => time: 22915 clock cycles
...
```

(b) Screen output

Figure 4.67: Typical distribution of the histogram processing time for a single event on the FPGA with regard to the parallel instantiated layers

Besides this, the analysis can be turned on with the parameter 'analysisInitHitReadoutDistribution' in the configuration file (see appendix A.2 on page A-4). Further on, the output is controlled by setting the corresponding bits, which are labeled with 'HITREADOUTDISTRI-

BUTION', 'HITREADOUTAVERAGEDISTRIBUTION' and 'FPGAHISTOGRAMPROCESSINGTIMING', in the parameters 'analysisInitAnalysisResultWarnings' and 'analysisInitAnalysisResultDisplays' (see appendix A.3 on page A-79). Furthermore it is self-evident that the first parameter enables the results to be printed to the standard output screen, while the second one displays graphical front-ends.

4.9.2 Statistic of the Number of Track candidates per Histogram Layer Column and Row

The analysis, which is presented in this section, offers the possibility to compute the necessary amount of memory for the Serialization process. As furthermore this step of the algorithm forms obviously a general interface between the parallel and serial computation, it is commonly suggested to be a bottleneck in the processing pipeline including a stall potentiality, which can be naturally avoided by using enough decoupling memory applying the FIFO strategy. Thence this analysis features not only an interesting statistic, because upon detailed inspection, it is clear that it can be used to estimate the amount of memory, which is necessary to avoid such a pipeline stall during the histogram read-out.
But since the time of the arriving data at the input of the Serialization unit is obviously very critical for such an estimation, it is suggested to produce two figures, whose content belongs either to the row-wise histogram read-out at the bottom or to the column-wise histogram read-out at right (see section 3.3.1 on page 166 and section 3.3.1 on page 171).
However it is self-evident that these two figures exhibit not enough statistics for a detailed analysis. Therefore each of these read-out strategies require three different figures. While the first one contains the minimal number of track candidates, which occur in the column or row of all histogram layers, the second one consists of the maximal number and finally the third figure shows of course the average.
For this purpose, the figures 4.68, 4.69 and 4.70 show a typical distribution of the minimal, maximal and average occurring track candidates with regard to the column-wise histogram read-out at right. Besides this, it has to be noticed that the figures for the row-wise histogram read-out at the bottom are apparently not shown here, because they are about the same.

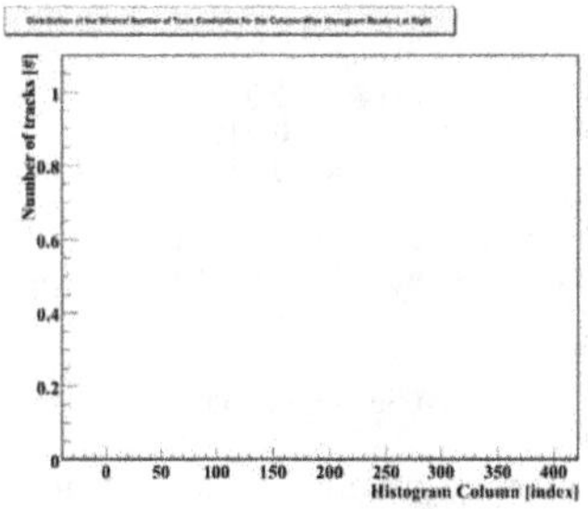

(a) Graphical front-end

AnalysisLIB WARNING: Distribution of the minimal number of the found tracks per histogram column in all layers:
No track is found!!!

(b) Screen output

Figure 4.68: Resulting distribution for the minimal number of track candidates per histogram column

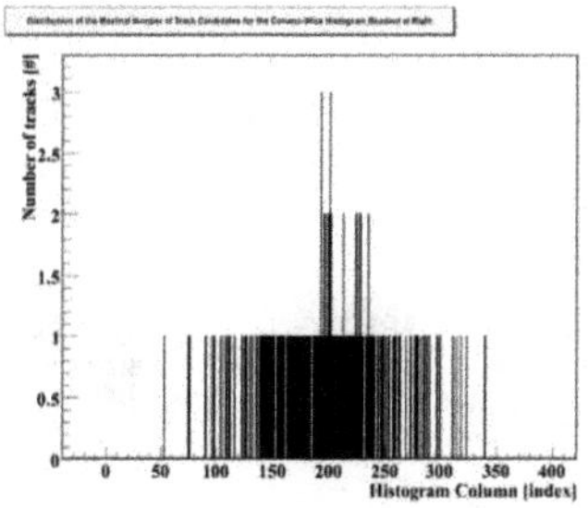

(a) Graphical front-end

AnalysisLIB WARNING: Distribution of the maximal number of the found tracks per histogram column in all layers:
...
Index: 195 => #Tracks: 1
Index: 196 => #Tracks: 2
Index: 197 => #Tracks: 1
Index: 198 => #Tracks: 2
Index: 199 => #Tracks: 1
Index: 200 => #Tracks: 2
Index: 201 => #Tracks: 3
There are 173 tracks found in the worst case in one layer.
...

(b) Screen output

Figure 4.69: Resulting distribution for the maximal number of track candidates per histogram column

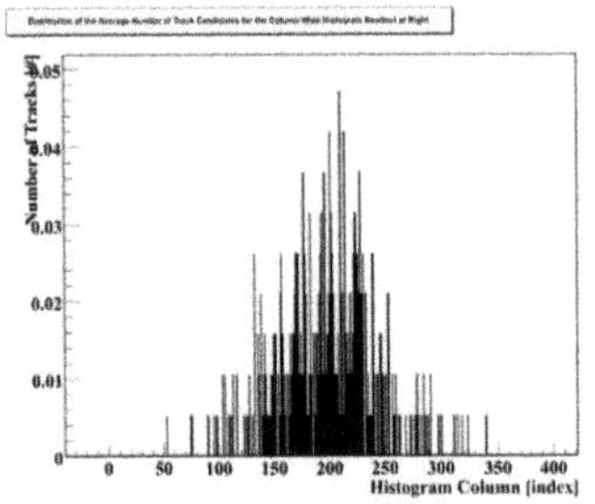

(a) Graphical front-end

```
AnalysisLIB WARNING: Distribution of the
average number of the found tracks per
histogram column in all layers:
...
Index: 195 => #Tracks: 0.031414
Index: 196 => #Tracks: 0.036649
Index: 197 => #Tracks: 0.010471
Index: 198 => #Tracks: 0.015707
Index: 199 => #Tracks: 0.015707
Index: 200 => #Tracks: 0.020942
Index: 201 => #Tracks: 0.041885
...
```

(b) Screen output

Figure 4.70: Resulting distribution for the average number of track candidates per histogram column

Further on, a more precise look to the Serialization process shows immediately that the systolic processed histogram read-out would require only supplementary FIFO memory, if multiple track candidates arrive on the final output stage at the same time, because these ones have to be then written to a memory, which accepts just a single track candidate at the same time. Thence the first track candidate can be processed on-line, while the others have to be delayed in such a memory. For this purpose, figure 4.71 depicts now the result of an analysis, which takes additionally a typical data arrival into account.

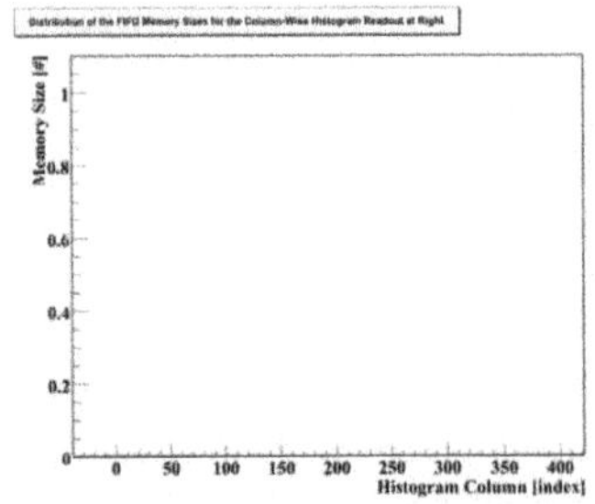

(a) Graphical front-end

```
AnalysisLIB WARNING: Distribution of the
assumed histogram column FIFO sizes for
the separator of the found tracks:
No Fifos are needed!!!
```

(b) Screen output

Figure 4.71: Resulting information for the size of the buffer memory needed for the Serialization process

With regard to this figure, it is impressing that no supplementary memory is needed for the Serialization process, because there are at no time track candidates detected, which arrive on the final output stage at the same time. Beyond that, it should be mentioned here that the production of these figures is turned on by the parameter 'analysisInitNumberOfTracksPerColumn' for the column and 'analysisInitNumberOfTracksPerRow' for the row in the configuration file (see appendix A.2 on page A-4). In addition to this, the output of the analysis is controlled by setting the corresponding bits, which are labeled with 'MINIMALTRACKSDISTRIBUTION', 'MAXIMALTRACKSDISTRIBUTION', 'AVERAGETRACKSDISTRIBUTION' and 'FIFOSDISTRIBUTION', in the parameters 'analysisInitAnalysisResultWarnings' and 'analysisInitAnalysisResultDisplays' (see appendix A.3 on page A-79). By the way, it is self-evident that the first parameter enables the results to be printed to the standard output screen, while the second one displays graphical front-ends.

4.9.3 Statistic of the Number of Track candidates per Histogram Layer

The analysis, which is presented in this section, offers the possibility to compute the necessary amount of memory for the LBuffer unit. Thence this analysis features again not just an interesting statistic, because the detailed inspection of each introduced implementation of the Hough transform algorithm clarifies rapidly that the number of track candidates per histogram layer is equal to the number of entries, which must be accommodated by the LBuffer unit.

Further on, this analysis can be used for two purposes at the same time, because the first one is obviously determined by the computation of the necessary amount of memory in total as well as separately for each corresponding histogram layer, while the second purpose comprises the determination of a pipeline strategy for the memory usage in all diverse implementations. However a closer look to these two features evidently suggests furthermore that both must be strongly related to each other, because the memory usage strategy influences usually also the required amount of total memory to ensure a secure processing.

For example, if a memory is used in combination with a ring buffer strategy for the read and write access, there should apparently be a security region, which ensures that the write process can not overtake the read process and vice versa.

Coming now to the point, figure 4.72 depicts a typical distribution of the occurring number of track candidates per histogram layer in a single event.

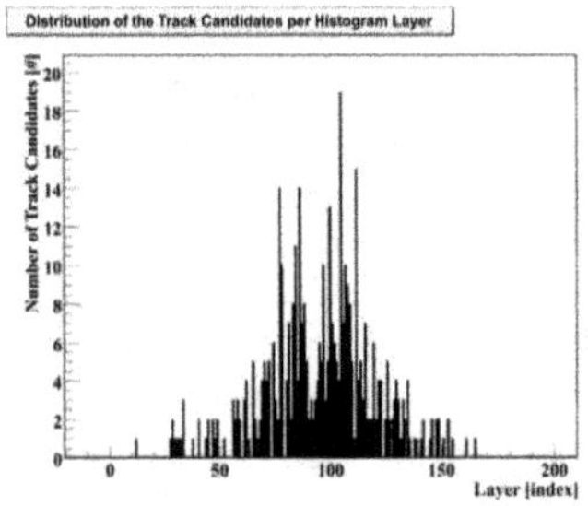

(a) Graphical front-end

```
AnalysisLIB WARNING: Distribution of the
number of track candidates per histogram layer:
...
Index: 104 => #Tracks:   4
Index: 105 => #Tracks:  19
Index: 106 => #Tracks:   7
Index: 107 => #Tracks:  10
Index: 108 => #Tracks:   9
Index: 109 => #Tracks:   8
Index: 110 => #Tracks:   5
Index: 111 => #Tracks:   1
Index: 112 => #Tracks:  15
...
```

(b) Screen output

Figure 4.72: Typical distribution for the number of track candidates per histogram layer

Having yet the example of the ring buffer in mind, it is clear that the distribution of the number of read and write accesses represents the major important information to avoid the overtaking instead of the absolute numbers. Thus the additional figure 4.73 shows a typical distribution of the difference number of track candidates for consecutive histogram layers.

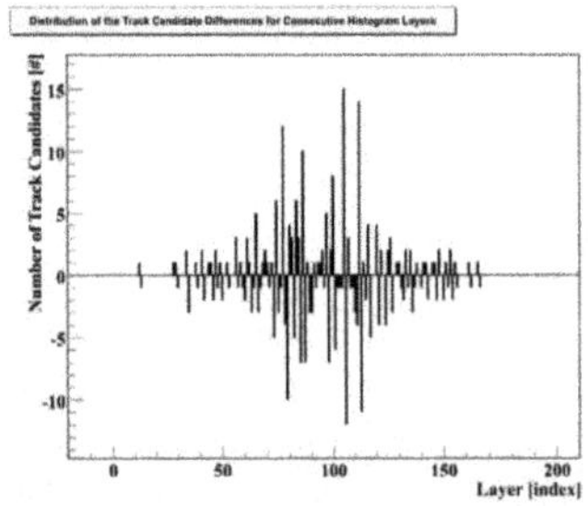

(a) Graphical front-end

```
AnalysisLIB WARNING: Distribution of the
difference number of track candidates between
consecutive histogram layers:
...
Index: 104 => #Tracks: -  1
Index: 105 => #Tracks:   15
Index: 106 => #Tracks: - 12
Index: 107 => #Tracks:    3
Index: 108 => #Tracks: -  1
Index: 109 => #Tracks: -  1
Index: 110 => #Tracks: -  3
Index: 111 => #Tracks: -  4
Index: 112 => #Tracks:   14
...
```

(b) Screen output

Figure 4.73: Typical distribution for the variation between the number of track candidates for consecutive histogram layers

So this means more precisely that the layer i contains now the information, which is computed by the following equation.

$$\text{difference number}[\mathit{layer}\ i] = \text{number}[\mathit{layer}\ i] - \text{number}[\mathit{layer}\ i-1]$$

Returning again to the ring buffer example, a final problem is obviously defined by the circumstance that a write access must not be consecutively followed by a read access. Hence the security region can not be determined by the maximum number of track candidates per histogram layer, but rather by the amount of memory, which is necessary to process the maximum number of write accesses. But since such a theoretical construct would be really specific to the event and additionally very hard to evaluate, it is apparently easier to calculate a worst-case amount of memory by simply adding the quantity of the maximum numbers for the track candidates per layer. For this purpose, figure 4.74 depicts this summed number of maximum track candidates with regard to a certain number of parallel instantiated histogram layers. As however the time to read these track candidates out of the memory is furthermore equal to the summed amount, because each read-out takes a single clock cycle, the following figure is created in analogousness to figure 4.67 of section 4.9.1 on page 287.

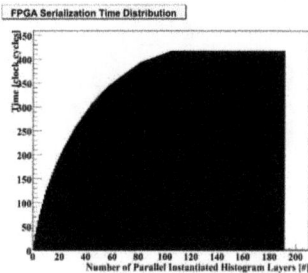

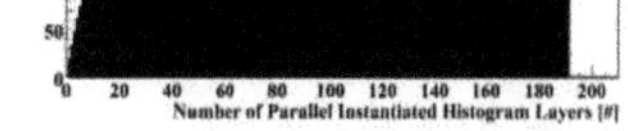

(a) Graphical front-end

```
AnalysisLIB WARNING: Distribution of the time needed for
the Serialization process, if a certain number of parallely
implemented histogram layers are used:
Parallely implemented layers: 1 => time: 19 clock cycles
Parallely implemented layers: 2 => time: 34 clock cycles
...
Parallely implemented layers: 14 => time: 156 clock cycles
...
Parallely implemented layers: 20 => time: 196 clock cycles
...
Parallely implemented layers: 24 => time: 218 clock cycles
...
Parallely implemented layers: 104 => time: 416 clock cycles
Parallely implemented layers: 105 => time: 417 clock cycles
...
Parallely implemented layers: 191 => time: 417 clock cycles
```

(b) Screen output

Figure 4.74: Typical distribution for the worst-case estimation of the FPGA Serialization time with regard to the parallel instantiated histogram layers

So based on this concept, it is obvious that the first entry, which represents a single instantiated histogram layer, has to be equal to the maximum number of track candidates per layer. By remembering now figure 4.72, it is self-evident that this circumstance can be easily approved by the search of this maximum. Moreover it is additionally clear that the second entry has to be equal to the sum of this value and the next smaller maximum number of track candidates per layer, and so on.

While finally recapitulating the introduced fragments of this analysis, the complexity can be guessed by the summary of the first figure 4.72, which offers the possibility to diagnose the total amount of necessary memory for the track candidates per histogram layer, the second figure 4.73, which helps to determine the required amount of memory related to the 3D Peak-finding strategy, and the third figure 4.74, which enables the worst-case estimation of the FPGA Serialization time with regard to the number of parallel instantiated histogram layers.
Further on, it should be also mentioned that the analysis is turned on by the parameter 'analysisInitNumberOfTracksPerLayer' in the configuration file (see appendix A.2 on page A-4). To this, the output of the analysis is as usual controlled by setting the corresponding bits, which are labeled with 'TRACKSPERLAYERDISTRIBUTION', 'TRACKDENSITYPERLAYERDISTRIBUTION' and 'FPGASERIALIZATIONPROCESSINGTIME', in the parameters 'analysisInitAnalysisMoreResultWarnings' and 'analysisInitAnalysisMoreResultDisplays' (see appendix A.3 on page A-79). Furthermore it is clear that the first parameter enables the results to be printed to the standard output screen, while the second one displays graphical front-ends.

Chapter 5

Results

This section demonstrates the assimilation of the implementations, which are introduced in section 3 on page 95, to the CBM experiment, which is illustrated in section 1.2 on page 7 and section 1.3 on page 10, by presenting the results for the varying parameters of the algorithm and their consequences. For this purpose, the first part of this section can be understood as customization or optimization recipe by the sequenced usage of the analyses, which are described in section 4 on page 213, while the second part illustrates the resulting tracking and implementation specific performance.

Besides this, it has to be supplementary noticed here that even if the results are attained by the application of the analytic formula in combination with the averaged magnetic field integrate, all consulted analyses can be surely also applied to each other introduced computational kernel like the Runge-Kutta approach for instance. Moreover the only reason for the selection of the analytic formula originates in the proof that the novel developed concept of a parametric description for the motion trajectory of a charged particle in the occurring inhomogeneous magnetic field is really feasible.

5.1 Searching for Well-Defined Parameters of the Algorithm

Due to the usage of the analytic formula for the computational kernel in the implemented Hough transform, the first objective in the focus is evidently required to be the magnetic field. This means more precisely that the analyses, which are presented in section 4.4 on page 223, have to be consulted to check whether the model representing the real inhomogeneous magnetic field in the formula is feasible or not. For this purpose, the two produced figures 5.1a and 5.1b have to be compared, because the first one shows the optimal constant magnetic field factor of each detector station, which is evaluated under the constraint that most hits of the MC Tracks encounter the correct histogram cell, while the second figure displays a cross-section of the averaged integrated magnetic field along the beam axis.

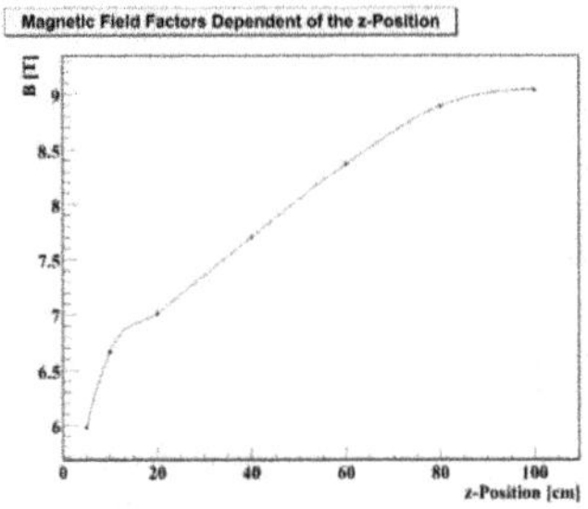

(a) Automatically evaluated optimal constant magnetic field factors for the detector stations

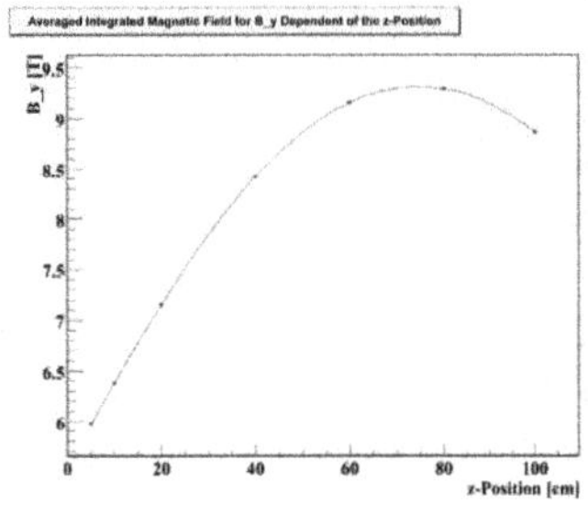

(b) Y component of the averaged integrated magnetic field in dependency of the beam axis

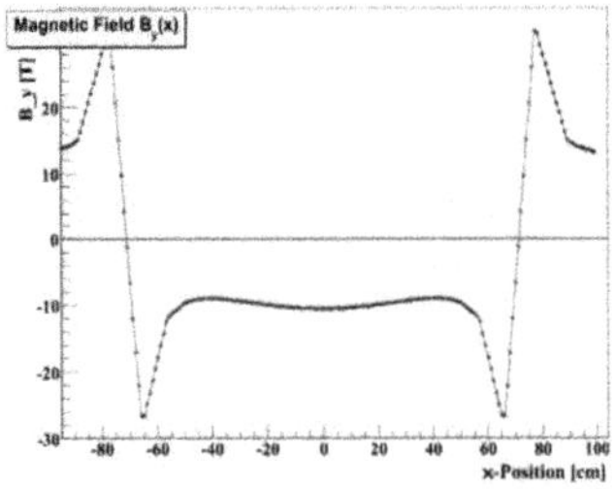

(c) Y component of the magnetic field in dependency of the x direction

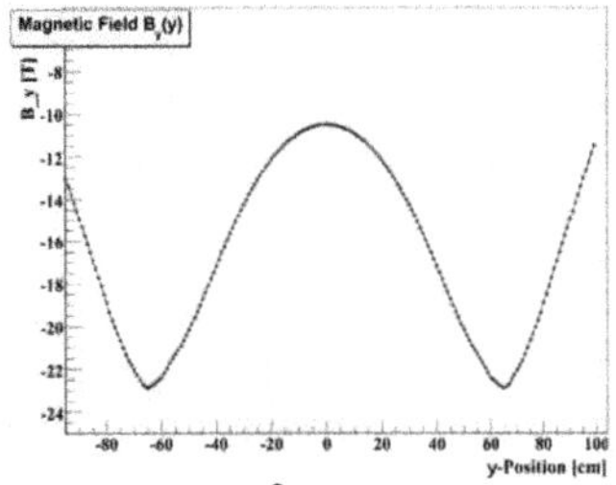

(d) Y component of the magnetic field in dependency of the y direction

Figure 5.1: Visualization of the similarity prerequisite for the analytic formula due to the inhomogeneous magnetic field

Even if these figures are not identical, the similarity should be sufficient for an efficient Hough tracking. Furthermore they can be actually not identical, because the constant factors are fuzzy due to the accumulation process representing a uniformity in the x and y direction of each detector station, whereas the magnetic field contrarily varies in these directions as the figures

5.1c and 5.1d depict. In addition to this, a real motion trajectory of a charged particle track is of course also not influenced by only a cross-section of the magnetic field.

So based on these facts neither the constant magnetic field factors nor the averaged integrated magnetic field provide absolute correct information for a concrete motion trajectory. Nevertheless they exhibit both good approximations, which just converge from different starting aspects. Hence the similarity of the two produced figures advises the fulfilled prerequisite and the focus can be set in the following to the parameters of the algorithm.

5.1.1 Dimensions of the Hough Space

Having now the prerequisites of the implemented Hough transform clarified, the next target to survey is obviously the Hough space. In this examination, it is rapidly distinct that the main duty belongs to the feasible configuration of the Hough space dimensions.
So since the region of interest is determined by the physics of the experiment, which is introduced in section 1.2 on page 7, it is self-evident that the minimum and the maximum of each dimension is fixed to dependent values as it is stated in section 2.2 on page 55. Moreover the usage of an accumulator for the implementation of the histogram, as it is presented in section 2.1 on page 45, claims further the commitment of a quantization configuration, which is generally also not a big deal, because the quantization is commonly set uniform to the original coordinate space. However this approach is apparently not feasible here, because the huge amount of steps in the coordinate space will explode the amount of memory, which is required for the accumulator. Nevertheless the number of steps must be at least big enough that multiple tracks will not encounter the same cell and all tracks will be as far from each other as needed to cope with the cluster problem, which is introduced in section 2.1 on page 45. Collaterally a further reason for the disqualification of the common approach is determined by the dependency of the Hough space dimensions from each other as deduced in section 2.2 on page 55 and section 2.3 on page 58. Beyond that, this is not just yet the total difficulty, because the restriction for the Hough curve slopes, which is caused by the planned efficient implementation of a systolic processable histogram on FPGAs (see section 3.3.1 on page 152) and already advertised in section 3.1 on page 96, complicates the situation certainly additionally.

Thence it is clear that multiple aspects are required to be able to cope with each fragment of this complex of problems. Thus the consecutively following

part of this section covers the major important constraint, which is constituted by the slope restriction, while the others inspect the quality of the peaks related to the quantization.
But before starting, it has to be finally mentioned that the subsequent results are not special to a single dedicated event, but rather averaged or accumulated for an amount of 100 ordinary events, which are characterized by the data of central Au+Au collisions with the beam energy of 25 AGeV and the standard detector setup of the STS.

Histogram Quantization Related to the First and Second Dimension Ratio

Since the Hough curve slopes are only allowed to realize 45 to 90 degrees in the actual Hough transform implementation (see section 3.1 on page 96), it is a common approach to cope with such a restriction by the selection of a well configured scaling for the affected dimensions. For that reason, a good quantization ratio for the first and the second dimension eases the slope enforcement, which is realized by the introduction of a LUT computation error. For this purpose, the analysis presented in section 4.8.1 on page 265 can be used to evaluate such a feasible ratio, because it is able to compute the percentage of LUT corrections, which are necessary to fulfill the requirement with regard to a specified ratio and given data. Hence figure 5.2 depicts now the percentage of corrected Hough curves for different histogram layer dimension ratios based on some common sample data sets. Moreover it is needless to say that the optimal ratio is defined by no LUT correction, which

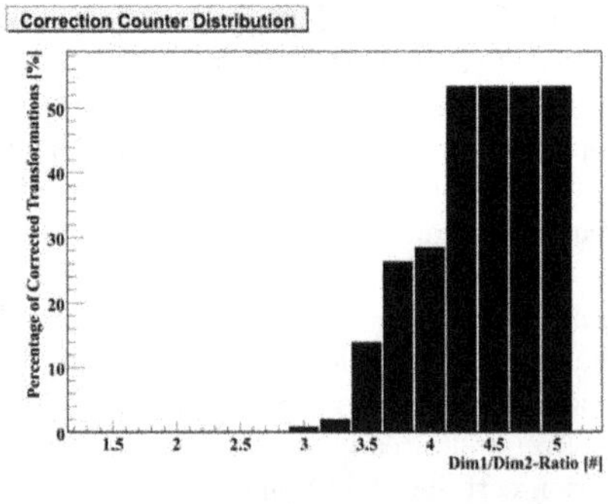

Figure 5.2: Resulting percentage distribution for the corrected Hough curves with regard to diverse given histogram layer dimension ratios

means also no error.

Further on, this figure advices now the ratio 2.5 or more precisely $\frac{319}{127}$ to be optimal for the first and the second dimension of the histogram.
However this optimal ratio can not be specified strictly, because on the one hand the transformation should be as accurate as possible, but on the other hand the histogram should require as sparse memory as possible. So the major effective aspect influenced by these two antidromic circumstances is certainly defined by the impact of the ratio to the tracking quality, which can be evaluated by the analyses presented in section 4.7 on page 258. Therefore table 5.1 presents the correspondent tracking quality for the same histogram layer dimension ratios and the same sample data sets.

$\frac{\textbf{dim2}}{\textbf{dim1}}$ - Ratio	$\frac{191}{127}$	$\frac{223}{127}$	$\frac{255}{127}$	$\frac{287}{127}$	$\frac{319}{127}$	$\frac{351}{127}$	$\frac{381}{127}$	$\frac{413}{127}$
Efficiency [%]	85	84	84	83	86	84	86	85
Fake rate [%]	14	7	6	6	9	8	8	8
Ghost rate [%]	6	5	4	4	5	4	7	7
Clones rate [%]	14	12	11	10	8	7	7	5
Identification rate [%]	54	62	65	67	67	70	67	70
Reduction rate [%]	76	79	80	81	80	82	80	81

$\frac{\textbf{dim2}}{\textbf{dim1}}$ - Ratio	$\frac{445}{127}$	$\frac{477}{127}$	$\frac{509}{127}$	$\frac{541}{127}$	$\frac{573}{127}$	$\frac{605}{127}$	$\frac{635}{127}$
Efficiency [%]	84	68	59	58	55	42	41
Fake rate [%]	7	7	7	7	7	7	9
Ghost rate [%]	6	6	6	7	6	8	7
Clones rate [%]	5	3	2	2	2	1	1
Identification rate [%]	72	72	74	74	76	74	75
Reduction rate [%]	82	86	88	88	89	91	92

Table 5.1: Tracking quality results for different given histogram layer dimension ratios

With regard to this table, it is now obvious that all ratios higher than 3.5 or $\frac{445}{127}$ disqualify themselves, because the high number of corrected Hough curves end up in a dramatically decreased efficiency. Further on, the almost stable tracking quality of the rest in combination with the distribution of the Hough curve corrections advices the selection of the ratio 3 or more precisely $\frac{381}{127}$, because the number of corrected Hough curves is smaller than 1% and thus ostensibly negligible, while the quality is still good.

Histogram Quantization Because of the Track Separation

As the ratio of the first and the second dimension of the histogram is successfully determined, the focus is now set to the combination of these two dimensions with the third one, which is surely kept constant before.
For this purpose, there are three different key elements, which have to be taken into account. The first one is evidently defined by the amount of required histogram memory, which increases with the quantization, but should be kept as small as possible. In contrast to this, the antidromic counterpart is certainly identified by the quality of the momentum reconstruction, because a bigger histogram quantization enables also a better resolution for the reconstructed momentum. And last but not least, the clustering problem introduced in section 2.1 on page 45 necessitates a well configured quantization, which leads to almost tiny clusters.
Imaginably the combination of these aims is quite complex and can not be analyzed by a simple concept. Nevertheless the analysis presented in section 4.6.1 on page 240 can be used to get a first impression about the minimal needed quantization, which is defined by being at least able to separate the MC Tracks, and the corresponding momentum resolution for the reconstruction. For this purpose, figure 5.3a depicts the number of not-findable MC Tracks, which are characterized by encountering the same histogram cell than another one, with regard to different predefined quantization configurations summarized in table 5.2. In addition to this, figure 5.3b shows the corresponding reconstructed momentum distribution of a single event for the hand-picked quantization.

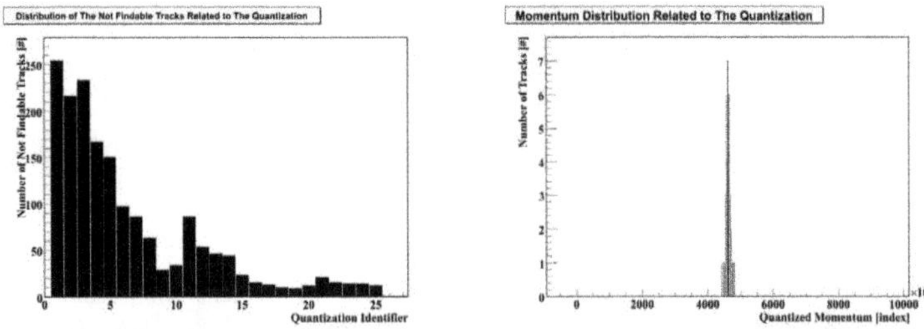

(a) Distribution of the not findable MC Tracks with regard to different quantization configurations

(b) Distribution of the MC Track momenta for s single hand-picked quantization

Figure 5.3: Visualization of the quantization effect to the ability of successful tracking

Identifier	1	2	3	4	5	6	7	8	9	10
Dim1	93	93	93	93	93	189	189	189	189	189
Dim2	31	31	31	31	31	63	63	63	63	63
Dim3	127	159	191	223	255	127	159	191	223	255

Identifier	11	12	13	14	15	16	17	18	19	20
Dim1	285	285	285	285	285	381	381	381	381	381
Dim2	95	95	95	95	95	127	127	127	127	127
Dim3	127	159	191	223	255	127	159	191	223	255

Identifier	21	22	23	24	25
Dim1	477	477	477	477	477
Dim2	159	159	159	159	159
Dim3	127	159	191	223	255

Table 5.2: Specification of all explored quantization identifier

Referring to figure 5.3a, there is apparently no optimal quantization at all, because each analyzed configuration exhibits MC Tracks, which are not findable, because they encounter the same histogram cell than another one. But as the amount of these tracks has to be also seen in comparison to the total number of findable MC Tracks, which is about 35,000 for 100 events, the quantization identifier 18, which represents the configuration $\begin{pmatrix} 381 \\ 127 \\ 191 \end{pmatrix}$, can be selected to be the best available one. In this connection, it should be additionally mentioned that even if smaller configurations are acceptable as well, this quantization is chosen with respect to the corresponding momentum resolution.

So the main conclusion, which can be derived from this result, is that a good resolution for the reconstructed momentum demands a bigger histogram quantization than the ability to separate the MC Tracks. However, methods, which are possibly able to cope with this discrepancy, are described in section 7 on page 435 and can be summarized by the application of simple algorithms following the Hough transform to improve the results.

Histogram Quantization Because of the Average Peak Distance

Although the previous section 5.1.1 on page 300 tries to manage the very complex problem of selecting an expedient quantization for the histogram by taking the peak separation of the MC Tracks and the corresponding reconstructed momentum resolution into account, it is obvious that this simple criterion is not satisfying here. Hence the analysis presented in section 4.6.2 on page 247 is applied additionally to evaluate figure 5.4, which depicts the average distance of the peaks representing the MC Tracks with regard to the same different quantization configurations than summarized in table 5.2 of section 5.1.1 on page 301.

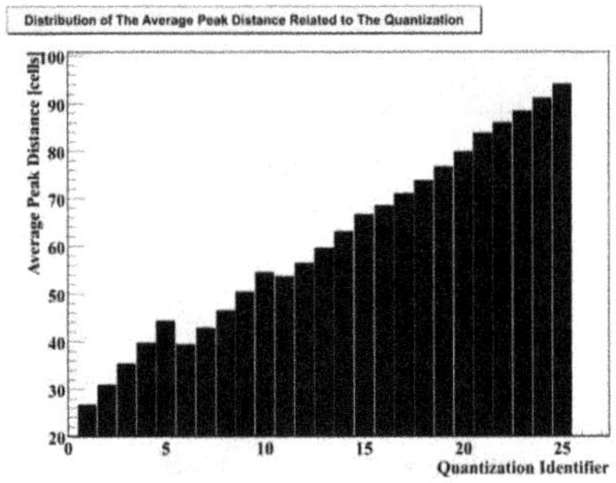

Figure 5.4: Resulting distribution for the average peak distance of the MC Tracks with regard to different quantizations

Relating this information now to the automatically generated Peak-finding geometry, which is exhibited by the analysis presented in section 4.8.4 on page 279, a very good imagination about the ability to manage the cluster problem introduced in section 2.1 on page 45 can be gained. In this connection, it has to be explicitly noted that the required amount of space avoids the displaying of each corresponding Peak-finding geometry for the quantization configurations. Furthermore even the corresponding geometry for the selected quantization is not depicted here, because this one can be found in section 5.1.4 on page 309.

Nevertheless another albeit less significant index for the quality of the quantization can be determined by the quotient $\frac{average\ distance}{dim1 \cdot dim2 \cdot dim3}$, because a higher one would suggest a better quality, if the clustering is assumed to be proportional to the quantization. Therefore table 5.3 summarizes this quotient for the analyzed quantization configurations.

Identifier	1	2	3	4	5	6	7
Quotient [%]	16.57	16.52	16.41	16.30	16.19	16.61	16.79

Identifier	8	9	10	11	12	13	14
Quotient [%]	16.85	16.80	16.80	16.39	16.61	16.75	16.81

Identifier	15	16	17	18	19	20	21
Quotient [%]	16.86	16.24	16.44	16.58	16.69	16.80	16.15

Identifier	22	23	24	25
Quotient [%]	16.29	16.44	16.54	16.67

Table 5.3: Survey report on the distance-quantization quotient for each analyzed quantization identifier of the histogram

With regard to this table, the almost constant quotient of all analyzed histogram quantization configurations implies that the relative peak distance, which is related to the histogram quantization, offers no additional information about the quality of each configuration. But even if the average peak distance is always appropriate, the absolute average peak distance shows that a smaller quantization causes smaller distances to occur more frequently. Thus the quantization identifier 18, whose exemplary entailed distance distribution is depicted in figure 5.5a, seems to represent here an adequate choice as well.

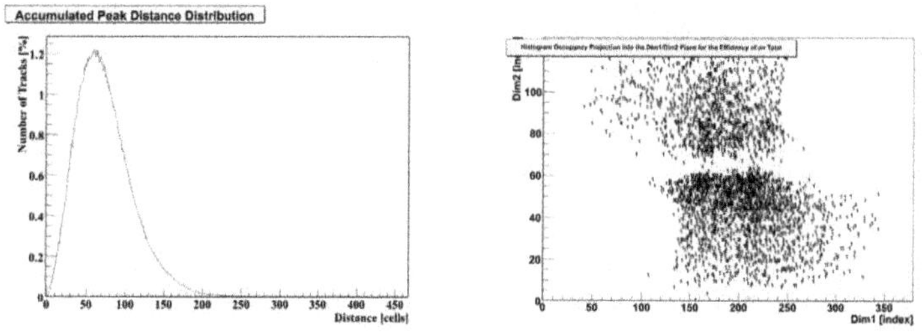

(a) Distribution for the accumulated peak distances of the MC Tracks

(b) Occupancy of the Histogram for the found MC Tracks

Figure 5.5: Peak distance visualizations with regard to the selected active quantization configuration

Moreover figure 5.5b shows that the MC Tracks of a common event are located in a cloud structure in the inner part of the histogram, which is naturally even more densely packed for smaller quantization configurations.
Nevertheless it is possible in all likelihood that this situation can be coped with the usage of non-uniform quantization unit intervals, which can be simply realized by an adaption of the LUT generation with regard to the dedicated intervals for the histogram cell regions. Although such an analysis is certainly far beyond this thesis, first speculations advice the quantization configuration $\begin{pmatrix} 95 \\ 31 \\ 191 \end{pmatrix}$ to perform quite well.
Besides this, it has to be finally mentioned here that such an approach affects surely all other aspects, which are surveyed in this section, as well.

Histogram Quantization Because of the Peak's Hit Signatures

As all previous sections care only about the ability to find the MC Tracks related to the quantization, it is clear that the final section has to offer some information about the resulting peak signatures representing such MC Tracks, because each one could be naturally only found, if it does not disappear due to an inadequate signature or another MC Track. For this purpose, the analysis, which is presented in section 4.5.1 on page 230, can be used to generate the two distributions shown in figure 5.6.

But before going into detail, it is important to realize at first that both distributions base apparently just on the results of the first LUT and are obviously related to the different quantization configuration identifiers, which are summarized in table 5.2 of section 5.1.1 on page 301.
Further on, it has to be additionally noticed that the analysis, which is presented in section 4.8.2 on page 268 and applied in section 5.1.2 on page 307, certainly influences these results.

Coming now to figure 5.6a, it is clear that this one illustrates the distribution of the percentage of MC Tracks, which correspond to a similar peak with a quality of at least 70 %. Thence a MC Track, which possesses for instance seven correct hits, would be only counted here, if one peak signature exhibits at least 70 % and thus five of these seven hits. Furthermore a MC Track owning four hits requires then obviously at least a signature containing three hits, and so on.
Besides this, figure 5.6b charts the distribution of the amount of peak signa-

tures, which are necessary to cover at least 75 % of the MC Tracks fulfilling the similarity requirement.

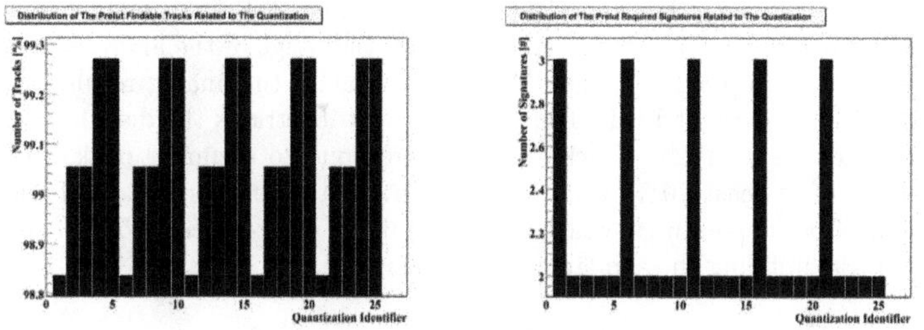

(a) Distribution of the findable MC Tracks due to the hit signature similarity of at least 70 %

(b) Distribution for the required amount of hit signatures to cover at least 75 % of the findable MC Tracks

Figure 5.6: Visualizations related to the hit signatures with regard to different quantizations and an assumed perfect second LUT

Having now a closer look to figure 5.6a, it is at first self-evident that the results depend only on the varying quantization of the third dimension. But even if the percentage increases for higher quantizations of this dimension, this circumstance could be neglected, because the total range is only about 0.5 %. Hence a bigger remark has to be set on figure 5.6b, because the smallest quantization requires one signature more than the others, which leads to the insight that a quantization of at least 159 seems to be adequate.

Besides this, table 5.4 presents supplementary results, which exhibit smaller quantizations of the third dimension than shown in figure 5.6, because it might be of additional interest to find the region of percentage degradation.

Dim1	127	127	127
Dim2	381	381	381
Dim3	31	63	95
Tracks	76.31 %	97.92 %	98.57 %
Signatures	5	3	3

Table 5.4: Additional report on the track-signature correlation for special quantization configurations and an assumed perfect second LUT

So with regard to these new results, a surprising fact is found, because the

expected degradation happens apparently not until the very small quantization of 31 is reached in the third dimension.
As these results depend however only on the first LUT, the focus has to be obviously changed to the dependency on the second LUT. For this purpose, the following figure 5.7, which is produced by the usage of the analysis presented in section 4.5.2 on page 233, depicts exactly the analog results for a perfect assumed first LUT. Thence figure 5.7a illustrates the distribution of the percentage of MC Tracks, which correspond to a similar peak with a quality of at least 70 %, while figure 5.7b charts the distribution of the amount of peak signatures, which are necessary to cover at least 75 % of the MC Tracks fulfilling the similarity requirement.

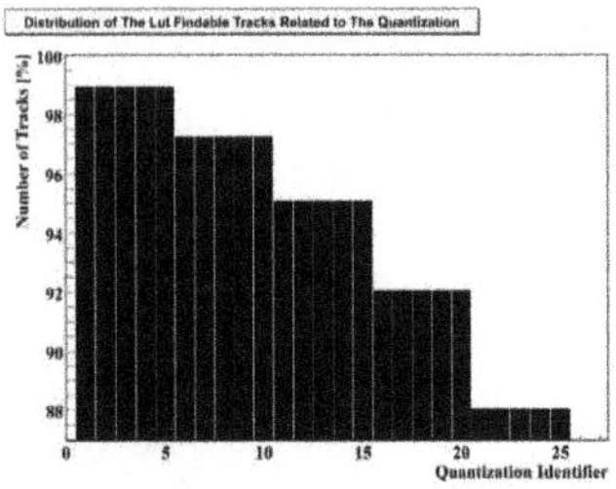

(a) Distribution of the findable MC Tracks due to the hit signature similarity of at least 70 %

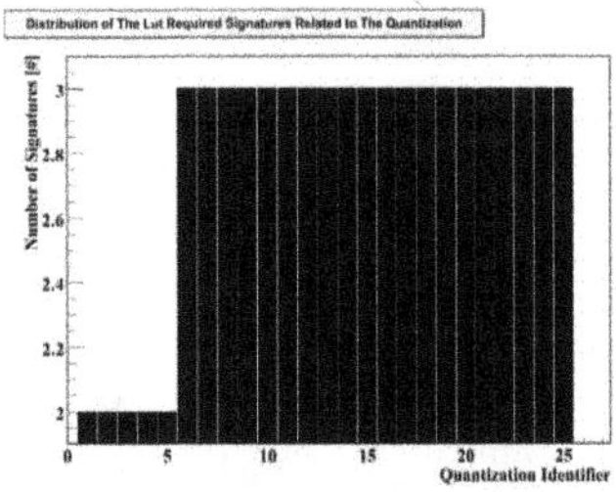

(b) Distribution for the required amount of hit signatures to cover at least 75 % of the findable MC Tracks

Figure 5.7: Visualizations related to the hit signatures with regard to different quantizations and an assumed perfect first LUT

Having again at first a closer look to figure 5.7a, it is clear that the results depend only on the varying quantization of the first and constrained second dimension. Further on, the assumption that the peaks get blurred for higher quantizations can be additionally confirmed, because this circumstance is apparently the reason for the percentage degradation.
However a bigger remark has to be set in turn on figure 5.7b, because the higher quantizations require obviously a constant amount of signatures, which leads to the insight that a quantization of less than 477 for the first and 159 for the second dimension seems to be adequate due to a volitional percentage of more than 90 %.

5.1.2 Optimal m in the Prelut Equation

Since the previous sections evaluate the feasible configuration of the Hough space dimensions for the CBM experiment to $\begin{pmatrix} 381 \\ 127 \\ 191 \end{pmatrix}$, it is obvious that the next step in the parameter search is determined by the optimization of the range m, which is used in the equation 2.13 of section 2.4.1 on page 73 to compute the first LUT by the application of the analytic formula. For this purpose, the analysis, which is presented in section 4.8.2 on page 268, is consulted to automatically compute an applicable range, which is afterwards defined to the resulting borders depicted in figure 5.8.

AnalysisLIB WARNING: There are 863 prelut ranges detected which fulfills the requirements!!!

AnalysisLIB WARNING: The prelut range causes an average of 2.7 entries per hit!!!

AnalysisLIB WARNING: The prelut range is actually set to [-0.015, 0.048]!!!

Figure 5.8: Detailed information about the automatically generated range m used in the analytic formula to compute the first LUT

So with regard to this figure, it is at a first glance surprising that the automatically evaluated range is asymmetrical around zero. But a second and more detailed look shows that this circumstance is not mysterious, because it depends certainly directly on the distribution of the momentum and especially p_y and p_z, which is featured by the applied configuring MC data.

5.1.3 Optimal Table Files

As the Encoding step of the Hough tracking algorithm, which is introduced in section 3.1 on page 96, is implemented by the application of a further LUT containing the connection from the hit signatures to the priority classes, it is clear that the content of the corresponding table file has to be also optimized. Furthermore the general concept of table files, which serve additionally a central role in the quality measurement, enables the need for analyses helping to create them automatically at best. Thus the following table 5.5 identifies each automatic algorithm, which is actually used to create the corresponding file for the tables 'codingTable', 'gradingPTable' and 'gradingRTable'.
However it has to be noticed in this connection that all files have to be generated with regard to the occurring peaks, because on the one hand many allowed degraded signatures in a table will lead to many wrong found peaks,

while on the other hand too less signatures inside will cause many missed correct but blurred peaks.

Table file	Algorithm
codingTable	LUTGOODNESSTABLE
gradingPTable	ONLINETABLE
gradingRTable	CHECKSUMTABLE

Table 5.5: Report on the chosen automatic generation algorithms for the table files

Moreover the results of these automatic algorithms, which are precisely described in appendix A.4 on page A-101, can be additionally visualized by the usage of the analysis presented in section 4.8.3 on page 272. Hence figure 5.9 reveals the table files for the recently selected quantization configuration $\begin{pmatrix}381\\127\\191\end{pmatrix}$ and the actually used configuring MC data.

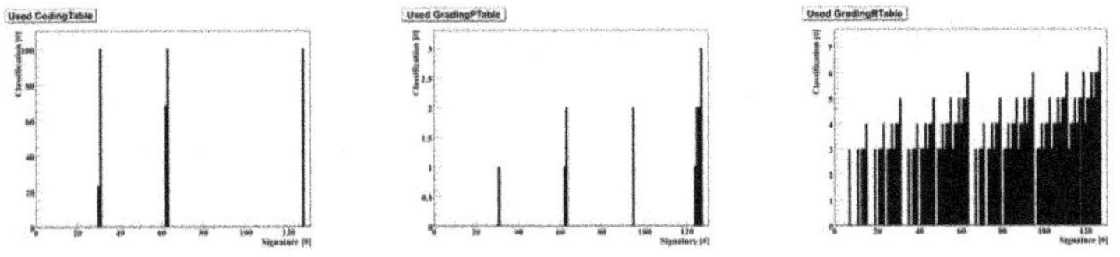

(a) CodingTable (b) GradingPTable (c) GradingRTable

Figure 5.9: Visualization of the automatically generated table files

With regard to this figure, it is immediately clear that the complexity of the table 'codingTable', which is used in the Encoding step of the algorithm, is really tiny and thus very good. Furthermore a detailed look demonstrates that it consists just of six hit signatures, which possess at least four of the maximal seven available hits. Even if the classifications represent further on the frequency of occurrence for each signature, these values would be naturally standardized in the following to the priority classes one, two and three due to the simplicity.

Besides this, the two other tables are evidently evaluated by a completely different type of algorithm, because they are both generated independent of any MC data, but dependent of the experimental setup and especially the

number of used detector stations.
So the table 'gradingPTable', which is used to determine the hit signatures of the findable MC Tracks, is created under the condition that a certain number of hits must be included in each signature, while the corresponding priority classes are generally set to the checksum of hits plus one minus the minimal number of hits. Thence the combination of this concept with the visualization accounts directly for the same 70 % rule as stated in section 5.1.1 on page 304, which restricts the signatures apparently to these ones possessing at least five of the maximal seven available hits, and the priority classes one, two and three.

Beyond that, the table 'gradingRTable', which defines the affiliation of the found peaks to the MC Tracks by the allowed hit signatures, contains simply the signatures in combination with its checksum of included hits and the minimal classification of three. So based on this concept, a peak would be only assumed to represent a certain MC Track, if this table includes a signature, which can be built only of hits corresponding to this certain MC Track.

5.1.4 Well-Defined Peak-finding Geometry

Since the Peak-finding step of the Hough tracking algorithm has to cope with the cluster problem, which is introduced in section 2.1 on page 45, the geometry of this process has to be surely also optimized. For this purpose, the automatic analysis, which is presented in section 4.8.4 on page 279, can be used to accumulate the affected histogram cells around the peak of each MC Track, which is taken into account. Thus figure 5.10 depicts the resulting geometry for the quantization configuration $\begin{pmatrix}381\\127\\191\end{pmatrix}$ and the applied configuring MC data of multiple events.

With regard to this figure, the first impressing fact is obviously determined by the circumstance that the geometry would surprisingly comprise only three histogram layers, even if the selected quantization configuration is that large. Furthermore the layer above and underneath the central layer are identical, which implies evidently a symmetry in the third dimension. In addition to this, another symmetry exists apparently around the center inside of each layer, which is even more surprising but just as satisfying. Moreover the only violation of this symmetry is detected by a single missing cell or a single cell too much in the histogram layer underneath and above the central layer. For that reasons, it has to be finally noted that such a Peak-finding geometry

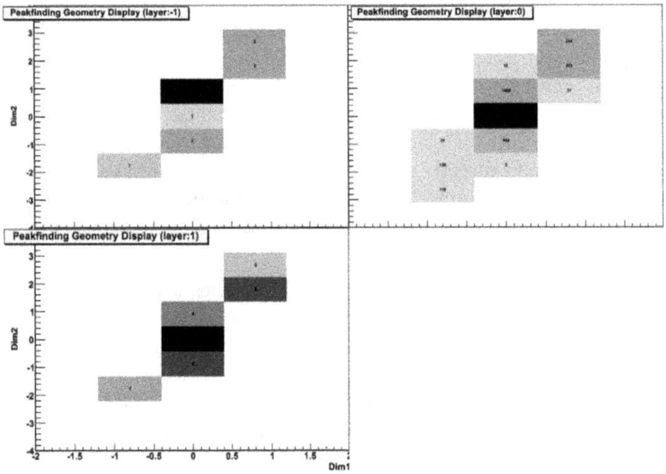

Figure 5.10: Visualization of the automatically generated Peak-finding geometry

is at its best, because it implies that the peak clusters are very small and defines no crucial problem.

5.2 Exactness of the Track Reconstruction

Since the parameters of the Hough tracking algorithm are successfully configured up to here, the focus of this section has to be obviously set to the evaluation of the consequent results for some applied sample events. For this purpose, there is at first a section presented, which illustrates the tracking performance, while the sections afterwards care about the momentum reconstruction and the resulting introduced error.

5.2.1 Tracking Performance

This section presents all important facts, which are necessary to benchmark the tracking performance. For this purpose, the analyses introduced in section 4.7.1 on page 258 and section 4.7.2 on page 260 are consulted to produce figure 5.11 and figure 5.12, which depict the absolute detailed and the relative summarized values for all available benchmarking aspects.

Absolute Analysis:
Peak Summary

Total number of peaks corresponding to no track:	739
Total number of peaks corresponding to one track:	4007
Total number of peaks corresponding to more tracks:	46
Total number of peaks:	4792
--	
Total number of peaks classified as track:	3165
Total number of peaks classified as clone:	327
Total number of peaks classified as fake:	402
Total number of peaks classified as ghost:	337
Total number of peaks classified as no track:	561
Total number of peaks:	4792
==	
Track Summary	
Total number of tracks corresponding to no peak:	503
Total number of tracks corresponding to one peak	2873
Total number of tracks corresponding to more peaks:	321
Total number of tracks:	3697
--	
Total number of well identified findable tracks:	2251
Total number of not well identified findable tracks:	943
Total number of identified not findable tracks:	508
Total number of identified tracks:	3702
--	
Total number of identified findable tracks:	3194
Total number of findable tracks:	3697

Figure 5.11: Exhibition of the absolute tracking performance for ten accumulated common events and an overall well-configured Hough transform

So with regard to figure 5.11, it is clear that such an analysis is evidently able to give precise information about each available detail of the Hough tracking, which is here also accumulated for ten evaluated sample events. Nevertheless the more essential figure is following, because this one summarizes the details with regard to the six most characterizing tracking aspects.

Relative Analysis:

Total efficiency [e] (aim 100%):	86%
Total fake rate [f] (aim 0%):	8%
Total ghost rate [g] (aim 0%):	7%
Total clones rate [c] (aim 0%):	7%
Total identification rate [i] (aim 100%):	67%
Total data reduction rate [r] (aim near 85%):	80%

Figure 5.12: Exhibition of the relative tracking performance for ten accumulated common events and an overall well-configured Hough transform

With regard to figure 5.12, it can be obviously asserted that the general aim of an optimal performance of the Hough tracking is achieved at its best with regard to the different quality aspects. Furthermore the degradation of around 10 % in each aspect except the identification rate is certainly acceptable because of the overall optimization and the dependency of each other. Moreover the small identification rate is not exciting, because it is simply caused by the accumulated amount of fake tracks, ghost tracks and clone tracks in combination with the MC Tracks, which are not found due to a blurred peak or the cluster problem.

Besides this, an essential fact, which is already stated in the sections presenting the above concerned analysis respectively, has to be reminded in this connection, because it is naturally impossible to define a simple aim for the Hough tracking algorithm, whose complexity is additionally worsened by the circumstance that an improvement of one aspect entails always a debasement of at least one another aspect. Hence it is needless to say that the configuration of the parameters, which are evaluated in the previous sections, is done with regard to the best available overall performance of the analytic formula. Thence a successive algorithm, which is totally independent of the Hough transform, can be applied to cope with the debasements induced by an improvement in the efficiency for instance. More information about such algorithms and possibilities can be found in section 7 on page 435.

5.2.2 Tracking Performance Related to the Momentum

Although the previous section seems to be enough to evaluate the performance of the tracking, a secondary characteristic, which is identified by the momentum of a findable MC Track, has to be obviously taken into account, because an imagination about the tracking performance related to this aspect might discover perhaps some special regions of interest, which exhibit a crucial performance and implicate thus a big resource for optimization resulting in an overall improvement.
Therefore this section presents evidently the tracking performance related to the momentum. For this purpose, the analysis introduced in section 4.7.3

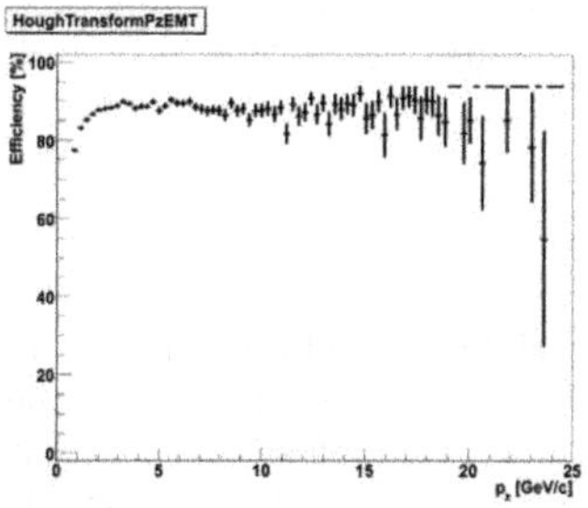

(a) Visualization of the tracking efficiency

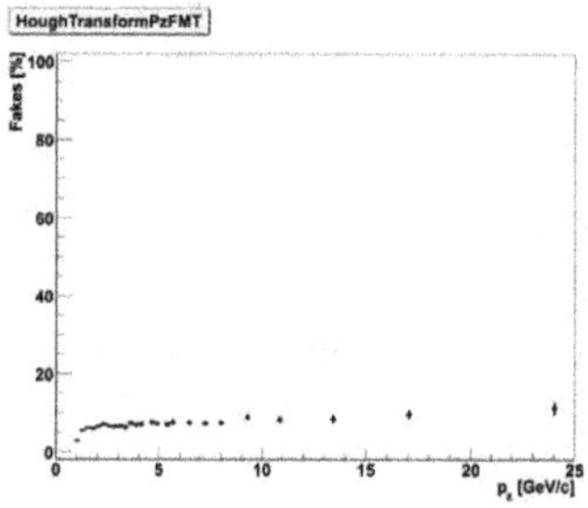

(b) Visualization of the tracking fake rate

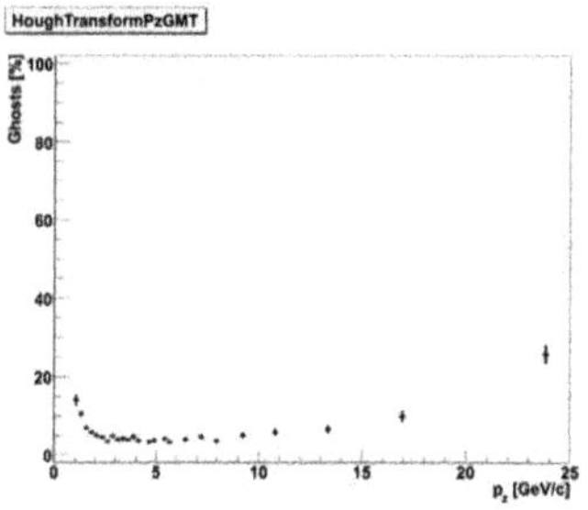

(c) Visualization of the tracking ghost rate

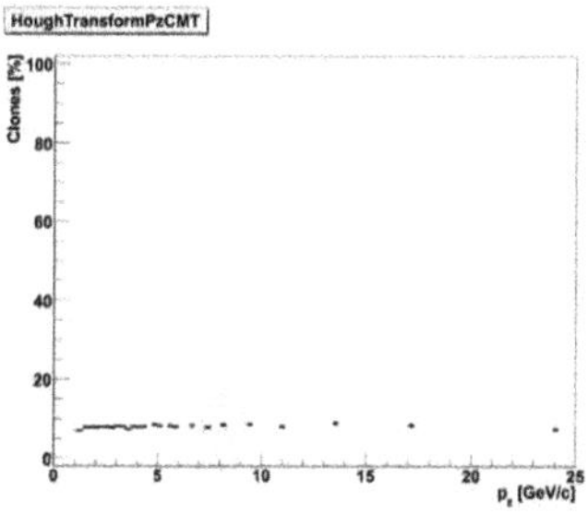

(d) Visualization of the tracking clones rate

Figure 5.13: Resulting distribution of each tracking performance aspect with regard to p_z

on page 262 is inquired to produce figure 5.13, which depicts the relative distribution for each available peak classification.

With regard to this figure, the tracking efficiency can be seen to be almost constant just with a small degradation at the minimal momentum cut of $1\,\frac{\text{GeV}}{\text{c}}$, which is apparently simply caused by a smearing effect due to the confinement of the Hough space.

Further on, the tracking clones rate is also almost constant, while the tracking fake rate is slightly rising towards higher momenta.

In contrast to this, the tracking ghost rate is not constant but rather increased for low and high momenta. However the behavior for low momenta is not surprising, because the highest track multiplicities occur evidently at low momenta, which causes thus the combinatorial coincidences to be more likely. In addition to that, the behavior for the high momenta is also not mysterious, because the less numerous tracks with high momenta correspond to almost straight lines near the beam pipe passing through areas with high track densities.

5.2.3 Momentum Error

As the previous sections care only about the ability to find each MC Track, it is obvious that there must be also a section, which informs about the quality of a found track. For this purpose, it is further on self-evident that the criterion, which is able to represent a measurement, is determined by the difference of the reconstructed momentum of the found track and the correct

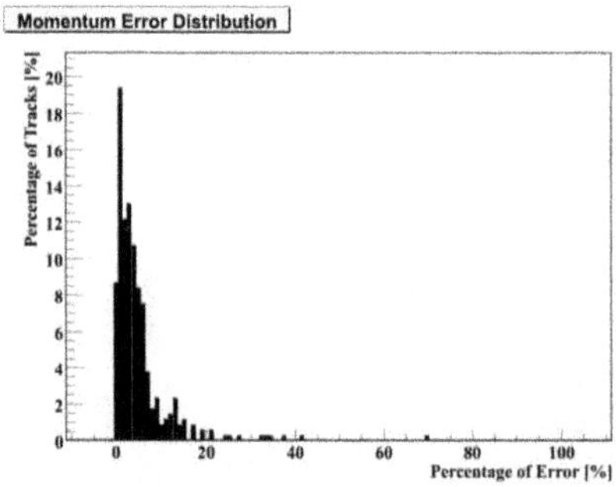

Figure 5.14: Distribution of the relative momentum error for the MC Tracks of a single sample event

one of the corresponding MC Track. Hence figure 5.14, which is produced by the application of the analysis illustrated in section 4.5.3 on page 236, depicts the distribution of such a momentum error evaluated for the found MC Tracks of a single sample event. Furthermore it has to be noticed in this connection that the error is evaluated relative to the correct momentum of each MC Track, because a small error for a small momentum should count as much as a big error for a big momentum.

Further on, the badness of the distribution shown in this figure suggests that the reconstructed momentum can not be used for any further physics analyses, which connotes also that the Hough tracking is not applicable for the CBM experiment. However before this consequence can be confirmed, the reason for the badness has to be analyzed, because there might be a workaround.
So since the error corresponds surely directly to the distance of the correct histogram cell, which is defined by the momentum of the MC Track, and the cell containing the corresponding found peak, the average distance might give additional information. For this purpose, the figure 5.15, which is generated by the analysis introduced in section 4.5.3 on page 236, informs about the average distance accumulated for all found MC Tracks of multiple events.

```
AnalysisLIB WARNING:
The average distance of the closest histogram cell, which contain the maximum
number of hits, to the correct histogram cell of the MC Track is: 1.88!!!
The average distance of the farest histogram cell, which contain the maximum
number of hits, to the correct histogram cell of the MC Track is: 2.84!!!
The average distance of the histogram cell, which contain the maximum
number of hits, to the correct histogram cell of the MC Track is: 2.29!!!
```

Figure 5.15: Exhibition of the universally accumulated average distance between the histogram cell identifying a MC Track and the one featuring the corresponding found peak

As this figure evidences now the average distance to be really small, the badness of the relative momentum error can be only caused by the quantization of the Hough space, because neighboring histogram cells represent simply a momentum difference, which is too high. Thence this situation can be only improved by using more quantization steps. However it has to be noted in this connection that such a solution will go evidently hand in hand with all other problems illustrated earlier. Moreover a different solution seems to be also given by the usage of another computational kernel or an applied successive algorithm as proposed in section 7 on page 435

5.3 Impact of the Experimental Setup on the Tracking Algorithm

Since the experimental setup defines the measuring instrument for the input of the Hough tracking due to the elements to measure, it is clear that this setup is able to influence the performance. Hence it is self-evident that an optimal setup, which is related to the Hough tracking, would be imaginable, if the corresponding effects are not homogeneous. But as the optimization of this setup due to the Hough tracking performance is far beyond this thesis, the next two sections present just some general impacts, which are gained during the development phase of the algorithm.

5.3.1 Micro Vertex Detector System Setup

While remembering section 1.3.1 on page 12 and especially section 1.3.1 on page 16, it is conspicuous that the read-out rate of the MAPS, which is actually tested to be 20 μs but planned to be less than 10 μs, is nevertheless two orders of magnitudes higher than the required read-out rates of about 25 ns, which are necessary to accommodate reaction rates of 10 MHz. Accordingly such a situation typically ends up in an event pileup in the affected detector stations, which is obviously not negligible and not avoidable.
Thence it is evidently important to analyze the consequences of this pileup with regard to the Hough tracking. And since just a more crowded histogram is further on expected in this connection, such a simple circumstance can be easily explored by the usage of the histogram layer occupancy analysis, which is presented in section 4.6.4 on page 255. For this purpose, the produced figure 5.16 depicts the result for a pileup of only ten events in a central histogram layer.

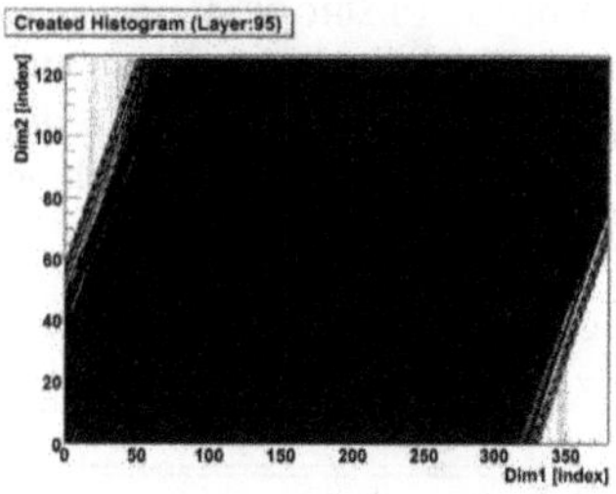

(a) Total scale of the layer

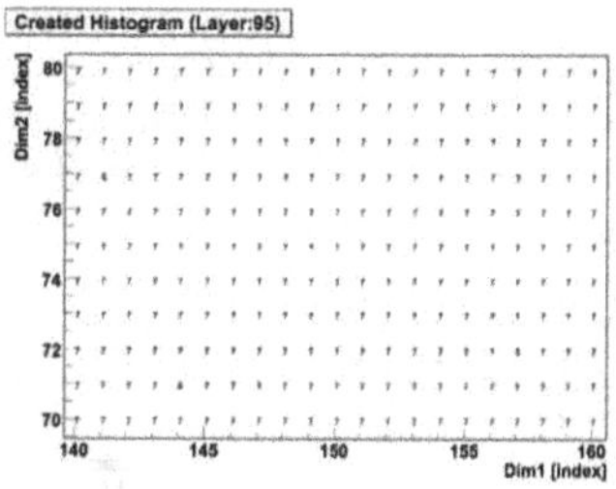

(b) Zoomed scale of the layer

Figure 5.16: Visualization of a single inner histogram layer after Histogramming the hits of three MAPS detector stations including the pileup of ten events

With regard to this figure, it can be easily seen that the histogram is not only more crowded, but totally jam-packed. Consequently such a histogram layer would contain either a track in almost all cells, if the signature is accepted, or no track at all, if the signature is not accepted. Thence the MAPS detector stations are not anymore able to deliver any useful information for the Hough tracking, because almost all signatures of the peaks are identical.

5.3.2 Silicon Tracking System Setup

In contrast to the MVD stations, each available type for the STS does obviously not implicate such a crucial pileup situation as described in section 5.3.1 on page 316. However it is clear that this pileup problem can be also not totally avoided here, because the triggering of the read-out causes evidently two consecutive events to be not split exactly, but the overlap seems to be quite small and the effect might be thus omitted.
Nevertheless it is of course as well important as for the MVD stations to analyze the consequences of this situation to the Hough tracking, which can be apparently realized in its simplest form by the study of varying Encoding signatures. For this purpose, the produced table 5.6 depicts the resulting relative performance of the Hough tracking algorithm with regard to the selected signatures in combination with the position of the corresponding detector stations by using the analysis, which is presented in section 4.7.2 on page 260.
Moreover it has to be additionally highlighted in this connection that these

results are evaluated by a very early version of the CBMROOT framework, which applies a simple binning algorithm for the MC Points to generate hits instead of realistic hit producers simulating hybrid pixel or micro-strip detector stations. Hence these results are certainly not representative in these days, but the conclusion can be generalized anyhow.

Detector station positions [cm] & Most degraded Encoding signature						**Tracking efficiency**	**Tracking ghost rate**	**Tracking clones rate**
20		40	60	80	100	91 %	25 %	1 %
20		40	60	80	100	92 %	45 %	1 %
		40	60	80	100	90 %	68 %	1 %
20	30	40	60	80	100	93 %	12 %	6 %
20	30	40	60	80	100	83 %	7 %	3 %
20	30	40	60		100	93 %	14 %	2 %

Table 5.6: Survey report on the effect of different detector setups in combination with binning hit producers and the most degraded Encoding signature to the relative tracking performance

With regard to this table, it turns immediately out that the dependency of the tracking efficiency to the detector setup in conjunction with the used Encoding signatures is tiny. The main differences related to this aspect occur apparently at the momentum cut of 1 $\frac{\text{GeV}}{\text{c}}$, because MC Tracks, which possess such a low momentum, do commonly not pass all detector stations and are further challenging the analytic formula. Furthermore the decreased tracking efficiency in row five originates obviously in this situation, because the most degraded Encoding signature requires hits in the first five detector stations.
In contrast to this, the tracking ghost rate decreases of course directly by the application of more appropriate Encoding signatures and more detector stations, which are also most suitable close to the target.
Besides this, the tracking clones rate is evidently independent of both parameters. But the main characteristic of this aspect seems to be designed by the Peak-finding geometry and the related arithmetic operation, because both are actually optimized for five detector stations and have to be thus adapted to the usage of six ones.

Finally the conclusion can be summarized by the usage of at least five detector stations, because the tracking efficiency is almost equal and less stations increase the tracking ghost rate dramatically. Further on, the movement of

the detector station four from the position at 80 cm to the position at 30 cm in the standard setup promises better results.
As however the actual development takes two different types of detector stations, which are introduced in section 1.3.1 on page 17 and section 1.3.1 on page 19, into account, this circumstance has to be apparently included in the analysis. For that reason, the following table 5.7 shows now the resulting relative performance by reusing the analysis presented in section 4.7.2 on page 260 of an early version of the CBMROOT framework, but with the difference of applying realistic hit producers for the varying number of hybrid pixel detector stations and micro-strip detector stations to compose the standard setup.

Detector station positions [cm] & Micro-strip stations					**Tracking efficiency**	**Tracking ghost rate**	**Tracking fake rate**
20	40	60	80	100	89 %	27 %	0 %
20	40	60	80	100	88 %	28 %	0 %
20	40	60	80	100	85 %	17 %	25 %
20	40	60	80	100	76 %	10 %	51 %
20	40	60	80	100	73 %	4 %	78 %

Table 5.7: Survey report on the effect of varying real hit producers to the relative tracking performance

With regard to this table, it is naturally not surprising that the detector setup using just hybrid pixel stations lead to comparable results than the corresponding setup in table 5.6, which employs a simple binning instead of realistic hit producers, because the only difference is defined by the nature of binning.
Besides this, the recent table exposes also that the application of more micro-strip detector stations result in a lower tracking efficiency, while many ghost tracks turn into fake tracks. Furthermore the tracking with more than two of such stations seems to be impossible, because more of them cause nearly as many fake tracks and ghost tracks as found MC Tracks.

So based on this situation, the decomposition of the performance results for the standard detector setup, which contains a single hybrid pixel and four micro-strip detector stations, might be of further interest. For that reason, the next table 5.8 demonstrates the resulting relative performance by using still the analysis presented in section 4.7.2 on page 260 of an early version of the CBMROOT framework, but by categorizing the results due to the number of findable MC Tracks.

Number of findable MC Tracks	Tracking efficiency	Tracking ghost rate	Tracking fake rate	Tracking clones rate
< 101	94 %	1 %	6 %	0 %
$101 - 200$	88 %	2 %	34 %	1 %
$201 - 300$	79 %	3 %	69 %	0 %
> 300	73 %	4 %	78 %	0 %

Table 5.8: Survey report on the effect of the micro-strip detector stations to the relative tracking performance

As this table reveals now the dependency of the tracking performance to the number of findable MC Tracks, it is obvious that the tracking in such a detector setup is impossible for more than 200 findable MC Tracks in a single event. However it is not the number, which is interesting, but the dependency itself, because the combination of the fake hit generation problem, which is illustrated in section 1.3.1 on page 19, with the circumstance, which counts for the MVD detector stations in section 5.3.1 on page 316, leads to the perfect explanation of this dependency. So within this context, it is immediately clear that a few MC Tracks are simply able to generate a very crowded histogram. Thus a reduction of the probability for the fake hit generation, which might be possibly realizable by shortening the micro-strip sensor length, will minimize this problem and the usage of such detector stations will produce better tracking performance results.

5.3.3 Failing Detector Stations

Since all detector stations, which are introduced in section 1.3.1 on page 17 and section 1.3.1 on page 19, are very complex building blocks at the leading edge of technology and additionally located in the harsh CBM radiation environment, it is clear that some fragments of a station or even the total one can possibly easily fail. Thence such a circumstance has to be evidently accounted in the tracking, because the effect of missing measurements is generally proportional to the size of failing fragments. But as such a detailed investigation is obviously far beyond this thesis, it should be only noticed here that the total failure of a detector station can be simply handled in the Encoding step of the Hough transform algorithm by adapting the table 'codingTable', because it does certainly not matter, whether a missing hit in a peak signature originates in a wrong transformation or a failed measuring instrument.

So coming now to the point, the following table 5.9 gives a short imagination about the tracking performance with regard to degraded peak signatures, which originate originally in wrong transformations, but can also represent totally failing stations. Moreover it is also important in this connection that these results are gained by the usage of an early version of the CBMROOT framework and an early implementation of the Hough transform. However the conclusion can be apparently in turn generalized.

CodingTable [Signature:Priority class]	111111:3	111111:3 101111:2 011111:2	111111:3 101111:2 011111:2 001111:1
Tracking efficiency	56 %	74 %	76 %
Tracking fake rate	23 %	49 %	63 %
Tracking ghost rate	3 %	8 %	13 %
Tracking identification rate	45 %	37 %	21 %

Table 5.9: Survey report on the effect of failing detector stations to the tracking performance

With regard to this table, the first column shows evidently that the tracking efficiency breaks dramatically down due to a detector station failure, because there are actually only signatures encoded, which contain a hit in each station.

But as the affected stations are suspected to be the fifth and sixth in order, the second column illustrates that the accreditation of degraded hit signatures at these positions increase the tracking efficiency to almost the original performance. However it is clear that such a less restrictive Encoding step causes also unavoidable side-effects, because these degraded signatures lead additionally to more fake tracks and ghost tracks in combination with a decreased tracking identification rate.

Moreover the third column suggests that the price to cope with a number of failures is unaffordable, because the side-effects are easily able to dominate the tracking. Nevertheless it is important to mention finally that this fact is not that strict, because such a bad situation might be improved by the usage of multiple priority classes, which can help to define a more complex and better deletion strategy in the Peak-finding process.

5.3.4 Beam Energy

Due to the planned manifoldness of investigations in the CBM experiment, which is summarized in section 1.2 on page 7, there are actually three different beam energies considered. As these different beam energies produce furthermore surely a different amount of hits, an effect on the Hough tracking is evidently expectable.
For this purpose, the following table 5.10, which is already presented in [GSM05], shows the performance of the Hough tracking with regard to two different detector setups and the beam energies of 15 AGeV, 25 AGeV and 35 AGeV. Moreover it is again important to realize that even if these results are gained by the usage of an early version of the CBMROOT framework and an early implementation of the Hough transform, the conclusion could be nonetheless generalized as mentioned in the previous section section 5.3.3 on page 320.

Beam energy	15 AGeV	25 AGeV	35 AGeV
Number of hits	3,500	4,000	5,000
Tracking efficiency	91 %	92 %	92 %
Tracking ghost rate	9 %	24 %	36 %
Tracking clones rate	13 %	14 %	14 %

(a) Detector stations at 20 cm, 40 cm, 60 cm, 80 cm and 100 cm

Beam energy	15 AGeV	25 AGeV	35 AGeV
Number of hits	3,500	4,000	5,000
Tracking efficiency	94 %	94 %	95 %
Tracking ghost rate	4 %	11 %	17 %
Tracking clones rate	9 %	10 %	10 %

(b) Detector stations at 20 cm, 30 cm, 40 cm, 60 cm, 80 cm and 100 cm

Table 5.10: Survey report on the effect of different beam energies to the tracking performance

With regard to this table, it is at a first glance surprising that the tracking efficiency and the tracking clones rate are only influenced by the detector setup, but not by the beam energy, while the tracking ghost rate varies dependent on both. However a closer look evidences that the reason for this behavior is not mysterious, because the number of hits, which are constituted by the beam energy via the number of MC Tracks, are proportional to the accidentally created peaks. And as such peaks can be only identified as ghost tracks, it is clear that the tracking efficiency and the tracking clones rate are not affected. Besides this, the impact of the detector setup is not further analyzed here, because a detailed inspection is presented in section 5.3.2 on page 317.

5.4 Results for the Software Implementation on a Personal Computer

This section presents the results for the personal computer implementation, which is introduced in section 3.2 on page 102, with regard to the usual benchmarking criteria time and memory consumption. However it has to be noticed before starting that this software implementation is especially developed for simulation purpose, which implies that no optimization is realized in any case due to the circumstance that the software will possibly run on each imaginable platform. Far from it, many computations are installed, which do not effect the tracking output, but consume time and memory to gain additional information about the quality of the algorithm for instance. Nevertheless the result is reported for each necessary step of the algorithm, because these summaries allow further a detailed step rating and thus optimization nomination.

As further on all results have to be obviously seen related to the input data, the key indicators can be summarized by seven detector stations, 15,658 hits, three entries in the table 'codingTable', three priority classes and nine elements for the Peak-finding geometry. In addition to this, the quantization of the histogram is set to 95 cells in the first dimension and 31 cells in the second dimension in conjunction with 191 layers.
Besides this, it has to be also noticed that the analyses, which are used to gain these results, are introduced in section 4.3 on page 221.

Before proceeding now with the detailed results, it is also important to highlight at first the overall performance, which is evaluated to a required time of 2.36 s and a consumed memory of 11.86 MB. Furthermore it is very interesting that this memory includes 5.41 MB for the input data, which can be further split into 4.99 MB for the MC data and 425.47 kB for additional information, and 677.14 kB for the output data, because these numbers form about 51 % of the total amount. So since the overall performance is successfully presented, the focus of the following sections can be set to the details of each algorithm step.

5.4.1 Look-up-tables

The first important fact to notice for the LUTs is obviously in turn defined by the circumstance that the results can be only estimated, because the digital information of the hits is recently not available and even not fixed as well (see section 3.2.3 on page 132).
Thence the total amount of required memory for the actual algorithm parameters can be only calculated to $2^{29}\,\text{addresses} \cdot (8+8)\,\frac{\text{bit}}{\text{address}} = 1.00\,\text{GB}$ for the first LUT and $2^{29}\,\text{addresses} \cdot (7+30)\,\frac{\text{bit}}{\text{address}} = 2.31\,\text{GB}$ for the second LUT.
As these sizes are furthermore apparently quite huge, it is additionally advised to use the same scheme as introduced in section 5.5.1 on page 328, which effects the tracking efficiency insignificantly but reduces the needed memory greatly.

5.4.2 HBuffer

Since the previous section exhibits both LUTs to be actually just theoretical due to the unavailable and unfixed digital information of the hits, it is clear that the application of the respective math formulas is required to be even able to process data. Furthermore this circumstance defines then apparently the reason for the impossibility to differentiate between the time, which is consumed for the evaluation of the LUTs and the saving of the resulting data in the HBuffer unit. Hence simply the combined time is measured, which ends up in 1.50 s or surprisingly about 63 % of the total time. Further on, it is consequently obvious that an optimization of this algorithm part, which might be for instance the usage of real existing LUTs instead of the math formulas, will lead directly to an improved result for the overall time consumption.

Besides this, the memory consumption is obviously more complex, because some memory is reserved statically, while the rest is allocated in blocks. By by referring to section 3.2.4 on page 136, the table 5.11 is able to illustrate the corresponding numbers.

Information	Reserved memory	Allocated memory	Used memory
First LUT	0.00 B	86.29 kB	61.17 kB
Second LUT	0.00 B	2.21 MB	1.90 MB
Hit	0.00 B	73.73 kB	61.17 kB
Total	4.00 B	2.39 MB	2.02 MB

Table 5.11: Resulting memory consumption of the HBuffer implemented in software

So with regard to this table, it is self-evident that the consumed memory, which is required to store all necessary information, is about 17 % of the total amount.

5.4.3 Histogram

In consideration of the details presented in section 3.2.5 on page 138, the necessary time for the Histogramming step of the algorithm is evaluated to 0.33 s, which is about 14 % of the total time and thus the second most consuming fragment. Further on, another essential fact, which does not have to be forget, is apparently determined by the circumstance that it takes additionally a total of about 0.16 s to reset the instantiated histogram layer before the next slice can be processed.

Besides this, the memory consumption is shown in table 5.12 in accordance with the previous section.

Information	Reserved memory	Allocated memory	Used memory
Signature	11.50 kB	0.00 B	2.88 kB
Hit	2.97 MB	0.00 B	2.97 MB
Total	2.98 MB	0.00 B	2.97 MB

Table 5.12: Resulting memory consumption of the histogram implemented in software

So with regard to this table, it is self-evident but really surprising that the consumed memory for the histogram itself is only about 0.03 % of the total amount, while the additional information, which is just needed to evaluate quality aspects of the algorithm, requires contrarily about 25 %.

5.4.4 Filtering

Since section 3.2.6 on page 144 defines the entire process of reducing the background noise in the histogram to comprise the separate algorithm steps Encoding, Diagonalization and Peak-finding including the Serialization, the following sections are naturally used to present the timing and memory results of each step individually, because this procedure offers the possibility to have a look at the results of each step independent of all others.

Encoding

Having section 3.2.6 on page 145 in mind, it is clear that the major important characteristic of the Encoding step is the time consumption. Consequently the rating of this algorithm step is measured to yield 0.27 s, which is surprisingly the third most consuming one due to taking about 11 % of the total time.

Diagonalization

Analog to the previous section, the major important characteristic of the Diagonalization concept introduced in section 3.2.6 on page 147 is obviously also described by the time consumption, which exhibits here the fewest one due to the measurement of only 0.02 s.

Peak-finding and Serialization

This section presents the timing results for the algorithm steps 2D Peak-finding, Serialization and 3D Peak-finding, which are all introduced in section 3.2.6 on page 148. So keeping now the precise implementations in mind, it is very impressive within this context that the 2D Peak-finding and the Serialization consumes nearly the same time, which is about 0.03 s and 0.03 s, while the 3D Peak-finding requires only the approximated half, which is rounded to 0.02 s.
For that reason, such a result advices the used platform to be highly memory optimized.

5.4.5 LBuffer

While remembering the concept of section 3.2.7 on page 150, the statically reserved memory is evaluated to 14 B, the allocated amount to 827 kB and the used number to 825 kB. So based on these results, it is for sure that the memory consumption of the LBuffer unit is just about 7 % of the total.

5.5 Results for the FPGA Implementation

This section features the results for the FPGA based implementations of each functional unit, which is introduced in section 3.3.1, which starts on page 152. Furthermore these results are evidently grouped into one section for each unit, which presents naturally simulation waveforms and synthesis results for all different available implementations including special attention payed to the required resources and the maximal possible clock frequency. Moreover it is assuredly essential to note in this connection that a Xilinx Virtex V FPGA device of the type XC5VLX110 with speedgrade -3 and package type FF1760, whose available resources are listed in table 5.13, is used to produce all synthesis results.

Array (*row* × *col*)	160 × 54
Virtex-5 slices	17,280
Maximum distributed RAM [kb]	1,120
DSP48E slices	64
Block RAM blocks [18 kb]	256
Block RAM blocks [36 kb]	128
Maximum Block RAM [kb]	4,608
Clock Management Tiles	6
Total I/O banks	23
Max user I/O	800
Clock technology [MHz]	550

Table 5.13: Overview about the resources of the FPGA device XC5VLX110 with speedgrade -3 and package type FF1760 (Source: [Vird])

Before proceeding now with the results for each functional unit, it has to be mentioned that more detailed information about the synthesis process itself can be found in appendix C on page C-1.

5.5.1 Look-up-tables

The first important circumstance to notice here is stated in section 3.3.1 on page 158, because both LUTs are actually not implemented. Due to this lack, it is further on certainly impossible to present real data results, but an assumption for the order of magnitude can be given at least.

So if a data input of 20 bit (17 bit for the x coordinate and 3 bit for the z coordinate) and a data output of 16 bit (two γ values: $dim3_{min} = dim3_{max} =$ 8 bit) is assumed for the first LUT, this one would obviously require 1 M $\cdot$ 2 Byte = 2 MB of RAM. But as this size is quite large for an implementation on the FPGA, I recommend to use external SRAM with just a connection interface inside the device.

While reflecting now the second LUT, it seems to be also imaginable to assume 20 bit (17 bit for the y coordinate and 3 bit for the z coordinate) for the data input. But in contrast to the first LUT, the data output is evidently determined by the quantization of the histogram dimensions $dim1 = \theta$ and $dim2 = -\frac{q}{p_{xz}}$ as well as the encoding style of the Hough curve. As furthermore this combination is recently supposed to be realized by 95 levels with a binary encoded starting index and 31 levels with a one-hot encoded histogram command due to the systolic processing, a data output of $log_2(95) + (31 - 1) = 7 + 30 = 37$ bit results. Besides this, it has to be mentioned in this connection that the minus one in the second bracket of this equation originates in the circumstance that the entry in the first row is made by the starting index and a further change after the last row is obviously senseless.
So based on these numbers, the second LUT requires apparently 1 M$\cdot$ 37 bit = 4.63 MB of RAM. Since this size is however even bigger than the one of the first LUT, I recommend again to use external SRAM with a connection interface inside the FPGA.

Keeping this memory consumption further on in mind, careful made estimations for both connection interfaces have shown that approximately 2,500 Virtex V slices are sufficient.

Hence the final conclusion for both LUTs is formed by the requirement of an external SRAM with about 6.63 MB and a FPGA occupancy of about 2,500 Virtex V slices.

5.5.2 HBuffer

Since the HBuffer unit defines a central part in the Hough transform implementation, the minimal example, which is presented in section 3.3.1 on page 160, has to be remembered to understand figure 5.17, which depicts the corresponding waveform of the simulated testbench including the following four input data samples.

1. Input data sample one:

$$\begin{aligned} \gamma_{min} &= index\_min\_in &= 4 \\ \gamma_{max} &= index\_max\_in &= 6 \\ indexLUT2 &= data\_in &= 02 \end{aligned}$$

2. Input data sample two:

$$\begin{aligned} \gamma_{min} &= index\_min\_in &= 3 \\ \gamma_{max} &= index\_max\_in &= 5 \\ indexLUT2 &= data\_in &= 08 \end{aligned}$$

3. Input data sample three:

$$\begin{aligned} \gamma_{min} &= index\_min\_in &= 2 \\ \gamma_{max} &= index\_max\_in &= 3 \\ indexLUT2 &= data\_in &= 10 \end{aligned}$$

4. Input data sample four:

$$\begin{aligned} \gamma_{min} &= index\_min\_in &= 5 \\ \gamma_{max} &= index\_max\_in &= 7 \\ indexLUT2 &= data\_in &= 20 \end{aligned}$$

Having now a closer look to this waveform, the four different data input samples can be easily identified by the high triggered 'we' signal in the beginning phase directly following the released high-active 'reset' signal. Further on, it is additionally important to realize in this connection that the data input has not to be delivered in a continuous stream, but just marked by the 'we' signal.
Afterwards the applied 'start_processing' signal switches the internal mode of the unit from receiving to processing. Thence the activated 'processing' signal shows obviously that no additional data can be further delivered.

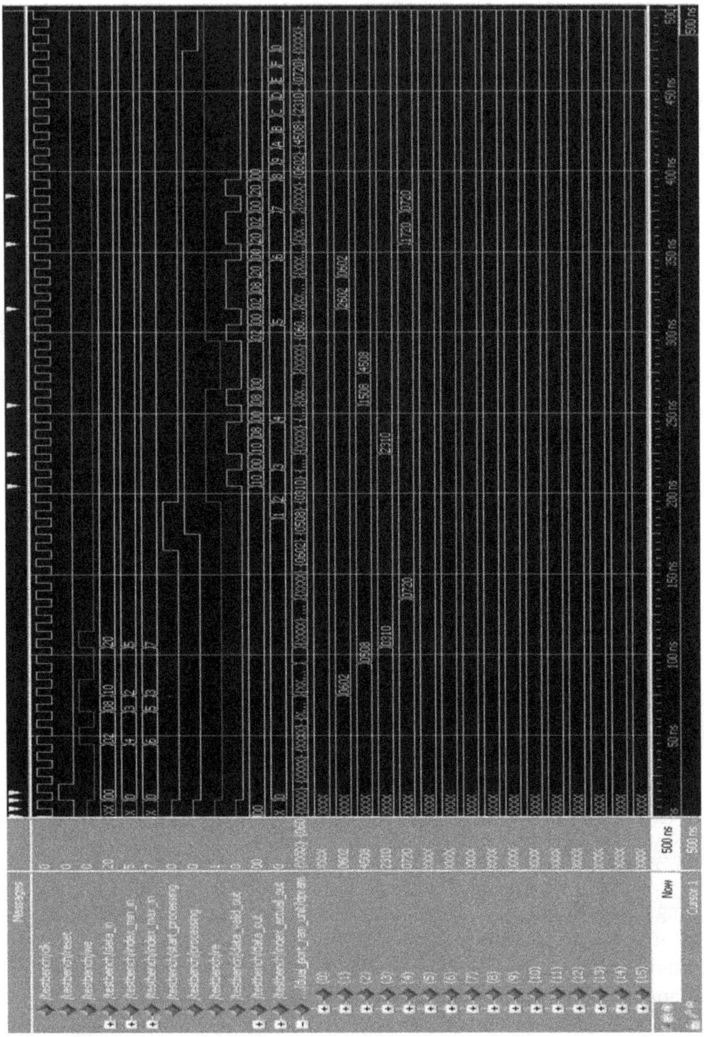

Figure 5.17: Example waveform for the HBuffer unit

By the way, it has to be noticed that all samples are written into consecutive addresses of the 'dpram' memory ordered by the arrival and independent of the index values, which represents the layers or lists.
Moreover a remark has to be set to the fragments of an entry in the 'dpram' memory. So by addressing this fact, the first fragment, which is identified by the first hexadecimal numeric character, defines naturally the essential link address to the previous entry in the corresponding list or zero in case of the first entry. Consequently the second one, which is also a hexadecimal numeric character, represents surely the γ_{max} or 'index_max_in' value, while the final fragment, which comprises the last two hexadecimal numeric characters, forms the 'indexLUT2' value or 'data_in' word.

Besides this, the internal activation of the 're' signal enables the unit later to deliver the stored data samples in the correct order, which is described in section 3.3.1 on page 160. In this connection, it has to be in turn noted that this data delivery does also not require a continuous stream, because the 'data_valid_out' signal marks correct data supplied on the 'data_out' signal, while the 'index_actual_out' signal determines the actual list or layer. Further on, it is clear that the 'processing' signal returns automatically to the inactive signal state after all data is read to show that the unit is ready again and has returned into the receiving mode.

Finally an important circumstance, which has to be also mentioned here, is determined by the rearrangement of data samples to consecutive lists, because this process is realized by just updating the link address entry in the 'dpram' memory. For this purpose, the processing phase in the waveform shows explicitly that the locating address of the concerned data samples has not to be touched.

As the functional description is extensively explained up to here, it is furthermore interesting in which way this functionality can be realized, because

Id	**Type**	**DPRAM ramstyle**
1	HBuffer	registers
2	HBuffer	block_ram
3	HBuffer	select_ram
4	HBuffer	no_rw_check

Table 5.14: Summary of the available synthesis possibilities for the HBuffer unit

table 5.14 shows different possibilities to implement such a HBuffer unit in theory.

A closer look to this table shows now that the column 'DPRAM ramstyle' specifies an attribute for the 'type' HBuffer, which defines the style of an inferred RAM. Hence the same source code, which is dictated by the used synthesis tool describing a DPRAM, can be used to produce different and sometimes hardware specific FPGA implementations.

Focusing currently the characteristics, the first value, which is called 'registers', turns RAM inferencing completely off. This means more precisely that the RAM is mapped to registers (flip-flops and logic) and not to technology-specific RAM resources.
In contrast to this, the value 'block_ram' arranges the inferred RAM to be mapped to the appropriate vendor-specific memory blocks, which are called Block SelectRAM+ for the Virtex families.
Moreover the value 'select_ram', which defines the default value, configures RAMs using the distributed RAMs in the CLBs.
And finally the value 'no_rw_check' establishes the Virtex Block SelectRAM+ like the 'block_ram' value but without inserting bypass logic for DPRAMs that would prevent a simulation mismatch between RTL and post-synthesis. However this value should be only used when simultaneously reading and writing to the same RAM location is not possible and overhead logic should be minimized, because reading and writing to the same Block SelectRAM+ address will lead to an indeterminate value at the output.
Since this information about the values are moreover apparently only summaries, more information can be found in the synthesis manual [Syn00] on page 7-84 to 7-87.

Anyway table 5.15 presents now the corresponding resource consumption for each implementation, which is produced by the synthesis tool chain in dependency of the following key data:

- Number of bits for indexing the second LUT (hbuffer::data_bits $= 20:$ 17 bit for the y coordinate and 3 bit for the z coordinate)
- Number of bits for indexing the histogram layer (hbuffer::index_bits $=$ $8:$ represents the $dim3_{stop}$ value)
- Number of bits for the link address required for the DPRAM memory usage (hbuffer::ramSize_bits $= 15:$ represents addresses for $2^{15} = 32\,\mathrm{k}$ memory entries)

Id	Clock	Constraint	RAMB36_EXP	Flip-Flops	LUTs	LUT-Flip-Flop pairs
1	-	-	-	-	-	-
2	136 MHz	7.3 ns	43	4,054	7,306	7,371
3	-	-	-	-	-	-
4	136 MHz	7.3 ns	43	4,054	7,306	7,371

Table 5.15: Survey report for the Place-and-Route results of the HBuffer unit

With regard to this table, it is at first interesting that the results for option one and three are missing. But a closer look to option one shows immediately that the reason is self-evident, because nobody should even try to instantiate 32 k registers with 43 bit, which results in a total amount of $1.41 \cdot 10^6$ flip-flops, in todays FPGA devices. Moreover the failing of option three can be explained just as easy, because table 5.16, which summarizes the mapping result with regard to the slice logic utilization, shows a fatal error based on an overcharged device (Number used as Memory: 245 %).

Number of Slice Flip Flops	4,103 out of 69,120	5 %
Number of Slice LUTs	55,182 out of 69,120	79 %
Number used as logic	11,148 out of 69,120	16 %
Number using O6 output only	11,092	
Number using O5 output only	16	
Number using O5 and O6	40	
Number used as Memory	44,032 out of 17,920	245 %
Number used as Dual Port RAM	44,032	
Number using O6 output only	44,032	
Number used as exclusive route-thru	2	
Number of route-thrus	18 out of 138,240	1 %
Number using O6 output only	18	

Table 5.16: Report for the Mapping results of the HBuffer unit option three

5.5.3 Histogram

Before presenting the results for the histogram, this section begins obviously also with a small testbench to explain the simulation waveform of all possible histogram layer implementations, which are introduced in section 3.3.1 on page 165.
However it has to be immediately noticed in this connection that the differences between these waveforms are certainly of major interest and are thus highlighted. Besides this, it should be also noted in this context that the concerned histogram layers are resized to 6×5 cells containing each an eight bit signature, because this reduced complexity enables surely a much easier comparison without changing or losing the general nature.

So coming now to the point, the next table 5.17 summarizes the three data samples of the testbench, which consists respectively of three input signals.

Order	**1**	**2**	**3**
Cmd_in [hex/bin]	F / 1111	0 / 0000	A / 1010
Id_in [hex/bin]	02 / 00000010	20 / 00100000	08 / 00001000
Startpos_in / cell	0 / 1	4 / 5	1 / 2

Table 5.17: Outline of the testbench input for the histogram generator unit

A closer look to these data samples shows now that the first one represents a hit in detector station two in combination with a histogram curve, which has a slope of 45 degrees due to the right-shift in front of each row starting at the first column in the first row. Further on, the second sample contains a hit in the sixth station with a vertical straight line as curve beginning at the fifth column in the first row. And finally the third data sample covers a hit in station four and a curve starting at the second column in the first row with a slope of about 67.5 degrees, which means an alternating right-shift in front of the rows.

So based on these samples, the resulting filled histogram layer is shown in table 5.18.

5	00000000	00000000	00000000	00001000	00100010	00000000
4	00000000	00000000	00001000	00000010	00100000	00000000
3	00000000	00000000	00001010	00000000	00100000	00000000
2	00000000	00001010	00000000	00000000	00100000	00000000
1	00000010	00001000	00000000	00000000	00100000	00000000
	1	2	3	4	5	6

Table 5.18: Configuration of the filled histogram layer based on the testbench input

By applying now either the row-wise read-out at the bottom or the column-wise read-out at the right side to this testbench filled histogram layer (see section 3.3.1 on page 165), the following two ordered data sets, which are shown in table 5.19 for the row and table 5.20 for the column, will result.

order	6	5	4	3	2	1
1	00	20	00	00	08	02
2	00	20	00	00	0A	00
3	00	20	00	0A	00	00
4	00	20	02	08	00	00
5	00	22	08	00	00	00

Table 5.19: Specification of the resulting data set for the row-wise ordered histogram read-out at the bottom

order	5	4	3	2	1
1	00	00	00	00	02
2	00	00	00	0A	08
3	00	08	0A	00	00
4	08	02	00	00	00
5	22	20	20	20	20
6	00	00	00	00	00

Table 5.20: Specification of the resulting data set for the column-wise ordered histogram read-out at the right side

Before presenting now the details, it has to be noticed at first that only four figures, which are determined by figure 5.18, 5.19, 5.20 and 5.21, are used to depict exemplary the resulting waveform for a register based implementation with a row-wise read-out at the bottom (see section 3.3.1 on page 166) and the waveform for a RAM based implementation with a column-wise read-out at the right side (see section 3.3.1 on page 171), because the other ones feature just negligible additional information but would fill many pages. Nevertheless it should be noted here that these waveforms are of course also generated and inspected.

Even if these figures seem to be really complex at a first glance, the transportation of the forecasts, which are illustrated in the tables 5.19 and 5.20, simplifies the depicted waveforms immediately.

Coming now to a more detailed look to figure 5.18 and figure 5.20, it is clear that the three data input samples can be again identified by the high triggered 'we_in' signal in the beginning phase directly following the released high-active 'reset' signal. Moreover it is in turn important to realize that the data input has not to be delivered in a continuous stream, but just marked by the 'we' signal.

Besides this, the first major difference between the register and RAM based implementations have already been happened in front of the data input delivery, because the length of the 'reset' signal is quite different. Furthermore the length in the register version is apparently caused by the circumstance that this version is able to reset all registers in the histogram in just a single clock cycle, while a single cycle is certainly impracticable to reset an entire memory block completely. Therefore the RAM version requires as many clock cycles as necessary to address each histogram cell in a row once, because the memory banks in the rows can be reset in parallel. For that reason, the internal 'reset' signal in the testbench remains six, which is equal to the number of columns, plus one clock cycles high before turning low. So based on this context, it is self-evident that this status has to be shown to the outer world by setting the 'processing' signal, because no data input can be taken in the meantime.
Another less significant but anyhow remarkable difference is apparently defined by the encoding style of the 'id_in' signal, which is binary in figure 5.18 and one-hot in figure 5.20.
Moreover it is also interesting that there is no need for a 'start_processing' signal as for the HBuffer unit, which is presented in section 5.5.2 on page 329, because the 'we_in' signal triggers the entire histogram processing.

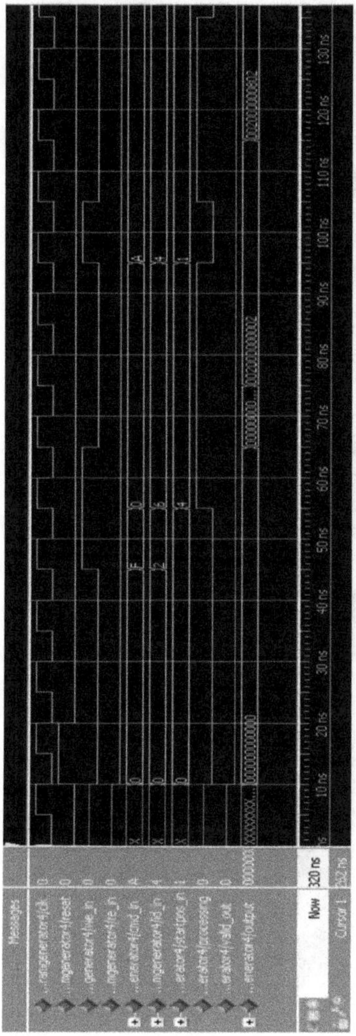

Figure 5.18: First part of the example waveform for the histogram implementations four, five and six in combination with the corresponding generator unit

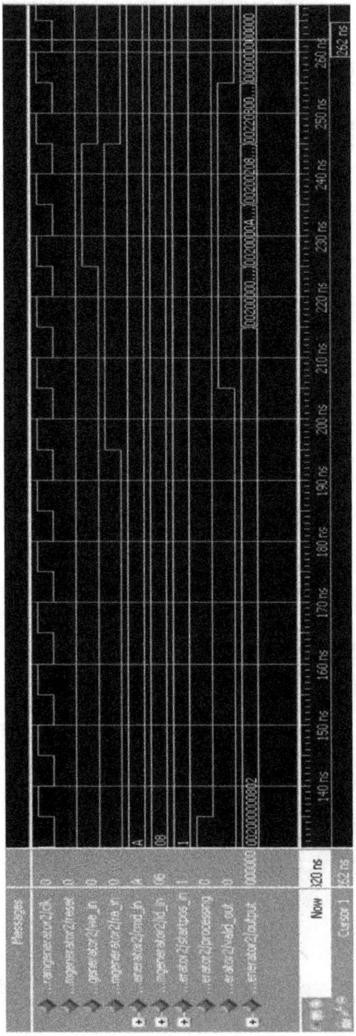

Figure 5.19: Final part of the example waveform for the histogram implementations one, two, three, four, five and six in combination with the corresponding generator unit

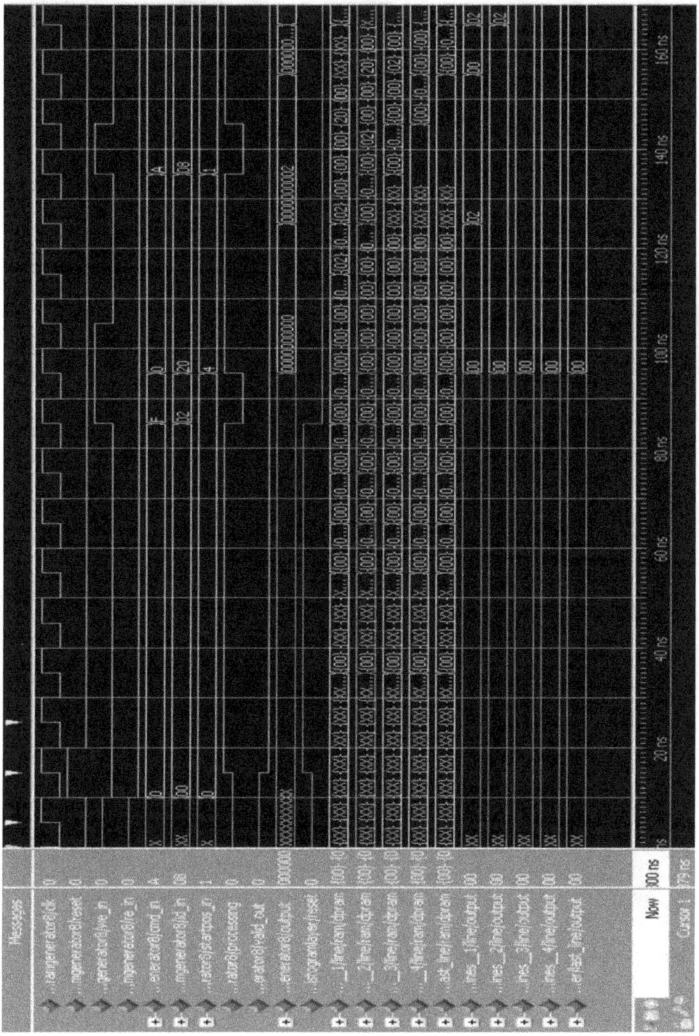

Figure 5.20: First part of the example waveform for the histogram implementation eight in combination with the corresponding generator unit

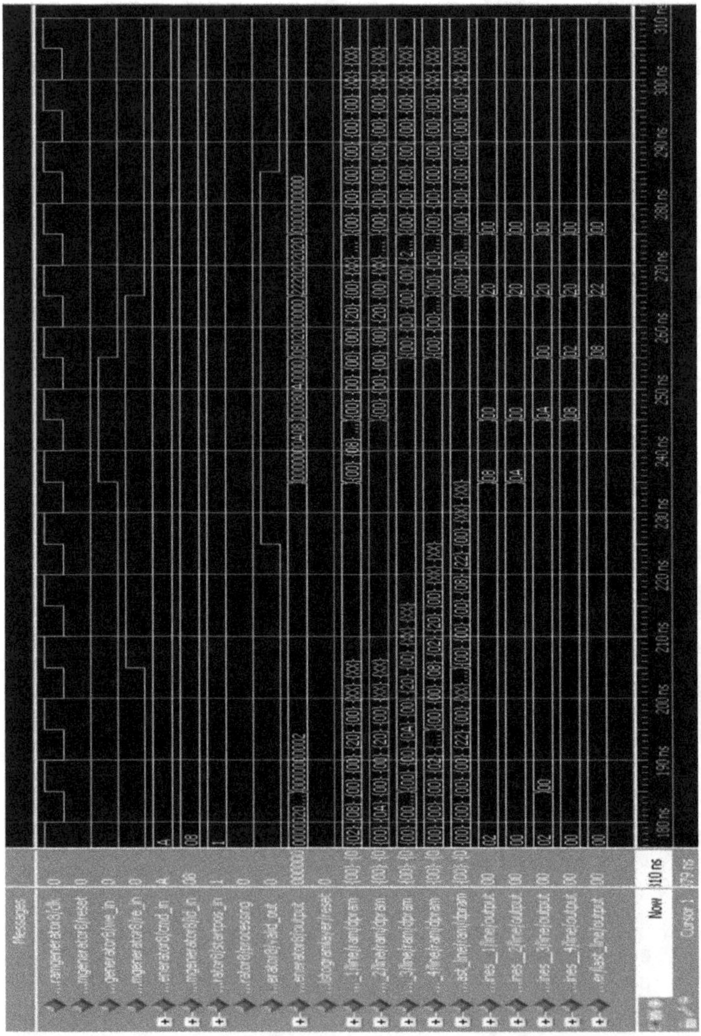

Figure 5.21: Final part of the example waveform for the histogram implementation eight in combination with the corresponding generator unit

Returning now to the waveforms, it has to be mentioned next that figure 5.20 does not completely depict the data injection based on the delivery into the 'dpram' memory, which is even totally hidden in figure 5.18, because the complexity is obviously too high.

Nevertheless this successful process evidently triggers the start of the histogram read-out afterwards. For this purpose, the unit is able to deliver the stored data in the correct order, which is either row-wise starting at the bottom or column-wise starting at the right side, by applying the 're' signal high active after the last entry is tagged by the 'we_in' signal and the 'processing' signal turns low.
But although all implementations create the same filled histogram layer, the two available read-out strategies require certainly different consecutive functional units, which can obviously lead to a discriminative timing and hardware requirement. So with regard to the testbench, the row-wise read-out strategy possesses naturally a data width of $6\,\text{data set members} \cdot 8\,\frac{\text{bit}}{\text{data set member}} = 48\,\text{bit}$ and needs 5 clock cycles, whereas the column-wise one manifests $5\,\text{data set members} \cdot 8\,\frac{\text{bit}}{\text{data set member}} = 40\,\text{bit}$ in conjunction with 6 clock cycles. Transferring this knowledge now to the given example waveforms, the corresponding condition can be seen respectively in figure 5.19, which depicts the row-wise read-out, and figure 5.21, which shows the column-wise read-out.
Having furthermore a closer look to both figures, it can be supplementary recognized that a data write access during the read-out process, which is marked by a high active 'we_in' signal, is always blocked to avoid data collisions.
Besides this, it has to be additionally noticed that the entire content would have also not to be read out in a continuous stream on the signal 'data_out', even if it is not figured out here, but in turn just marked by the 'valid_out' signal.

As the functional description is extensively explained up to here, it is furthermore interesting in which way this functionality can be realized, because different possibilities to implement such a unit are in turn theoretically available.
But before having a closer look to the histogram itself, it is obviously important to analyze the generator unit in front of the histogram, which shifts the parallel input data into the correct time slots with regard to the systolic processing. By the way, it is also clear that each histogram version requires an adequate generator unit, which comprises then in theory again different possibilities for the concrete implementation. For this purpose, table 5.21

shows all of these possibilities in combination with each generator unit.

Id	Type	Encoding [startpos/id]	Cntrl	Queue srlstyle
1	oneHot	1hot/1hot	0	registers
2	oneHot	1hot/1hot	0	noextractff_srl
3	oneHot	1hot/1hot	0	select_srl
4	oneHot	1hot/binary	0	registers
5	oneHot	1hot/binary	0	noextractff_srl
6	oneHot	1hot/binary	0	select_srl
7	number1	binary/1hot	1	registers
8	number1	binary/1hot	1	noextractff_srl
9	number1	binary/1hot	1	select_srl
10	number2	binary/1hot	2	registers
11	number2	binary/1hot	2	noextractff_srl
12	number2	binary/1hot	2	select_srl

Table 5.21: Summary of the available synthesis possibilities for the histogram generator units

A closer look to this table illustrates now that the two columns 'Encoding' and 'Cntrl' specify details of the implementation type, which are very important, because they mark the possible connections of such a generator unit to its allowed histograms implementation. Hence it is immediately clear that an expedient processing could be only guaranteed, if these characteristics match for both implementations in combination with the gathered information from the LUTs.
Coming next to a short explanation, the column 'Encoding' informs apparently about the encoding style, which is used to interpret the start position of the Hough curve (startpos) and the detector station of the hit (id). In addition to this, the following column 'Cntrl' determines evidently the control strategy, which is applied to the data path.
Thence each generator unit, which is characterized by control strategy '0' and encoding '1hot/1hot', can be combined with each histogram unit, which is also characterized by control strategy '0' and encoding '1hot/1hot'.

By keeping this classification in mind, the column 'Queue srlstyle' specifies finally an attribute, which defines the implementation style of an inferred queue. Hence the same source code, which is also dictated by the used synthesis tool, can be used as well as the attribute 'ramstyle' of the HBuffer

unit to produce different and sometimes hardware specific FPGA implementations.
So focusing currently the characteristics, the first value, which is called 'registers', turns the technology-specific resource inferencing completely off and maps the functionality to sequential connected registers (flip-flops) in shifting mode.
In contrast to this, the value 'noextractff_srl' maps the functionality to the appropriate vendor-specific Shift Register Look-up-table (SRL) elements, which are called 'SRL16' in the Virtex V family and imply thus 16 bit shift register LUT primitives.
And finally the value 'select_srl' instantiates a special case of 'noextractff_srl', which includes as unique difference a register inserted ahead of the output buffer to improve timing.
Since this information about the values are apparently only summaries, more information can be found in [Syn02] on page 7-102 to 7-104.

Anyway table 5.22 presents now the corresponding resource consumption for each implementation, which is produced by the synthesis tool chain in dependency of the following key data:

- Number of bits for the shift command of the histogram curve (generator::cmd_bits = 30: always set to histogramlayer::histogram_dim2 - 1)
- Number of bits for the detector station index (generator::id_bits = 7: for a histogram layer with one-hot encoding; generator::id_bits = 3: for a histogram layer with binary encoding)
- Number of bits for the start position of the histogram curve (generator::startPos_bits = 7: for a histogram layer with binary encoding; generator::startPos_bits_in = 7 and generator::startPos_bits_out = 95: for a histogram layer with one-hot encoding, because the value must be converted from binary to one-hot)

Id	Clock	Constraint	Flip-Flops	LUTs	LUT-Flip-Flop pairs
1	833 MHz	1.2 ns	525 (1 %)	71 (1 %)	596 (1 %)
2	833 MHz	1.2 ns	61 (1 %)	SR: 30 (1 %) + 101 (1 %) = 131 (1 %)	189 (1 %)
3	833 MHz	1.2 ns	94 (1 %)	SR: 28 (1 %) + 99 (1 %) = 127 (1 %)	193 (1 %)
4	909 MHz	1.1 ns	525 (1 %)	71 (1 %)	594 (1 %)
5	833 MHz	1.2 ns	61 (1 %)	SR: 30 (1 %) + 101 (1 %) = 131 (1 %)	187 (1 %)
6	833 MHz	1.2 ns	94 (1 %)	SR: 28 (1 %) + 99 (1 %) = 127 (1 %)	192 (1 %)
7	400 MHz	2.5 ns	506 (1 %)	29 (1 %)	527 (1 %)
8	384 MHz	2.6 ns	72 (1 %)	SR: 29 (1 %) + 58 (1 %) = 87 (1 %)	150 (1 %)
9	370 MHz	2.7 ns	104 (1 %)	SR: 27 (1 %) + 56 (1 %) = 83 (1 %)	152 (1 %)
10	400 MHz	2.5 ns	507 (1 %)	30 (1 %)	524 (1 %)
11	370 MHz	2.7 ns	102 (1 %)	SR: 28 (1 %) + 58 (1 %) = 86 (1 %)	178 (1 %)
12	370 MHz	2.7 ns	133 (1 %)	SR: 26 (1 %) + 56 (1 %) = 82 (1 %)	180 (1 %)

Table 5.22: Survey report for the Place-and-Route results of the different histogram generator units

With regard to this table, the results show immediately that just one percent of the available hardware resources are required for a generator unit independent of the implementation.
Notwithstanding the combined consumption of LUT and flip-flop resources for the different 'srlstyle' attributes is expected, the general nature between both values can be easily illustrated by the LUT factor of 1.7 to 3.0 and the flip-flop factor of 3.8 to 8.6. So for that reason, it is clear that the versions with the attributes 'noextractff_srl' or 'select_srl' would be favored over the 'registers' version, if multiple generator units are required.

Besides this, a not less important result is apparently determined by the achieved maximum clock frequency for each implementation, because it is really impressive that the first six implementations, which are convenient for the histogram units based on registers, can theoretically be clocked with a frequency twice as high as the last six implementations.
However it is obvious that these frequencies are not practicable, because the used FPGA device features just a clock technology of 550 MHz. Thence the obtained frequencies can only be that high, because the technology check is made in the clock management tiles section of the synthesis, which is not instantiated here. Hence the synthesis output will not work on such a device, but the results will give a good imagination about the theoretical limits. Moreover it should be supplementary mentioned here that an optimization of the last six implementations to close the frequency gap is senseless, because the clock technology restriction will certainly do that job.

Since all different implementations of the generator unit are satisfyingly analyzed, the focus is moved in the following to the histogram unit itself. For this purpose, section 3.3.1 on page 165 has to be remembered, because all available implementations, which base either on registers or on RAMs, are exhaustively illustrated there. Further on, table 5.23 summarizes now an enormous number of possible histogram implementations, which base on registers. However it has to be noted in this connection that only the first three implementations differ in strategy, because all others are simply derived from these ones especially the types 'cell4', 'cell5' and 'cell6', which incorporate the same principles than 'cell1', 'cell2' and 'cell3'. Furthermore the only reason for the existence of these additional types is the requirement of different data bus sizes, which are caused by the encoding styles.

Id	Type	Encoding start/hit	Cntrl/ Read	DPRAM ramstyle	Queue srlstyle
1	cell1	1hot/1hot	0/↓	-	-
2	cell2	1hot/1hot	0/↓	-	-
3	cell3	1hot/1hot	0/↓	-	-
4	cell4	1hot/bin	0/↓	-	-
5	cell5	1hot/bin	0/↓	-	-
6	cell6	1hot/bin	0/↓	-	-
7	cell1	1hot/1hot	0/→	-	-
8	cell2	1hot/1hot	0/→	-	-
9	cell3	1hot/1hot	0/→	-	-
10	cell4	1hot/bin	0/→	-	-
11	cell5	1hot/bin	0/→	-	-
12	cell6	1hot/bin	0/→	-	-

Table 5.23: Summary of the available synthesis possibilities for the histogram unit implemented with register based cells

By remembering now the two columns 'Encoding' and 'Cntrl' in the table 5.21 of section 5.5.3 on page 342, which presents the diverse concrete implementations of the generator unit, it is not surprising that table 5.23 contains the corresponding information, because an efficient connection of a generator implementation and a histogram implementation require a precise match as stated earlier in this section.

As furthermore the histogram unit allows two different strategies for the read-out, it is obvious that the additional information 'Read' specifies the correspondingly supported one, which affects naturally the implementation of the consecutive functional units.

And finally the last two columns offer in turn the well-known concept of synthesis attributes, which are already introduced in table 5.14 and table 5.21. But even if these attributes are unused here, they are still kept because of the consistency to table 5.24, which utilizes them for the RAM based histogram implementations.

Id	Type	Encoding start/hit	Cntrl/ Read	DPRAM ramstyle	Queue srlstyle
13	linecell7	bin/1hot	1/→	registers	registers
14	linecell7	bin/1hot	1/→	registers	noextractff_srl
15	linecell7	bin/1hot	1/→	registers	select_srl
16	linecell7	bin/1hot	1/→	block_ram	registers
17	linecell7	bin/1hot	1/→	block_ram	noextractff_srl
18	linecell7	bin/1hot	1/→	block_ram	select_srl
19	linecell7	bin/1hot	1/→	select_ram	registers
20	linecell7	bin/1hot	1/→	select_ram	noextractff_srl
21	linecell7	bin/1hot	1/→	select_ram	select_srl
22	linecell7	bin/1hot	1/→	no_rw_check	registers
23	linecell7	bin/1hot	1/→	no_rw_check	noextractff_srl
24	linecell7	bin/1hot	1/→	no_rw_check	select_srl
25	linecell8	bin/1hot	2/→	registers	-
26	linecell8	bin/1hot	2/→	block_ram	-
27	linecell8	bin/1hot	2/→	select_ram	-
28	linecell8	bin/1hot	2/→	no_rw_check	-

Table 5.24: Summary of the available synthesis possibilities for the histogram unit implemented with RAM based cells

Before presenting now the results, it is obviously important to note that the synthesis has to be made more careful than for any other unit, because the histogram implementations feature different theoretic hand-made optimizations, which may be killed during the automatic optimization process of the synthesis tool.
In this connection, an easy example of such a theoretic optimization can be found in section 3.3.1 on page 166 by comparing the introduced histogram implementation one with version two, because the only difference between them is determined by the number of registers per histogram row, which contain the actual detector station index to be set for the actual histogram curve. Furthermore a closer look even exhibits that implementation one combines a minimal number of such registers with a maximum fanout, while version two realizes the maximal number of these registers with a minimum fanout. So since these two implementations form apparently the two maximal contrary conditions, which are identified by the less required resources or the superior wiring and thus timing due to the easier Place-and-route Report (PAR) process, the optimum is naturally located somewhere in between. Thence these hand-made optimizations are just of theoretic nature, because the real optimum is evidently defined by the best trade-off between these two conditions, which depends surely on the capabilities of the used FPGA.
As however the results of these two contrary conditions are very important to set the limits for the trade-off, it is clear that the optimization process of the synthesis tool has to be turned off, which entails the modification of some tool dependent files.

So based on this circumstance, the first file to modify is obviously defined by the one, which contains the architecture body of the design's top entity, because there is the possibility to set the synthesis attributes 'syn_replicate' and 'syn_preserve' for the synthesis tool SYNPLIFY PRO or 'register_duplication' and respectively 'equivalent_register_removal' for the synthesis tool XST. Thusly within this concept, the value 'false' for the attribute 'syn_replicate' or 'register_duplication' disables the register replication, while setting the attribute 'syn_preserve' to the value 'true' or 'equivalent_register_removal' to the value 'false' prevents sequential optimizations across a flip-flop boundary and preserves the signal.
Furthermore exhaustive information about the attributes 'syn_replicate' and 'syn_preserve' of the actual used tool SYNPLIFY PRO can be found in [Syn02] on page 7-92 to 7-94 and 7-167 to 7-171 respectively.

Collaterally the second affected file would be the option file of the synthesis tool SYNPLIFY PRO, which is introduced in appendix C.3 on page C-5, if it is used, because the sequential optimization has to be turned off by setting the synthesis parameter 'no_sequential_opt to the value '1' instead of the default value '0'.

Coming now to the evaluation of the synthesis results including these modified files, table 5.25 and 5.26 present the corresponding resource consumption for each implementation, which is produced by the synthesis tool chain in dependency of the following key data:

- Number of quantized histogram layer cells in the first dimension (histogramlyer::dim1 = 95)
- Number of quantized histogram layer cells in the second dimension (histogramlyer::dim2 = 31)
- Number of bits for the detector station index (histogramlyer::id_bits = 7 : for a histogram layer with one-hot encoding; histogramlyer::id_bits = 3 in combination with histogramlyer::id_bits_oneHot = 7 : for a histogram layer with binary encoding)

Id	**Clock**	**Constraint**	**Flip-Flops**	**LUTs**	**LUT-Flip-Flop pairs**
1	500 MHz	2.0 ns	23,675 (34 %)	26,319 (38 %)	26,562 (38 %)
2	476 MHz	2.1 ns	43,415 (62 %)	26,319 (38 %)	45,076 (65 %)
3	625 MHz	1.6 ns	40,565 (58 %)	40,569 (58 %)	40,569 (58 %)
4	434 MHz	2.3 ns	23,555 (34 %)	26,319 (38 %)	26,460 (38 %)
5	526 MHz	1.9 ns	32,015 (46 %)	26,319 (38 %)	34,263 (49 %)
6	500 MHz	2.0 ns	29,165 (42 %)	29,591 (42 %)	34,273 (49 %)
7	526 MHz	1.9 ns	23,675 (34 %)	26,319 (38 %)	26,570 (38 %)
8	555 MHz	1.8 ns	43,414 (62 %)	26,319 (38 %)	45,201 (65 %)
9	666 MHz	1.5 ns	40,565 (58 %)	40,569 (58 %)	40,569 (58 %)
10	476 MHz	2.1 ns	23,555 (34 %)	26,319 (38 %)	26,459 (38 %)
11	555 MHz	1.8 ns	32,015 (46 %)	26,319 (38 %)	34,301 (49 %)
12	526 MHz	1.9 ns	29,165 (42 %)	29,634 (42 %)	34,150 (49 %)

Table 5.25: Survey report for the Place-and-Route results of the different histogram units based on register cells and disabled synthesis parameters for register replication and sequential optimization

With regard to this table, the resource comparison forecast between the histogram implementation one and two can be proven, because version two requires many more flip-flops, while the number of LUTs are equal. Moreover the prediction for the timing can be contrarily not verified, because the result for implementation one is surprisingly better than the timing of version two. However this situation would be not that confusing, if the very good wiring capability of the used FPGA is taken into account, because the PAR process does not face routing limits in this case.

Comparing these two implementations further on to the next version, the nearly constant amount of flip-flops and the highly increased number of LUTs are both expected due to the basic structure of the implementation three, which is introduced in section 3.3.1 on page 166.

Besides this, the results of the implementations four to six are also not astonishing, because the changed encoding style for the detector station index from one-hot to binary requires evidently less flip-flops for storing and not more but content changed LUTs.

Beyond that, this table exhibits supplementary the information that the read-out direction of the histogram, which is remembered to be row-wise at the bottom for id one to six and column-wise at the right for id seven to twelve, causes generally no mentionable difference between the resource consumption of the corresponding partner implementations, which are obviously identified by one and seven, two and eight and so on. However it is also an interesting fact that the timing of a version with column-wise read-out at the right is always better than the timing of the corresponding partner, which is read-out row-wise at the bottom.

Finally it should be also noticed here that the achieved maximum clock frequency of all units is not practicable, because the used FPGA device features just a clock technology of 550 MHz. Thence the obtained frequencies can only be that high, because the technology check is made in the clock management tiles section of the synthesis, which is not instantiated here. Hence the synthesis output will not work on such a device, but the results will give a good imagination about the theoretical limits.

So since all implementations, which utilize registers, are successfully analyzed up to here, the focus can be evidently set to the implementations based on RAM. Therefore table 5.26 presents currently the synthesis results for these histogram implementations including all different available synthesis possibilities.

Id	Clock	Constraint	RAMB18X2s	Flip-Flops	LUTs	LUT-Flip-Flop pairs
13	322 MHz	3.1 ns	-	32,264 (46 %)	13,971 (20 %)	33,484 (48 %)
14	322 MHz	3.1 ns	-	32,264 (46 %)	13,971 (20 %)	33,484 (48 %)
15	322 MHz	3.1 ns	-	32,264 (46 %)	13,971 (20 %)	33,484 (48 %)
16	384 MHz	2.6 ns	-	4,488 (6 %)	RAM: 1,984 (11 %) + 1,323 (1 %) = 3,307 (4 %)	6,790 (9 %)
17	384 MHz	2.6 ns	-	4,488 (6 %)	RAM: 1,984 (11 %) + 1,323 (1 %) = 3,307 (4 %)	6,790 (9 %)
18	384 MHz	2.6 ns	-	4,488 (6 %)	RAM: 1,984 (11 %) + 1,323 (1 %) = 3,307 (4 %)	6,790 (9 %)
19	416 MHz	2.4 ns	-	4,488 (6 %)	RAM: 868 (4 %) + 672 (1 %) = 1,540 (2 %)	5,236 (7 %)
20	416 MHz	2.4 ns	-	4,488 (6 %)	RAM: 868 (4 %) + 672 (1 %) = 1,540 (2 %)	5,236 (7 %)
21	416 MHz	2.4 ns	-	4,488 (6 %)	RAM: 868 (4 %) + 672 (1 %) = 1,540 (2 %)	5,236 (7 %)
22	384 MHz	2.6 ns	-	4,488 (6 %)	RAM: 1,984 (11 %) + 1,323 (1 %) = 3,307 (4 %)	6,790 (9 %)
23	384 MHz	2.6 ns	-	4,488 (6 %)	RAM: 1,984 (11 %) + 1,323 (1 %) = 3,307 (4 %)	6,790 (9 %)
24	384 MHz	2.6 ns	-	4,488 (6 %)	RAM: 1,984 (11 %) + 1,323 (1 %) = 3,307 (4 %)	6,790 (9 %)
25	294 MHz	3.4 ns	-	28,977 (41 %)	14,098 (20 %)	32,136 (46 %)
26	344 MHz	2.9 ns	-	1,201 (1 %)	RAM; 1,984 (11 %) + 1,479 (2 %) = 3,463 (5 %)	3,744 (5 %)
27	384 MHz	2.6 ns	-	1,202 (1 %)	RAM: 868 (4 %) + 828 (1 %) = 1,696 (2 %)	2,704 (3 %)
28	344 MHz	2.9 ns	-	1,201 (1 %)	RAM: 1,984 (11 %) + 1,479 (2 %) = 3,463 (5 %)	3,744 (5 %)

Table 5.26: Survey report for the Place-and-Route results of the different histogram units based on RAM cells and disabled synthesis parameters for register replication and sequential optimization

With regard to this table, the first interesting fact is obviously defined by the resource consumption of the implementations, which use flip-flops to offer the RAM functionality, because these results are absolutely comparable to the result of the implementations based on register cells, although the realization strategy is quite different. However it has to be additionally instantly noticed that this strategy constitutes surely the reason for the diverse timing of these implementations.

Moreover the results, which are shown in this table, suggest that the implementations based on RAM cells suit universally much better to the given FPGA device than the histogram implementations based on register cells, because the overall resource consumption is generally less than 10 % in contrast to more than 35 %.

Aside from this, the major surprising circumstance is apparently determined by the results for the implementation groups, which are alike except for the values of the attributes 'DPRAM ramstyle' and 'Queue srlstyle', because the resource consumption is regardless of that completely identical. Although this outcome is unexpected, the reason is quite simple, because the two earlier introduced synthesis attributes 'syn_replicate' and 'syn_preserve' avoid the correct vendor-specific RAM and SRL extraction during the synthesis process. Thence it is clear that the resource consumption has to be respectively even identical.
But as this situation is of course unintentional, the next two tables 5.27 and 5.28 present the resource consumption, which is produced by the synthesis tools with an enabled optimization process and the same key data.

Id	Clock	Constraint	Flip-Flops	LUTs	LUT-Flip-Flop pairs
1	454 MHz	2.2 ns	24,309 (35 %)	26,330 (38 %)	27,171 (39 %)
2	434 MHz	2.3 ns	24,309 (35 %)	26,330 (38 %)	27,174 (39 %)
3	666 MHz	1.5 ns	40,565 (58 %)	40,569 (58 %)	40,569 (58 %)
4	476 MHz	2.1 ns	29,128 (42 %)	26,330 (38 %)	31,618 (45 %)
5	500 MHz	2.0 ns	29,113 (42 %)	26,330 (38 %)	31,615 (45 %)
6	526 MHz	1.9 ns	29,234 (42 %)	29,635 (42 %)	34,362 (49 %)
7	333 MHz	3.0 ns	26,930 (38 %)	28,951 (41 %)	29,789 (43 %)
8	370 MHz	2.7 ns	26,930 (38 %)	28,951 (41 %)	29,800 (43 %)
9	714 MHz	1.4 ns	40,565 (58 %)	40,569 (58 %)	40,569 (58 %)
10	400 MHz	2.5 ns	31,887 (46 %)	28,951 (41 %)	34,370 (49 %)
11	370 MHz	2.7 ns	31,881 (46 %)	28,951 (41 %)	34,317 (49 %)
12	526 MHz	1.9 ns	29,255 (42 %)	29,717 (42 %)	34,520 (49 %)

Table 5.27: Survey report for the Place-and-Route results of the different histogram units based on register cells and enabled synthesis parameters for register replication and sequential optimization

While remembering now the synthesis conditions in combination with the two different available optimization possibilities from the beginning of this section, which are characterized by handmade by the developer and automatic by the synthesis tool, it is not surprising that the results for the resource consumption of the histogram implementation one and two are such similar. Furthermore this circumstance causes evidently all results, which are presented in table 5.27, to become closer to each other.
Although this situation is certainly expected, it is nevertheless very hard to retrace the behavior of the automatic optimization, which is offered by the synthesis tool.

Aside from this, the most interesting alterations can be naturally found in table 5.28, because this one presents the results for the histogram implementations based on RAM cells.

Id	Clock	Constraint	RAMB18X2s	Flip-Flops	LUTs	LUT-Flip-Flop pairs
13	322 MHz	3.1 ns	-	33,981 (49 %)	15,015 (21 %)	35,257 (51 %)
14	322 MHz	3.1 ns	-	30,733 (44 %)	SR: 204 (1 %) + 15,219 (22 %) = 15,423 (22 %)	32,096 (46 %)
15	303 MHz	3.3 ns	-	30,958 (44 %)	SR: 190 (1 %) + 15,205 (21 %) = 15,395 (22 %)	32,534 (47 %)
16	294 MHz	3.4 ns	31 (12 %)	4,271 (6 %)	614 (1 %)	4,442 (6 %)
17	294 MHz	3.4 ns	31 (12 %)	1,023 (1 %)	SR: 204 (1 %) + 818 (1 %) = 1,022 (1 %)	1,668 (2 %)
18	277 MHz	3.6 ns	31 (12 %)	1,248 (1 %)	SR: 190 (1 %) + 804 (1 %) = 994 (1 %)	1,694 (2 %)
19	370 MHz	2.7 ns	-	4,620 (6 %)	RAM: 868 (4 %) + 678 (1 %) = 1,546 (2 %)	5,406 (7 %)
20	384 MHz	2.6 ns	-	1,372 (1 %)	SR: 204 (1 %) + RAM: 868 (4 %) + 882 (1 %) = 1,954 (2 %)	2,533 (3 %)
21	384 MHz	2.6 ns	-	1,597 (2 %)	SR: 190 (1 %) + RAM: 868 (4 %) + 868 (1 %) = 1,926 (2 %)	2,526 (3 %)
22	294 MHz	3.4 ns	31 (12 %)	4,271 (6 %)	614 (1 %)	4,442 (6 %)
23	294 MHz	3.4 ns	31 (12 %)	1,023 (1 %)	SR: 204 (1 %) + 818 (1 %) = 1,022 (1 %)	1,668 (2 %)
24	277 MHz	3.6 ns	31 (12 %)	1,248 (1 %)	SR: 190 (1 %) + 804 (1 %) = 994 (1 %)	1,694 (2 %)
25	243 MHz	4.1 ns	-	29,535 (42 %)	14,124 (20 %)	32,164 (46 %)
26	285 MHz	3.5 ns	31 (12 %)	936 (1 %)	769 (1 %)	1,299 (1 %)
27	370 MHz	2.7 ns	-	1,176 (1 %)	RAM: 868 (4 %) + 799 (1 %) = 1,667 (2 %)	2,176 (3 %)
28	285 MHz	3.5 ns	31 (12 %)	936 (1 %)	769 (1 %)	1,299 (1 %)

Table 5.28: Survey report for the Place-and-Route results of the different histogram units based on RAM cells and enabled synthesis parameters for register replication and sequential optimization

With regard to this table, it has to be highlighted at first that the optimization effects of the attributes 'ramstyle' for the DPRAM and 'srlstyle' for the queue can be recognized very well.

Further on, the best result related to the timing, which is surprisingly also the best one concerning the resource consumption, is obviously described by the implementations with id 20 and 21. Furthermore a closer look to both implementations shows that they feature a maximum clock frequency of about 384 MHz, while their resource consumption is about 1 % to 2 % of the available flip-flops and 2 % of the obtainable LUTs.
However even if the overall FPGA occupancy is 3 %, the limitation for the number of parallel units on a single-chip would be apparently dominated by the RAM inferring, because it utilizes currently about 4 % of the available resources. Moreover the reason for this high percentage number is quite simple, because the used FPGA can not configure each LUT to be usable as RAM.
Besides this, it is also not astonishing that the results of these two implementations are so close to each other, because the only minor difference is identified by the adjustment of the attribute 'srlstyle', which is anyway almost equal.

Beyond that, a major remark has to be set on the implementations, which employ the technology specific Block RAM, because the utilization shown in table 5.28 is much bigger than expected and especially more than three times higher than the comparatively consumed LUT RAM resources.
So while having now a more detailed look to this situation, it is clear that the estimation of the required histogram memory can be easily computed to $95\,\frac{\text{cells}}{\text{dim1}} \cdot 31\,\frac{\text{cells}}{\text{dim2}} \cdot 7\,\frac{\text{bit}}{\text{cell}} = 20{,}615\,\frac{\text{bit}}{\text{histogram layer}}$, which leads then to an anticipated utilization of $\frac{20{,}615\,\frac{\text{bit}}{\text{histogram layer}}}{4{,}608 \cdot 1{,}024\,\frac{\text{maximum available bits}}{\text{FPGA}}} \cdot 100\,\% \approx 0.44\,\%$. But as this estimation is quite different compared to the evaluated result of about 12 %, there must be a crucial design aspect, which seems to be really mysterious at a first glance.
Anyway the word 'Block' in the name of the concerned resource is immediately able to bring light into the darkness, because the Block RAM is of course just available in memory blocks. Thence the utilization is suggested to be $\frac{31\,\text{histogram rows} \cdot 1\,\frac{\text{inferred block}}{\text{histogram row}}}{256\,\text{available 18 kb blocks}} \cdot 100\,\% \approx 12.11\,\%$.
Moreover it has to be also directly noted in this connection that this special form of resource consumption implies that most capabilities of such a single memory block are unused. Furthermore a more precise look evinces even that the occupancy of such a single memory block can be computed to only $\frac{95\,\frac{\text{cells}}{\text{dim1}} \cdot 7\,\frac{\text{bit}}{\text{cell}}}{18 \cdot 1{,}024\,\text{bits in block}} \cdot 100\,\% \approx 3.61\,\%$. Hence it is really essential to real-

ize that no additional memory would be required, if more histogram cells in the first dimension have to be implemented. However it has to be supplementary noticed that the requirement of two access ports, which are needed both at the same time for the communication between the rows, avoids the implementation of multiple histogram rows in a single block.

5.5.4 Filtering

Since section 3.3.1 on page 174 defines the entire process of reducing the background noise in the histogram to comprise the separate algorithm steps Encoding, Diagonalization and Peak-finding including the Serialization, the following sections are naturally used to individually present the simulation waveforms and synthesis results of all different implementations including special attention payed to the required resources and the maximal possible clock frequency, because this procedure offers the possibility to have a look at the results of each step independent of all others.

Encoding

This section starts analog to all other sections with a simple waveform, which illustrates a small example for the implemented Encoding unit.
But before coming to the details, it has to be explicitly highlighted that just a single waveform would be depicted in figure 5.22, even if multiple implementations are presented in section 3.3.1 on page 174. However this circumstance is not that surprising, because the shown waveform, which corresponds evidently to the LUT implementation, is apparently the most complex one. Furthermore the other waveform can be obviously easily deduced from this one by just removing the signals 'clk', 'load', 'load_encode' and 'load_input', because the loading functionality has to be certainly removed due to the hard-coded capabilities, while the operating mode has to be surely identical at the end.
Besides this, it has to be also noticed in this connection that the Encoding unit does naturally not feature a 'reset' signal at all. Thus the one shown in the waveform is just used to initialize the testbench into a defined initial state.

Having now a closer look to the waveform, the loading functionality can be immediately identified by the signal 'load_encode', which contains the LUT addresses or value identifiers, and the signal 'load_input', which represents the value for the actual address. In addition to this, the signal 'load' has to be

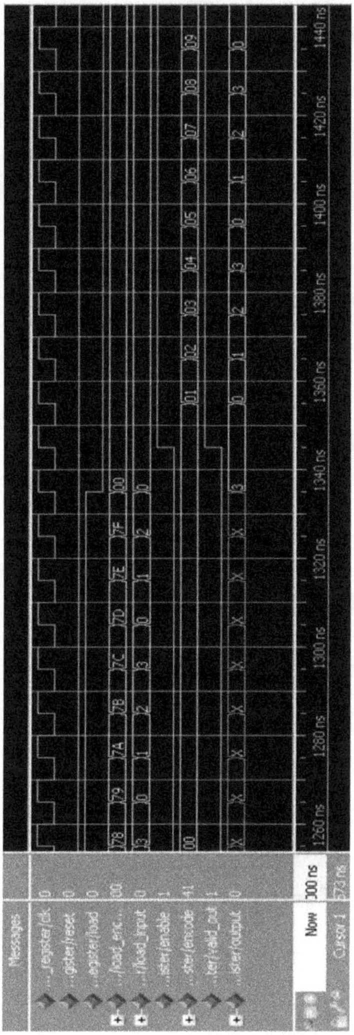

Figure 5.22: Example waveform for the priority encoder of the histogram cells based on a LUT implementation with Block RAM or registers

once again simply used to mark valid data in the initial configuration phase of the LUT at startup, which features obviously in turn just the possibility for a discontinuous data stream.
Thence the first part of the waveform depicts the end of such a configuration phase, which writes consecutively repetitive data values in the range of zero to three into successive addresses starting with zero. Moreover it has to be mentioned in this connection that this configuration phase should obviously not end until each possible address contains valid data.

Further on, it is clear that the unit has to turn into the operating mode to realize the Encoding after the configuration is successful. For this purpose, the active signals change in the middle of this figure to the signal 'encode', which delivers certainly the value to encode, and the signal 'enable', which triggers the Encoding. Furthermore the corresponding output is then obviously determined by the signal 'output' and marked by the signal 'valid_out'. Consequently the following part of the waveform shows finally some Encoding examples.

As the functional description is now extensively explained, it is furthermore interesting in which way this functionality can be realized, because table 5.29 presents different possibilities to implement such an Encoding unit in theory.

Id	**Type**	**Read**	**SPRAM ramstyle**
1	logic	↓	-
2	register	↓	-
3	ram	↓	registers
4	ram	↓	block_ram
5	ram	↓	select_ram
6	ram	↓	no_rw_check
7	logic	→	-
8	register	→	-
9	ram	→	registers
10	ram	→	block_ram
11	ram	→	select_ram
12	ram	→	no_rw_check

Table 5.29: Summary of the available synthesis possibilities for the Encoding unit

By remembering now the earlier introduced general concept for the efficient

connection of the different available functional unit implementations, the third column of this table is certainly not surprising, because the read-out direction of the actual unit must of course precisely match the one of the correspondingly selected histogram implementation, whose direction is provided in the 'Read' part of the fourth column of table 5.23 of section 5.5.3 on page 346 and table 5.24 of section 5.5.3 on page 347.
In addition to this, the column afterwards offers apparently in turn the well-known concept of a synthesis attribute, which is similarly introduced in table 5.14 of section 5.5.2 on page 331. Furthermore the only negligible difference between the accordingly affected fragment of the two implementations is obviously determined by the instantiation of a Single-Ported Random Access Memory (SPRAM) instead of a Dual-Ported Random Access Memory (DPRAM), which leads then naturally regardless of that to the usage of the the identical memory related attribute.

So coming now to the evaluation of the results, table 5.30 presents the corresponding resource consumption for each implementation, which is produced by the synthesis tool chain in dependency of the following key data:

- Number of parallel processing units depending on the histogram read-out and especially on the size of the corresponding histogram dimension (priority_encoder::parallelity $= 95:$ size of the first histogram dimension for the row-wise read-out at the bottom [$\downarrow$]; priority_encoder::parallelity $= 31:$ size of the second histogram dimension for the column-wise read-out at right [$\rightarrow$])
- Number of bits for the Encoding input of a single processing unit (priority_encoder::encoding_bits) $= 7:$ contains one bit for each detector station such as the histogram cells
- Number of bits for the encoded output of a single processing unit (priority_encoder::priority_bits $= 2:$ represents three priority classes)

Id	Clock	Constraint	RAMB18X2s	Flip-Flops	LUTs	LUT-Flip-Flop pairs
1	769 MHz	1.3 ns	-	813 (1 %)	285 (1 %)	851 (1 %)
2	212 MHz	4.7 ns	-	24,320 (35 %)	20,710 (29 %)	31,085 (44 %)
3	208 MHz	4.8 ns	-	24,573 (35 %)	22,185 (32 %)	31,626 (45 %)
4	232 MHz	4.3 ns	48 (37 %)	0 (0 %)	363 (1 %)	363 (1 %)
5	322 MHz	3.1 ns	-	665 (1 %)	RAM: 760 (4 %) + 363 (1 %) = 1,123 (1 %)	1,759 (2 %)
6	232 MHz	4.3 ns	48 (37 %)	0 (0 %)	363 (1 %)	363 (1 %)
7	769 MHz	1.3 ns	-	237 (1 %)	93 (1 %)	250 (1 %)
8	217 MHz	4.6 ns	-	7,934 (11 %)	6,964 (10 %)	10,369 (15 %)
9	196 MHz	5.1 ns	-	7,759 (11 %)	7,899 (11 %)	10,808 (15 %)
10	263 MHz	3.8 ns	16 (12 %)	0 (0 %)	134 (1 %)	134 (1 %)
11	312 MHz	3.2 ns	-	217 (1 %)	RAM: 248 (1 %) + 134 (1 %) = 382 (1 %)	555 (1 %)
12	263 MHz	3.8 ns	16 (12 %)	0 (0 %)	134 (1 %)	134 (1 %)

Table 5.30: Survey report for the Place-and-Route results of the different Encoding units

With regard to this table, it can be immediately seen that the implementations with row-wise read-out at the bottom (id 1 - 6), which use naturally 95 subunits for parallel processing, require approximately three times more hardware resources than their corresponding versions with column-wise read-out at right (id 7 - 12), which utilize in contrast only 31 subunits. Consequently the major interesting result of this table is obviously determined by the circumstance that all implementations of the Encoding unit scale with their number of subunits, which are needed to process the parallel data.

Besides this, the best result for each read-out direction with regard to the timing and resource consumption is evidently defined by option one and seven. Furthermore this fact is of course not surprising, because the implementations, which apply all time fixed logic gates for a special Encoding model, offer naturally less flexibility and require thus less resources than the LUT versions. In addition to this, the differences in the results are so enormous, because the actual Encoding table contains just three entries, which enable a tremendous logic gate optimization and implicate many entries with the priority class zero in the LUT versions.
However if the implementations with logic gates are skipped, it would be impressing that the best unit, which offers the total flexibility, is given by the versions five and eleven. Thence these two units, which apply both distributed RAM, combine the flexibility and the best timing with the fewest resource consumption.

Beyond that, it is also not astonishing that the results for option two and three or eight and nine are quite similar, because the implementation pairs employ all flip-flops. By the way, it is clear that this circumstance counts as well for the ids four and six or ten and twelve with the only difference of applying Block RAM instead of flip-flops.

Diagonalization

While remembering the testbench of section 3.3.1 on page 176, it is quite easy to explain the following two waveforms 5.23 and 5.24, which base on the following three input data samples specialized to each read-out strategy:

1. row-wise = hex 1B = $bin\,011011$; column-wise = $hex\,39$ = $bin\,111001$
2. row-wise = $hex\,06$ = $bin\,000110$; column-wise = $hex\,24$ = $bin\,100100$
3. row-wise = $hex\,31$ = $bin\,110001$; column-wise = $hex\,13$ = $bin\,010011$

Having now a closer look to both waveforms, these three different data input samples can be easily identified at the 'input' signal by the high triggered 'enable' signal in the beginning phase directly following the released high-active 'reset' signal. In this connection, it has to be once again noticed that the data input has not to be delivered in a continuous stream, but just marked by the 'enable' signal. Moreover the correct delivered data on the 'output' signal is naturally marked by the signal 'valid_out'.

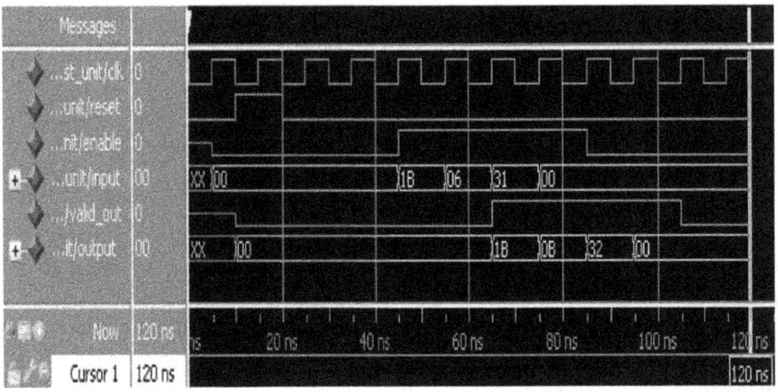

Figure 5.23: Example waveform for the Diagonalization unit suitable for histogram implementations with row-wise read-out at the bottom

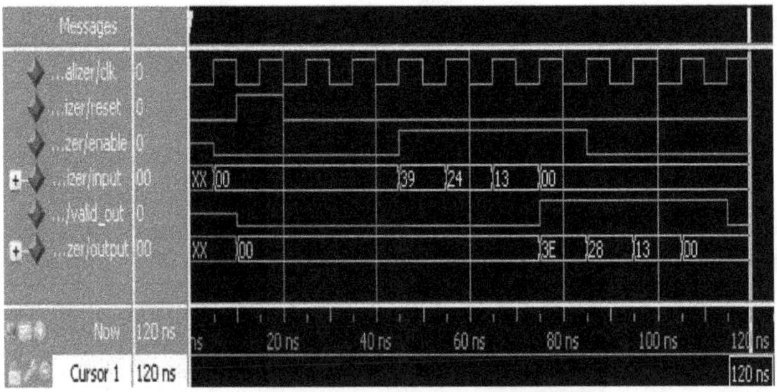

Figure 5.24: Example waveform for the Diagonalization unit suitable for histogram implementations with column-wise read-out at right

As the functional description of the Diagonalization unit is clarified up to here, it is interesting in which way this functionality can be realized. For this purpose, table 5.31 lists all available implementation possibilities.

Id	Type	Read
1	row	↓
2	column	→

Table 5.31: Summary of the available synthesis possibilities for the Diagonalization unit

In contrast to the previous sections, it is conspicuous that this table contains only two entries, which represent just a different read-out strategy respectively identified in turn by the 'Read' column.

Further on, table 5.32 presents now the corresponding resource consumption for each implementation, which is produced by the synthesis tool chain in dependency of the following key data:

- Number of parallel processing units depending on the histogram read-out and especially on the size of the corresponding histogram dimension (row_diagonalizer::parallelism $= 95:$ size of the first histogram dimension for the row-wise read-out at the bottom [↓]; column_diagonalizer::parallelism $= 31:$ size of the second histogram dimension for the column-wise read-out at right [→])
- Number of concerned diagonal histogram cells (row_diagonalizer::size = column_diagonalizer::size $= 2$)
- Index of the centric diagonal histogram cell (row_diagonalizer::index $= 1$; column_diagonalizer::index $= 0$)
- Number of bits for the data (row_diagonalizer::size = column_diagonalizer::size $= 2:$ must be equal to the number of bits for the encoded priority classes)

Id	Clock	Constraint	Flip-Flops	LUTs	LUT-Flip-Flop pairs
1	833 MHz	1.2 ns	568 (1%)	188 (1%)	568 (1%)
2	909 MHz	1.1 ns	184 (1%)	60 (1%)	184 (1%)

Table 5.32: Survey report for the Place-and-Route results of the different Diagonalization units

With regard to this table, the implementation with row-wise read-out at the bottom (id 1), which uses 95 subunits for parallel processing, requires obviously again approximately three times more hardware resources than the version with column-wise read-out at right (id 2), which utilizes only 31 subunits. Consequently the major interesting result is analog to the Encoding unit presented in section 5.5.4 on page 356, which means that the Diagonalization unit scales with its number of subunits.

Finally it should be also noticed here that the achieved maximum clock frequencies are not practicable, because the used FPGA device features just a clock technology of 550 MHz. Thence these frequencies can only be that high, because the technology check is not made here. Hence the synthesis output will not work on such a device, but the results will give a good imagination about the theoretical limits.

Peak-finding and Serialization

While remembering the testbench of section 3.3.1 on page 179, the following two waveforms 5.25 and 5.26, which base on the subsequent four input data samples specialized to each read-out strategy, can be quite easily explained as well as in all previous sections.

1. row-wise = $hex\,17 = bin\,010111$; column-wise = $hex\,1B = bin\,011011$
2. row-wise = $hex\,1A = bin\,011010$; column-wise = $hex\,19 = bin\,011001$
3. row-wise = $hex\,35 = bin\,110101$; column-wise = $hex\,35 = bin\,110101$
4. row-wise = $hex\,00 = bin\,000000$; column-wise = $hex\,00 = bin\,000000$

Having now a closer look to both waveforms, these four different data input samples can be once again easily identified at the 'input' signal by the high triggered 'enable' signal in the beginning phase directly following the released high-active 'reset' signal. In this connection, it has to be also in turn noticed that the data input has not to be delivered in a continuous stream, but just marked by the 'enable' signal. Moreover the correct delivered data on the 'output' signal is naturally marked by the signal 'valid_out'.

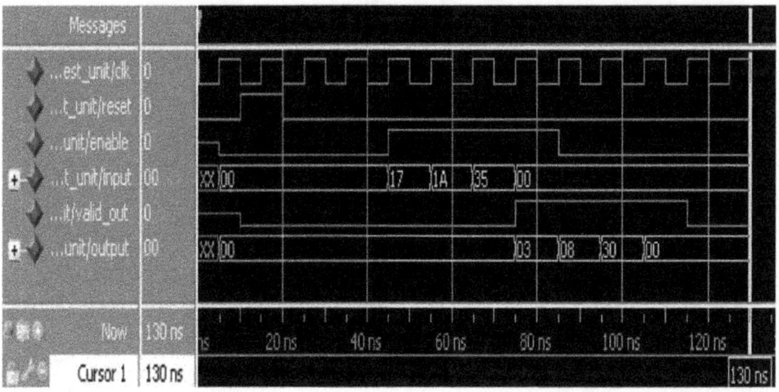

Figure 5.25: Example waveform for the 2D Peak-finding unit suitable for histogram implementations with row-wise read-out at the bottom

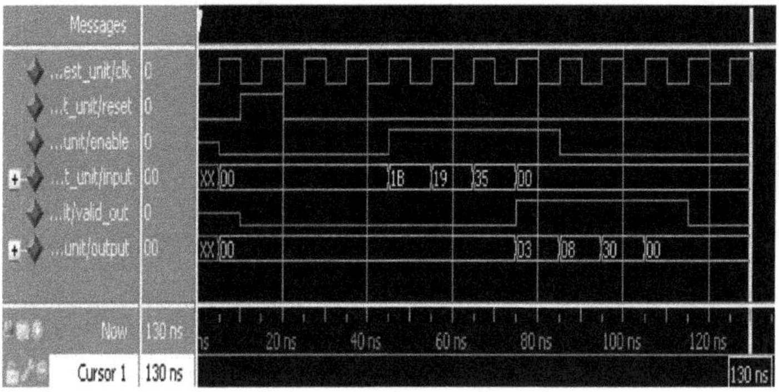

Figure 5.26: Example waveform for the 2D Peak-finding unit suitable for histogram implementations with column-wise read-out at right

As the functional description of the 2D Peak-finding unit is clarified up to here, it is interesting in which way this functionality can be realized. For this purpose, table 5.33 lists all available implementation possibilities.

Id	Type	Read
1	row	↓
2	column	→

Table 5.33: Summary of the available synthesis possibilities for the 2D Peak-finding unit

As well as in the previous section and in contrast to the sections before that one, there are only two entries in the synthesis possibility table 5.33, which represent just a different read-out strategy respectively identified in turn by the 'Read' column.

Further on, table 5.34 presents now the corresponding resource consumption for each implementation, which is produced by the synthesis tool chain in dependency of the following key data:

- Number of parallel processing units depending on the histogram read-out and especially on the size of the corresponding histogram dimension (row_peakfinder2d::parallelism $= 95:$ size of the first histogram dimension for the row-wise read-out at the bottom [↓]; column_peakfinder2d::parallelism $= 31:$ size of the second histogram dimension for the column-wise read-out at right [→])
- Number of concerned histogram cells in the first dimension (row_peakfinder2d::size_dim1 = column_peakfinder2d::size_dim1 = 3)
- Number of concerned histogram cells in the second dimension (row_peakfinder2d::size_dim2 = column_peakfinder2d::size_dim2 = 3)
- Number of bits for the data (row_peakfinder2d::size = column_peakfinder2d::size $= 2:$ must be equal to the number of bits for the encoded priority classes)

Id	Clock	Constraint	Flip-Flops	LUTs	LUT-Flip-Flop pairs
1	454 MHz	2.2 ns	769 (1 %)	494 (1 %)	998 (1 %)
2	588 MHz	1.7 ns	312 (1 %)	132 (1 %)	324 (1 %)

Table 5.34: Survey report for the Place-and-Route results of the different 2D Peak-finding units

With regard to this table, it can be once more immediately seen that the implementation with row-wise read-out at the bottom (id 1), which uses naturally 95 subunits for parallel processing, requires approximately three times more hardware resources than its corresponding version with column-wise read-out at right (id 2), which utilizes in contrast only 31 subunits. Consequently the major interesting result of this table is again determined by the circumstance that the 2D Peak-finding unit scales with its number of subunits as well as for the Diagonalization and Encoding unit, which are presented in section 5.5.4 on page 361 and in section 5.5.4 on page 356.

Finally it should be also noticed again that the achieved maximum clock frequency of each unit is not practicable, because the used FPGA device features just a clock technology of 550 MHz. Thence the obtained frequencies can only be that high, because the technology check is made in the clock management tiles section of the synthesis, which is not instantiated here. Hence the synthesis output will not work on such a device, but the results will give a good imagination about the theoretical limits.

So since the 2D Peak-finding unit is satisfyingly analyzed up to here, the focus is moved in the following to the implementation of the Serialization unit.
But before continuing, special thanks go to Andreas Wurz, because he has realized the functionality, which is introduced in section 3.3.1 on page 179. Moreover the circumstance, that the developed description is not done by myself, leads apparently to the lack of a testbench produced waveform and further to the direct presentation of the available implementation possibilities in table 5.35.

Id	Type	Read
1	trig_multiplex_96_1	↓
2	trig_multiplex_32_1	→

Table 5.35: Summary of the available synthesis possibilities for the Serialization unit

As well as in multiple previous sections, this synthesis possibility table contains obviously also just two entries featuring a different read-out strategy, which can be identified in turn by the 'Read' column.

Further on, it has to be obviously mentioned here that the size of the composing memories, which form evidently the almost unique element of the Serialization unit, can be determined with the assistance of the analysis introduced in section 4.9.2 on page 288.

Besides this, it is additionally important to notice that the description in HDL is fixed, which means that there is no generic parameter to modify the unit in any case.

Anyway table 5.36 presents now the corresponding resource consumption for each implementation, which is produced by the synthesis tool chain in dependency of the following key data:

- Number of parallel processing units depending on the histogram read-out and especially on the size of the corresponding histogram dimension (trig_multiplex_96_1 $= 96:$ ceil to the next power of two based on the size of the first histogram dimension for the row-wise read-out at the bottom [↓]; trig_multiplex_32_1 $= 32:$ ceil to the next power of two based on the size of the second histogram dimension for the column-wise read-out at right [→])

Id	Clock	Constraint	RAMB18X2s	Flip-Flops	LUTs	LUT-Flip-Flop pairs
1	277 MHz	3.6 ns	2 (1 %)	572 (1 %)	SR: 517 (2 %) + 1,225 (1 %) = 1,742 (2 %)	1,789 (2 %)
2	294 MHz	3.4 ns	2 (1 %)	363 (1 %)	SR: 538 (3 %) + 555 (1 %) = 1,093 (1 %)	1,114 (1 %)

Table 5.36: Survey report for the Place-and-Route results of the different Serialization units

So with regard to this table, it can be evidently only attested that the implementation with row-wise read-out at the bottom (id 1), which uses here 96 subunits for parallel processing due to the power of two scalability, requires at least more hardware resources than its corresponding version with column-wise read-out at right (id 2), which utilizes by contrast 32 subunits. Moreover it is directly clear that the expected well-known scalability with the number of subunits can not be verified here such as in the other sections. However this discrepancy is furthermore not surprising, because the needed memory resources depend obviously on the amount of arising data, which is certainly identical for both implementations, as well as on the timing assimilation of the data, which is surely different but can be precisely investigated by the usage of the analysis presented in section 4.9.2 on page 288.

5.5.5 LBuffer

In contrast to all other sections, this one is obviously not able to present results, because the concept of the LBuffer and the 3D Peak-finding unit, which is introduced in section 3.3.1 on page 183, is not hypercritical and thus not implemented yet. Furthermore the required resources of both units are additionally assumed to be insignificant, because the implementation is planned to be realized on the PPC, which exists always inside the FPGA device whether used or not.
Nevertheless the analysis, which is introduced in section 4.9.3 on page 291, can be used to evaluate the number of entries in the LBuffer unit, which enables further the theoretical calculation for the amount of necessary memory. Finally it has to be also mentioned here that CBMROOT simulations evince that a lack of the 3D Peak-finding unit has even not an impressive consequence to the performance of the tracking.

5.6 Results for the Sony Playstation III Implementation

This section presents the results for the Sony Playstation III implementation, which is introduced in section 3.3.2 on page 184, with regard to the benchmarking criteria time and memory consumption.

Even if the alteration of the criteria for quantitative design in contrast to section 5.5 on page 327 is surprising at a first glance, the reason for this modification would become naturally quite obvious when taking the fix hardware of such an embedded system into account just as for the personal computer implementation, which is presented in section 5.4 on page 323. However it has to be additionally mentioned in this connection that the implementation for the Sony Playstation III is unlike the personal computer one very optimized and does not contain computations, which do not effect the tracking output. Nevertheless the result is surely again separated for each necessary overhead and algorithm step to feature a detailed rating and thus optimization nomination.

So for this purpose, it is evidently perspicuous that the focus has to be initially set to each overhead step to clarify the corresponding importance. So within this intention, the first nominee is certainly identified by the initialization, which consumes time to read the data files containing the hits, the coding table, the Peak-finding geometry and both LUTs. Afterwards another group of steps is apparently characterized by the thread handling, because these ones consume time for the setup, the configuration and the termination of the SPE usage. Moreover the configuration step is in particular noticeable within the context of the Sony Playstation III system, because it is required to automatically detect the available size in the local storage of each SPE, which defines as mentioned in section 3.3.2 on page 203 a limit for the job creation and for the amount of parallel realizable histogram layers.

Further on, it has to be finally explicitly highlighted that these recently introduced overhead steps are not presented in detail, because they are naturally negligible due to the circumstance that they arise only once independent of the number of computed events. Furthermore the memory consumption, which is presented in the following, is likewise restricted to the local storage of the SPEs, because this resource forms a limit as mentioned earlier, while the XDRDRAM is uncritical.

Coming now to more details, it is clear that the results have to be of course seen related to the same input data as utilized for the software implementation on a personal computer, which is already presented in section 5.4 on page 323. Therefore the key indicators can be in turn summarized by seven detector stations, 15,658 hits, three entries in the table 'codingTable', three priority classes and nine elements for the Peak-finding geometry. In addition to this, the quantization of the histogram is obviously also again set to 95 cells in the first dimension and 31 cells in the second dimension in conjunction with 191 layers.
Moreover it has to be noticed in this connection that sample input data is generally produced by running the software simulation package, which is introduced in section 3.2 on page 102, on a common personal computer.

Besides this, it should be also noted here that the correspondingly used makefile (see Make) for the program compilation and linking can be found in appendix D.2 on page D-3, while exhaustive information about the fastest algorithm configuration can be found in appendix D.3 on page D-11.

Before proceeding now with the detailed results for each step of the algorithm, the tables 5.37, 5.38, 5.39 and 5.40 present the overall results for all different available algorithm configurations, which are introduced in section 3.3.2 on page 184.
So based on this context, it is self-evident that the column 'CV' in these tables represents the code version, which defines the job creation strategy, while the column 'MV' identifies the memory version, which determines the memory access type.
Further on, the column 'CPU' configures apparently the processing elements, which are used to evaluate the results. By the way, it is clear that the Sony Playstation III system offers two major differentiable configurations to consider, which are defined by the usage of multiple SPEs controlled by the PPE or the PPE uniquely. Furthermore it has to be explicitly highlighted at this juncture that if the unique PPE is taken, the whole algorithm would be evaluated only there and the SPEs would be unused at all, whereas the other version uses contrarily all available SPEs in combination with the PPE. Although it is collaterally obvious in this context that the SPE version allows also the usage of just some SPEs, the detailed result with regard to the number of used SPEs is shifted to [Str09], because such a topic is far beyond this thesis but indeed not less interesting.
Moreover the final configuration column 'ARCH' identifies evidently the computation mode, which is either determined by 'scalar' for SISD computation or 'vector' for SIMD computation. To dispel some misunderstandings in this connection, it should be emphasized that it is really possible to access the memory vectorial and compute scalar or vice versa. However it should be not discussed any further, if such a feature makes sense or not, because the results tell their own tale.

CV	MV	CPU	ARCH	Time $[ms]$	Local storage $[kB]$
1	1	ppe	scalar	162.13	47.64 - 121.46
1	1	ppe	vector	122.08	47.64 - 121.46
1	1	spe	scalar	43.71	47.64 - 121.46
1	1	spe	vector	25.03	47.64 - 121.46
1	2	ppe	scalar	151.08	21.76 - 119.10
1	2	ppe	vector	1,539.69	21.76 - 119.10
1	2	spe	scalar	41.32	21.76 - 119.10
1	2	spe	vector	102.15	21.76 - 119.10
1	3	ppe	scalar	155.93	24.64 - 121.46
1	3	ppe	vector	136.94	24.64 - 121.46
1	3	spe	scalar	43.16	24.64 - 121.46
1	3	spe	vector	29.72	24.64 - 121.46

Table 5.37: Overview of the time and memory consumption for the Sony Playstation III system with regard to the configuration parameter CODEVERSION 1

CV	MV	CPU	ARCH	Time $[ms]$	Local storage $[kB]$
2	1	ppe	scalar	164.10	47.64 - 121.46
2	1	ppe	vector	122.69	47.64 - 121.46
2	1	spe	scalar	45.10	47.64 - 121.46
2	1	spe	vector	26.70	47.64 - 121.46
2	2	ppe	scalar	152.43	21.76 - 119.10
2	2	ppe	vector	1,547.50	21.76 - 119.10
2	2	spe	scalar	42.45	21.76 - 119.10
2	2	spe	vector	104.05	21.76 - 119.10
2	3	ppe	scalar	157.58	24.64 - 121.46
2	3	ppe	vector	139.46	24.64 - 121.46
2	3	spe	scalar	45.18	24.64 - 121.46
2	3	spe	vector	31.82	24.64 - 121.46

Table 5.38: Overview of the time and memory consumption for the Sony Playstation III system with regard to the configuration parameter CODEVERSION 2

CV	MV	CPU	ARCH	Time $[ms]$	Local storage $[kB]$
3	1	ppe	scalar	163.42	47.64 - 121.46
3	1	ppe	vector	123.70	47.64 - 121.46
3	1	spe	scalar	40.19	47.64 - 121.46
3	1	spe	vector	22.74	47.64 - 121.46
3	2	ppe	scalar	151.78	21.76 - 119.10
3	2	ppe	vector	1,545.98	21.76 - 119.10
3	2	spe	scalar	38.58	21.76 - 119.10
3	2	spe	vector	101.07	21.76 - 119.10
3	3	ppe	scalar	157.13	24.64 - 121.46
3	3	ppe	vector	138.15	24.64 - 121.46
3	3	spe	scalar	39.57	24.64 - 121.46
3	3	spe	vector	27.65	24.64 - 121.46

Table 5.39: Overview of the time and memory consumption for the Sony Playstation III system with regard to the configuration parameter CODEVERSION 3

CV	MV	CPU	ARCH	Time $[ms]$	Local storage $[kB]$
4	1	ppe	scalar	166.36	47.64 - 121.46
4	1	ppe	vector	121.90	47.64 - 121.46
4	1	spe	scalar	50.66	47.64 - 121.46
4	1	spe	vector	32.34	47.64 - 121.46
4	2	ppe	scalar	155.83	21.76 - 119.10
4	2	ppe	vector	1,542.57	21.76 - 119.10
4	2	spe	scalar	52.37	21.76 - 119.10
4	2	spe	vector	107.10	21.76 - 119.10
4	3	ppe	scalar	160.72	24.64 - 121.46
4	3	ppe	vector	142.20	24.64 - 121.46
4	3	spe	scalar	50.66	24.64 - 121.46
4	3	spe	vector	35.35	24.64 - 121.46

Table 5.40: Overview of the time and memory consumption for the Sony Playstation III system with regard to the configuration parameter CODEVERSION 4

With regard to these tables, the first interesting fact to notice is obviously only related to the two tables 5.37 and 5.38, because the corresponding time consumption of each configuration is quite similar. Even if this circumstance is certainly expectable, such a condition implies collaterally that the memory access to the jobs and especially the hits is realized very close to the optimum.

In addition to this, the next major conspicuous fact is evidently determined by the circumstance that the used memory in the local storage depends just on the chosen memory version and not on the code version.

Further on, an unsurprising time tendency inside of a code version and memory version block can be apparently identified by the order PPE scalar, PPE vectorial, SPE scalar and then the fastest one SPE vectorial. Such a tendency is obviously expected, because the vector version processes multiple data with a single instruction and the PPE is just a single engine compared to multiple SPEs. The only timing violation, which can be found in this order, exists for memory version two in all code versions, because the algorithm strategy induces the vector variables of the computation to be built by accessing the memory minimally. So for example, if the data parallelism level is 16, there would be surely 16 memory accesses needed to load or store a vectorial value from or to the histogram memory. And as the applied vector arithmetics are furthermore quite easy and consume thus sparse time, the memory access naturally dominates the arithmetics, which causes the exceptional consumption of time.

Finally it should be explicitly pointed out here that the timing comparison of the worst implementation, which is determined by code version two and memory version two in conjunction with vector processing on the unique PPE, with the best one, which is defined by code version three and memory version one in combination with vector processing on all available SPEs, yields an improvement factor of approximately 68. Consequentially the Sony Playstation III system capabilities are suggested to be successfully explored.

As the overall result is clarified now, the following sections present evidently the details for each algorithm step.

5.6.1 Look-up-tables

The major important fact to realize here at first is determined by the circumstance that the result for both LUTs can be only estimated, because the data input is unfixed as stated in section 3.3.2 on page 190.
Further on, there is additionally a remarkable difference in the delivery of this data input, because the hits have to be actually read from a file in place of the network interface due to the fact that the Sony Playstation III platform is unsupported by the CBMROOT framework. Furthermore this file contains just the detector station index in conjunction with the LUT index for each occurring hit instead of the in reality produced information of the experimental setup. Hence the software simulation, which is introduced in section 3.2 on page 102, is obviously required to compute such a hit file in combination with the dependent LUT files related to the indexes.

So consequently the information in these files exhibits in this context an amount of 244.74 kB for the hits, 30.72 kB for the first and 122.88 kB for the second LUT. However it has to be additionally noted in this connection that this very small amount of memory originates in the circumstance that all files contain just usable data for about 15,658 determined hits, while the calculation of the correct amount of reserved memory for the actual algorithm parameters can be found in section 5.4.1 on page 324.
Nevertheless both planned and both actually implemented LUTs are obviously located in the XDRDRAM memory of the Sony Playstation III system as introduced in section 3.3.2 on page 201. For that reason, it is clear that the initialization phase must be anyway applied to load the hit information and the content of each LUT from different files into the XDRDRAM memory.

5.6.2 HBuffer

Before presenting the result for the HBuffer unit, it has to be noted that the real implementation is a little bit more complex than stated in section 3.3.2 on page 202, because it has to support obviously all different available code versions, which realize possible job creation strategies, in addition to the common functionality of decoupling the LUT evaluation from the Histogramming. Therefore the actual unit has evidently to organize the data input of each SPE as well as the overhead information to manage the job processing on the SPEs.
For that reason, the total data input comprises for each job an array of 64 bit elements, which consist of 8 bit for the relative start index and 8 bit for the relative stop index of the histogram layer, 8 bit for the station index, 8 bit for

the start position as well as 32 bit for the histogram command. Furthermore the main characteristic of these arrays, which is caused by a prerequisite of the chosen DMA transfer mode, is a 128 bit memory alignment, although the bit width of a common transformed hit is just 64 bit. However this requirement is not crucial, because it could be easily satisfied by adding a neutral element, if the number of elements is odd.
Besides this, the overhead information to manage the job processing includes apparently an offset value, which features the conversion from the relative start and stop indexes into the absolute indexes of the corresponding histogram layers, the memory and data parallelism level as well as the amount of input data.

So based on these facts, table 5.41 shows now the resulting size of each job, which can be summed up to a total amount of 612.23 kB.

Job	0	1	2	3	4	5	6
Data [kB]	12.36	21.52	27.22	51.91	72.81	74.24	71.97

Job	7	8	9	10	11	12	13
Data [kB]	74.08	75.17	61.56	35.49	17.94	14.61	1.36

Table 5.41: Survey report for the size of the input data related to each job

With regard to this table, it is at a first glance surprising that this result is generally admitted for all different available configurations, which are introduced in the beginning of section 5.6 on page 370. But a second view shows further that this circumstance can be easily explained by a closer look to the effective job restriction of each configuration, because it is simply always identical, whether the requirements are different. Furthermore this situation is apparently enforced to feature the same data input for all configurations to enable the results to be more comparable. So based on this context, it is clear that the most curious restriction takes effect for the implementation, which uses the PPE uniquely, because this one does naturally not require a limitation for the amount of local storage, because the SPEs are unused. Moreover it has to be noticed in this connection that most restrictions are related to the workload balancing, which can be found detailed in [Str09]. So it should be just mentioned here that the selective restriction is actually defined by the limitation of the maximal number of hits for a job.

Besides this, it is certainly of additional interest which implementation takes how much time to generate the job prerequisites and to create all jobs. For this purpose, it is obvious that neither the SPE usage nor the scalar/vector utilization has an effect, because the jobs are always created in the same fashion on the PPE. Consequently the time consumption for the generation of the prerequisites is expected to be almost constant for all different versions and is thus universally evaluated to be around 1.23 ms.

Before coming now to the details for the job creation, it has to be recognized that the results shown in table 5.42 contains the total time, which means that the parallelization of the job creation and the job processing is not accounted in any case.

	CV1	**CV2**	**CV3**	**CV4**
MV1	4.59 ms	6.34 ms	5.86 ms	6.89 ms
MV2	4.49 ms	6.06 ms	5.42 ms	6.39 ms
MV3	4.54 ms	6.24 ms	5.60 ms	6.40 ms

Table 5.42: Inquiry report of the time consumption for the job creation related to the configuration parameters CODEVERSION and MEMORYVERSION

With regard to this table, it is at a first glance surprising that the time, which is required for the job creation, depends not only on the code version, but also on the memory version. However the reason of this dependency is quite simple, because the memory version affects obviously the memory limit of the local storage, which defines further a restriction to the jobs.
So coming now to the point, the time consumption ranking of the different memory versions is as expected MV2, MV3 and finally MV1, because this order is equal to the available amount of histogram layers, which determine a further restriction to comply. Moreover the behavior of the code versions is also anticipated, because CV1 implements the optimal timing strategy for the job creation, while CV2 realizes an intermediate step to the parallelized optimum CV3. Finally CV4 is the slowest one, because it specializes each job for a dedicated SPE.

5.6.3 Histogram

In contrast to the HBuffer unit, the histogram unit is naturally implemented straight forward. This means more precisely that the parallel layers are either instantiated in the local storage of each SPE or in the XDRDRAM dependent on the respectively chosen specification of the processing elements SPE or PPE uniquely. Furthermore it is clear that the available memory access types or memory versions, which are already introduced in section 3.3.2 on page 203, influence the amount of parallel instantiated histogram layers for each job directly, whereas the code version has only to cope with this limitation. Thus the major interesting characteristic is obviously determined by the amount of allocated memory for each job with regard to the selected memory version. For this purpose, table 5.43 summarizes the results.

Job	**Histogram layers**					
	MV1		**MV 2**		**MV3**	
	[#]	[kB]	[#]	[kB]	[#]	[kB]
0 - 4	16	46.02	16	46.02	16	46.02
5	16	46.02	9	25.88	16	46.02
6	16	46.02	9	25.88	16	46.02
7	16	46.02	8	23.01	8	23.01
8	16	46.02	14	40.26	16	46.02
9 - 12	16	46.02	16	46.02	16	46.02
13	16	46.02	7	20.13	16	46.02

Table 5.43: Survey report for the amount of histogram layers and the corresponding memory consumption related to each job and the configuration parameter MEMORYVERSION

A closer look to the result of memory version one shows now immediately that the concept of this access type, which is determined by allocating always the maximal amount of histogram layers for each job only with regard to the hardware restriction, is successfully implemented.

Further on, the result of memory version two exhibits contrarily the exact adjustment of the number of parallel histogram layers for each job related to all restrictions.

Finally the result of memory version three describes obviously also the detailed adjustment of the amount of layers for each job with regard to all restrictions but only in the power of two.

Besides this, another major interesting fact, which is substantiated by this

table, is surely defined by the circumstance that the most frequently taken job restriction is defined by the ALU width and not determined by the size of the local storage as assumed, because almost all jobs require 16 histogram layers. In this connection, it has to be also noticed that such a conclusion can be easily made, because the enforced job restrictions are identical for all available configurations of the algorithm to feature the same data input, which enables thus the results to be certainly more comparable (see section 5.6.2 on page 376).
However it is clear that the time consumption of the Histogramming step varies, although the memory consumption is quite constant for all configurations. For this purpose, the following tables 5.44 and 5.45 list the required time of each thread with regard to each possible configuration, which is naturally characterized by the memory version in conjunction with the processing elements PPE or SPE and the architecture scalar or vector.

Time [ms]					
Scalar			**Vector**		
MV1	**MV2**	**MV3**	**MV1**	**MV2**	**MV3**
87.74	86.33	86.58	51.52	1,424.77	53.16

Table 5.44: Inquiry report of the time consumption for the Histogramming step on the PPE related to the configuration parameter MEMORYVERSION

Thread	**Time** [ms]					
	Scalar			**Vector**		
	MV1	**MV2**	**MV3**	**MV1**	**MV2**	**MV3**
1	17.55	17.55	15.19	9.43	61.69	14.02
2	17.16	17.16	19.52	9.24	64.49	16.02
3	18.18	18.18	18.18	9.76	75.85	14.51
4	20.23	20.23	20.23	10.82	69.84	16.08
5	19.27	16.14	19.27	8.65	58.79	15.35
6	16.38	19.50	16.38	10.46	87.85	13.30

Table 5.45: Inquiry report of the time consumption for the Histogramming step on the SPE related to each thread and the configuration parameter MEMORYVERSION

So with regard to these tables, it has to be at first noticed that the general

assumptions of SPE being faster than PPE processing and the vector being faster than the scalar architecture can be proven. Furthermore it is very interesting that the data distribution to all available SPEs leads to a higher speed gain than the data vectorization. In addition to this, the resulting order of the memory versions is obviously also predictable, because memory version one, which realizes the easiest memory access with regard to the ALU computation, forms the fastest configuration, while memory version three constitutes a compromise between this one and memory version two, which describes the slowest configuration.
However the most surprising result in these tables can be found for the vector architectures of memory version two, because the general rule of vector being faster than scalar is violated there. But a more detailed look to this violation shows immediately that the reason originates apparently in the combination of the memory access and the architecture, because they are as far away from each other as possible. This means more precisely that the combination of the memory access in the smallest supported fraction with the architecture usage in vector mode causes naturally a heavy memory access to build the variables at any time dedicated for each ALU computation, which leads thus evidently to the required huge amount of time. Moreover this circumstance is certainly also the reason for the reversed speed order when comparing the scalar architectures of the different memory versions.
Finally it has to be explicitly highlighted that this algorithm step would imply the best capability for a further enhancement, even if the maximal speed gain between the slowest and the fastest configuration of memory version one is only about eight, because it requires the most time compared to all other steps.

As the result of the Histogramming step is successfully presented now, the focus has to be currently set to a special aspect of the histogram usage, which consumes additional time. So within that context, this means more precisely that a closer look has to be taken to the allocation and deallocation of the memory implementing the instantiated histogram layers, because these two needed duties require evidently time for any evaluation loop, which is of course independent of the code version and the scalar or vector usage but dependent of the memory version. Hence the following tables 5.46 and 5.47 list the required time of each thread with regard to the memory versions in conjunction with the processing units PPE or SPE.

With regard to these tables, it is at first remarkable that the memory allocation and deallocation is approximately five to six times faster on the SPE than on the PPE. Moreover the time consumption of the memory versions is ordered as expected to MV2, MV1 and finally the slowest one MV3, because

Thread	Time [μs]					
	PPE			SPE		
	MV1	MV2	MV3	MV1	MV2	MV3
1	158.78	134.14	169.02	5.03	4.52	5.23
2	-	-	-	5.00	4.42	2.98
3	-	-	-	3.45	3.30	3.97
4	-	-	-	3.77	3.43	3.91
5	-	-	-	3.43	2.78	3.56
6	-	-	-	3.43	3.28	4.61

Table 5.46: Inquiry report of the time consumption for the histogram layer allocation related to each thread and the configuration parameter MEMORYVERSION

Thread	Time [μs]					
	PPE			SPE		
	MV1	MV2	MV3	MV1	MV2	MV3
1	12.06	10.99	15.56	1.13	1.10	1.12
2	-	-	-	1.07	1.07	0.70
3	-	-	-	0.74	0.73	0.73
4	-	-	-	0.73	0.69	0.74
5	-	-	-	0.75	0.75	0.74
6	-	-	-	0.73	0.73	1.12

Table 5.47: Inquiry report of the time consumption for the histogram layer deallocation related to each thread and the configuration parameter MEMORYVERSION

memory version two uses just few big blocks, whereas memory version one takes many of the smallest fraction and memory version three needs time for the decision about the optimal block size.
Further on, a more precise look to the histogram allocation and deallocation is obviously senseless, because both waste so sparse time that it does not count for the total consumption. Nevertheless it is noticeable at the end that the deallocation is approximately ten times faster than the allocation on the PPE and five times on the SPE.

5.6.4 Filtering

Since section 3.3.2 on page 207 defines the entire process of reducing the background noise in the histogram to comprise the separate algorithm steps Encoding, Diagonalization and Peak-finding including the Serialization, the following sections are once again used to present the timing and memory results of each step individually, because this procedure offers the possibility to have a look at the results of each step independent of all others.

Encoding

Having section 3.3.2 on page 207 in mind, it is clear that the major important characteristic of the Encoding step is identified by the time consumption, which is thus listed in table 5.48 for the unique PPE processing and in table 5.49 for each thread of the SPE processing with regard to each possible configuration determined by the memory version in conjunction with the architecture scalar or vector.

Time [ms]					
Scalar			**Vector**		
MV1	**MV2**	**MV3**	**MV1**	**MV2**	**MV3**
10.12	8.60	9.49	41.11	56.52	54.24

Table 5.48: Inquiry report of the time consumption for the Encoding step on the PPE related to the configuration parameter MEMORYVERSION

So with regard to these tables, the common assumption of SPE being faster than PPE processing apparently still counts, while the general assumption of the vector being faster than the scalar architecture can not be acknowledged. However the reason for this contrary behavior is not mysterious, because the vector capability of the ALU can not be entirely used due to the circumstance that the Encoding step solely substitutes the signature of each histogram cell exclusively for a LUT based priority class, which is merely available scalar. Consequentially the vectorized histogram cells have to be naturally split into scalar signatures to be able to evaluate the table look-up. Hence the scalar architecture has even to be faster in this special case, because the data segmentation and vectorization is needless.
Moreover it has to be additionally noticed in this connection that this situation could easily change, if a vectorized LUT can be applied for the En-

Thread	Time [ms]					
	Scalar			Vector		
	MV1	MV2	MV3	MV1	MV2	MV3
1	1.51	1.29	1.26	3.36	2.33	2.55
2	1.51	0.97	1.26	2.55	2.28	1.28
3	1.00	0.94	1.00	2.46	1.69	1.85
4	1.00	1.00	1.00	2.62	1.79	1.96
5	1.00	1.00	1.01	2.62	1.33	1.96
6	1.00	0.79	1.00	2.05	1.43	1.85

Table 5.49: Inquiry report of the time consumption for the Encoding step on the SPE related to each thread and the configuration parameter MEMORYVERSION

coding instead of the scalar one. But a closer look to this idea shows immediately that the augmented amount of memory, which rises easily from $2^8 \cdot 8\,\text{bit} = 256\,\text{Byte}$ for the scalar LUT to $2^{128} \cdot 128\,\text{bit}$ for the vectorized one, avoids evidently the employment of a vectorized LUT. Furthermore even a two-word vectorized LUT is not applicable, because such a version takes already $2^{16} \cdot 16\,\text{bit} = 128\,\text{kByte}$ of memory, which is obviously also too much for the local storage of a SPE.
Further on, the resulting order of the memory versions is surely predictable as well, because memory version two, which realizes the fewest memory access with regard to the reasonable histogram cells, forms the fastest configuration, while memory version three constitutes a compromise between this one and memory version one, which describes the slowest configuration.

Anyway the most surprising result in these tables can be found for the vector architectures of the PPE processing, because the performance order of the memory versions is reversed. By investigating this circumstance, a potential explanation of this fact might be that the compiler is able to optimize the table look-up process by caching the LUT, because then it is not necessary to access the XDRDRAM memory for each look-up. Furthermore this explanation implies evidently also the correct performance order for the vector architectures of the SPE processing, because the local storage of the SPEs is not cacheable in the strict sense.

Diagonalization

Analog to the previous section, the major important characteristic of the Diagonalization step introduced in section 3.3.2 on page 209 is certainly again identified by the time consumption, which is therefore listed in table 5.50 for the unique PPE processing and in table 5.51 for each thread of the SPE processing with regard to each possible configuration determined by the memory version in conjunction with the architecture scalar or vector.

Time [ms]					
Scalar			Vector		
MV1	MV2	MV3	MV1	MV2	MV3
5.82	4.89	5.34	2.58	30.82	2.44

Table 5.50: Inquiry report of the time consumption for the Diagonalization step on the PPE related to the configuration parameter MEMORYVERSION

Thread	Time [ms]					
	Scalar			Vector		
	MV1	MV2	MV3	MV1	MV2	MV3
1	1.32	1.12	1.10	0.18	1.48	0.26
2	1.32	0.85	1.10	0.18	1.45	0.18
3	0.88	0.82	0.88	0.12	1.07	0.17
4	0.88	0.88	0.88	0.12	1.13	0.17
5	0.88	0.88	0.88	0.12	0.86	0.17
6	0.88	0.69	0.88	0.12	0.92	0.26

Table 5.51: Inquiry report of the time consumption for the Diagonalization step on the SPE related to each thread and the configuration parameter MEMORYVERSION

With regard to these tables, all tendencies, which are introduced in section 5.6.3 on page 379 for the Histogramming step of the algorithm, can be obviously recovered. Thence a short summary exhibits that the general assumptions of SPE being faster than PPE processing and the vector being faster than the scalar architecture still count. In addition to this, the tendency for the speed gain related to the data distribution and the data vectorization is valid as well. Moreover the resulting order of the memory versions is also

identically defined by memory version one, which is followed by version two and finally version three.
Besides this, it is collaterally not surprising that the results for the vector architectures of memory version two violate in turn the general rule of vector being faster than scalar, because the same reason as stated in section 5.6.3 on page 379 for the Histogramming step prevails here.

Peak-finding and Serialization

This section presents the timing results for the algorithm steps 2D Peak-finding, Serialization and 3D Peak-finding, which are all introduced in section 3.3.2 on page 210. So while remembering now the implementation details, the following tables 5.52 and 5.53 as well as the tables 5.54 and 5.55 list analog to the previous sections the required time for the unique PPE processing and for each thread of the SPE processing with regard to each possible con-

Time [ms]					
Scalar			**Vector**		
MV1	**MV2**	**MV3**	**MV1**	**MV2**	**MV3**
44.53	37.91	41.10	4.37	19.45	4.97

Table 5.52: Inquiry report of the time consumption for the 2D Peak-finding step on the PPE related to the configuration parameter MEMORYVERSION

Thread	**Time** [ms]					
	Scalar			**Vector**		
	MV1	**MV2**	**MV3**	**MV1**	**MV2**	**MV3**
1	11.88	10.16	9.92	0.82	1.53	1.05
2	11.90	7.72	9.93	0.84	1.57	0.80
3	7.96	7.47	7.96	0.64	1.34	0.85
4	7.93	7.93	7.93	0.60	1.06	0.76
5	7.94	7.94	7.94	0.61	1.15	0.80
6	7.97	6.26	7.97	0.62	1.38	1.14

Table 5.53: Inquiry report of the time consumption for the 2D Peak-finding step on the SPE related to each thread and the configuration parameter MEMORYVERSION

figuration, which is determined by the memory version in conjunction with the architecture scalar or vector.

Time [ms]					
Scalar			Vector		
MV1	**MV2**	**MV3**	**MV1**	**MV2**	**MV3**
4.71	3.98	4.39	13.60	4.20	13.07

Table 5.54: Inquiry report of the time consumption for the Serialization step on the PPE related to the configuration parameter MEMORYVERSION

Thread	**Time** [ms]					
	Scalar			**Vector**		
	MV1	**MV2**	**MV3**	**MV1**	**MV2**	**MV3**
1	1.89	1.64	1.60	1.70	1.74	1.76
2	1.89	1.28	1.60	1.33	1.70	1.03
3	1.26	1.19	1.26	1.23	1.26	1.27
4	1.26	1.26	1.26	1.31	1.34	1.35
5	1.26	1.26	1.26	1.31	0.99	1.35
6	1.26	1.01	1.26	1.05	1.07	1.42

Table 5.55: Inquiry report of the time consumption for the Serialization step on the SPE related to each thread and the configuration parameter MEMORYVERSION

With regard to these tables, it is obvious that these results can not offer a novel insight compared to the previous sections. The only interesting difference is evidently determined by the circumstance that the vectorization of the algorithm leads to a higher speed gain than the SPE processing. But nevertheless the optimum is assembled by applying certainly both strategies.

Besides this, the consumed time for the 3D Peak-finding is of course universally evaluated to 3.21 ms, because all versions implement this step on the PPE without any optimization and the job packages are naturally identical as illustrated in section 5.6.2 on page 376.

5.6.5 LBuffer

Since the implementation of the LBuffer unit, which is introduced in section 3.3.2 on page 211, is certainly analog to the implementation of the HBuffer unit, it is clear that the adjustment to the overhead data, which is needed to manage the job processing on the SPEs, has to be also taken into account similar to section 5.6.2 on page 376.
Thus the total content of the LBuffer unit comprises for each job an array of 32 bit elements, which consist of 8 bit for the absolute index in the first histogram dimension, 8 bit for the absolute index in the second histogram dimension, 8 bit for the relative index of the histogram layer and 8 bit for the priority class. Furthermore it has to be additionally noted that the main characteristic of these arrays, which is apparently caused as well as for the HBuffer unit by the identical prerequisite of the chosen DMA transfer mode, is obviously in turn identified by a 128 bit memory alignment, although the bit width of a common track candidate is just 32 bit. However this requirement is once again not crucial, because it can be likewise easily satisfied by adding up to three neutral elements depending on the amount of data, which must be finally surely restless dividable through 128 bit.
Beyond that, it has to be evidently also noticed that the offset value of the overhead information in the HBuffer unit has to be used to reconvert the relative index of the histogram layer into the absolute index before the 3D Peak-finding is able to process the data, which is stored in the LBuffer unit.

So based on these facts, table 5.56 shows now the resulting size of each job, which can be summed up to a total amount of 3.19 kB.

Job	0	1	2	3	4	5	6
Data [kB]	0.02	0.02	0.08	0.14	0.52	0.66	0.41

Job	7	8	9	10	11	12	13
Data [kB]	0.44	0.52	0.30	0.06	0.03	0.02	0.00

Table 5.56: Survey report for the size of the output data related to each job

With regard to this table, it is at a first glance surprising that the job distribution for the total amount of output data is generally admitted, because different implementation configurations are introduced in the beginning of section 5.6 on page 370. However this circumstance can be easily explained by the identical data input for all configurations, which is assured in section 5.6.2 on page 376.

Besides this, it should be finally highlighted that the data, which is stored in the LBuffer unit, is further reduced to 2.55 kB by the 3D Peak-finding step of the algorithm.

5.6.6 Internal Multi-Core Overhead

As the Sony Playstation III system exhibits an embedded multi-core chip as introduced in section 3.3.2 on page 185, there is obviously the need for internal overhead to configure each core and to transport the data. For that reason, the following tables 5.57 and 5.58 as well as 5.59 summarize exemplarily the results for each concerned criterion of the fastest implementation, which is identified by the application of vector processing on all available SPEs in combination with code version three and memory version one.

Process	**Time** [ms]
Setup threads	8.61
Configuration	0.69

Table 5.57: Inquiry report of the time consumption for the internal multi-core overhead on the PPE related to each overhead process

Process	**Time** [μs]		
	Thread 1	**Thread 2**	**Thread 3**
Configuration	24.70	14.80	10.92
Standby	13,434.01	24,196.84	32,600.73
Receive table	9.26	9.10	9.10
Receive data	15.45	15.99	14.95
Send data request	1.14	1.19	0.83
Send data standby	364.09	216.18	410.34
Send data	2.09	1.79	1.35

Table 5.58: Inquiry report of the time consumption for the internal multi-core overhead on the SPE threads one, two and three related to each overhead process

Process	**Time [μs]**		
	Thread 4	**Thread 5**	**Thread 6**
Configuration	10.72	9.80	9.25
Standby	39,762.48	49,979.02	57,489.85
Receive table	9.10	9.10	9.09
Receive data	15.00	12.73	14.57
Send data request	0.74	0.77	0.75
Send data standby	61.92	89.86	58.23
Send data	1.29	1.29	1.40

Table 5.59: Inquiry report of the time consumption for the internal multi-core overhead on the SPE threads four, five and six related to each overhead process

So with regard to these tables, it is obvious that a closer look to the internal overhead is dispensable, because the measured times are either that sparse that they are insignificant or just consumed once.

Chapter 6

Conclusions

6.1 Algorithm

With regard to the results presented in section 5.1 on page 295 and section 5.2 on page 310 as well as in section 5.3 on page 316, the described adapted and customized Hough transform applying the analytic formula in combination with the averaged magnetic field integrate defines apparently a feasible approach for the tracking in the very demanding environment of the CBM experiment, because the performance is certainly sufficient.
Furthermore a detailed look exhibits the tracking efficiency to be around 86 %, the tracking reduction rate about 80 % and the tracking identification rate approximately 67 %, whereas the unwanted side-effects can be kept below 10 %.
Since each tracking performance aspect depends moreover on the multiplicity of the event, the beam energy, the number of detector stations and the resolution of the Hough space, even better results can be obviously expected for events with less tracks, for detector setups with more stations and for improvements to the histogram, which might be a non-uniform resolution for instance.

Besides this, speculations suggest supplementary that the Runge-Kutta method utilized as computational kernel is possibly able to increase the tracking efficiency about 5 %. In addition to this, another improvement of approximately 5 % for the tracking efficiency might be gained by the application of a Hough transform following algorithm, which is exemplarily introduced in section 7 on page 435, because such an algorithm is able to cope with the unwanted side-effects and enables thus a less restrictive Hough transform parameter set.

Finally it has to be once again explicitly highlighted that each addressed optimization is regardless of the timing, because the processing speed of the adapted and customized Hough transform is generally proportional to the number of hits and particularly independent of the computational kernel due to the application of LUTs. Furthermore the pipelining strategy on the FPGA leads naturally just to an increased latency without effects on the throughput.

6.2 Suitable systems for the CBM experiment

Remembering now the statement from the beginning of this thesis, the CBM experiment requires a first level tracking system, which is able to process high reaction rates of up to $10^7 \, \frac{\text{events}}{\text{s}}$ with a high number of multiplicities of up to 1,000 particles per central collision. However the research and development phase of the experiment reveals such a very high complexity that the reduction of the reaction rate by the number of four to $25 \cdot 10^5 \, \frac{\text{events}}{\text{s}}$ without a change to the multiplicities defines currently the only chance to cope with it.
Besides this, the situation is also not simplified by the circumstance that general simulations of central Au+Au collisions at energies of 35 AGeV have shown that the amount of hits in the recent STS detector system can be up to 20,000 per event.

So based on this information, the following indicators for the first level tracking system result by additionally assuming the encoding of a single hit with 32 bit.

- Required processing time of a single event: $\frac{1}{25 \cdot 10^5 \, \frac{\text{events}}{\text{s}}} = 0.4 \, \frac{\mu\text{s}}{\text{event}}$
- Occurring data size of a single event: $20{,}000 \, \frac{\text{hits}}{\text{event}} \cdot 32 \, \frac{\text{bit}}{\text{hit}} \approx 80 \, \frac{\text{kByte}}{\text{event}}$
- Network load: $25 \cdot 10^5 \, \frac{\text{events}}{\text{s}} \cdot 80 \, \frac{\text{kByte}}{\text{event}} \approx 200 \, \frac{\text{GByte}}{\text{s}}$

But since these universal key indicators describe obviously an estimation, which is evidently to simple to be used to benchmark the implementations presented in the following sections, it is clear that a generated sample event has to be utilized with all its versatility for that purpose. Consequentially all implementations are thus analyzed in detail by using a sample event, which is determined by:

- Number of detector stations: 7
- Number of input hits: 15,658
- Number of entries in the coding table: 3
- Number of priority classes: 3
- Number of Peak-finding geometry elements: 9
- Size of the histogram: 95×31 cells with 191 layers

Before proceeding now with the detailed results, it has to be finally noted that this sample event is surely identical to the one, which is used to evaluate the results for the Sony Playstation III system presented in section 5.6 on page 370 and for the personal computer system exhibited in section 5.4 on page 323.

6.2.1 Personal Computer

Within this context, the result, which is presented in section 5.4 on page 323, has to be remembered at first, because it demonstrates that the CBMROOT framework implementation of the adapted and customized Hough transform evaluates the sample event in 2.36 s on a common personal computer. However it has to be once again explicitly highlighted in this connection that this implementation ascertains overhead information for analysis purpose and is furthermore in particular not optimized to architecture specific features.

Further on, the combination of this timing with the required processing time of an event from the CBM experiment, which is $0.40 \frac{\mu s}{\text{event}}$, leads then immediately to a necessary amount of $\frac{2.36 \frac{s}{\text{sample event}}}{0.40 \frac{\mu s}{\text{event}}} = 5.9 \cdot 10^6$ parallel used personal computer systems in a computation factory. Consequentially the total costs could be easily computed to about $5.9 \cdot 10^6 \cdot \$0.5 \cdot 10^3 \approx \$2.8 \cdot 10^9$, if the price of such a common personal computer system, which can be found in [Int], is taken into account.

Besides this, the required bandwidth of the CBM experiment, which is $200 \frac{\text{GByte}}{s}$ or $1.6 \cdot 10^3 \frac{\text{Gbit}}{s}$, defines obviously not a restriction for such a computation factory, because common personal computers feature a 1-Gigabit ethernet connection. Furthermore a more detailed analysis exposes that the necessary time to receive the data of the sample event is $\frac{15.66 \cdot 10^3 \frac{\text{hits}}{\text{sample event}} \cdot 32 \frac{\text{bit}}{\text{hit}}}{1 \frac{\text{Gbit}}{s}} \approx$

$0.47\,\frac{\text{ms}}{\text{event}}$, which is evidently about $5.1 \cdot 10^3$ times smaller than the processing time.
Nevertheless the physical network connection is naturally a demanding task, because each personal computer system requires an own network link. So if one considers cascaded 48-port NETGEAR GS748AT network switches (see [NETa] and [NETb]), there would be an amount of 128,261 articles, which cost $128.3 \cdot 10^3 \cdot \$\,1.3 \cdot 10^3 \approx \$\,168.3 \cdot 10^6$.

In addition to these purchasing costs, a lack of such a factory will be apparently also the operating costs, which are caused by the power consumption. As furthermore a common personal computer system consumes about 200 Watt (see [Sch]) in combination with 67 Watt (see [NETa]) for the network switch, the operating costs of the entire computation factory including its network connection can be computed for a single year and a electricity tariff of $0.2\,\frac{\$}{\text{kWh}}$ to:

$$
\begin{aligned}
\text{operating costs} &= \quad 5.9 \cdot 10^6 \cdot 0.20\,\text{kW} \cdot 365\,\text{d} \cdot 24\,\tfrac{\text{h}}{\text{d}} \cdot 0.2\,\tfrac{\$}{\text{kWh}} \\
&+ \ 128.3 \cdot 10^3 \cdot 0.07\,\text{kW} \cdot 365\,\text{d} \cdot 24\,\tfrac{\text{h}}{\text{d}} \cdot 0.2\,\tfrac{\$}{\text{kWh}} \\
&\approx \quad \$\,2.1 \cdot 10^9
\end{aligned}
$$

So due to the complexity and costs of such a computation factory, this implementation is obviously unfeasible to realize an on-line first level tracking system for the CBM experiment. However a single personal computer system can be certainly applied to simulate and analyze the event performance of the adapted and customized Hough transform algorithm off-line and is thus very useful.

6.2.2 FPGA

While remembering the results, which are presented in section 5.5 on page 327, it is immediately clear that the required amount of resources for the different synthesis possibilities of the Hough transform units and especially the ones of the histogram layer implementation presume that a single FPGA can contain just a basic algorithm realization. As however the CBM first level tracking system indicators, which are described at the beginning of section 6.2 on page 392, necessitate obviously really fast processing elements, a multichip FPGA platform has certainly to be considered as well, because more available hardware resources offer naturally a bigger range in the algorithm's realization flexibility concerning the amount of usable synthesis possibilities and the applicable parallelism level of each unit.
For this purpose, a very good imagination about the theoretical amount

of synthesis possibility combinations, which are evidently spanned by the possibilities of each necessary Hough transform unit respectively, is given in table 6.1.

Id	Prelut	HBuffer	Lut	Generator	Histogram	Encoder	Diagonalizer	Peakfinder2d	Serializer	LBuffer	Peakfinder3d
1 - 216	-	1 - 4	-	1 - 3	1 - 3	1 - 6	1	1	1	-	-
217 - 433	-	1 - 4	-	4 - 6	4 - 6	1 - 6	1	1	1	-	-
434 - 650	-	1 - 4	-	1 - 3	7 - 9	7 - 12	2	2	2	-	-
651 - 867	-	1 - 4	-	4 - 6	10 - 12	7 - 12	2	2	2	-	-
868 - 1,732	-	1 - 4	-	7 - 9	13 - 24	7 - 12	2	2	2	-	-
1,733 - 2,021	-	1 - 4	-	10 - 12	25 - 28	7 - 12	2	2	2	-	-

Table 6.1: General view of all conceivable synthesis possibilities for the Hough transform

By having further the aim of a fast implementation in mind, it is evident that the combination strategy of the Hough transform units in association with their applied parallelism level require a general analyzing scheme for the actual algorithm's timing behavior, which takes surely the result of each single unit presented in section 5.5 on page 327 into account.
For this purpose, the capability to decouple the three major algorithm steps of the adapted and customized Hough transform, which is introduced in section 3.3.1 on page 156, by the application of memories is naturally utilized in combination with the general pipelining structure, which is depicted in the following figure 6.1, to determine the latency of an event and the pipelined throughput.

So consequentially the common benchmarking criteria of such a platform are apparently defined by the required amount of resources, the latency of an event, which represents the time delay to entirely process a single unique data set, and the throughput, which distinguishes the amount of processed data sets in a brief span of time.

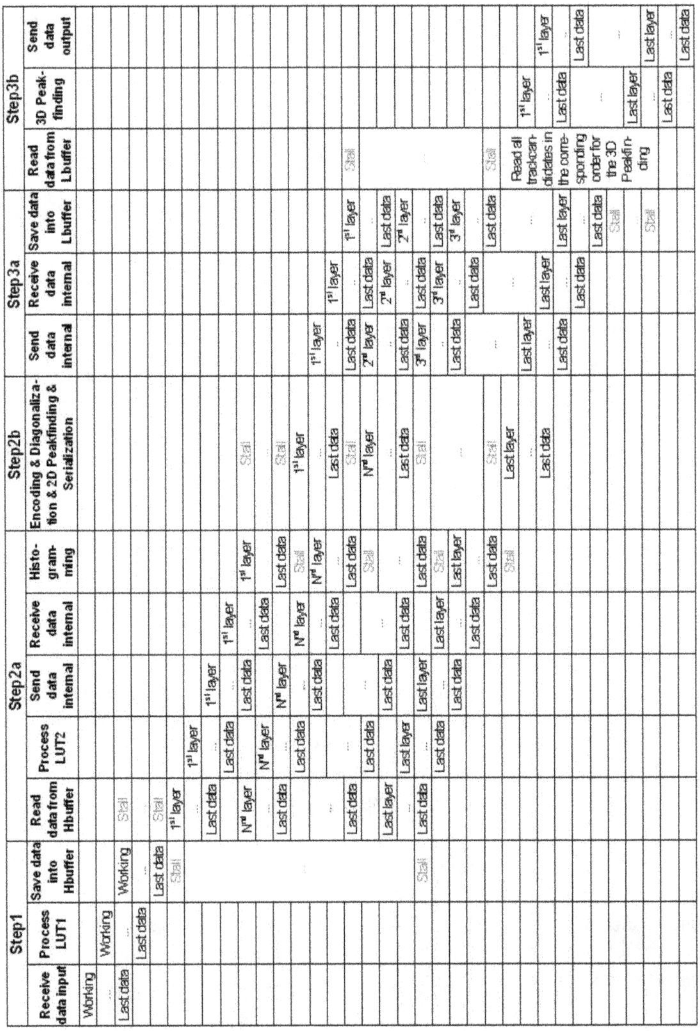

Figure 6.1: Illustration of the general pipelining strategy for the FPGA implementation

Coming now to a precise exploration of these major steps, the figure illustrates directly that the algorithm step 'Step1' is quite simple, because it contains just the data receiving from the input network, the first LUT transform and the saving of the data in the HBuffer unit.

Besides this, the next step 'Step2a', which evidently locks the previous step until the successful read-out of the entire HBuffer unit, implements obviously the second LUT transform and the Histogramming as well as the HBuffer read-out.
Moreover it has to be additionally mentioned within this context that the last data value in each Histogramming iteration certainly violates the systolic processing, which leads thus to an extra stall of the histogram. A closer look to this stall shows furthermore that it takes as many clock cycles as necessary to propagate this data value into all histogram rows.
By the way, it is also important to notice that this step contains a platform internal chip-to-chip network for the multi-chip version, which is of course not applicable for the single-chip version.

Continuing now with the exploration, the subsequent algorithm step 'Step2b' realizes apparently the entire histogram read-out pipeline, which consists of the histogram read-out itself, the Encoding, the Diagonalization, the 2D Peak-finding and the Serialization.
Notwithstanding many more or less complex algorithm fragments are combined in this pipeline, it is immediately clear that just two of them can lead to an effective stall.
Since this fact is really interesting, the analysis of the first one, which is surely defined by the histogram read-out, results in the knowledge that the stall possibility has two key elements. Having now an even closer look, the first one is evidently determined by the circumstance that the common read-out stall depends on the size of the histogram dimension in the read-out direction. Furthermore the second key element takes surely the situation into account that multiple histogram layers can be instantiated in parallel, but read out serially by the utilization of only a single read-out chain.
Anymore the second effective pipeline stall is commonly expected by the Serialization stall possibility, which can be in contrast easily avoided to the costs of latency by dimensioning the FIFO memory of the Serialization unit as deep as necessary. For this purpose, the analysis, which is presented in section 4.9.2 on page 288, features doubtless aid to determine this depth precisely.

Afterwards it is not surprising that the subsequent step 'Step3a' is determined by the second chip-to-chip network and the merge of the actual track candidates into the LBuffer unit.

In this connection, it is also clear that the network connection is as well as earlier just needed for the multi-chip version.

Coming now to the end, the last step 'Step3b' consists finally of the LBuffer read-out, the 3D Peak-finding and the delivery of the data to the output network.

So based on this general structure, the following equations 6.1, 6.2 and 6.3 can be henceforth used to rate all implementations against each other. However it is clear that such a rating requires additionally the definition of a main focus, which is here expectedly set to the throughput, while the latency is subordinated to the second level of interest.

$$
\begin{aligned}
\text{time}(\mathit{step1}) &= \text{time}(\mathit{Receive\ hits}) + \text{time}(\mathit{Process\ LUT1}) \\
&+ \text{time}(\mathit{Save\ HBuffer}) \\
\text{time}(\mathit{step2a}) &= \text{time}(\mathit{Read\ HBuffer}) + \text{time}(\mathit{Process\ LUT2}) \\
&+ \text{time}(\mathit{Histograming}) \\
&+ \left\langle \frac{\mathit{existing\ layers}}{\mathit{instantiated\ layers}} \right\rangle \cdot \text{time}(\mathit{Last\ data\ histograming}) \\
\text{time}(\mathit{step2b}) &= \left\langle \frac{\mathit{existing\ layers}}{\mathit{instantiated\ layers}} \right\rangle \cdot \left\langle \frac{\mathit{instantiated\ layers}}{\mathit{read-out\ chains}} \right\rangle \\
&\cdot \text{time}(\mathit{Read\ Histogram}) \\
&+ \text{time}(\mathit{Encoding}) + \text{time}(\mathit{Diagonalization}) \\
&+ \text{time}(\mathit{2D\ Peakfinding}) + \text{time}(\mathit{Serialization}) \\
&+ \text{time}(\mathit{Save\ Serialization\ buffer}) \\
\text{time}(\mathit{step3a}) &= \text{time}(\mathit{Read\ Serialization\ buffer}) \\
&+ \text{time}(\mathit{Save\ LBuffer}) \\
\text{time}(\mathit{step3b}) &= \text{time}(\mathit{Read\ LBuffer}) + \text{time}(\mathit{3D\ Peakfinding}) \\
&+ \text{time}(\mathit{Deliver\ tracks})
\end{aligned}
\tag{6.1}
$$

$$
\begin{aligned}
\text{latency} &= \text{time}(\mathit{step1}) + \text{time}(\mathit{step2a}) + \text{time}(\mathit{step2b}) \\
&+ \text{time}(\mathit{step3a}) + \text{time}(\mathit{step3b})
\end{aligned}
\tag{6.2}
$$

$$
\begin{aligned}
\text{throughput} = \max(\ & \text{time}(\mathit{step1}) + \text{time}(\mathit{Read\ HBuffer}); \\
& \text{time}(\mathit{step2a}) + \text{time}(\mathit{step2b}) \\
& + \text{time}(\mathit{Read\ Serialization\ buffer}); \\
& \text{time}(\mathit{step3a}) + \text{time}(\mathit{step3b})\ )
\end{aligned}
\tag{6.3}
$$

Even if these equations seem to be now satisfying, a really big problem would be certainly identified by the not yet developed Hough transform steps 'Step3a' and 'Step3b' due to the reasons, which are already stated in section 5.5.5 on page 370. However it is nevertheless easily possible to cope with

the missing timing behavior of these steps by simply assuming that they require at the utmost as much time as the slowest one of the other steps in the pipeline.
So based on this supposition, which is evidently practical here because of the pipeline balancing with regard to the throughput, the equations turn into:

$$\begin{aligned} \text{latency} &= \text{time}(\textit{step1}) + \text{time}(\textit{step2a}) + \text{time}(\textit{step2b}) \\ &+ \max(\ \text{time}(\textit{step1}); \text{time}(\textit{step2a}) + \text{time}(\textit{step2b})\ ) \\ \Rightarrow \text{latency} &= \max(\ 2 \cdot \text{time}(\textit{step1}) + \text{time}(\textit{step2a}) + \text{time}(\textit{step2b}); \\ &\quad \text{time}(\textit{step1}) + 2 \cdot (\text{time}(\textit{step2a}) + \text{time}(\textit{step2b}))\ ) \end{aligned} \tag{6.4}$$

$$\begin{aligned} \text{throughput} &= \max(\ \text{time}(\textit{step1}) + \text{time}(\textit{Read HBuffer}); \\ &\quad \text{time}(\textit{step2a}) + \text{time}(\textit{step2b}) \\ &\quad +\text{time}(\textit{Read Serialization buffer}); \\ &\quad \max(\ \text{time}(\textit{step1}) + \text{time}(\textit{Read HBuffer}); \\ &\quad \text{time}(\textit{step2a}) + \text{time}(\textit{step2b}) \\ &\quad +\text{time}(\textit{Read Serialization buffer})\ )\ ) \\ \Rightarrow \text{throughput} &= \max(\ \text{time}(\textit{step1}) + \text{time}(\textit{Read HBuffer}); \\ &\quad \text{time}(\textit{step2a}) + \text{time}(\textit{step2b}) \\ &\quad +\text{time}(\textit{Read Serialization buffer})\ ) \end{aligned} \tag{6.5}$$

Due to the fact that all different Hough transform implementations can be now benchmarked by the usage of this general analyzing scheme, the following sections are of course utilized to introduce optimal realizations, which fulfill the requirements of the CBM experiment bearing different aspects in mind. Although detailed information about the combination strategy of the synthesis possibilities for each building block is presented together with their corresponding resource consumption, it is clear that the main focus is set to the aspects, which brand the optimal implementation by either using just a single FPGA, by being the fastest implementation or by being the cheapest one.
Further on, the main design parameter of the presented results is naturally determined by the balanced pipelined throughput, which leads to the task of evaluating a well-defined parallelism level for each unit.

But before proceeding with the details, it should be finally noticed here that all given conclusions are based on the FPGA device, which is already used to produce the results of section 5.5 on page 327. Albeit it is obvious that the application of future FPGA devices will allow to implement a more optimized Hough transform algorithm with a higher histogram resolution for instance, or will lead to a better result with regard to the timing behavior or resource allocation.

A Single-Chip Version Fulfilling the CBM Requirements

Coming now to a more precise analysis of the optimal realization, which is branded by the aspect of utilizing just a single FPGA to implement the entire algorithm, this section presents a Hough transform based first level tracking system for the CBM experiment, which parallelizes such devices to cope with the data rate. Consequently the maximal usable resources for one algorithm instance are apparently limited by the resources, which are available on a single FPGA device.

So the combination of this limit with the strategy of the Hough transform building blocks, which are exposed in figure 3.30 of section 3.3.1 on page 158, leads of course to the implementation startup of choosing these blocks without alternative options at first.
Hence the HBuffer unit with id two is chosen at the beginning, because there is no other possibility. In this connection, it has to be also mentioned that the HBuffer unit is only needed, because not all histogram layers can be instantiated in parallel.
Moreover it is clear that the subsequent units must be separated for the row-wise histogram read-out at the bottom and the column-wise histogram read-out at right. Thence the concerned units for the row-wise read-out at the bottom are the Diagonalization unit with id one, the Peak-finding unit with id one and the Serialization unit with id one. The corresponding units for the column-wise read-out at right are contrarily determined by the Diagonalization unit with id two, the Peak-finding unit with id two and the Serialization unit with id two.
Further on, there are just three units missing up to now, which are obviously identified by the histogram with the corresponding generator and the Encoding unit.
Since the flexibility of the loadable Encoding unit should be certainly used, but the flip-flop and LUT resources preserved, the selected implementation is thus naturally labeled with id four for the row-wise read-out at the bottom and id ten for the column-wise read-out at right.
Moreover a closer look to the occupancy of the FPGA device shows yet that there is no restriction for the choice of the histogram implementation, because each single one fits. As however the utilization of the different FPGA resources should be balanced here, the histogram unit with id one and the corresponding generator unit with id two is chosen for the row-wise read-

out at the bottom. For the same reason, the histogram unit with id 27 and the corresponding generator with id twelve are picked for the column-wise read-out at right.

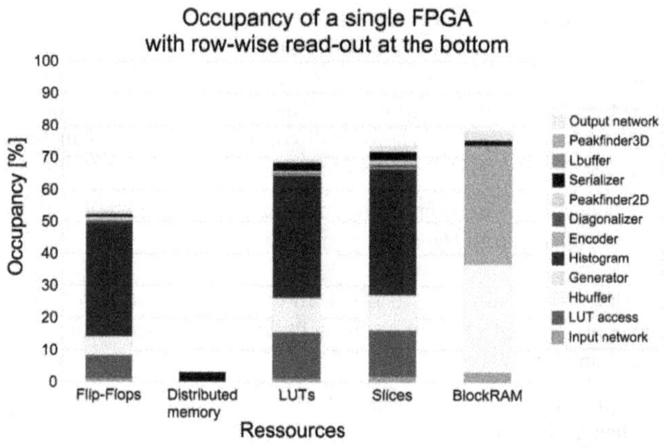

(a) Row-wise histogram read-out at the bottom

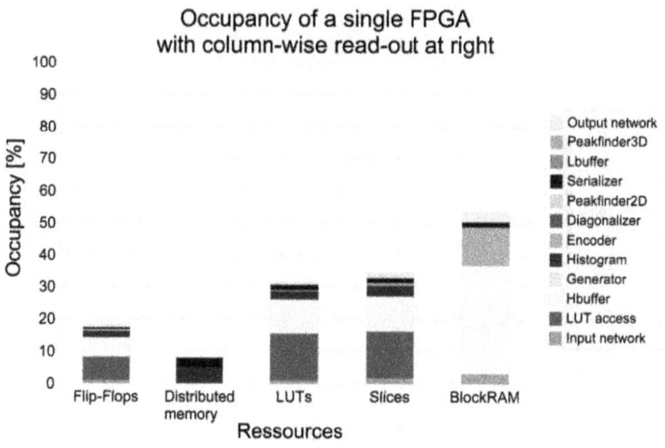

(b) Column-wise histogram read-out at right

Figure 6.2: Occupancy of the FPGAs for the implementation with a single-chip

So having in the end two successfully configured Hough transform implementations, their generated FPGA resource occupancies are depicted in figure 6.2, while the absolute values can be found in appendix C.9.1 on page C-17.

Although figure 6.2b shows that the current occupancy allows multiple parallel instantiated histogram layers, it has to be explicitly highlighted that this alternative is not investigated here, because the multi-chip version, which is introduced in section 6.2.2 on page 408, offers even more capabilities. In addition to this, it has to be also noticed that some resources on the FPGA device have to be reserved for the overall control unit, which can be easily implemented by using the state machine concept.

Since the resource consumption is now successfully analyzed in detail, the focus is naturally changed to the corresponding timing behavior of both single-chip versions.
Therefore the general equations 6.1, 6.4 and 6.5 are evaluated for each specific implementation with regard to the sample event, which is introduced in section 6.2.2 on page 394.

But before proceeding, the analysis, which is presented in section 4.9.1 on page 284, has to be obviously consulted to get detailed information about the hit read-out of the HBuffer unit, because each hit contributes to a certain

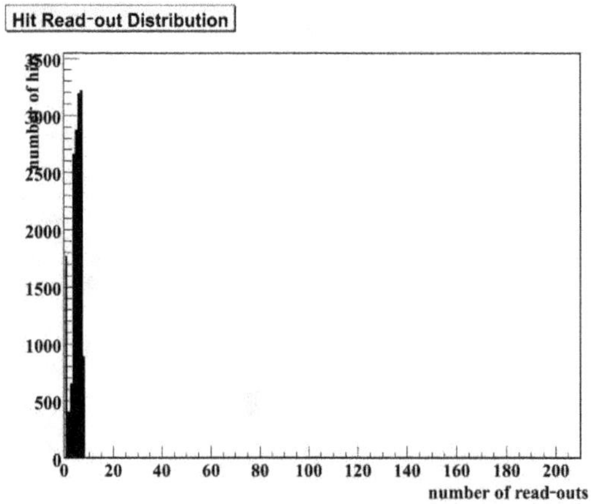

(a) Standard distribution

Figure 6.3: Distribution of the number of hit read-outs for the sample event and the amount of used histogram layers

amount of histogram layers.
As figure 6.3 depicts furthermore the result of this analysis based on the sample event, the distribution of the number of hits over the number of read-outs, which is surely equal to the number of hits over the number of affected histogram layers for each hit, is illustrated and can be consequently utilized for the computation of the timing behavior.

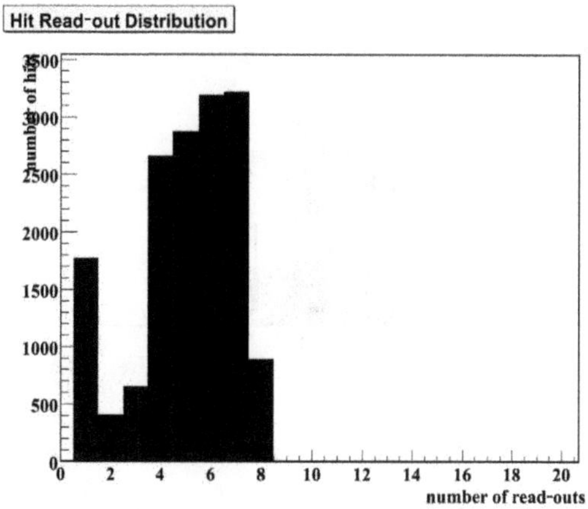

(b) Region of interest

Figure 6.3: Distribution of the number of hit read-outs for the sample event and the amount of used histogram layers

So applying this distribution to the instantiation of a single histogram layer, it is clear that the number of hits, which have to be read out of the HBuffer unit, can be easily computed by the equation:

$$\begin{aligned}\text{number of read} - \text{outs} &= \sum_{i=1}^{\text{max. layers}} \left(i \cdot \text{number of hits}_{(i\ layers)}\right) \\ \text{number of read} - \text{outs} &= \sum_{i=1}^{191} \left(i \cdot \text{number of hits}_{(i\ layers)}\right) \\ &= 78.4 \cdot 10^3\end{aligned}$$

In addition to this, the analysis, which is introduced in section 4.9.3 on page 291, has to be apparently inquired to get extensive information about the track candidate read-out of the buffer, which is used in the Serialization unit. Consequentially figure 6.4 shows the result, which reveals the distribution of the number of track candidates over the corresponding histogram layers for the sample event.

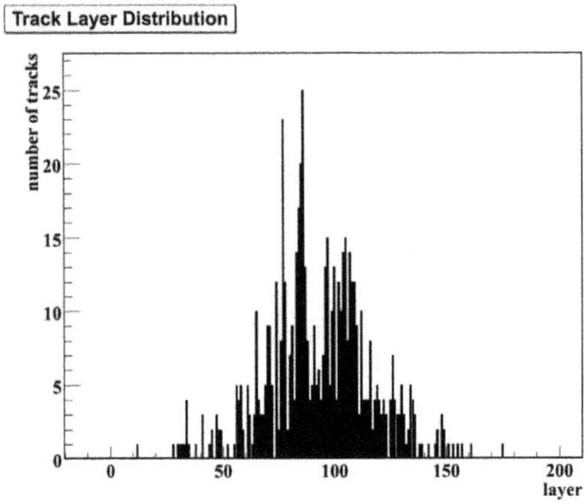

Figure 6.4: Distribution of the number of track candidate read-outs for the sample event and the amount of used histogram layers

Compared to the previous computation of the number of hit read-outs, the necessary amount of track candidate read-outs has to be obviously evaluated in a different way, because this processing step is located at the last pipeline stage. Moreover the main impact on this number of read-outs is apparently determined by the number of parallel instantiated histogram layers, which fixes also the number of serial iterations to process all layers, and the number of track candidates, which exist in these layers. So at a first glance it is self-evident that the number of track candidate read-outs varies with the number of serial iterations, but as just a single histogram layer is instantiated in the actual considered Hough transform implementations, the worst-case number of read-outs for the serial iterations can be easily defined by the equation:

$$\begin{aligned} \text{number of read}-\text{outs} &= \max\left(\bigcup_{i=1}^{\text{max. layers}} \text{number of tracks}_{(layer\ i)}\right) \\ \text{number of read}-\text{outs} &= \max\left(\bigcup_{i=1}^{191} \text{number of tracks}_{(layer\ i)}\right) \\ &= 25 \end{aligned}$$

Keeping these values in mind, the timing in clock cycles for each step can be evaluated.

$$
\begin{aligned}
\text{time}_{[\text{clock cycles}]}(\mathit{step1}) &= 15.7 \cdot 10^3 + 5 + 2 \\
&= 15.7 \cdot 10^3 \\
\text{time}_{[\text{clock cycles}]}(\mathit{step2a}) &= 78.4 \cdot 10^3 + 5 + 2 + 191 \cdot 30 \\
&= 84.1 \cdot 10^3 \\
\text{time}^{\mathit{row-wise}}_{[\text{clock cycles}]}(\mathit{step2b}) &= \langle \tfrac{191}{1} \rangle \cdot \langle \tfrac{1}{1} \rangle \cdot 31 + 1 + 1 + 3 + 3 + 1 \\
&= 5.9 \cdot 10^3 \\
\text{time}^{\mathit{column-wise}}_{[\text{clock cycles}]}(\mathit{step2b}) &= \langle \tfrac{191}{1} \rangle \cdot \langle \tfrac{1}{1} \rangle \cdot 95 + 1 + 2 + 3 + 3 + 1 \\
&= 18.2 \cdot 10^3 \\
\text{time}_{[\text{clock cycles}]}(\mathit{step3a}) &= 25 + 2 \\
&= 27
\end{aligned}
\tag{6.6}
$$

By using these numbers, the latency and the throughput can be further computed in clock cycles.

$$
\begin{aligned}
\text{latency}^{\mathit{row-wise}}_{[\text{clock cycles}]} &= \max(\, 2 \cdot 15.7 \cdot 10^3 + 84.1 \cdot 10^3 + 5.9 \cdot 10^3; \\
&\quad 15.7 \cdot 10^3 + 2 \cdot (84.1 \cdot 10^3 + 5.9 \cdot 10^3)\,) \\
&= 195.7 \cdot 10^3 \\
\text{latency}^{\mathit{column-wise}}_{[\text{clock cycles}]} &= \max(\, 2 \cdot 15.7 \cdot 10^3 + 84.1 \cdot 10^3 + 18.2 \cdot 10^3; \\
&\quad 15.7 \cdot 10^3 + 2 \cdot (84.1 \cdot 10^3 + 18.2 \cdot 10^3)\,) \\
&= 220.2 \cdot 10^3
\end{aligned}
\tag{6.7}
$$

$$
\begin{aligned}
\text{throughput}^{\mathit{row-wise}}_{[\text{clock cycles}]} &= \max(\, 15.7 \cdot 10^3 + 78.4 \cdot 10^3; \\
&\quad 84.1 \cdot 10^3 + 5.9 \cdot 10^3 + 25\,) \\
&= 94.0 \cdot 10^3 \\
\text{throughput}^{\mathit{column-wise}}_{[\text{clock cycles}]} &= \max(\, 15.7 \cdot 10^3 + 78.4 \cdot 10^3; \\
&\quad 84.1 \cdot 10^3 + 18.2 \cdot 10^3 + 25\,) \\
&= 102.3 \cdot 10^3
\end{aligned}
\tag{6.8}
$$

Taking now the maximum achievable clock frequency of each handpicked Hough transform unit presented in section 5.5 on page 327 into account, it is reasonable that an overall frequency of 125 MHz ($8\,\frac{\text{ns}}{\text{clock cycle}}$) is practicable. Thence the equations 6.7 for the latency and 6.8 for the throughput can be evaluated to the absolute timing for the sample event.

$$
\begin{aligned}
\text{latency}^{row-wise}_{\text{[time]}} &= 195.7 \cdot 10^3 \, \frac{\text{clock cycles}}{\text{sample event}} \cdot 8 \, \frac{\text{ns}}{\text{clock cycle}} \\
&= 195.7 \cdot 10^3 \, \frac{\text{clock cycles}}{\text{sample event}} \cdot \frac{8 \, \mu\text{s}}{10^3 \, \text{clock cycle}} \\
&\approx 1.57 \cdot 10^3 \, \mu\text{s} \\
\text{latency}^{column-wise}_{\text{[time]}} &= 220.2 \cdot 10^3 \, \frac{\text{clock cycles}}{\text{sample event}} \cdot 8 \, \frac{\text{ns}}{\text{clock cycle}} \\
&= 220.2 \cdot 10^3 \, \frac{\text{clock cycles}}{\text{sample event}} \cdot \frac{8 \, \mu\text{s}}{10^3 \, \text{clock cycle}} \\
&\approx 1.76 \cdot 10^3 \, \mu\text{s}
\end{aligned} \tag{6.9}
$$

$$
\begin{aligned}
\text{throughput}^{row-wise}_{\text{[time]}} &= 94.0 \cdot 10^3 \, \frac{\text{clock cycles}}{\text{sample event}} \cdot 8 \, \frac{\text{ns}}{\text{clock cycle}} \\
&= 94.0 \cdot 10^3 \, \frac{\text{clock cycles}}{\text{sample event}} \cdot \frac{8 \, \mu\text{s}}{10^3 \, \text{clock cycle}} \\
&\approx 752.18 \, \mu\text{s} \\
\text{throughput}^{column-wise}_{\text{[time]}} &= 102.3 \cdot 10^3 \, \frac{\text{clock cycles}}{\text{sample event}} \cdot 8 \, \frac{\text{ns}}{\text{clock cycle}} \\
&= 102.3 \cdot 10^3 \, \frac{\text{clock cycles}}{\text{sample event}} \cdot \frac{8 \, \mu\text{s}}{10^3 \, \text{clock cycle}} \\
&\approx 818.20 \, \mu\text{s}
\end{aligned} \tag{6.10}
$$

Further on, the combination of these throughput timing behavior results with the required processing time of an event from the CBM experiment, which is $0.40 \, \frac{\mu\text{s}}{\text{event}}$, leads then immediately to a necessary amount of $\frac{752.18 \, \frac{\mu\text{s}}{\text{sample event}}}{0.40 \, \frac{\mu\text{s}}{\text{event}}} = 1.9 \cdot 10^3$ parallel used systems for the row-wise read-out version and $\frac{818.20 \, \frac{\mu\text{s}}{\text{sample event}}}{0.40 \, \frac{\mu\text{s}}{\text{event}}} = 2.0 \cdot 10^3$ systems for the column-wise read-out version in a computation factory. Consequently the total device purchasing costs could be easily calculated to about $1.9 \cdot 10^3 \cdot \$\,1.6 \cdot 10^3 \approx \$\,3.0 \cdot 10^6$ for the row-wise read-out version and about $2.0 \cdot 10^3 \cdot \$\,1.6 \cdot 10^3 = \$\,3.3 \cdot 10^6$ for the column-wise read-out version, if the common FPGA price for the currently used device, which can be found in section 3.3.2 on page 184, is taken into account.

Since the timing of both implementations is recently successfully matched to the CBM experiment, the network load, which defines evidently also an important benchmark, has to be analyzed. By remembering now the introduced design goal, it is clear that the network must be able to provide as peak performance a single hit in every clock cycle, which entails apparently a peak network traffic of $\frac{1 \, \frac{\text{hit}}{\text{clock cycle}} \cdot 32 \, \frac{\text{bit}}{\text{hit}}}{8 \, \frac{\text{ns}}{\text{clock cycle}}} \approx 3.73 \, \frac{\text{Gbit}}{\text{s}}$.
Matching this result in the following to the network capability of the used FPGA device, which is introduced in section 3.3.1 on page 156, it is obvious that two MGTs are enough to cover the needs. But as the platform is not

able to work with a continuous data stream due to the pipeline stalls, the average performance can be computed to $\frac{15.66 \cdot 10^3 \frac{\text{hits}}{\text{sample event}} \cdot 32 \frac{\text{bit}}{\text{hit}}}{752.18 \cdot 10^{-6} \frac{\text{s}}{\text{sample event}}} \approx 0.62 \frac{\text{Gbit}}{\text{s}}$ for the row-wise read-out version and $\frac{15.66 \cdot 10^3 \frac{\text{hits}}{\text{sample event}} \cdot 32 \frac{\text{bit}}{\text{hit}}}{818.20 \cdot 10^{-6} \frac{\text{s}}{\text{sample event}}} \approx 0.57 \frac{\text{Gbit}}{\text{s}}$ for the column-wise read-out version.
So if one considers further the combination of a network bridge for these two links to the common 10-Gigabit ethernet specification and cascaded 24-port BLADE G8124 network switches (see [BLAb] and [Blaa]), there would be an amount of 86 articles for the row-wise read-out version and 93 articles for the column-wise read-out version, which generate consequential purchasing costs of either $86 \cdot \$\,12.0 \cdot 10^3 \approx \$\,1.0 \cdot 10^6$ for the row-wise read-out version or $93 \cdot \$\,12.0 \cdot 10^3 \approx \$\,1.1 \cdot 10^6$ for the column-wise read-out version.

In addition to these purchasing costs, a lack of such a factory will be apparently also defined by the operating costs, which are caused by the power consumption. As furthermore a closer look to the technical device specification details exhibits a single FPGA to consume roughly estimated about 10 Watt (see [Virf]) in combination with 168 Watt (see [BLAb]) for the network switch, the operating costs of the entire computation factory including its network connection can be computed for a single year and a electricity tariff of $0.2 \frac{\$}{\text{kWh}}$ to:

$$\begin{aligned}
\text{operating costs}^{row-wise} &= 1.9 \cdot 10^3 \cdot 0.01\,\text{kW} \cdot 365\,\text{d} \cdot 24\,\tfrac{\text{h}}{\text{d}} \cdot 0.2\,\tfrac{\$}{\text{kWh}} \\
&+ \quad 86 \quad \cdot 0.17\,\text{kW} \cdot 365\,\text{d} \cdot 24\,\tfrac{\text{h}}{\text{d}} \cdot 0.2\,\tfrac{\$}{\text{kWh}} \\
&\approx \$\,58.3 \cdot 10^3
\end{aligned}$$

$$\begin{aligned}
\text{operating costs}^{column-wise} &= 2.0 \cdot 10^3 \cdot 0.01\,\text{kW} \cdot 365\,\text{d} \cdot 24\,\tfrac{\text{h}}{\text{d}} \cdot 0.2\,\tfrac{\$}{\text{kWh}} \\
&+ \quad 93 \quad \cdot 0.17\,\text{kW} \cdot 365\,\text{d} \cdot 24\,\tfrac{\text{h}}{\text{d}} \cdot 0.2\,\tfrac{\$}{\text{kWh}} \\
&\approx \$\,63.2 \cdot 10^3
\end{aligned}$$

By taking now the complexity and costs of these computation factories into account, both implementations do evidently not define a practicable modality to realize an on-line first level tracking system for the CBM experiment. However the FPGA technology seems to be the correct choice.

A Multi-Chip Version Fulfilling the CBM Requirements

In contrast to the previous section, this one introduces evidently a Hough transform based first level tracking system for the CBM experiment, which parallelizes a multi-chip FPGA platform implementing the entire algorithm to cope with the occurring data rate. Hence the main discrepancy to the previous section is obviously determined by the circumstance that the available resources on a single-chip does not longer limit the usable resources for the algorithm, but the algorithm resource consumption dictates the necessary amount of FPGA devices for a single platform. Thence it is clear that the number of expedient FPGA devices for such a platform is investigated in the following in combination with their corresponding activities.

So for this purpose, it is not surprising that the activity or job concepts, which are devised for the Sony Playstation III implementation and already presented in section 3.3.2 on page 191, are transferred to the FPGA by the application of the developed functional Hough transform units, which are introduced in section 3.3.1 on page 156.
Consequentially these concepts suggest furthermore immediately that the platform utilizes at least three FPGA devices, which are connected by a platform internal chip-to-chip network. Moreover the functionality of these three devices is apparently distinguished by forming the input interface, the output interface and the processing part in between. Further on, it can be directly derived from this principle that both interfaces should be not replicated, whereas the processor is explicitly wanted to be parallelized as often as useful. However it is nevertheless obvious that the total number of applied FPGAs should be kept as small as possible.

Going further into the activity details of each specialized device, a more precise look to the functionality of the input interface shows that the corresponding FPGA has to implement the connection unit to the input network, which delivers the hits, the first LUT access unit, the HBuffer unit, the second LUT access unit and a platform internal chip-to-chip network connection unit, which is able to send the transformed hits to the FPGA processor devices.
According to this, the output interface FPGA has evidently to realize also a platform internal chip-to-chip network connection unit, which is able to receive the track candidates from the processor devices, the LBuffer unit, the 3D Peak-finding unit and the connection unit to the output network, which delivers finally the found tracks.
In contrast to this, the job of the processor FPGA is quite more complex, because it has to offer the full histogram processing pipeline, which contains

the Histogramming, the Encoding, the Diagonalization, the 2D Peak-finding and the Serialization step of the algorithm. By the way, it has to be also noticed in this context that such a conceptual FPGA is naturally capable of featuring a secondary level of parallelism, which is on-chip and analyzed later in this section, in addition to the recently introduced chip parallelism, because these steps form apparently the heart of the algorithm.

Coming now to the point, the focus of this thesis is set subsequently to the input interface FPGA and exceptionally to the processor FPGAs, because the units, which have to be instantiated on the output interface FPGA, are not developed yet due to the reasons introduced in section 5.5.5 on page 370.

So related to the single-chip version, which is presented in section 6.2.2 on page 400, the implementation startup is in turn determined by the selection of these building blocks, which feature no alternative options.
Therefore the first choice for the input interface FPGA is characterized again by the HBuffer unit with id two. But as there is yet no other possibility to select one of the other functional units for the input interface FPGA, because not any is actually implemented, the spotlight is set to the processor FPGAs. By remembering further the central part of the implementation, which is naturally defined by the evaluation of an optimal amount of parallel processor FPGA devices as already mentioned earlier, it is obviously at first important to explore the parallelism level, which can be realized by the usage of a unique processor FPGA. For this purpose, the aim of the processor FPGA implementation turns to the realization of the maximal possible parallelism level on the device, which changes also the focus for the selection. Hence the histogram implementations, which base on register cells (id 1 - 12), disqualify themselves, because if one will consider them, one would directly think about multi-chip FPGA platforms, which would apply $\frac{191\,\text{layers}}{2\,\frac{\text{layers}}{\text{chip}}} \approx 96\,\text{chips}$ to instantiate just the histogram layers. And as such an amount of FPGA devices is surely too much, the following investigations are focused on the histogram units, which are implemented with memory based cells (id 13 - 28). Further on, a consequence of this circumstance is apparently defined by the fact that only histogram units with a column-wise read-out at right are considered. Thence all consecutive units of the histogram must also apply the column-wise read-out at right. However this restriction is certainly not a disadvantage, because these implementations require in general less resources with no effect on the maximum clock frequency.
So starting now the concrete selection process, the concerned units for this read-out strategy are determined by the Diagonalization unit with id two, the Peak-finding unit with id two and the Serialization unit with id two.
Moreover the last three missing units, which are the histogram with the

corresponding generator and the Encoding unit, must be handpicked more careful, because the parallelism level of such a unit can require a lot of unintentional resources.
So beginning now with the Encoding unit, the loadable one with id ten, which is used in the single-chip version, must be evidently exchanged by the fixed Encoding unit with id seven, because the multi-chip version has to be very fast and should employ as sparse hardware as possible.
Remembering further the constraints for the histogram implementation, it is also patent that the one to use is the one, which requires the fewest resources. As furthermore table 5.28 of section 5.5.3 on page 354 summarizes the resource consumption of the applicable histogram implementations, a quick search leads directly to the selection of the one labeled with the id 26. But even if this implementation requires the fewest LUT-flip-flop pairs, the utilization of the Block RAM, which is a special feature of the used FPGA device family, would naturally limit the parallelism level. Hence another histogram implementation has to be considered additionally. So because of that reason, it is clear that the implementation with id 27 is accessorily chosen. Collaterally this condition offers certainly the extra advantage that both implementations can be exchanged with each other without changing the generator unit.
Last but not least the chosen generator unit is defined by the implementation with id ten because of the easy and straight implementation scheme.

Since the appropriate implementation of each Hough transform unit is evidently selected up to here, it is clear that the next interesting and essential characteristic of the platform to analyze is defined by the on-chip parallelism capability of a single processor FPGA. For this purpose, the main focus is naturally set to the combination of the required units for the Histogramming, the Encoding, the Diagonalization, the 2D Peak-finding and the Serialization as well as the number of parallel instantiated entities with regard to the available resources of the FPGA device.

So having this analysis in mind, a closer look to the first unit in the processing pipeline, which is surely identified by the generator unit, shows at the beginning that this one is generally marginal, because it is needed just once in any case, because the transformed hits arrive always serial. Furthermore it is additionally important to notice that the generator unit forms no restriction to the consecutive units, because as many simple extra 'enable' signals as parallel instantiated histogram layers are apparently able to arrange the correct operation for the Histogramming.

Besides this, the most manifest strategy for the parallelism level of all other units in the pipeline is quite simple, because the entire rest is initially just parallelized as often as the FPGA resources allow. Thus the resulting level can be easily evaluated to 14 $\frac{\text{layers}}{\text{chip}}$ of the histogram, which are read out by 14 $\frac{\text{read-outs}}{\text{chip}}$ arranged in chains (1 $\frac{\text{layer}}{\text{read-out}}$). Consequently figure 6.5a depicts now the FPGA resource occupancy, while the absolute values can be found in table C.6 of appendix C.9.2 on page C-22.

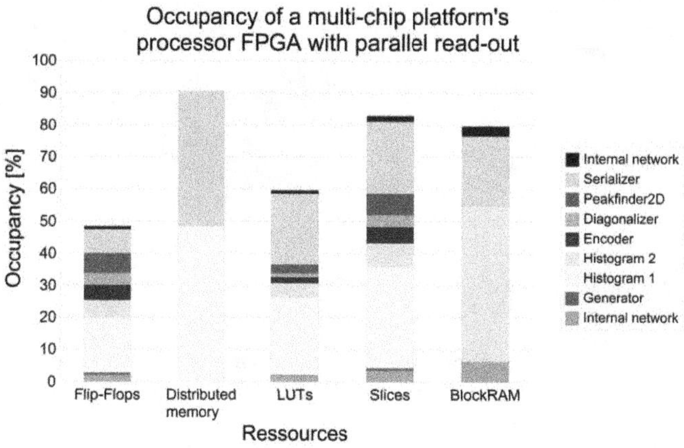

(a) Parallelism level of 14 $\frac{\text{layers}}{\text{chip}}$ and 14 $\frac{\text{read-out chains}}{\text{chip}}$ processing all layers in parallel

Figure 6.5: Occupancy of the available processor FPGAs for the implementation with a multi-chip platform

However it is clear that this parallelism level demonstrates just a first attempt, which may be improved to receive better results. Thence the next considered parallelism level is expectedly looked straight forward, because it is determined by the one, which defines the exact oppositional strategy resulting in the instantiation of just one of the read-out pipeline chains but as many histogram layers as possible. So this idea leads then obviously to the parallelism level of 24 $\frac{\text{layers}}{\text{chip}}$ and 1 $\frac{\text{read-out}}{\text{chip}}$ (24 $\frac{\text{layer}}{\text{read-out}}$), which ends in the resource occupancy shown in figure 6.5b and the absolute values found in table C.8 of appendix C.9.2 on page C-24. Moreover it has to be consequentially noted in this connection that the timing behavior of such a device is certainly changed, because the histogram layers can be filled in parallel, but has to be read out serially.

As these two versions realize evidently the most opposed situations of the available parallelism levels, because more read-out pipelines than histogram

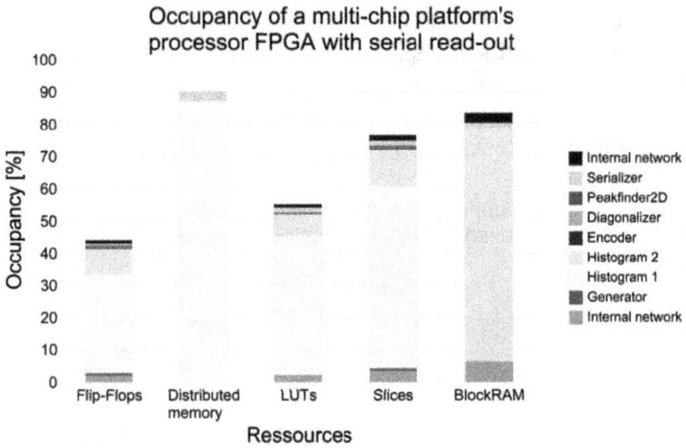

(b) Parallelism level of 24 $\frac{\text{layers}}{\text{chip}}$ and a single read-out chain per chip processing all layers serially

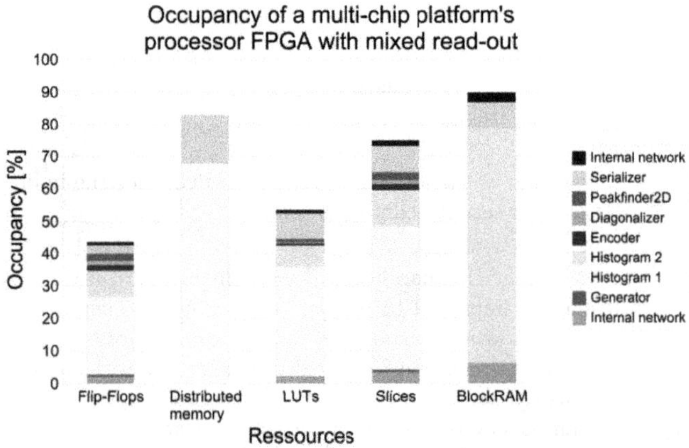

(c) Parallelism level of 20 $\frac{\text{layers}}{\text{chip}}$ and five read-out chains per chip processing the parallel layers four times serially

Figure 6.5: Occupancy of the available processor FPGAs for the implementation with a multi-chip platform layers are surely senseless, it is assumed that the optimum might be located somewhere in between. For that reason, another well-defined and FPGA related parallelism level is considered, which combines $20\,\frac{\text{layers}}{\text{chip}}$ and $5\,\frac{\text{read-outs}}{\text{chip}}$ ($4\,\frac{\text{layer}}{\text{read-out}}$), because then five parallel histogram layers have to be read-out four times serially to process all layers. Consequently figure 6.5c gives an imagination about the resulting FPGA resource occupancy, while the absolute values can be once again found in table C.10 of appendix C.9.2 on page C-26.

Beyond the implementation of these different parallelism levels, it has to be additionally explicitly highlighted that all versions naturally have to reserve some resources on the FPGA device to realize the overall control unit, which can be obviously easily implemented by using the state machine concept.
Further on, these different realization schemes for the parallelism capability of a single processor FPGA have to be evidently onwards analyzed to evaluate the next interesting objective, which is apparently determined by the integration of these schemes into the platform system resulting in the FPGA device parallelism level. For this purpose, the following two paragraphs investigate possible multi-chip FPGA platforms, which differ just in the amount of parallel realized histogram layers. So as it is more precisely mentioned in section 3.3.1 on page 165, the first paragraph introduces a platform, which instantiates all histogram layers in parallel, while the second one presents a platform, which offers the advantage of reading each hit maximally twice out of the HBuffer unit by instantiating as many histogram layers in parallel as the maximum number of affected layers for a single hit.

The Fastest Version Fulfilling the CBM Requirements This paragraph presents the results for an implementation, which is characterized by offering enough hardware resources to instantiate all histogram layers in parallel. So this characteristic enables thus the removal of the HBuffer unit, because all hits, which are received via the input network, can be immediately transformed and afterwards directly passed to the Histogramming step of the algorithm. Furthermore it has to be explicitly noted that this circumstance defines the major difference to the single-chip version, which is exhibited in section 6.2.2 on page 400, or to the version, which is shown in the next paragraph.
Before proceeding now with the details of this implementation, figure 6.6 depicts the modified platform pipeline by featuring a different timing at the beginning.

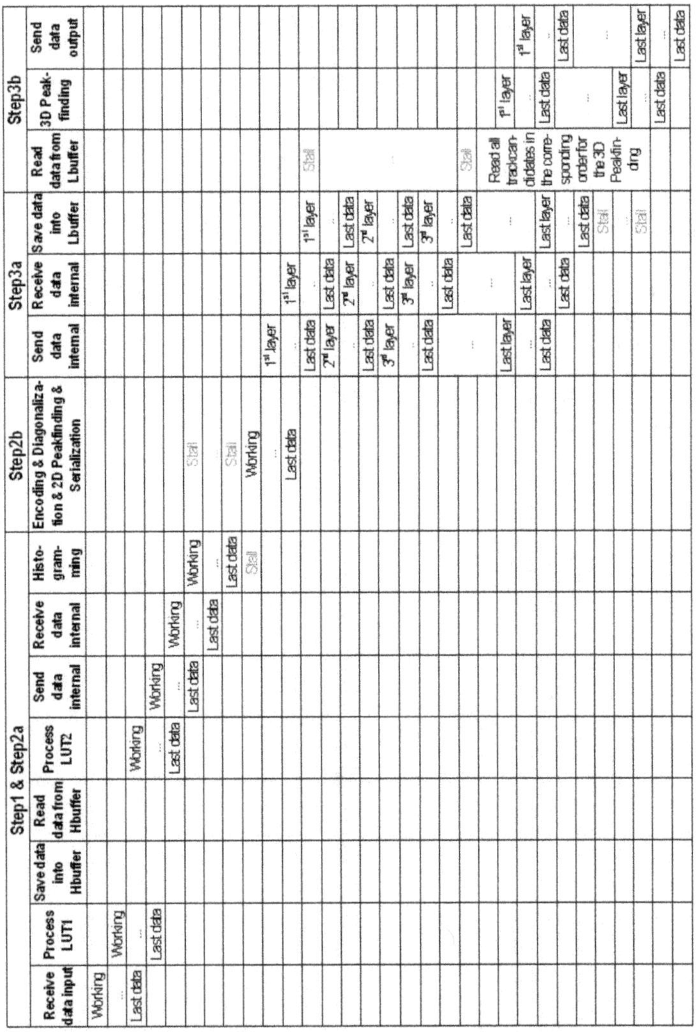

Figure 6.6: Illustration of the modified pipelining strategy for the specialized multi-chip FPGA implementation

With regard to this figure, it is self-evident that the introduced alteration in the pipeline affects especially the timing of the first two algorithm steps, whose general nature is already described in equation 6.1 of section 6.2.2 on page 398. Hence it is clear that modifications must be installed, which entail apparently further consequences to the equations 6.4 and 6.5. Consequently the adapted formulas turn into:

$$\begin{aligned} \text{time}(\textit{step1 \& step2a}) \quad = & \quad \text{time}(\textit{Receive hits}) \\ + & \quad \text{time}(\textit{Process LUT1}) \\ + & \quad \text{time}(\textit{Process LUT2}) \\ + & \quad \text{time}(\textit{Histograming}) \\ + & \quad 1 \cdot \text{time}(\textit{Last data histograming}) \end{aligned} \tag{6.11}$$

$$\begin{aligned} \text{latency} \quad = \quad & \max(\, 2 \cdot \text{time}(\textit{step1 \& step2a}) + \text{time}(\textit{step2b}); \\ & \text{time}(\textit{step1 + step2a}) + 2 \cdot \text{time}(\textit{step2b}) \,) \end{aligned} \tag{6.12}$$

$$\begin{aligned} \text{throughput} \quad = \quad & \max(\, \text{time}(\textit{step1 \& step2a}); \\ & \text{time}(\textit{step2b}) + \text{time}(\textit{Read Serialization buffer}) \,) \end{aligned} \tag{6.13}$$

Besides this, the removal of the HBuffer unit leads naturally also to an alteration in the occupancy of the input interface FPGA, because this one has now only to implement the connection unit to the input network, which delivers the hits, the first LUT access unit, the second LUT access unit and a platform internal chip-to-chip network connection unit, which is able to send the transformed hits to the processor FPGAs. As such a resulting occupancy is furthermore surely quite low, some histogram layers can be apparently implemented already there. Thence the only supplementary required feature is certainly determined by a secondary platform internal chip-to-chip network connection unit, which is able to send the track candidates directly to the output interface FPGA, in addition to the established platform internal chip-to-chip network connection unit to the processor FPGAs. However this capability is only used in the following to extend the number of parallel instantiated histogram layers on the processor FPGAs to the total amount.

So for instance, if a platform system is built by using the processor FPGAs, which are determined earlier by the parallelism level of 14 $\frac{\text{layers}}{\text{chip}}$ and 14 $\frac{\text{read-outs}}{\text{chip}}$ (1 $\frac{\text{layer}}{\text{read-out}}$), the input interface FPGA would have to additionally implement evidently 9 $\frac{\text{layers}}{\text{chip}}$ and 9 $\frac{\text{read-outs}}{\text{chip}}$, because 13 processor FPGAs could realize just 182 histogram layers. Hence such a platform requires a total of 15 FPGAs including one input interface chip, one output interface chip and 13 processor chips for the realization of the algorithm with the characteristic of these parallelism levels.

In contrast to this, the other introduced parallelism level of $24\,\frac{\text{layers}}{\text{chip}}$ and $1\,\frac{\text{read-out}}{\text{chip}}$ $(24\,\frac{\text{layer}}{\text{read-out}})$ leads obviously to a platform, which requires ten chips for the realization of the algorithm, because eight processor FPGAs can realize 192 histogram layers. So based on this insight, it is also immediately clear that the input interface chip contains no processing unit in this version and is therefore quite empty.

Further on, the last developed parallelism level concept results finally in a platform, which employs eleven chips, because nine processor FPGAs implementing $20\,\frac{\text{layers}}{\text{chip}}$ and $5\,\frac{\text{read-outs}}{\text{chip}}$ $(4\,\frac{\text{layer}}{\text{read-out}})$ can certainly instantiate 180 histogram layers. As this system apparently requires then the input interface FPGA to contain eleven supplemental histogram layers, the occupancy is assumed to be the highest within this context.
However a closer look to the occupancy of the input interface chip, which belongs to the first introduced platform, shows that this one is bigger. Furthermore the reason for that is not mysterious, because this chip has to implement nine read-out chains in addition to the nine histogram layers. Therefore figure 6.7 depicts now the occupancy of the most occupied input interface FPGA.

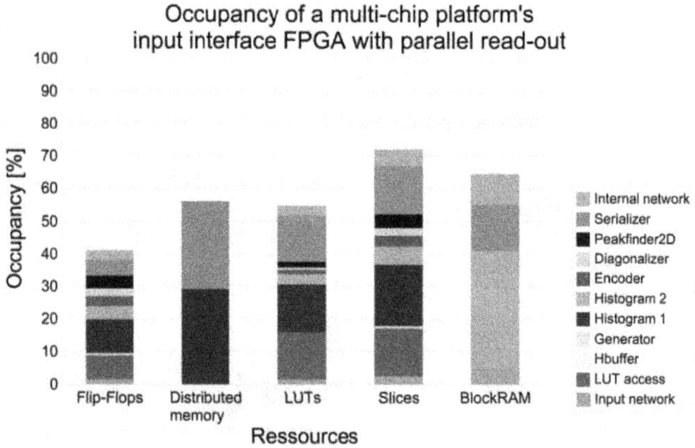

Figure 6.7: Occupancy of the input interface FPGA for the implementation with a multi-chip platform characterized by a parallelism level of $9\,\frac{\text{layers}}{\text{chip}}$ and $9\,\frac{\text{read-out chains}}{\text{chip}}$

Since the resource consumption is successfully and extensively analyzed up to here, the focus is now obviously changed to the corresponding timing behavior of the diverse implementations.
For this purpose, the general equation 6.1, which is partially modified to 6.11, and the adapted equations 6.12 as well as 6.13 are evaluated with regard to the sample event, which is introduced in section 6.2.2 on page 394.

But before proceeding, it is as important as in the single-chip version to think about the hit read-out of the HBuffer unit and the track candidate read-out of the buffer, which is used in the Serialization unit. However it is in contrast to the single-chip version neither necessary to consult the analysis for the hit read-out, which is presented in section 4.9.1 on page 284, nor needed to inquire the analysis for the track candidate read-out, which is introduced in section 4.9.3 on page 291, because the information details are quite simple.
Beginning with a closer look to the number of hit read-outs of the HBuffer unit, it is immediately clear that this value is zero, because the HBuffer unit has been removed.
In addition to this, a precise view of the read-out strategy for the track candidates, which is introduced in the single-chip version, results instantly in the parallel read-out of all track candidates, because all histogram layers are instantiated in parallel. Consequently the following simple formula can be used to compute the amount for the sample event to:

$$\begin{aligned}
\text{number of read} - \text{outs} &= \sum_{i=1}^{\text{max. layers}} \text{number of tracks}_{(\mathit{layer}\ i)} \\
\text{number of read} - \text{outs} &= \sum_{i=1}^{191} \text{number of tracks}_{(\mathit{layer}\ i)} \\
&= 604
\end{aligned}$$

Keeping these two values in mind, the timing in clock cycles for each step can be evaluated.

$$\begin{aligned}
\text{time}_{[\text{clock cycles}]}(\mathit{step1}\ \&\ \mathit{step2a}) &= 15.7 \cdot 10^3 + 5 + 5 + 2 + 1 \cdot 30 \\
&= 15.7 \cdot 10^3 \\
\text{time}_{[\text{clock cycles}]}^{\mathit{1}\ \frac{\text{layer}}{\text{read-out}}}(\mathit{step2b}) &= 1 \cdot \langle \tfrac{191}{191} \rangle \cdot 95 + 1 + 2 + 3 + 3 + 1 \\
&= 105 \\
\text{time}_{[\text{clock cycles}]}^{\mathit{4}\ \frac{\text{layer}}{\text{read-out}}}(\mathit{step2b}) &= 1 \cdot \langle \tfrac{191}{48} \rangle \cdot 95 + 1 + 2 + 3 + 3 + 1 \\
&= 390 \\
\text{time}_{[\text{clock cycles}]}^{\mathit{24}\ \frac{\text{layer}}{\text{read-out}}}(\mathit{step2b}) &= 1 \cdot \langle \tfrac{191}{8} \rangle \cdot 95 + 1 + 2 + 3 + 3 + 1 \\
&= 2.3 \cdot 10^3
\end{aligned} \tag{6.14}$$

$$\text{time}_{\text{[clock cycles]}}(\mathit{step3a}) = 604 + 2 = 606 \tag{6.14}$$

By using these numbers further, the latency and the throughput can be computed in clock cycles.

$$\begin{aligned}
\text{latency}^{\mathit{1}\,\frac{\text{layer}}{\text{read-out}}}_{\text{[clock cycles]}} &= \max(\ 2 \cdot 15.7 \cdot 10^3 + 105; \\
&\qquad 15.7 \cdot 10^3 + 2 \cdot 105\ ) \\
&= 31.5 \cdot 10^3 \\
\text{latency}^{\mathit{4}\,\frac{\text{layer}}{\text{read-out}}}_{\text{[clock cycles]}} &= \max(\ 2 \cdot 15.7 \cdot 10^3 + 390; \\
&\qquad 15.7 \cdot 10^3 + 2 \cdot 390\ ) \\
&= 31.8 \cdot 10^3 \\
\text{latency}^{\mathit{24}\,\frac{\text{layer}}{\text{read-out}}}_{\text{[clock cycles]}} &= \max(\ 2 \cdot 15.7 \cdot 10^3 + 2.3 \cdot 10^3; \\
&\qquad 15.7 \cdot 10^3 + 2 \cdot 2.3 \cdot 10^3\ ) \\
&= 33.7 \cdot 10^3
\end{aligned} \tag{6.15}$$

$$\begin{aligned}
\text{throughput}^{\mathit{1}\,\frac{\text{layer}}{\text{read-out}}}_{\text{[clock cycles]}} &= \max(\ 15.7 \cdot 10^3; 105 + 606\ ) \\
&= 15.7 \cdot 10^3 \\
\text{throughput}^{\mathit{4}\,\frac{\text{layer}}{\text{read-out}}}_{\text{[clock cycles]}} &= \max(\ 15.7 \cdot 10^3; 390 + 606\ ) \\
&= 15.7 \cdot 10^3 \\
\text{throughput}^{\mathit{24}\,\frac{\text{layer}}{\text{read-out}}}_{\text{[clock cycles]}} &= \max(\ 15.7 \cdot 10^3; 2.3 \cdot 10^3 + 606\ ) \\
&= 15.7 \cdot 10^3
\end{aligned} \tag{6.16}$$

Taking now the maximum achievable clock frequency of each handpicked Hough transform unit presented in section 5.5 on page 327 into account, it is reasonable that an overall frequency of 250 MHz ($4\,\frac{\text{ns}}{\text{clock cycle}}$) is practicable. In this connection, it has to be explicitly highlighted that this doubled clock frequency is certainly attainable, because the removed HBuffer unit has formed the limitation. Thence the equations 6.15 for the latency and 6.16

for the throughput can be evaluated to the absolute timing for the sample event.

$$\begin{aligned}
\text{latency}_{[\text{time}]}^{1\,\frac{\text{layer}}{\text{read-out}}} &= 31.5 \cdot 10^3\,\frac{\text{clock cycles}}{\text{sample event}} \cdot 4\,\frac{\text{ns}}{\text{clock cycle}} \\
&= 31.5 \cdot 10^3\,\frac{\text{clock cycles}}{\text{sample event}} \cdot \frac{4\,\mu\text{s}}{10^3\,\text{clock cycle}} \\
&\approx 126.02\,\mu\text{s} \\
\text{latency}_{[\text{time}]}^{4\,\frac{\text{layer}}{\text{read-out}}} &= 31.8 \cdot 10^3\,\frac{\text{clock cycles}}{\text{sample event}} \cdot 4\,\frac{\text{ns}}{\text{clock cycle}} \\
&= 31.8 \cdot 10^3\,\frac{\text{clock cycles}}{\text{sample event}} \cdot \frac{4\,\mu\text{s}}{10^3\,\text{clock cycle}} \\
&\approx 127.16\,\mu\text{s} \\
\text{latency}_{[\text{time}]}^{24\,\frac{\text{layer}}{\text{read-out}}} &= 33.7 \cdot 10^3\,\frac{\text{clock cycles}}{\text{sample event}} \cdot 4\,\frac{\text{ns}}{\text{clock cycle}} \\
&= 33.7 \cdot 10^3\,\frac{\text{clock cycles}}{\text{sample event}} \cdot \frac{4\,\mu\text{s}}{10^3\,\text{clock cycle}} \\
&\approx 134.76\,\mu\text{s}
\end{aligned} \tag{6.17}$$

$$\begin{aligned}
\text{throughput}_{[\text{time}]} &= 15.7 \cdot 10^3\,\frac{\text{clock cycles}}{\text{sample event}} \cdot 4\,\frac{\text{ns}}{\text{clock cycle}} \\
&= 15.7 \cdot 10^3\,\frac{\text{clock cycles}}{\text{sample event}} \cdot \frac{4\,\mu\text{s}}{10^3\,\text{clock cycle}} \\
&\approx 62.80\,\mu\text{s}
\end{aligned} \tag{6.18}$$

Further on, the combination of this throughput timing behavior, which is surprisingly independent of the read-out parallelism level, with the required processing time of an event from the CBM experiment, which is $0.40\,\frac{\mu\text{s}}{\text{event}}$, leads then obviously to the necessary amount of $\frac{62.80\,\frac{\mu\text{s}}{\text{sample event}}}{0.40\,\frac{\mu\text{s}}{\text{event}}} = 157$ parallel used systems in a computation factory. Consequently the total device purchasing costs could be easily calculated in the following, if the common FPGA price for the currently used device, which can be found in section 3.3.2 on page 184, is taken into account.

$$\begin{aligned}
\text{purchasing costs}^{1\,\frac{\text{layer}}{\text{read-out}}} &= 157 \cdot 15 \cdot \$\,1.6 \cdot 10^3 \approx \$\,3.8 \cdot 10^6 \\
\text{purchasing costs}^{4\,\frac{\text{layer}}{\text{read-out}}} &= 157 \cdot 11 \cdot \$\,1.6 \cdot 10^3 \approx \$\,2.8 \cdot 10^6 \\
\text{purchasing costs}^{24\,\frac{\text{layer}}{\text{read-out}}} &= 157 \cdot 10 \cdot \$\,1.6 \cdot 10^3 \approx \$\,2.5 \cdot 10^6
\end{aligned}$$

Since the timing of all implementations is successfully matched to the CBM experiment, it is evidently again necessary to analyze the network load, because this aspect defines in turn an important benchmark as well as for the single-chip version. By remembering now the same introduced design goal, it is also immediately clear that the network must be again able to provide as peak performance a single hit in every clock cycle, which entails thus a

peak network traffic of $\frac{1\,\frac{\text{hit}}{\text{clock cycle}}\cdot 32\,\frac{\text{bit}}{\text{hit}}}{4\,\frac{\text{ns}}{\text{clock cycle}}} \approx 7.45\,\frac{\text{Gbit}}{\text{s}}$.
Matching this result in the following to the network capability of the used FPGA device, which is introduced in section 3.3.1 on page 156, it is obvious that three MGTs are enough to cover the needs. But as all platforms are, like the single-chip versions, not able to work with a continuous data stream due to the pipeline stalls, the average performance can be easily computed to $\frac{15.66\cdot 10^3\,\frac{\text{hits}}{\text{sample event}}\cdot 32\,\frac{\text{bit}}{\text{hit}}}{62.80\cdot 10^{-6}\,\frac{\text{s}}{\text{sample event}}} \approx 7.43\,\frac{\text{Gbit}}{\text{s}}$.
So if one considers further the combination of a network bridge for these three links to the common 10-Gigabit ethernet specification and cascaded 24-port BLADE G8124 switches (see [BLAb] and [Blaa]), there would be naturally an amount of 4 articles, which generate consequential purchasing costs of $4 \cdot \$\,12.0 \cdot 10^3 \approx \$\,47.8 \cdot 10^3$.

In addition to these purchasing costs, a lack of such a factory will be apparently also defined by the operating costs, which are caused by the power consumption. As furthermore a closer look to the technical device specification details exhibits once again a single FPGA to consume roughly estimated about 10 Watt (see [Virf]) in combination with 168 Watt (see [BLAb]) for the network switch, the operating costs of the entire computation factory including its network connection can be computed for a single year and a electricity tariff of $0.2\,\frac{\$}{\text{kWh}}$ to:

$$\begin{aligned}
\text{operating costs}^{\,1\,\frac{\text{layer}}{\text{read-out}}} &= 157 \cdot 15 \cdot 0.01\,\text{kW} \cdot 365\,\text{d} \cdot 24\,\tfrac{\text{h}}{\text{d}} \cdot 0.2\,\tfrac{\$}{\text{kWh}} \\
&+ \quad 4 \cdot 0.17\,\text{kW} \cdot 365\,\text{d} \cdot 24\,\tfrac{\text{h}}{\text{d}} \cdot 0.2\,\tfrac{\$}{\text{kWh}} \\
&\approx \$\,42.4 \cdot 10^3
\end{aligned}$$

$$\begin{aligned}
\text{operating costs}^{\,4\,\frac{\text{layer}}{\text{read-out}}} &= 157 \cdot 11 \cdot 0.01\,\text{kW} \cdot 365\,\text{d} \cdot 24\,\tfrac{\text{h}}{\text{d}} \cdot 0.2\,\tfrac{\$}{\text{kWh}} \\
&+ \quad 4 \cdot 0.17\,\text{kW} \cdot 365\,\text{d} \cdot 24\,\tfrac{\text{h}}{\text{d}} \cdot 0.2\,\tfrac{\$}{\text{kWh}} \\
&\approx \$\,31.4 \cdot 10^3
\end{aligned}$$

$$\begin{aligned}
\text{operating costs}^{\,24\,\frac{\text{layer}}{\text{read-out}}} &= 157 \cdot 10 \cdot 0.01\,\text{kW} \cdot 365\,\text{d} \cdot 24\,\tfrac{\text{h}}{\text{d}} \cdot 0.2\,\tfrac{\$}{\text{kWh}} \\
&+ \quad 4 \cdot 0.17\,\text{kW} \cdot 365\,\text{d} \cdot 24\,\tfrac{\text{h}}{\text{d}} \cdot 0.2\,\tfrac{\$}{\text{kWh}} \\
&\approx \$\,28.7 \cdot 10^3
\end{aligned}$$

Besides this platform external network, it has to be collaterally remarked that the multi-chip versions require obviously two extra platform internal chip-to-chip network connections, which feature the transfer of the transformed hits from the input interface FPGA to the processor devices and the track candidates from the processor devices to the output interface FPGA (see

section 6.2.2 on page 408).
But before analyzing now the loads of each network, it is understood that the content of a single data package has to be clarified respectively before.

So according to a single job data package of the Playstation III implementation, which is introduced in section 5.6.2 on page 376, a single data package for the first network takes 60 bit, because it comprises 8 bit for the start index and 8 bit for the stop index of the histogram layer, 7 bit for the station index, 7 bit for the start position and 30 bit for the histogram command (see section 5.5.1 on page 328).
By matching this package size now to the introduced design goal, which is in turn determined by the processing of a single hit in every clock cycle, and further to the capability of the used FPGA device, which is introduced in section 3.3.1 on page 156, it is clear that a network bandwidth of $\frac{1\,\frac{\text{hit}}{\text{clock cycle}}\cdot 60\,\frac{\text{bit}}{\text{hit}}}{4\,\frac{\text{ns}}{\text{clock cycle}}} \approx 13.97\,\frac{\text{Gbit}}{\text{s}}$ or the requirement of at least four MGTs on the input interface FPGA and on each processor FPGA result.

Moreover the second network, which transfers the track candidates, requires apparently 22 bit for a single data package, because it has to provide 7 bit for the coordinate in the first dimension, 5 bit for the coordinate in the second dimension, 8 bit for the coordinate in the third dimension, which represents naturally the histogram layer, and actually 2 bit for the priority class.
Hence a bandwidth of $\frac{1\,\frac{\text{hit}}{\text{clock cycle}}\cdot 22\,\frac{\text{bit}}{\text{hit}}}{4\,\frac{\text{ns}}{\text{clock cycle}}} \approx 5.12\,\frac{\text{Gbit}}{\text{s}}$ is needed, which leads further to the reservation of two MGTs on the output interface FPGA and on each processor FPGA.

So finally it has to be explicitly highlighted that the bottleneck of the network concept is determined by the input interface FPGA, which contains also some histogram layers, because there is the requirement for connections to all available networks. But as the total number of calculated MGTs evaluates to nine, this bottleneck is uncritical, because the used FPGA device features up to 16 (see [Vird]).

By taking now the complexity and costs of such computation factories into account, these implementations would describe naturally a good solution to realize an on-line first level tracking system for the CBM experiment, even if each one is respectively quite expensive.

Cheapest Version Fulfilling the CBM Requirements This paragraph presents the results for an implementation, which is characterized by offering the best trade-off between the hardware resource consumption and the timing. While striving furthermore for this goal, section 3.3.1 on page 165 strongly advices this implementation to be defined by instantiating as many parallel histogram layers as the maximum number of layer entries, which have to be considered for a single hit, because then all hits have to be obviously processed maximally twice.
So based on this concept, the application of this strategy to the common sample event, whose distribution for the number of hit read-outs, which can be surely renamed to histogram layer entries, is shown earlier in figure 6.3 of section 6.2.2 on page 402, leads evidently directly to the instantiation of eight parallel layers. But as the general processor FPGA device, whose occupancy is depicted in figure 6.5, supports already more histogram layers and the result counts apparently just for the sample event, it is certainly better to use the amount of layers, which fit into a single chip. Hence the considered FPGA platform is quite simple and contains just three devices, which comprise one for the input interface, one for the processor and one for the output interface.

As this result can however only claim to realize the best trade-off between the required hardware resources and the timing, it is clear that the concept has to be analyzed more precisely. For this purpose, the analysis, which is introduced in section 4.9.1 on page 284, can be used to substantiate this suspicion by the evaluation of the average number of hit read-outs for a certain number of parallel instantiated histogram layers, because it is well-known from previous implementations that the hit read-out of the HBuffer unit and the hit processing have both a really huge effect on the timing.
Consequently a closer look to figure 6.8, which depicts the result of this analysis evaluated for the sample event, illustrates for instance that the instantiation of two histogram layers instead of a single one decreases the average number of hit read-outs by 40 %. Furthermore the result for eight layers and especially the result for the presented processor FPGA device exhibit that the made assumption leads to a quite good implementation, because the advantage of multiple parallel instantiated histogram layers decreases with the number of layers as expected. Thence the utilization of a unique processor FPGA forms a really effective trade-off between the resource consumption and the hit read-out performance.

Since the available parallelism levels, which are once again identified by the number of histogram layers and the number of processor FPGAs, are clarified up to here, the focus can be naturally changed in the following to the

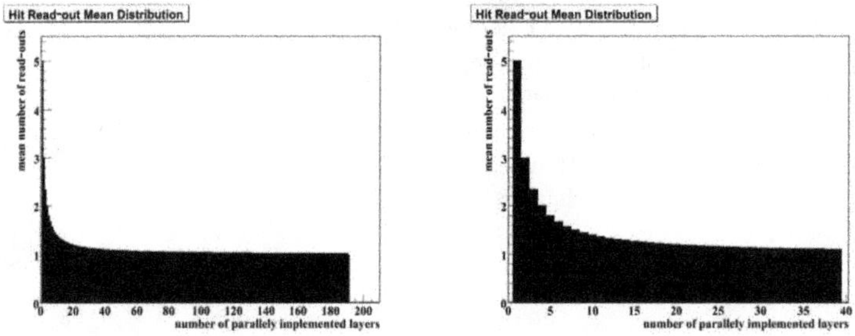

(a) Standard distribution　　(b) Region of interest

Figure 6.8: Distribution of the average number of read-outs for a single hit with regard to the parallel instantiated histogram layers

input interface device. For this purpose, the general timing of the Hough transform implementation, which is depicted in figure 6.1 of section 6.2.2 on page 396, has to be remembered, because not all histogram layers are anymore instantiated in parallel, and the HBuffer unit can be thus not removed from the input interface FPGA in contrast to the previous multi-chip version. Nevertheless a comparably fast implementation can be achieved by paying special attention to the restrictions of the HBuffer unit, which are surely determined by the pipeline stall and the maximum clock frequency. As however the maximum clock frequency leaves commonly no room for any easy optimization, the pipeline stall is examined in more detail at first.
So while remembering equation 6.5 of section 6.2.2 on page 399, it is immediately clear that the pipeline stall belongs only to the first fragment in the maximum function. But as the timing of the related algorithm parts is evidently fixed, it is clear that an improvement is impossible without changing the general structure. Nonetheless an easy and well-known concept is able to provide a solution regardless of this fact, because the quite low occupancy of the input interface FPGA allows apparently the instantiation of two identical HBuffer units in parallel, which can be then utilized simultaneously in alternating processing modes. Consequentially the mode of both units is of course swapped permanently, which means more precisely that one unit is always able to save new data, whereas the other one is ready to deliver its data to the consecutive units.

Even if a pipeline stall can be surely not generally avoided within this concept, it could offer anyhow a big advantage with regard to the throughput, because the related equation turns obviously from time(*step1*) +

time(*Read HBuffer*) into max(time(*step1*); time(*Read HBuffer*)).
Further on, this new equation demonstrates now that the time, which is consumed by the faster fragment of the maximum function, is completely hidden. This circumstance leads furthermore to the even more simplified formula time(*Read HBuffer*), because it is commonly senseless to save more data into the HBuffer unit than read-out again. Thence equation 6.5, which describes the throughput, changes collaterally into:

$$\begin{aligned}\text{throughput} \;=\; \max(\quad & \text{time}(\mathit{Read\ HBuffer}); \\ & \text{time}(\mathit{step2a}) + \text{time}(\mathit{step2b}) \\ + \;& \text{time}(\mathit{Read\ Serialization\ buffer}) \;\;)\end{aligned} \tag{6.19}$$

Moreover the resulting occupancy of the input interface FPGA, which implements this concept, is depicted in the following figure 6.9.

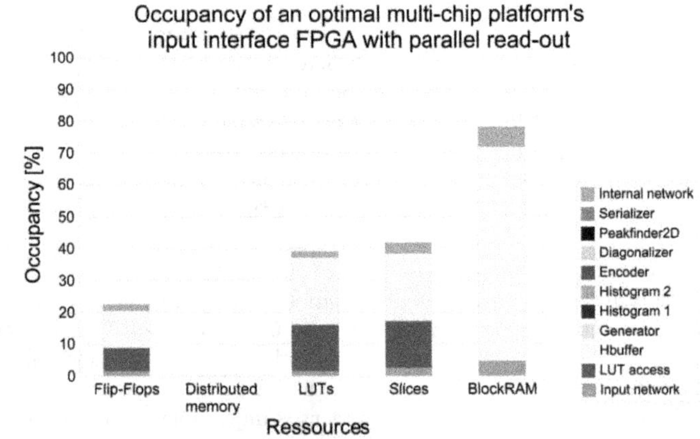

Figure 6.9: Occupancy of the input interface FPGA for the implementation with an optimal multi-chip platform

Beyond that, the center of interest is currently set to the maximum clock frequency, although its increasing is in the majority of cases very hard and commonly even not possible without changing the general structure of the algorithm. However a related improvement can be easily realized by applying another well-known general concept of memory usage, which is called memory bank interleaving, because it enables the HBuffer unit to just multiply the bandwidth. For this purpose, the existing memory is simply divided

into multiple banks with each containing the uniform distributed amount of entries from each list, because this circumstance enables then each bank to deliver data at the same time during the read-out process, which entails thus a multiplied data rate at the same clock frequency.
Coming now to further implementation details, a closer look to this modification shows immediately that no additional memory is needed, because the existing amount is just split into multiple separate banks. But in contrast to this circumstance, the list register set, which contains the link addresses of the previous entry in the lists, can not be retained, because these memory addresses are commonly not identical for different banks. So as each memory bank needs obviously its own list register set, the corresponding bit width can be easily determined by reducing the bit width of the existing list register set by the number of bits, which are necessary to differentiate between the banks. Hence the remaining amount of bits form the basement of the list register multiplication. In addition to this, the removed number of bits can be evidently used to build registers for the alternate access of the banks, because their usage will then guaranty the required prerequisite of a uniform distribution of the entries from the lists.

However this alternate access of the HBuffer memory introduces also a not negligible problem, which occurs during the data saving, because it is possible within this process that two consecutive data samples must be written into the same memory bank at the same time. Nonetheless an easy concept to soften this problem is naturally characterized by the initialization of the access registers with links to different memory banks for consecutive lists at startup. But as this care defines obviously not a sufficient solution, there must be an additional FIFO memory in front of each memory bank, which is able and big enough to buffer concurrent data. Moreover it is clear that the final topic in this concept is determined by a small state machine, which controls this interleaving process.

While relating now this interleaving strategy to the timing results of the previous implementations, it is self-evident that a twofold interleaving, which accomplishes two memory banks, is assumed to fit best. In spite of that, it is also clear that all these modifications can of course not exactly duplicate the bandwidth, but they come hopefully very close.

Anyway the focus is changed in the following to the corresponding timing behavior, because the resource consumption is successfully analyzed in detail. Therefore the general equations 6.1 and 6.4 in combination with the adapted equation 6.19 are evaluated with regard to the sample event, which is introduced in section 6.2.2 on page 394.
But before starting, the analysis, which is presented in section 4.9.1 on page

284, has to be consulted once again to get additional information about the hit read-out of the HBuffer unit as well as in the single-chip versions. However in contrast to these versions, it is apparently not adequate here to compute just the sum over the number of hits for all histogram layers, because a hit can be currently spread over multiple layers in the way that these layers are located in either the same or in different serial computation steps, which entails obviously a differing amount of read-outs for such hits. Although this problem is of course implicitly solved for the single-chip versions and the fast multi-chip versions, because a hits is either spread with its number of occurrence due to the existence of a single histogram layer or simply not spreadable due to the existence of all histogram layers in parallel, it is evidently quite hard to cope with this circumstance here.
Nevertheless figure 6.8, which is depicted earlier in this section, offers a simple solution, because the total number of hit read-outs for the single-chip versions can be for instance approximated by the multiplication of the average number of hit read-outs for a single instantiated histogram layer, which is about five, with the number of existing hits, which is 15,658 . As this result, which includes evidently just a small rounding error caused by the quantization of the figure, is furthermore certainly not surprising, this circumstance can be naturally used to deduce the following equation.

$$\begin{aligned}\text{number of read} - \text{outs} &= \sum_{i=1}^{\text{max. layers}} \left(1 + \frac{i-1}{\text{number of instantiated layers}}\right) \\ &\cdot \text{number of hits}_{(i\ layers)} \\ \text{number of read} - \text{outs} &= \sum_{i=1}^{191} \left(1 + \frac{i-1}{\text{number of instantiated layers}}\right) \\ &\cdot \text{number of hits}_{(i\ layers)}\end{aligned}$$

So based on this formula, the corresponding number of hit read-outs can be computed for the introduced processor FPGAs and the sample event to:

$$\begin{aligned}\text{number of read} - \text{outs}^{1\ \frac{\text{layer}}{\text{read-out}}} &= 20.1 \cdot 10^3 \\ \text{number of read} - \text{outs}^{4\ \frac{\text{layer}}{\text{read-out}}} &= 18.8 \cdot 10^3 \\ \text{number of read} - \text{outs}^{24\ \frac{\text{layer}}{\text{read-out}}} &= 18.3 \cdot 10^3\end{aligned}$$

In addition to this, the analysis, which is introduced in section 4.9.3 on page 291, has to be once again inquired as well to get more information about the track candidate read-out of the buffer, which is used in the Serialization unit. For that reason, figure 6.10 shows the distribution for the worst-case estimation of the FPGA Serialization time with regard to the parallel instantiated histogram layers based on the sample event.

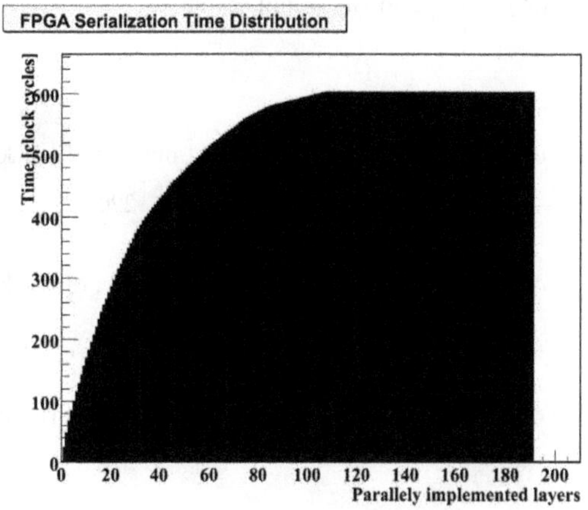

Figure 6.10: Distribution for the worst-case estimation of the FPGA Serialization time with regard to the parallel instantiated histogram layers based on the sample event

While having further section 6.2.2 on page 400 and in particular the definition of the number of track candidate read-outs in the single-chip versions in mind, it is self-evident that this strategy can be also used here. Thus the worst-case number of read-outs can be easily computed by the subsequent modified equation.

$$\text{number of read} - \text{outs} = \sum_{j=1}^{\text{inst. layers}} \max_j \left(\bigcup_{i=1}^{\substack{\text{max.}\\ \text{layers}}} \text{number of tracks}_{(layer\ i)} \right)$$

$$\text{number of read} - \text{outs} = \sum_{j=1}^{\text{inst. layers}} \max_j \left(\bigcup_{i=1}^{191} \text{number of tracks}_{(layer\ i)} \right)$$

So with regard to this concept, the major lack is obviously determined by the circumstance that not all maxima are located in consecutive histogram layers. Nonetheless the presented estimation is naturally applicable to compute a worst-case behavior, because there is commonly a track candidate concentration in some adjacent layers. For that reason, the corresponding worst-case number of track candidate read-outs can be computed for the introduced processor FPGAs and the sample event to:

$$\begin{aligned}
\text{number of read}-\text{outs}\ ^{1\ \frac{\text{layer}}{\text{read-out}}} &= \sum_{j=1}^{14} \max_j \left(\bigcup_{i=1}^{191} \text{number of tracks}_{(layer\ i)} \right) \\
&= 220 \\
\text{number of read}-\text{outs}\ ^{4\ \frac{\text{layer}}{\text{read-out}}} &= \sum_{j=1}^{20} \max_j \left(\bigcup_{i=1}^{191} \text{number of tracks}_{(layer\ i)} \right) \\
&= 286 \\
\text{number of read}-\text{outs}^{24\ \frac{\text{layer}}{\text{read-out}}} &= \sum_{j=1}^{24} \max_j \left(\bigcup_{i=1}^{191} \text{number of tracks}_{(layer\ i)} \right) \\
&= 323
\end{aligned}$$

Keeping these values now in mind, the timing in clock cycles for each step can be evaluated.

$$\begin{aligned}
\text{time}_{[\text{clock cycles}]}(step1) &= 15.7 \cdot 10^3 + 5 + 2 \\
&= 15.7 \cdot 10^3 \\
\text{time}_{[\text{clock cycles}]}^{1\ \frac{\text{layer}}{\text{read-out}}}(step2a) &= 20.1 \cdot 10^3 + 5 + 2 + \langle \tfrac{191}{14} \rangle \cdot 30 \\
&= 20.6 \cdot 10^3 \\
\text{time}_{[\text{clock cycles}]}^{4\ \frac{\text{layer}}{\text{read-out}}}(step2a) &= 18.8 \cdot 10^3 + 5 + 2 + \langle \tfrac{191}{20} \rangle \cdot 30 \\
&= 19.1 \cdot 10^3 \\
\text{time}_{[\text{clock cycles}]}^{24\ \frac{\text{layer}}{\text{read-out}}}(step2a) &= 18.3 \cdot 10^3 + 5 + 2 + \langle \tfrac{191}{24} \rangle \cdot 30 \\
&= 18.5 \cdot 10^3 \\
\text{time}_{[\text{clock cycles}]}^{1\ \frac{\text{layer}}{\text{read-out}}}(step2b) &= \langle \tfrac{191}{14} \rangle \cdot \langle \tfrac{14}{14} \rangle \cdot 95 + 1 + 2 + 3 + 3 + 1 \\
&= 1.3 \cdot 10^3 \\
\text{time}_{[\text{clock cycles}]}^{4\ \frac{\text{layer}}{\text{read-out}}}(step2b) &= \langle \tfrac{191}{20} \rangle \cdot \langle \tfrac{20}{5} \rangle \cdot 95 + 1 + 2 + 3 + 3 + 1 \\
&= 3.8 \cdot 10^3 \\
\text{time}_{[\text{clock cycles}]}^{24\ \frac{\text{layer}}{\text{read-out}}}(step2b) &= \langle \tfrac{191}{24} \rangle \cdot \langle \tfrac{24}{1} \rangle \cdot 95 + 1 + 2 + 3 + 3 + 1 \\
&= 18.3 \cdot 10^3
\end{aligned} \tag{6.20}$$

$$\begin{aligned}
\text{time}_{[\text{clock cycles}]}^{\mathit{1}\,\frac{\text{layer}}{\text{read-out}}}(\mathit{step3a}) &= 220+2 \\
&= 222 \\
\text{time}_{[\text{clock cycles}]}^{\mathit{4}\,\frac{\text{layer}}{\text{read-out}}}(\mathit{step3a}) &= 286+2 \\
&= 288 \\
\text{time}_{[\text{clock cycles}]}^{\mathit{24}\,\frac{\text{layer}}{\text{read-out}}}(\mathit{step3a}) &= 323+2 \\
&= 325
\end{aligned} \tag{6.20}$$

By using these numbers, the latency and the throughput can be further computed in clock cycles.

$$\begin{aligned}
\text{latency}_{[\text{clock cycles}]}^{\mathit{1}\,\frac{\text{layer}}{\text{read-out}}} &= \max(\,2\cdot 15.7\cdot 10^3+20.6\cdot 10^3+1.3\cdot 10^3; \\
&\quad 15.7\cdot 10^3+2\cdot(20.6\cdot 10^3+1.3\cdot 10^3)\,) \\
&= 59.5\cdot 10^3 \\
\text{latency}_{[\text{clock cycles}]}^{\mathit{4}\,\frac{\text{layer}}{\text{read-out}}} &= \max(\,2\cdot 15.7\cdot 10^3+19.1\cdot 10^3+3.8\cdot 10^3; \\
&\quad 15.7\cdot 10^3+2\cdot(19.1\cdot 10^3+3.8\cdot 10^3)\,) \\
&= 61.5\cdot 10^3 \\
\text{latency}_{[\text{clock cycles}]}^{\mathit{24}\,\frac{\text{layer}}{\text{read-out}}} &= \max(\,2\cdot 15.7\cdot 10^3+18.5\cdot 10^3+18.3\cdot 10^3; \\
&\quad 15.7\cdot 10^3+2\cdot(18.5\cdot 10^3+18.3\cdot 10^3)\,) \\
&= 89.2\cdot 10^3
\end{aligned} \tag{6.21}$$

$$\begin{aligned}
\text{throughput}_{[\text{clock cycles}]}^{\mathit{1}\,\frac{\text{layer}}{\text{read-out}}} &= \max(\,20.1\cdot 10^3; 20.6\cdot 10^3+1.3\cdot 10^3+220\,) \\
&= 22.1\cdot 10^3 \\
\text{throughput}_{[\text{clock cycles}]}^{\mathit{4}\,\frac{\text{layer}}{\text{read-out}}} &= \max(\,18.8\cdot 10^3; 19.1\cdot 10^3+3.8\cdot 10^3+286\,) \\
&= 23.2\cdot 10^3 \\
\text{throughput}_{[\text{clock cycles}]}^{\mathit{24}\,\frac{\text{layer}}{\text{read-out}}} &= \max(\,18.3\cdot 10^3; 18.5\cdot 10^3+18.3\cdot 10^3+323\,) \\
&= 37.1\cdot 10^3
\end{aligned} \tag{6.22}$$

Taking now the maximum achievable clock frequency of each handpicked Hough transform unit presented in section 5.5 on page 327 into account, the reasonable overall frequency decreases in contrast to the fast versions obviously again to 125 MHz, because the HBuffer unit reappears. However the introduced optimizations related to the HBuffer unit enable the application of two different clock domains for the design, which are determined by a frequency of 250 MHz ($4 \frac{\text{ns}}{\text{clock cycle}}$) for the general system and 125 MHz ($8 \frac{\text{ns}}{\text{clock cycle}}$) with a duplicated bandwidth for the HBuffer unit. Thence the equations 6.21 for the latency and 6.22 for the throughput can be evaluated to the absolute timing for the sample event.

$$\begin{aligned}
\text{latency}_{[\text{time}]}^{1\,\frac{\text{layer}}{\text{read-out}}} &= 59.5 \cdot 10^3 \,\frac{\text{clock cycles}}{\text{sample event}} \cdot 4\,\frac{\text{ns}}{\text{clock cycle}} \\
&= 59.5 \cdot 10^3 \,\frac{\text{clock cycles}}{\text{sample event}} \cdot \frac{4\,\mu\text{s}}{10^3\,\text{clock cycle}} \\
&\approx 237.81\,\mu\text{s} \\
\text{latency}_{[\text{time}]}^{4\,\frac{\text{layer}}{\text{read-out}}} &= 61.5 \cdot 10^3 \,\frac{\text{clock cycles}}{\text{sample event}} \cdot 4\,\frac{\text{ns}}{\text{clock cycle}} \\
&= 61.5 \cdot 10^3 \,\frac{\text{clock cycles}}{\text{sample event}} \cdot \frac{4\,\mu\text{s}}{10^3\,\text{clock cycle}} \\
&\approx 245.83\,\mu\text{s} \\
\text{latency}_{[\text{time}]}^{24\,\frac{\text{layer}}{\text{read-out}}} &= 89.2 \cdot 10^3 \,\frac{\text{clock cycles}}{\text{sample event}} \cdot 4\,\frac{\text{ns}}{\text{clock cycle}} \\
&= 89.2 \cdot 10^3 \,\frac{\text{clock cycles}}{\text{sample event}} \cdot \frac{4\,\mu\text{s}}{10^3\,\text{clock cycle}} \\
&\approx 356.80\,\mu\text{s}
\end{aligned} \tag{6.23}$$

$$\begin{aligned}
\text{throughput}_{[\text{time}]}^{1\,\frac{\text{layer}}{\text{read-out}}} &= 22.1 \cdot 10^3 \,\frac{\text{clock cycles}}{\text{sample event}} \cdot 4\,\frac{\text{ns}}{\text{clock cycle}} \\
&= 22.1 \cdot 10^3 \,\frac{\text{clock cycles}}{\text{sample event}} \cdot \frac{4\,\mu\text{s}}{10^3\,\text{clock cycle}} \\
&\approx 88.46\,\mu\text{s} \\
\text{throughput}_{[\text{time}]}^{4\,\frac{\text{layer}}{\text{read-out}}} &= 23.2 \cdot 10^3 \,\frac{\text{clock cycles}}{\text{sample event}} \cdot 4\,\frac{\text{ns}}{\text{clock cycle}} \\
&= 23.2 \cdot 10^3 \,\frac{\text{clock cycles}}{\text{sample event}} \cdot \frac{4\,\mu\text{s}}{10^3\,\text{clock cycle}} \\
&\approx 92.78\,\mu\text{s} \\
\text{throughput}_{[\text{time}]}^{24\,\frac{\text{layer}}{\text{read-out}}} &= 37.1 \cdot 10^3 \,\frac{\text{clock cycles}}{\text{sample event}} \cdot 4\,\frac{\text{ns}}{\text{clock cycle}} \\
&= 37.1 \cdot 10^3 \,\frac{\text{clock cycles}}{\text{sample event}} \cdot \frac{4\,\mu\text{s}}{10^3\,\text{clock cycle}} \\
&\approx 148.36\,\mu\text{s}
\end{aligned} \tag{6.24}$$

Further on, the combination of these throughput timing behavior results with the required processing time of an event from the CBM experiment, which is $0.40\,\frac{\mu s}{\text{event}}$, leads obviously immediately to the following necessary amount of parallel used systems in a computation factory.

$$\begin{aligned}
\text{number of systems}^{\mathit{1}\,\frac{\text{layer}}{\text{read-out}}} &= \frac{88.46\,\frac{\mu s}{\text{sample event}}}{0.40\,\frac{\mu s}{\text{event}}} \\
&= 222 \\
\text{number of systems}^{\mathit{4}\,\frac{\text{layer}}{\text{read-out}}} &= \frac{92.78\,\frac{\mu s}{\text{sample event}}}{0.40\,\frac{\mu s}{\text{event}}} \\
&= 232 \\
\text{number of systems}^{\mathit{24}\,\frac{\text{layer}}{\text{read-out}}} &= \frac{148.36\,\frac{\mu s}{\text{sample event}}}{0.40\,\frac{\mu s}{\text{event}}} \\
&= 371
\end{aligned}$$

Consequently the total device purchasing costs could be easily calculated subsequently, if the common FPGA price for the currently used device, which can be found in section 3.3.2 on page 184, is taken into account.

$$\begin{aligned}
\text{purchasing costs}^{\mathit{1}\,\frac{\text{layer}}{\text{read-out}}} &= 222 \cdot 3 \cdot \$\,1.6 \cdot 10^3 \approx \$\,1.1 \cdot 10^6 \\
\text{purchasing costs}^{\mathit{4}\,\frac{\text{layer}}{\text{read-out}}} &= 232 \cdot 3 \cdot \$\,1.6 \cdot 10^3 \approx \$\,1.1 \cdot 10^6 \\
\text{purchasing costs}^{\mathit{24}\,\frac{\text{layer}}{\text{read-out}}} &= 371 \cdot 3 \cdot \$\,1.6 \cdot 10^3 \approx \$\,1.8 \cdot 10^6
\end{aligned}$$

Since the timing of all implementations is successfully matched to the CBM experiment, it is clear that the network load, which defines evidently in turn an important benchmark as well as for all other versions, has to be set into the focus. However the checking is quite easy here, because the peak performance is naturally identical to the fast versions. Therefore the approximated $7.45\,\frac{\text{Gbit}}{\text{s}}$ leads apparently again to the usage of three MGTs. Moreover the average network load for the different versions is computed consecutively.

$$\begin{aligned}
\text{average network load}^{\mathit{1}\,\frac{\text{layer}}{\text{read-out}}} &= \frac{15.66 \cdot 10^3\,\frac{\text{hits}}{\text{sample event}} \cdot 32\,\frac{\text{bit}}{\text{hit}}}{88.46 \cdot 10^{-6}\,\frac{\text{s}}{\text{sample event}}} \approx 5.28\,\frac{\text{Gbit}}{\text{s}} \\
\text{average network load}^{\mathit{4}\,\frac{\text{layer}}{\text{read-out}}} &= \frac{15.66 \cdot 10^3\,\frac{\text{hits}}{\text{sample event}} \cdot 32\,\frac{\text{bit}}{\text{hit}}}{92.78 \cdot 10^{-6}\,\frac{\text{s}}{\text{sample event}}} \approx 5.03\,\frac{\text{Gbit}}{\text{s}} \\
\text{average network load}^{\mathit{24}\,\frac{\text{layer}}{\text{read-out}}} &= \frac{15.66 \cdot 10^3\,\frac{\text{hits}}{\text{sample event}} \cdot 32\,\frac{\text{bit}}{\text{hit}}}{148.36 \cdot 10^{-6}\,\frac{\text{s}}{\text{sample event}}} \approx 3.15\,\frac{\text{Gbit}}{\text{s}}
\end{aligned}$$

So if one considers further the combination of a network bridge for these three links to the common 10-Gigabit ethernet specification and cascaded 24-port

BLADE G8124 switches (see [BLAb] and [Blaa]), there would be naturally an amount of ten, eleven and 17 articles, which generate the following consequential purchasing costs.

$$\begin{aligned}
\text{switch purchasing costs}^{\mathit{1}\ \frac{\text{layer}}{\text{read-out}}} &= 10 \cdot \$\,12.0 \cdot 10^3 \approx \$\,119.5 \cdot 10^3 \\
\text{switch purchasing costs}^{\mathit{4}\ \frac{\text{layer}}{\text{read-out}}} &= 11 \cdot \$\,12.0 \cdot 10^3 \approx \$\,131.5 \cdot 10^3 \\
\text{switch purchasing costs}^{\mathit{24}\ \frac{\text{layer}}{\text{read-out}}} &= 17 \cdot \$\,12.0 \cdot 10^3 \approx \$\,203.2 \cdot 10^3
\end{aligned}$$

In addition to these purchasing costs, a lack of such a factory will be once again also defined by the operating costs, which are caused by the power consumption. As furthermore a closer look to the technical device specification details exhibits in turn a single FPGA to consume roughly estimated about 10 Watt (see [Virf]) in combination with 168 Watt (see [BLAb]) for the network switch, the operating costs of the entire computation factory including its network connection can be computed for a single year and a electricity tariff of $0.2\,\frac{\$}{\text{kWh}}$ to:

$$\begin{aligned}
\text{operating costs}^{\mathit{1}\ \frac{\text{layer}}{\text{read-out}}} &= 222 \cdot 3 \cdot 0.01\,\text{kW} \cdot 365\,\text{d} \cdot 24\,\tfrac{\text{h}}{\text{d}} \cdot 0.2\,\tfrac{\$}{\text{kWh}} \\
&+ \quad 10 \ \ \cdot 0.17\,\text{kW} \cdot 365\,\text{d} \cdot 24\,\tfrac{\text{h}}{\text{d}} \cdot 0.2\,\tfrac{\$}{\text{kWh}} \\
&\approx \$\,14.6 \cdot 10^3
\end{aligned}$$

$$\begin{aligned}
\text{operating costs}^{\mathit{4}\ \frac{\text{layer}}{\text{read-out}}} &= 232 \cdot 3 \cdot 0.01\,\text{kW} \cdot 365\,\text{d} \cdot 24\,\tfrac{\text{h}}{\text{d}} \cdot 0.2\,\tfrac{\$}{\text{kWh}} \\
&+ \quad 11 \ \ \cdot 0.17\,\text{kW} \cdot 365\,\text{d} \cdot 24\,\tfrac{\text{h}}{\text{d}} \cdot 0.2\,\tfrac{\$}{\text{kWh}} \\
&\approx \$\,15.4 \cdot 10^3
\end{aligned}$$

$$\begin{aligned}
\text{operating costs}^{\mathit{24}\ \frac{\text{layer}}{\text{read-out}}} &= 371 \cdot 3 \cdot 0.01\,\text{kW} \cdot 365\,\text{d} \cdot 24\,\tfrac{\text{h}}{\text{d}} \cdot 0.2\,\tfrac{\$}{\text{kWh}} \\
&+ \quad 17 \ \ \cdot 0.17\,\text{kW} \cdot 365\,\text{d} \cdot 24\,\tfrac{\text{h}}{\text{d}} \cdot 0.2\,\tfrac{\$}{\text{kWh}} \\
&\approx \$\,24.5 \cdot 10^3
\end{aligned}$$

Besides this, the supplementary required two platform internal chip-to-chip network connections, which are already introduced in section 6.2.2 on page 408 to feature the transfer of the transformed hits from the input interface FPGA to the processor devices and the track candidates from the processor devices to the output interface FPGA, are indicated according to section 6.2.2 on page 413 by 60 bit for a single data package of the first network and 22 bit for a single data package of the second network.
So based on this analogousness, it is self-evident that the necessary bandwidth of approximately $13.97\,\frac{\text{Gbit}}{\text{s}}$ results again in the application of at least

four MGTs for the first network on the input interface FPGA and on each processor FPGA, whereas the bandwidth of about $5.12\,\frac{\text{Gbit}}{\text{s}}$ for the second network ends in the reservation of two MGTs on the output interface FPGA and on each processor FPGA.

Beyond that, it has to be finally noted that the bottleneck, which is introduced in section 6.2.2 on page 413, is not existing anymore, because the input interface FPGA is quite different in the functionality, even though the platform internal chip-to-chip network connections are identical. Admittedly the recent bottleneck is now located on the processor FPGA, because this device is the only one, which needs access to both networks. But as the total number of calculated MGTs evaluates to six, this bottleneck is in turn uncritical, because the used FPGA device features up to 16 (see [Vird]).

By taking now the complexity and costs of such computation factories into account, these implementations define obviously the best available group to realize an on-line first level tracking system for the CBM experiment.

6.2.3 Sony Playstation III

In contrast to the earlier described systems, the one presented in this section is naturally not that complex, because the simple recapitulation of the result, which is revealed in section 5.6 on page 370, exhibits directly that the fastest implementation of the adapted and customized Hough transform algorithm on a Sony Playstation III evaluates the sample event in less than 23 ms.
As this version would furthermore use only vector arithmetics, even if the scalar Encoding step is faster (see section 5.6.4 on page 383) and a mixture of both is apparently possible due to the well-known memory endianness, the required time to process the sample event can be obviously reduced onwards to less than 21.5 ms.

So the combination of this timing with the required processing time of an event from the CBM experiment, which is $0.40\,\frac{\mu\text{s}}{\text{event}}$, entails evidently further a exigency of $\frac{21.50\,\frac{\text{s}}{\text{sample event}}}{0.40\,\frac{\mu\text{s}}{\text{event}}} = 53.8 \cdot 10^3$ parallel used Sony Playstation III systems in a computation factory. Consequentially the total costs could be easily calculated to $53.8 \cdot 10^3 \cdot \$\,0.4 \cdot 10^3 \approx \$\,21.4 \cdot 10^6$, if the price of such a Sony Playstation III system, which can be found in section 3.3.2 on page 184, is taken into account.

Besides this, another important benchmark would be certainly determined once again by the network load, if the statement in section 3.3.2 on page 190 is withdrawn. So when assuming this circumstance, the usability of such a

computation factory has to be surely checked against the required bandwidth of the CBM experiment, which is 200 $\frac{\text{GByte}}{\text{s}}$ or $1.6 \cdot 10^3 \, \frac{\text{Gbit}}{\text{s}}$. For that reason, the network specification of a Sony Playstation III, which can be found in section 3.3.2 on page 190, results in a need of at least $\frac{1.6 \cdot 10^3 \, \frac{\text{Gbit}}{\text{s}}}{1 \, \frac{\text{Gbit}}{\text{s}}} = 1.6 \cdot 10^3$ parallel used systems to fulfill this requirement. But as the number of systems, which are necessary because of the processing time, is much higher and the time to receive the data of the sample event is just $\frac{15.66 \cdot 10^3 \, \frac{\text{hits}}{\text{sample event}} \cdot 32 \, \frac{\text{bit}}{\text{hit}}}{1 \, \frac{\text{Gbit}}{\text{s}}} \approx 0.47 \, \frac{\text{ms}}{\text{event}}$, which is also approximately 50 times smaller than the processing time, the only decisive significant evaluation criterion is of course uniquely defined by the processing time.
Nevertheless the physical network connection is naturally an essential and demanding task, because each Sony Playstation III system requires evidently an own network link. So if one considers further cascaded 48-port NETGEAR GS748AT network switches (see [NETa] and [NETb]), there would be an amount of 1,169 articles, which generate additional purchasing costs of $1.2 \cdot 10^3 \cdot \$\,1.3 \cdot 10^3 \approx \$\,1.5 \cdot 10^6$.

Beyond that, a lack of such a factory will be in turn collaterally determined by the operating costs, which are caused by the power consumption. As furthermore a closer look to the technical system specification details exhibits a single Sony Playstation III system to consume about 140 Watt (see [Grä08]) in combination with 67 Watt (see [NETa]) for the network switch, the operating costs of the entire computation factory including its network connection can be computed for a single year and a electricity tariff of $0.2 \, \frac{\$}{\text{kWh}}$ to:

$$\begin{aligned} \text{operating costs} &= 53.8 \cdot 10^3 \cdot 0.14\,\text{kW} \cdot 365\,\text{d} \cdot 24\,\tfrac{\text{h}}{\text{d}} \cdot 0.2\,\tfrac{\$}{\text{kWh}} \\ &+ 1.2 \cdot 10^3 \cdot 0.07\,\text{kW} \cdot 365\,\text{d} \cdot 24\,\tfrac{\text{h}}{\text{d}} \cdot 0.2\,\tfrac{\$}{\text{kWh}} \\ &\approx \$\,13.3 \cdot 10^6 \end{aligned}$$

So due to the complexity and costs of such a computation factory, such an implementation seems to be as unfeasible to realize an on-line first level tracking system for the CBM experiment as the implementation, which bases on common personal computers. But in contrast to this one, a single Sony Playstation III system features additionally the possibility to analyze the applicability of the algorithm's parallelism capabilities in combination with their levels.

Chapter 7

Outlook

This section presents a short prospect about the not yet touched improvement reservoirs of the recently introduced Hough tracking algorithm in combination with the resulting assumed benefits.
But before going into details, the conclusions, which are extensively illustrated in section 6.2 on page 392 and briefly summarized in table 7.1, have to be remembered, because all improvements are naturally required to suit at least the most applicable tracking implementation for the planned CBM fixed-target experiment at the FAIR, which is obviously advised to be the three chip platform FPGA architecture.
By the way, it has to be also mentioned here that a prototype of such a system is actually not constructed and describes thus an additional interesting topic.

Being now more specific, section 6.1 on page 391 has already refined the most simple realizable forms of improvement to the Hough tracking performance by using more appropriate LUTs as introduced exemplarily in section 3.2.3 on page 134, by utilizing a non-equidistant quantization as addressed in section 4.6.3 on page 251 as well as by applying successive improvement algorithms. Incidentally the Hough tracking was suggested in a first proposal to generate the initial track candidates or so-called seeds for a Kalman filter or the Cellular Automaton, which is outlined in section 1.4.3 on page 40, but the quite huge effort for only seed generation rapidly overcomes this idea.

Anyway the ability to use fast successive algorithms, which offer the possibility to improve the performance of the Hough tracking based on the computed results, is evidently given without loss of generality. But since the major prerequisite of all these algorithms is obviously specified by the affiliation of hits to each final Hough tracking track candidate and/or vice versa, which can

Architecture	Time for a single event	Required platforms for CBM	Purchasing costs for the platforms
PC: P4@3GHz and 2GB RAM; not optimized CBMROOT simulation	2.36 s	$5.9 \cdot 10^6$	\$ $2.8 \cdot 10^9$
Playstation III: CellBE@3.1GHz, 256MB RAM and 256kB LS; vectorized, 6SPUs	21.50 ms	$53.8 \cdot 10^3$	\$ $21.4 \cdot 10^6$
1 FPGA: xc5vlx110-ff1760-3@125MHz; implementing 1 histogram layer	752.18 μs	$1.9 \cdot 10^3$	\$ $3.0 \cdot 10^6$ **(only chips)**
15 FPGAs: xc5vlx110-ff1760-3@250MHz; implementing all histogram layers	62.80 μs	157	\$ $2.5 \cdot 10^6$ **(only chips)**
3 FPGAs: xc5vlx110-ff1760-3@250MHz; implementing 14 histogram layers	88.46 μs	222	\$ $1.1 \cdot 10^6$ **(only chips)**

Table 7.1: Executive summary to the conclusions of all diverse implementations

be basically not satisfied by the recently implemented Hough transform, a supporting procedure must be certainly evaluated before the optimizing algorithm can proceed. Hence two different models are considered for this purpose, which can be characterized by LUT usage or peak reconstruction.

In this connection, a closer look shows now immediately that the easier one is formed by the application of an extra LUT, whose content is defined by a computational propagated hit for each detector station of each potential Hough tracking track candidate. Consequently each that way evaluated hit of a final track candidate has to be additionally utilized for a subsequent hit

search in the sorted input data to identify the corresponding real existing hit and to affiliate this one onward to the actually processed final track candidate.
So based on this concept, it is directly clear that the two main disadvantages of this elemental model are determined by the definition of the hit correlation in the search process and the LUT size, because this one can easily explode due to the dependency on the number of detector stations and the quantization of the histogram. However the generation of this LUT is surprisingly facile, because the initial step of the Hough transform LUT generation based on the Runge-Kutta approach, which is presented in section 3.2.3 on page 134, already includes all required computations.

In contrast to this, the second model avoids exactly these disadvantages by the computational reconstruction of the concrete peak composition for each histogram cell, which corresponds to a final track candidate. So consequentially the hits, which are attached to such a track candidate, are intrinsically assured to contribute to the peak in the corresponding cell, but the existence of a unique hit for each detector station can be evidently not guaranteed, because the quality of the peak influences directly the affiliation. In particular, reality leads often to multiple hits, which are attached to a track candidate for a single detector station. Therefore another procedure seems to be needed, which is able to cope with this situation. But later on, it can be seen that an expedient solution can be easily integrated into the successive algorithms.
Contemplating now the hardware implementation, the design of the HBuffer unit, which is described in section 3.3.1 on page 160, apparently volunteers to be reused twice to store all hits in one unit and all final track candidates in the other unit, because both data sets are then sorted with regard to the histogram layer. Further on, this circumstance can be thereafter used to apply a fundamental check for the hit contribution to the final track candidate, which utilizes the following equation 7.1 to simply compute the intersection of the Hough curve for each hit in each layer with the histogram cell of all layer-correspondent track candidates in the original histogram geometry.

$$dim1_{track} == startIndex_{hit} + \sum_{i=0}^{dim2_{track}} cmd_{hit}(i) \qquad (7.1)$$

Since both models for the affiliation of hits to track candidates are basically illustrated, the focus has to be obviously moved now to the improvement algorithms themselves.

Consequently the first and most simple one to introduce realizes just a well-known thresholded smoothness computation for each track candidate trajec-

tory including the marking of the affiliated hits, which disables them naturally for other track candidates.
So within this conceptual design, it is immediately clear that such a process will probably improve the performance of the Hough tracking, because some track candidates will not own a smooth enough hit sampled trajectory or not any longer enough hits.

Besides this, a further enhancement to this algorithm leads to a second more developed approach, which is mainly characterized by the insertion of a usage frequency counter for each hit in the affiliation procedure, because such a counter is evidently able to give additional information about a separate quality aspect of a final track candidate. For instance, figure 7.1 depicts the distribution of the number of track candidates with regard to the sum of the smallest hit usage frequency for each detector station based on the typical sample event introduced in section 6.2 on page 392. However it has to be also noticed in this context that the lack of hits in a detector station defines surely an essential problem for the quality comparison of these sums, because the quality is commonly assumed to be better the smaller the sum is. But this circumstance is furthermore not crucial, because the utilization of a high predefined value for the imaginary frequency of such a missing hit, which depends apparently on the quantification of the allowed signatures for the tracks, describes certainly a pretty easy and universally known solution.

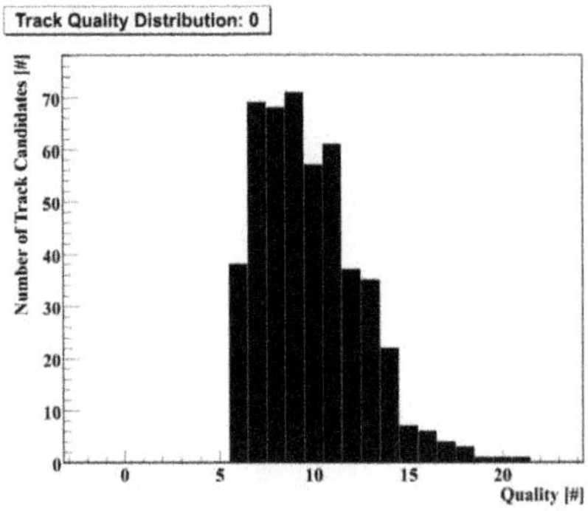

Figure 7.1: Illustration of a typical distribution for the accumulated quality aspect of each track candidate related to the sum of the smallest affiliated hit usage frequency for each detector station

So coming now to the description of the algorithm, it is obvious that the processing of these ordered track candidates has to be started with the ones possessing the smallest sum index, because these ones are assumed to own the best quality and are thus most probable to represent a real track. Despite or rather on account of that simplicity, it has to be also directly highlighted that even if these indexes are calculated incipiently, the index of each track candidate would have to be recomputed and checked for recent correctness before finishing, because the hit marking caused by a previously accepted track candidate can evidently invalidate hits, which are as well used to calculate the initial sum index of the actually examined track candidate. Thence one of two different successive data paths is taken by a surveyed track candidate in dependency on the binary decision for the correctness. Furthermore it is not really surprising in this context that the identity of the recomputed and the initial sum enables the data path, which is designated by accepting the track candidate and marking the correspondent hits, whereas the disparity leads to the reordering of this track candidate into the new computed subsequent sum index. Moreover the end of processing is consequentially identified by either lacking for track candidates or exceeding a threshold for the sum index, which is commonly related to the quality of the peaks.

Beyond that, another more complex but noticeable enhancement to this algorithm might be also given by the supplementary introduction of a dummy fake hit usage frequency counter, which is incremented for all available fake hit possibilities of an occurring real hit, because these dummy fake hits can be then matched to the real occurring hits and combined somehow with the ordinary hit usage frequency counter.

Anyway it has to be finally noted that the main problem of the approaches on the basis of the hit usage frequency counter is of course determined by the selection of the hit for a detector station, which is used to calculate the sum or quality of a track candidate, because the smallest hit usage frequency is apparently not related to any sampled smooth motion trajectory. However a simplified Kalman filter, which is implemented in [Tra06] for example, can be possibly applied to select an optimal hit for each single detector station with regard to such a motion trajectory before the evaluation of the sum index or quality is realized. Even though originating by this circumstance, it has to be additionally noticed that such an advancement would lead collaterally to another level of track candidate qualification, which is able to cope with the attachment of multiple hits to a single detector station.

Appendix A

The Package HTRACK

This chapter presents a detailed description of the CBMROOT package HTRACK (see also section 3.2.2 on page 120). For this purpose, a closer look to the interface is taken, before the parameters of the algorithm are exhibited. At this juncture, it has to be noticed yet that the parameters are further divided into three different categories. In addition to this, it has to be realized that the membership of a parameter to such a category is determined by the way of changing its value. So the first group is evidently formed by the parameters, which are accessible via a configuration file, the second set is determined by the availability via special table files and the third block is built by define statements in the source code. As this is obviously not the total story, the last section contains finally two example scripts for the tagged version FEB07 of the CBMROOT framework, which include respectively one simulation and one reconstruction chain applying the module HTRACK for tracking in the STS.

A.1 Class *CbmHoughStsTrackFinder*

At first it has to be noticed here that the general task, which is responsible for the STS tracking in the reconstruction chain of the CBMROOT framework is called *CbmStsFindTracks*. For that reason, it is clear that the input to this task comprises the detector geometry, the magnetic field and a *TClonesArray* object including the hits. In addition to this, another *TClonesArray* object, which has to contain finally the output tracks, exists obviously anyway. Furthermore this task uses an engine to implement the precise tracking algorithm, which must be derived from the abstract base class *CbmStsTrack-*

finder and setup by using the function *void UseFinder(CbmStsTrackFinder* finder)*.

So within this context, it is self-evident that the main STS tracking class, which is named *CbmHoughStsTrackFinder* and located in the folder htrack-/CbmHoughStsTrackfinder, is derived from this base class. Hence the three virtual functions, *void Init()* for the initialization, *int DoFind()* for the tracking algorithm itself returning the found number of tracks and *void Finish()* for cleaning up the task, have to be implemented. At this juncture, it is very important to realize that the initialization is done just once at startup, while the tracking and cleaning is done for each event (see event simulation).
Besides this, the major important parameter of the base class is used to set a verbosity level, which controls the displaying of information on the standard output screen. So this means more precisely that the level zero provokes the task to run in quite mode, the level one configures the task to show only messages on the event level, the level two represents the track level and all higher levels are reserved for debugging purpose.
Beyond these circumstances, which are caused by the inheritance from the abstract base class, it has to be also noticed here that the Hough tracking offers some additional functionality.
For instance, it would be of course imaginably good, if the three dimensional histogram, which identifies the key element of the algorithm, can be visualized somehow. Furthermore life is not made easier by the signature of a histogram cell, because it forms collaterally a fourth dimension. But as such a visualization is nevertheless essential, the same solution than for the memory resource problem of the histogram is applied here. Thus it is possible to draw a single histogram layer, which is specified by the parameters 'eventNumber' and 'layer', in a graphical window on the standard output screen by using the function named *void ShowHistogramLayer(unsigned int eventNumber, unsigned short layer, const char* drawOption)*. In this connection, it is obvious that all layers can not be shown in parallel, because they require too much display resources. Moreover it is of course self-evident that the specified layer must exist in the actual analysis environment, which implies that the analysis has to be turned on by using the corresponding parameters in the configuration file (see appendix A.2 on page A-4). By the way, it should be just touched here that this analysis can be used to display the modified histogram after each algorithm step, which determines then also the number of graphical windows on the screen displaying apparently the same layer.
Finally the last mentionable function is named *void SetOutputFile(const char* name)*, because it sets a separate file to place the output of the enabled analysis. However if this optional function call is missing, the output

of the analysis would be simply written in the same file than the output of the reconstruction, which consists of the tracking results.

Since all necessary functions have been successfully explained now, the focus can be set to the constructors of the class, which are summarized in the following. But before having a closer look, it has to be realized that there are four different ones due to the algorithm parameters, whose value can be modified by them. So the first common constructor is of course the default constructor, which uses all standard parameters being not able to change a single value.

```
CbmHoughStsTrackFinder();
```

Further on, the second constructor uses a simple configuration file, whose name has to be committed, to set the parameters of the algorithm.

```
CbmHoughStsTrackFinder(const char* name);
```

The next constructor increases in complexity, because it allows to overwrite the standard algorithm parameters for those ones, which are delivered to the constructor. Further on, it is clear that these parameter names are identical to the ones in the configuration file and their meaning can be thus found in appendix A.2 on page A-4.

```
CbmHoughStsTrackFinder(bitArray inputDetectorMask,
    int inputMinTracks, int inputMaxTracks,
    unsigned short inputMagneticFieldIntegration-
        StepwidthPerStation,
    int trackfinderGammaStep,
    int trackfinderThetaStep,
    int trackfinderLutRadiusStep,
    unsigned short trackfinderFirstFilterNeighborhood-
        Dim1ClearRadius,
    unsigned short trackfinderFirstFilterNeighborhood-
        Dim2ClearRadius,
    unsigned short trackfinderSecondFilterNeighborhood-
        Dim1ClearRadius,
    unsigned short trackfinderSecondFilterNeighborhood-
        Dim2ClearRadius,
    unsigned short trackfinderSecondFilterNeighborhood-
        Dim3ClearRadius,
```

```
    bool initStatus,
    const char* analysisOutputFileName);
```

Finally the last but not least constructor applies the parameter values, which are read from a configuration file, and overwrites the values of these ones, which are additionally passed to the constructor.

```
CbmHoughStsTrackFinder(const char* name,
    bitArray inputDetectorMask,
    int inputMinTracks, int inputMaxTracks,
    unsigned short inputMagneticFieldIntegration-
        StepwidthPerStation,
    int trackfinderGammaStep,
    int trackfinderThetaStep,
    int trackfinderLutRadiusStep,
    unsigned short trackfinderFirstFilterNeighborhood-
        Dim1ClearRadius,
    unsigned short trackfinderFirstFilterNeighborhood-
        Dim2ClearRadius,
    unsigned short trackfinderSecondFilterNeighborhood-
        Dim1ClearRadius,
    unsigned short trackfinderSecondFilterNeighborhood-
        Dim2ClearRadius,
    unsigned short trackfinderSecondFilterNeighborhood-
        Dim3ClearRadius,
    bool initStatus,
    const char* analysisOutputFileName);
```

Since the whole tracking interface is presented in detail, a closer look to the parameters of the algorithm is offered in the following.

A.2 Configuration File

This section presents all details, which are necessary to understand the concepts featuring the setting of the value for each parameter of the Hough tracking algorithm. For this purpose, general information is exhibited, before the actual parameters are described themselves.
So the first major important circumstance is obviously determined by the location of the file, which contains the corresponding class implementation, because this file located in the folder htrack/CbmHoughStsTrackfinder has to

be evidently adapted to be able to administrate the recently chosen parameter set.
Within this context, the next essential fact to mention is apparently identified by the complexity of this parameter set, because there are recently about 160 parameters, which can be divided into data flow parameters, parameters of the algorithm itself and parameters for analyses.

Besides this, the actual class implementing the configuration file offers two different concepts, whose priority is of course implicitly defined by the usage, to set the value of each parameter.
In this connection, the first one, which realizes commonly the standard way to backup unset parameters, is naturally identified by an existing default value in the source code. For this purpose, a define statement, which is characterized by the convention of concatenating the prefix 'defVal' with the parameter's name, has to exist for each available parameter to fix a default value.
But as this concept requires of course a recompilation for every change of a parameter value, the second one, which is characterized by the usage of an external ASCII formatted configuration file, should be preferred, because a recompilation is then unnecessary. At a first glance, this circumstance seems to play a less important role, but the more often the tracking has to run with different parameter values, the more favorably becomes the usage of such a configuration file.

Beyond that, the first two parameters in the source code, which are named 'defFileName' and 'writeDefaultHeader', enable the availability of the external ASCII formatted configuration file, because the first one would define the default file name, if it is not supplied by the program call, while the second one would enable the writing of a default configuration file, if the actually set file does not exist. Hence a startup from scratch, which means no existing external configuration file, will take always the default values of the source code. But this first run could offer the possibility to write a configuration file containing the default values for each parameter, if the parameter flag 'writeDefaultHeader' is enabled. Afterwards, it is clear that the values of this existing file would be used, if the file exists. Thus the default values in the source code are omitted except for the parameters, which are missing in the configuration file due to a deletion.
Finally it has to be mentioned here that each value of a parameter, which belongs to the specification of a file name, can be configured to contain an absolute path or a relative path. At this juncture, it is self-evident that a relative path has to start always with ./, ../ or without any slashes. However the base for a relative path without slashes is the current working direc-

tory, while the starting point for a relative path with slashes is set by the environment variable $VMCWORKDIR. But if this environment variable is furthermore not set, the current working directory would be also used for the base of this version.

Since the necessary general information is exhaustively illustrated, the next part of this section lists all parameters of the Hough tracking algorithm including their meaning, their standard value, their value range and their associated parameters. Further on, all of them can be apparently adapted in their complete range, but the existing analysis, which are introduced in section 4 on page 213, are advised to find an optimal value.

So coming now to the point, the data flow parameters are following.

Name	**inputFileName**
Meaning	This parameter holds the name of the file consisting of the input data like MC Points and/or hits and/or geometry. **(Deprecated API)**
See	inputFileFormat

Table A.1: Specification of the data flow parameter 'inputFileName'

Name	**inputFileFormat**
Meaning	This parameter sets the correct reader for the input file belonging to the format of the file. (Deprecated API)
Standard	ROOTFILEFORMAT
Range	{ ASCIIFILEFORMAT, ROOTFILEFORMAT, ROOT2-FILEFORMAT }
See	inputFileName

Table A.2: Specification of the data flow parameter 'inputFileFormat'

Name	**inputMinTracks**
meaning	This parameter sets the minimal number of MC Tracks, which one event (see event simulation) must have to be taken into account for tracking. If there are less tracks, the event would be skipped and the next one would be processed.
Standard	0
Range	$[-2^{31}; +2^{31}-1]$
Format	signed decimal number
See	inputMaxTracks

Table A.3: Specification of the data flow parameter 'inputMinTracks'

Name	**inputMaxTracks**
Meaning	This parameter sets the maximal number of MC Tracks, which one event (see event simulation) is allowed to have to be taken into account for tracking. If there are more tracks, the event would be skipped and the next one would be processed. If inputMaxTracks is not bigger than inputMinTracks, each event would be processed independent of the number of MC Tracks.
Standard	0
Range	$[-2^{31}; +2^{31}-1]$
Format	signed decimal number
See	inputMinTracks

Table A.4: Specification of the data flow parameter 'inputMaxTracks'

Name	**inputFileNameMagneticField**
Meaning	This parameter holds the name of the file consisting of the information for the magnetic field.
Standard	"../input/FieldActive.root"
See	inputMagneticFieldIsRootFile, inputMapNameMagneticField, inputMagneticFieldIntegrationStepwidthPerStation, inputMagneticFieldIntegrationFactor, inputDisableAutomaticMagneticField

Table A.5: Specification of the data flow parameter 'inputFileNameMagneticField'

Name	**inputMagneticFieldIsRootFile**
Meaning	This parameter defines the format of the file consisting of the magnetic field. If it is true, the file must be in ROOT format. Else the file is assumed to be in a special ASCII format. **(ASCII is deprecated, just ROOT is supported in the future)**
Standard	"true"
Range	{ "false"; "true" } or { 0; 1 }
Format	string or decimal
See	inputFileNameMagneticField, inputMapNameMagneticField, inputMagneticFieldIntegrationStepwidthPerStation, inputMagneticFieldIntegrationFactor, inputDisableAutomaticMagneticField

Table A.6: Specification of the data flow parameter 'inputMagneticFieldIsRootFile'

Name	**inputMagneticFieldIntegrationStepwidthPerStation**
Meaning	This parameter sets the number of factors per station, which are used to integrate the magnetic field numerically with the Simpson method. So if there is a certain number of factors used for the first station, there would be twice as much factors for the second one and so on. at this juncture, an odd number will be made even by adding one, because the Simpson method needs an even number of factors.
Standard	10
Range	$[\,0\,;\,+\,2^{16}-1\,]$
Format	unsigned decimal number
See	inputFileNameMagneticField, inputMagneticFieldIsRootFile, inputMapNameMagneticField, inputMagneticFieldIntegrationFactor, inputDisableAutomaticMagneticField

Table A.7: Specification of the data flow parameter 'inputMagneticFieldIntegrationStepwidthPerStation'

Name	**inputDisableAutomaticMagneticField**
Meaning	This parameter offers the possibility to disable the magnetic field, which is delivered from the framework, and to enable the usage of another one or no one instead.
Standard	"false"
Range	{ "false"; "true" } or { 0; 1 }
Format	string or decimal
See	inputFileNameMagneticField, inputMagneticFieldIsRootFile, inputMapNameMagneticField, inputMagneticFieldIntegrationStepwidthPerStation, inputMagneticFieldIntegrationFactor

Table A.8: Specification of the data flow parameter 'inputDisableAutomaticMagneticField'

Name	**inputMagneticFieldIntegrationFactor**
Meaning	This parameter offers the possibility to weight the magnetic field factors, which are used in the analytic tracking formula. This implies of course the direct weighting of the generated factors or the computed integral of the magnetic field.
Standard	1.0
Range	$[\,-1.7 \cdot 10^{308}\,;\,+1.7 \cdot 10^{308}\,]$
Format	signed decimal number
See	inputFileNameMagneticField, inputMagneticFieldIsRootFile, inputMapNameMagneticField, inputMagneticFieldIntegrationStepwidthPerStation, inputDisableAutomaticMagneticField

Table A.9: Specification of the data flow parameter 'inputMagneticFieldIntegrationFactor'

Name	**inputMapNameMagneticField**
Meaning	This parameter holds the name of the field map to identify the magnetic field in the ROOT based file.
Standard	"NewMap"
See	inputFileNameMagneticField, inputMagneticFieldIsRootFile, inputMagneticFieldIntegrationStepwidthPerStation, inputMagneticFieldIntegrationFactor, inputDisableAutomaticMagneticField

Table A.10: Specification of the data flow parameter 'inputMapNameMagneticField'

Name	**inputNumberOfVolumesInfrontOfSTS**
Meaning	This parameter defines the number of geometric volumes, which are established in front of the STS volumes. It is needed in an old version of the framework to compute the correct volumeID. **(Deprecated API)**
Standard	10
Range	$[\,-2^{31};\,+2^{31}-1\,]$
Format	signed decimal number
See	inputDetectorMask, inputFileNameDetector, inputDisableAutomaticDetector

Table A.11: Specification of the data flow parameter 'inputNumberOfVolumesInfrontOfSTS'

Name	**inputFileNameDetector**
Meaning	This parameter holds the name of the ASCII file consisting of the geometry for the detector **(Deprecated API)**
Standard	"../geometry/sts_station_2sts_4strips.dat"
See	inputDetectorMask, inputNumberOfVolumesInfrontOfSTS, inputDisableAutomaticDetector

Table A.12: Specification of the data flow parameter 'inputFileNameDetector'

Name	**inputDisableAutomaticDetector**
Meaning	This parameter offers the possibility to disable the detector geometry, which is committed from the framework, and to enable the usage of another one instead.
Standard	"false"
Range	{ "false"; "true" } or { 0; 1 }
Format	string or decimal
See	inputDetectorMask, inputFileNameDetector, inputNumberOfVolumesInfrontOfSTS

Table A.13: Specification of the data flow parameter 'inputDisableAutomaticDetector'

Name	**inputDetectorMask**
Meaning	This parameter can completely remove detector planes from tracking. That means that this value is seen binary with one bit representing one plane. Furthermore the position of the bit represents the position of the station. So if such a bit has the value one, this plane including all hits in this station would be removed and skipped from tracking. If there are more stations than bit positions, the assumed value for the missing bits is evidently always zero.
Standard	0
Range	$[\,0\,;\,2^{maxDimSH}-1\,]$ (see appendix A.3 on page A-79)
Format	radix format (see appendix B on page B-1)
See	inputFileNameDetector, inputNumberOfVolumesInfrontOfSTS, inputDisableAutomaticdetector

Table A.14: Specification of the data flow parameter 'inputDetectorMask'

Name	**inputCodingTableMode**
Meaning	This parameter sets the mode for the table used for the Encoding of the histogram (see appendix A.4 on page A-101).
Standard	FILETABLE
Range	{ NOTABLE, CROSSFOOTTABLE, ONLINETABLE, FILETABLE, LUTGOODNESSTABLE }
Format	signed decimal number
See	inputCodingTableFileName, inputCodingTableWrite

Table A.15: Specification of the data flow parameter 'inputCodingTableMode'

Name	**inputCodingTableFileName**
Meaning	This parameter holds the name of the file consisting of the table, which is used for the Encoding of the histogram. (see appendix A.4 on page A-101)
Standard	"../../input/codingTable.txt"
See	inputCodingTableMode, inputCodingTableWrite

Table A.16: Specification of the data flow parameter 'inputCodingTableFileName'

Name	**inputGradingPTableMode**
Meaning	This parameter sets the mode for the table, which is used for the grading of the findable MC Tracks (see appendix A.4 on page A-101).
Standard	FILETABLE
Range	{ NOTABLE, CROSSFOOTTABLE, ONLINETABLE, FILETABLE }
Format	signed decimal number
See	inputGradingPTableFileName, inputGradingPTableWrite

Table A.17: Specification of the data flow parameter 'inputGradingPTableMode'

Name	**inputGradingPTableFileName**
Meaning	This parameter holds the name of the file consisting of the table, which is used for the grading of the findable MC Tracks. (see appendix A.4 on page A-101)
Standard	"../../input/gradingPTable.txt"
See	inputGradingPTableMode, inputGradingPTableWrite

Table A.18: Specification of the data flow parameter 'inputGradingPTable-FileName'

Name	**inputGradingRTableMode**
Meaning	This parameter sets the mode for the table, which is used for the grading of the found tracks (see appendix A.4 on page A-101).
Standard	FILETABLE
Range	{ NOTABLE, CROSSFOOTTABLE, ONLINETABLE, FILETABLE, LUTGOODNESSTABLE }
Format	signed decimal number
See	inputGradingRTableFileName, inputGradingRTableWrite

Table A.19: Specification of the data flow parameter 'inputGradingRTable-Mode'

Name	**inputGradingRTableFileName**
Meaning	This parameter holds the name of the file consisting of the table, which is used for the grading of the found tracks. (see appendix A.4 on page A-101)
Standard	"../../input/gradingRTable.txt"
See	inputGradingRTableMode, inputGradingRTableWrite

Table A.20: Specification of the data flow parameter 'inputGradingRTable-FileName'

Name	**inputCodingTableWrite**
Meaning	This parameter enables the writing of the table, which is used for the Encoding of the histogram, into a file.
Standard	"false"
Range	{ "false"; "true" } or { 0; 1 }
Format	string or decimal
See	inputCodingTableMode, inputCodingTableFileName

Table A.21: Specification of the data flow parameter 'inputCodingTableWrite'

Name	**inputGradingPTableWrite**
Meaning	This parameter enables the writing of the table, which is used for the grading of the findable MC Tracks, into a file.
Standard	"false"
Range	{ "false"; "true" } or { 0; 1 }
Format	string or decimal
See	inputGradingPTableMode, inputGradingPTableFileName

Table A.22: Specification of the data flow parameter 'inputGradingPTableWrite'

Name	**inputGradingRTableWrite**
Meaning	This parameter enables the writing of the table, which is used for the grading of the found tracks, into a file.
Standard	"false"
Range	{ "false"; "true" } or { 0; 1 }
Format	string or decimal
See	inputGradingRTableMode, inputGradingRTableFileName

Table A.23: Specification of the data flow parameter 'inputGradingRTableWrite'

Having all data flow parameters successfully presented up to here, the subsequent part belongs obviously to the algorithm parameters.

Name	**trackfinderHitProducer**
Meaning	This parameter can define a special hit producer, which is able to compute hits based on points. **This API was used before realistic hit producers exist in the framework. The API is deprecated and not supported any longer.**
Standard	NONEHITPRODUCER
Range	{ NONEHITPRODUCER, BINNINGHITPRODUCER, SMEARINGHITPRODUCER }
Format	signed decimal number

Table A.24: Specification of the algorithm parameter 'trackfinderHitProducer'

Name	**trackfinderReadPointsFromFile**
Meaning	This parameter enables the manually reading of the STS Points from a file by using a special framework interface. If it is disabled, no STS Points would be taken. In this case the internal analysis, which base on the MC information, are not available. If there are additionally no hits, no tracking could be applied also, because there would be no data.
Standard	true
Range	{ "false"; "true" } or { 0; 1 }
Format	string or decimal
See	trackfinderHitProducer, trackfinderReadHitsFromFile

Table A.25: Specification of the algorithm parameter 'trackfinderReadPointsFromFile'

Name	**trackfinderPrelutFileName**
Meaning	This parameter holds the name of the file consisting of the pre-lut.
See	trackfinderLutsVersion, trackfinderLutFileName, trackfinder-RungeKuttaPdgCode

Table A.26: Specification of the algorithm parameter 'trackfinderPrelutFile-Name'

Name	**trackfinderLutFileName**
Meaning	This parameter holds the name of the file consisting of the lut.
See	trackfinderLutsVersion, trackfinderPrelutFileName, trackfind-erRungeKuttaPdgCode

Table A.27: Specification of the algorithm parameter 'trackfinderLutFile-Name'

Name	**trackfinderReadHitsFromFile**
Meaning	This parameter enables the manually reading of the STS Hits from a file by using a special framework interface. If it is disabled, the STS Hits delivered from the automatic framework interface would be taken. If there are no hits delivered, one of the deprecated hit producers would be applied to the points to produce hits.
Standard	"false"
Range	{ "false"; "true" } or { 0; 1 }
Format	string or decimal
See	trackfinderHitProducer, trackfinderReadPointsFromFile

Table A.28: Specification of the algorithm parameter 'trackfinderReadHits-FromFile'

Name	**trackfinderReadMapsHits**
Meaning	This parameter enables the reading of the MAPS Hits in an older framework version. In the new version it enables the reading of the MVD Hits.
Standard	"false"
Range	{ "false"; "true" } or { 0; 1 }
Format	string or decimal

Table A.29: Specification of the algorithm parameter 'trackfinderReadMapsHits'

Name	**trackfinderReadHybridHits**
Meaning	This parameter enables the reading of the Hybrid pixel Hits in an older framework version. In the new version it enables the reading of the STS Hits.
Standard	"false"
Range	{ "false"; "true" } or { 0; 1 }
Format	string or decimal
See	trackfinderReadStripHits

Table A.30: Specification of the algorithm parameter 'trackfinderReadHybridHits'

Name	**trackfinderReadStripHits**
Meaning	This parameter enables the reading of the Micro-strip Hits in an older framework version. In the new version it enables the reading of the STS Hits.
Standard	"false"
Range	{ "false"; "true" } or { 0; 1 }
Format	string or decimal
See	trackfinderReadHybridHits

Table A.31: Specification of the algorithm parameter 'trackfinderReadStripHits'

Name	**trackfinderWriteTracksToFile**
Meaning	This parameter enables the manually writing of the found STS Tracks into the standard file by using a special framework interface. If it is disabled, the found STS Tracks would be delivered to the automatic framework interface.
Standard	"false"
Range	{ "false"; "true" } or { 0; 1 }
Format	string or decimal

Table A.32: Specification of the algorithm parameter 'trackfinderWriteTracksToFile'

Name	**trackfinderLutsVersion**
Meaning	This parameter sets the version, which is used for both LUTs. Possible ones are identified by the generation of both LUTs via the Runge-Kutta approach or the analytic formula as well as the direct usage of a file or the analytic formula.
Standard	ANALYTICFORMULALUT
Range	{ GENERATERUNGEKUTTALUT, GENERATEANALYTICFORMULALUT, FILELUT, ANALYTICFORMULALUT }
Format	signed decimal number
See	trackfinderPrelutFileName, trackfinderLutFileName, trackfinderRungeKuttaPdgCode

Table A.33: Specification of the algorithm parameter 'trackfinderLutsVersion'

Name	**trackfinderRungeKuttaPdgCode**
Meaning	This parameter sets the PDG (see [Parb]) code, which is used in the Runge-Kutta approach to generate both LUTs. This code is further committed to GEANE, which propagates the specified particle with different momenta along motion trajectories through the detector setup of the experiment.
Standard	13
Range	see [MON]
Format	signed decimal number
See	trackfinderLutsVersion, trackfinderPrelutFileName, trackfinderLutFileName

Table A.34: Specification of the algorithm parameter 'trackfinderRungeKuttaPdgCode'

Name	**trackfinderGammaMin**
Meaning	This parameter sets the minimum value in the γ-dimension of the Hough space. That means that this value is the minimal γ, which can occur in the Hough space.
Standard	−0.467,64
Range	$[\,-1.7 \cdot 10^{308}\,;\,+1.7 \cdot 10^{308}\,]$
Format	signed decimal number
See	trackfinderGammaMax, trackfinderGammaStep

Table A.35: Specification of the algorithm parameter 'trackfinderGammaMin'

Name	**trackfinderGammaMax**
Meaning	This parameter sets the maximum value in the γ-dimension of the Hough space. That means that this value is the maximal γ, which can occur in the Hough space.
Standard	+0.467,64
Range	$[\,-1.7 \cdot 10^{308}\,;\,+1.7 \cdot 10^{308}\,]$
Format	signed decimal number
See	trackfinderGammaMin, trackfinderGammaStep

Table A.36: Specification of the algorithm parameter 'trackfinderGammaMax'

Name	**trackfinderGammaStep**
Meaning	This parameter sets the number of quantization steps in the γ-dimension of the Hough space. That means that this value sets the resolution of γ in the Hough space.
Standard	maxDimGamma (see appendix A.3 on page A-79)
Range	$[\ 0;\ +2^{16}-1\ ]$ **More limiting is the size of the needed memory, which is allocated for the histogram (one layer of the Hough space)**
Format	unsigned decimal number
See	trackfinderGammaMin, trackfinderGammaMax

Table A.37: Specification of the algorithm parameter 'trackfinderGammaStep'

Name	**trackfinderThetaMin**
Meaning	This parameter sets the minimum value in the Θ-dimension of the Hough space. That means that this value is the minimal Θ, which can occur in the Hough space.
Standard	−0.467,64
Range	$[\ -1.7\cdot 10^{308};\ +1.7\cdot 10^{308}\ ]$
Format	signed decimal number
See	trackfinderThetaMax, trackfinderThetaStep

Table A.38: Specification of the algorithm parameter 'trackfinderThetaMin'

Name	**trackfinderThetaMax**
Meaning	This parameter sets the maximum value in the Θ-dimension of the Hough space. That means that this value is the maximal Θ, which can occur in the Hough space.
Standard	+0.467,64
Range	$[\ -1.7\cdot 10^{308};\ +1.7\cdot 10^{308}\ ]$
Format	signed decimal number
See	trackfinderThetaMin, trackfinderThetaStep

Table A.39: Specification of the algorithm parameter 'trackfinderThetaMax'

Name	**trackfinderThetaStep**
Meaning	This parameter sets the number of quantization steps in the Θ-dimension of the Hough space. That means that this value sets the resolution of Θ in the Hough space.
Standard	maxDimTheta (see appendix A.3 on page A-79)
Range	$[\ 0\ ;\ +2^{16}-1\ ]$ **More limiting is the size of the needed memory, which is allocated for the histogram (one layer of the Hough space)**
Format	unsigned decimal number
See	trackfinderThetaMin, trackfinderThetaMax

Table A.40: Specification of the algorithm parameter 'trackfinderThetaStep'

Name	**trackfinderLutRadiusMin**
Meaning	This parameter sets the minimum value in the $\frac{q}{p_{xz}}$-dimension of the Hough space. That means that this value is the minimal $\frac{q}{p_{xz}}$, which can occur in the Hough space.
Standard	-1.11
Range	$[\ -1.7 \cdot 10^{308}\ ;\ +1.7 \cdot 10^{308}\ ]$
Format	signed decimal number
See	trackfinderLutRadiusMax, trackfinderLutRadiusStep

Table A.41: Specification of the algorithm parameter 'trackfinderLutRadius-Min'

Name	**trackfinderLutRadiusMax**
Meaning	This parameter sets the maximum value in the $\frac{q}{p_{xz}}$-dimension of the Hough space. That means that this value is the maximal $\frac{q}{p_{xz}}$, which can occur in the Hough space.
Standard	$+1.11$
Range	$[\ -1.7 \cdot 10^{308}\ ;\ +1.7 \cdot 10^{308}\ ]$
Format	signed decimal number
See	trackfinderLutRadiusMin, trackfinderLutRadiusStep

Table A.42: Specification of the algorithm parameter 'trackfinderLutRadius-Max'

Name	**trackfinderLutRadiusStep**
Meaning	This parameter sets the number of quantization steps in the $\frac{q}{p_{xz}}$-dimension of the Hough space. That means that this value sets the resolution of $\frac{q}{p_{xz}}$ in the Hough space.
Standard	maxDimRadius (see appendix A.3 on page A-79)
Range	$[\ 0\ ;\ +2^{16}-1\ ]$ **More limiting is the size of the needed memory, which is allocated for the histogram (one layer of the Hough space)**
Format	unsigned decimal number
See	trackfinderLutRadiusMin, trackfinderLutRadiusMax

Table A.43: Specification of the algorithm parameter 'trackfinderLutRadiusStep'

Name	**trackfinderPrelutRadiusMin**
Meaning	This parameter defines the minimum of the computation range for the prelut formula, which is used to build the corresponding LUT. This range is used to find the correct entries in the γ-dimension of the Hough space. It is just a supporting parameter.
Standard	-0.25
Range	$[\ -1.7\cdot 10^{308}\ ;\ +1.7\cdot 10^{308}\ ]$
Format	signed decimal number
See	trackfinderPrelutRadiusMax

Table A.44: Specification of the algorithm parameter 'trackfinderPrelutRadiusMin'

Name	**trackfinderPrelutRadiusMax**
Meaning	This parameter defines the maximum of the computation range for the prelut formula, which is used to build the corresponding LUT. This range is used to find the correct entries in the γ-dimension of the Hough space. It is just a supporting parameter.
Standard	+0.25
Range	$[-1.7 \cdot 10^{308} ; +1.7 \cdot 10^{308}]$
Format	signed decimal number
See	trackfinderPrelutRadiusMin

Table A.45: Specification of the algorithm parameter 'trackfinderPrelutRadiusMax'

Name	**trackfinderMinClassCoding**
Meaning	This parameter defines the minimal priority class for the coding table, which is taken into account. This value is strongly dependent from the used tables (see appendix A.4 on page A-101). It could be evidently set always to one, if the used tables just consist of the relevant classes starting with priority class one.
Standard	1
Range	$[0 ; +2^{16} - 1]$
Format	unsigned decimal number
See	trackfinderMinClassGradingP, trackfinderMinClassGradingR

Table A.46: Specification of the algorithm parameter 'trackfinderMinClassCoding'

Name	**trackfinderMinClassGradingP**
Meaning	This parameter defines the minimal priority class for the gradingP table, which is taken into account. This value is strongly dependent from the used tables (see appendix A.4 on page A-101). It could be evidently set always to one, if the used tables just consist of the relevant classes starting with priority class one.
Standard	1
Range	$[\,0;\, +2^{16}-1\,]$
Format	unsigned decimal number
See	trackfinderMinClassCoding, trackfinderMinClassGradingR

Table A.47: Specification of the algorithm parameter 'trackfinderMinClassGradingP'

Name	**trackfinderMinClassGradingR**
Meaning	This parameter defines the minimal priority class for the gradingR table, which is taken into account. This value is strongly dependent from the used tables (see appendix A.4 on page A-101). It could be evidently set always to one, if the used tables just consist of the relevant classes starting with priority class one.
Standard	4
Range	$[\,0;\, +2^{16}-1\,]$
Format	unsigned decimal number
See	trackfinderMinClassCoding, trackfinderMinClassGradingP

Table A.48: Specification of the algorithm parameter 'trackfinderMinClassGradingR'

Name	**trackfinderAutomaticFilterWrite**
Meaning	This parameter enables the manually writing of the automatic generated filter geometry into a file. Important is that this parameter would have no effect, if the parameter 'trackfinder-FilterType' is set to FILEFILTER.
Standard	"false"
Range	{ "false"; "true" } or { 0; 1 }
Format	string or decimal
See	trackfinderFilterType, trackfinderAutomaticFilterCover-Percentage, trackfinderAutomaticFilterDataPercentage, trackfinderAutomaticFilterFileName

Table A.49: Specification of the algorithm parameter 'trackfinderAutomaticFilterWrite'

Name	**trackfinderFirstFilterGeometry**
Meaning	This parameter allows to switch for the 2D Peak-finding between all different available geometries, which are defined for the type MAXMORPHSEARCHFILTER. More information can be found in the source code file HistogramTransformation-LIB/include/filterDef.h. However this value would be not considered, if the type NOFILTER or ERASERFILTER is selected. Additionally an automatic type enables the value FIRSTFINALMODGEOMETRY to produce a specialized processing.
Standard	FIRSTFINALGEOMETRY
Range	NOFIRSTGEOMETRY, FIRST21GEOMETRY, FIRST122GEOMETRY, FIRST12GEOMETRY, FIRST12MODGEOMETRY, FIRST121GEOMETRY, FIRSTFINALGEOMETRY, FIRSTFINALMODGEOMETRY
Format	unsigned decimal number
See	trackfinderFilterType, trackfinderFirstFilterArithmetic, trackfinderSecondFilterGeometry, trackfinderSecondFilter-Arithmetic

Table A.50: Specification of the algorithm parameter 'trackfinderFirstFilter-Geometry'

Name	**trackfinderFilterType**
Meaning	This parameter defines the filter type, which is used in the Peak-finding process. It is obvious that NOFILTER implements a filter, which does not change the data in any way. Further on the ERASERFILTER and MAXMORPHSEARCHFILTER implement static filter geometries in combination with static filter arithmetics. The main difference between both is that ERASERFILTER implements just one geometry, which can vary in the size, while using just one arithmetic. The next value FILEFILTER implements the reading of a file to receive the geometry, which should be used. But at this juncture, the filter arithmetics are still static. Moreover the final values, which are AUTOMATICFIRSTEVENTFILTER, AUTOMATICEACHEVENTFILTER and AUTOMATICUPDATEEVENTFILTER implements a filter geometry, which is automatically generated by using a special analysis applied to the MC Tracks and their hits. In addition to this, the filter arithmetics are again static.
Standard	MAXMORPHSEARCHFILTER
Range	NOFILTER, MAXMORPHSEARCHFILTER, ERASERFILTER, FILEFILTER, AUTOMATICFIRSTEVENTFILTER, AUTOMATICEACHEVENTFILTER, AUTOMATICUPDATEEVENTFILTER
Format	unsigned decimal number
See	trackfinderFirstFilterGeometry, trackfinderFirstFilterArithmetic, trackfinderSecondFilterGeometry, trackfinderSecondFilterArithmetic, trackfinderAutomaticFilterCoverPercentage, trackfinderAutomaticFilterDataPercentage, trackfinderAutomaticFilterWrite, trackfinderAutomaticFilterFileName

Table A.51: Specification of the algorithm parameter 'trackfinderFilterType'

Name	**trackfinderFirstFilterArithmetic**
Meaning	This parameter allows to switch for the 2D Peak-finding between all different available filter arithmetics except for the filter types NOFILTER and ERASERFILTER. More information can be found in the source code file HistogramTransformationLIB/include/filterDef.h.
Standard	FIRSTSIMPLEARITHMETIC
Range	NOFIRSTARITHMETIC, FIRSTSIMPLEARITHMETIC, FIRSTSIMPLEMODARITHMETIC, FIRSTCOMPLEXARITHMETIC, FIRSTCOMPLEXMODARITHMETIC, FIRSTSPECIALARITHMETIC
Format	unsigned decimal number
See	trackfinderFilterType, trackfinderFirstFilterGeometry, trackfinderSecondFilterGeometry, trackfinderSecondFilterArithmetic

Table A.52: Specification of the algorithm parameter 'trackfinderFirstFilterArithmetic'

Name	**trackfinderAutomaticFilterFileName**
Meaning	This parameter holds the name of the file, which is used to access the automatic filter geometry for writing and reading.
See	trackfinderFilterType, trackfinderAutomaticFilterCoverPercentage, trackfinderAutomaticFilterDataPercentage, trackfinderAutomaticFilterWrite

Table A.53: Specification of the algorithm parameter 'trackfinderAutomaticFilterFileName'

Name	**trackfinderSecondFilterGeometry**
Meaning	This parameter allows to switch for the 3D Peak-finding process between all different available filter geometries, which are defined for the filter type MAXMORPHSEARCHFILTER. More information can be found in the source code file HistogramTransformationLIB/include/filterDef.h. However this value would be not considered, if the filter type NOFILTER or ERASERFILTER is selected. Additionally an automatic filter type enables the value SECONDFINALMODGEOMETRY to produce a specialized processing.
Standard	SECONDFINALGEOMETRY
Range	NOSECONDGEOMETRY, SECOND3MODGEOMETRY, SECONDFINALGEOMETRY, SECONDFINALMODGEOMETRY, SECOND3GEOMETRY
Format	unsigned decimal number
See	trackfinderFilterType, trackfinderFirstFilterGeometry, trackfinderFirstFilterArithmetic, trackfinderSecondFilterArithmetic

Table A.54: Specification of the algorithm parameter 'trackfinderSecondFilterGeometry'

Name	**trackfinderFirstFilterNeighborhoodDim1ClearRadius**
Meaning	This parameter defines the size of the filter window, which is applied to one layer of the Hough space, in the first dimension before and after the actually considered filtering element.
Standard	3
Range	$[\,0\,;\,+2^{16}-1\,]$
Format	unsigned decimal number
See	trackfinderFirstFilterNeighborhoodDim2ClearRadius

Table A.55: Specification of the algorithm parameter 'trackfinderFirstFilterNeighborhoodDim1ClearRadius'

Name	**trackfinderSecondFilterArithmetic**
Meaning	This parameter allows to switch for the 3D Peak-finding process between all different available filter arithmetics except for the filter types NOFILTER and ERASERFILTER. More information can be found in the source code file HistogramTransformationLIB/include/filterDef.h.
Standard	SECONDSIMPLEARITHMETIC
Range	NOSECONDARITHMETIC, SECONDSIMPLEARITHMETIC, SECONDSIMPLEMODARITHMETIC, SECONDCOMPLEXARITHMETIC, SECONDCOMPLEXMODARITHMETIC, SECONDSPECIALARITHMETIC
Format	unsigned decimal number
See	trackfinderFilterType, trackfinderFirstFilterGeometry, trackfinderFirstFilterArithmetic, trackfinderSecondFilterGeometry

Table A.56: Specification of the algorithm parameter 'trackfinderSecondFilterArithmetic'

Name	**trackfinderFirstFilterNeighborhoodDim2ClearRadius**
Meaning	This parameter defines the size of the filter window, which is applied to one layer of the Hough space, in the second dimension before and after the actually considered filtering element.
Standard	1
Range	$[\,0\,;\,+2^{16}-1\,]$
Format	unsigned decimal number
See	trackfinderFirstFilterNeighborhoodDim1ClearRadius

Table A.57: Specification of the algorithm parameter 'trackfinderFirstFilterNeighborhoodDim2ClearRadius'

Name	**trackfinderSecondFilterNeighborhoodDim1Clear-Radius**
Meaning	This parameter defines the size of the filter window, which is applied to the track candidates of each layer, in the first dimension before and after the actually considered filtering element.
Standard	1
Range	$[\,0\,;\,+2^{16}-1\,]$
Format	unsigned decimal number
See	trackfinderSecondFilterNeighborhoodDim2ClearRadius, trackfinderSecondFilterNeighborhoodDim3ClearRadius

Table A.58: Specification of the algorithm parameter 'trackfinderSecondFilterNeighborhoodDim1ClearRadius'

Name	**trackfinderSecondFilterNeighborhoodDim2Clear-Radius**
Meaning	This parameter defines the size of the filter window, which is applied to the track candidates of each layer, in the second dimension before and after the actually considered filtering element.
Standard	1
Range	$[\,0\,;\,+2^{16}-1\,]$
Format	unsigned decimal number
See	trackfinderSecondFilterNeighborhoodDim1ClearRadius, trackfinderSecondFilterNeighborhoodDim3ClearRadius

Table A.59: Specification of the algorithm parameter 'trackfinderSecondFilterNeighborhoodDim2ClearRadius'

Name	**trackfinderSecondFilterNeighborhoodDim3Clear-Radius**
Meaning	This parameter defines the size of the filter window, which is applied to the track candidates of each layer, in the third dimension before and after the actually considered filtering element.
Standard	1
Range	$[\,0\,;\,+2^{16}-1\,]$
Format	unsigned decimal number
See	trackfinderSecondFilterNeighborhoodDim1ClearRadius, trackfinderSecondFilterNeighborhoodDim2ClearRadius

Table A.60: Specification of the algorithm parameter 'trackfinderSecondFilterNeighborhoodDim3ClearRadius'

Name	**trackfinderAutomaticFilterCoverPercentage**
Meaning	This parameter defines the coverage percentage, which is used in the automatic generated filter geometry, to decide, if a geometry element should be used or not during the Peak-finding process. At this juncture, this percentage is evidently related to the maximum frequent geometry element, which is obviously always located at the matching origin of the geometry. Besides this, the filter type FILEFILTER enables the used percentage in the Peak-finding process to be set by the file, if the actual value of this parameter is set to zero. Otherwise the file value is overwritten.
Standard	100
Range	$[\,0\,;\,100\,]$
Format	unsigned decimal number
See	trackfinderFilterType, trackfinderAutomaticFilterDataPercentage, trackfinderAutomaticFilterWrite, trackfinderAutomaticFilterFileName

Table A.61: Specification of the algorithm parameter 'trackfinderAutomaticFilterCoverPercentage'

Name	**trackfinderAutomaticFilterDataPercentage**
Meaning	This parameter defines the percentage of MC Tracks, which should be used to generate the automatic filter geometry for the Peak-finding process.
Standard	100
Range	[0 ; 100]
Format	unsigned decimal number
See	trackfinderFilterType, trackfinderAutomaticFilterCoverPercentage, trackfinderAutomaticFilterWrite, trackfinderAutomaticFilterFileName

Table A.62: Specification of the algorithm parameter 'trackfinderAutomaticFilterDataPercentage'

Due to the circumstance that all parameters of the algorithm are successfully exhibited up to now, the parameters for the analyses will follow.
But before continuing with these ones, there are two possible pitfalls, which have to be mentioned, because they can avoid the expected evaluation of the analyses. The nature of the first one is theoretical, because the internal analyses, which base commonly on the MC information, would be obviously not available, if there are no STS Points present during runtime. In addition to this, the second one is more practical, because the output of the analyses would be definitely prevented from displaying (see appendix A.1 on page A-1 and appendix A.5 on page A-108), if the used verbosity level in the running script is too small.
Nevertheless the successive parameters are very useful to determine a well suited configuration for the recent tracking purpose.

Name	**analysisOutputFileName**
Meaning	This parameter holds the name of the file, which should be accessed to place the write-enabled output of all analysis.
Standard	"analysisOutput.root"

Table A.63: Specification of the analysis parameter 'analysisOutputFileName'

Name	**analysisThresholdForP**
Meaning	This parameter defines the minimal momentum of a MC Track to be taken into account in the analysis section. Additionally it has an important role by defining the range of the Hough space, because the Hough space for track with a smaller momentum must not exist.
Standard	1.0
Range	$[\,-1.7 \cdot 10^{308}\,;\,+1.7 \cdot 10^{308}\,]$
Format	signed decimal number
See	trackfinderLutRadiusMin, trackfinderLutRadiusMax

Table A.64: Specification of the analysis parameter 'analysisThresholdForP'

Name	**analysisInitConfiguration**
Meaning	This parameter enables a summary output on the standard output screen, which displays information about the actually set configuration of the Hough tracking algorithm.
Standard	"false"
Range	{ "false"; "true" } or { 0; 1 }
Format	string or decimal

Table A.65: Specification of the analysis parameter 'analysisInitConfiguration'

Name	**analysisInitDetector**
Meaning	This parameter enables a summary output on the standard output screen, which displays information about the actually set detector configuration of the Hough tracking algorithm.
Standard	"false"
Range	{ "false"; "true" } or { 0; 1 }
Format	string or decimal

Table A.66: Specification of the analysis parameter 'analysisInitDetector'

Name	**analysisInitEvent**
Meaning	This parameter enables a summary output on the standard output screen, which displays information about the actually processed event by the Hough tracking algorithm.
Standard	"true"
Range	{ "false"; "true" } or { 0; 1 }
Format	string or decimal

Table A.67: Specification of the analysis parameter 'analysisInitEvent'

Name	**analysisInitClassPriority**
Meaning	This parameter enables a summary output on the standard output screen, which displays information about the priority classes of the actually processed MC Tracks by the Hough tracking algorithm.
Standard	"false"
Range	{ "false"; "true" } or { 0; 1 }
Format	string or decimal

Table A.68: Specification of the analysis parameter 'analysisInitClassPriority'

Name	**analysisInitMemory**
Meaning	This parameter enables a summary output on the standard output screen, which displays information about the actual memory consumption of the Hough tracking algorithm.
Standard	"false"
Range	{ "false"; "true" } or { 0; 1 }
Format	string or decimal

Table A.69: Specification of the analysis parameter 'analysisInitMemory'

Name	**analysisInitQualityEFGCEventAbsolute**
Meaning	This parameter enables a tracking summary output on the standard output screen, which displays information consisting of absolute numbers for the tracking efficiency, the tracking fake rate, the tracking ghost rate, the tracking clones rate, the tracking identification rate and the tracking reduction rate for each separate event processed by the Hough tracking algorithm.
Standard	"false"
Range	{ "false"; "true" } or { 0; 1 }
Format	string or decimal
See	analysisInitQualityEFGCEventRelative

Table A.70: Specification of the analysis parameter 'analysisInitQualityEFGCEventAbsolute'

Name	**analysisInitQualityEFGCEventRelative**
Meaning	This parameter enables a tracking summary output on the standard output screen, which displays information consisting of percentage numbers for the tracking efficiency, the tracking fake rate, the tracking ghost rate, the tracking clones rate, the tracking identification rate and the tracking reduction rate for each separate event processed by the Hough tracking algorithm.
Standard	"false"
Range	{ "false"; "true" } or { 0; 1 }
Format	string or decimal
See	analysisInitQualityEFGCEventAbsolute

Table A.71: Specification of the analysis parameter 'analysisInitQualityEFGCEventRelative'

Name	**analysisInitQualityEFGCTotalAbsolute**
Meaning	This parameter enables a tracking summary output on the standard output screen, which displays information consisting of absolute numbers for the tracking efficiency, the tracking fake rate, the tracking ghost rate, the tracking clones rate, the tracking identification rate and the tracking reduction rate accumulated for all events processed by the Hough tracking algorithm.
Standard	"false"
Range	{ "false"; "true" } or { 0; 1 }
Format	string or decimal
See	analysisInitQualityEFGCTotalRelative

Table A.72: Specification of the analysis parameter 'analysisInitQualityEFGCTotalAbsolute'

Name	**analysisInitQualityEFGCTotalRelative**
Meaning	This parameter enables a tracking summary output on the standard output screen, which displays information consisting of percentage numbers for the tracking efficiency, the tracking fake rate, the tracking ghost rate, the tracking clones rate, the tracking identification rate and the tracking reduction rate accumulated for all events processed by the Hough tracking algorithm.
Standard	"true"
Range	{ "false"; "true" } or { 0; 1 }
Format	string or decimal
See	analysisInitQualityEFGCTotalAbsolute

Table A.73: Specification of the analysis parameter 'analysisInitQualityEFGCTotalRelative'

Name	**analysisInitMomentumEFGCEventPzEFGC**
Meaning	This parameter enables a tracking output in the graphical display of a window, which shows information about the tracking efficiency, the tracking fake rate, the tracking ghost rate and the tracking clones rate in accordance to the absolute value of the momentum in z-direction for each separate event processed by the Hough tracking algorithm.
Standard	"false"
Range	{ "false"; "true" } or { 0; 1 }
Format	string or decimal
See	analysisInitMomentumEFGCTotalPzEFGC, analysisInitMomentumEvent, analysisInitMomentumToRoot

Table A.74: Specification of the analysis parameter 'analysisInitMomentumEFGCEventPzEFGC'

Name	**analysisInitMomentumEFGCTotalPzEFGC**
Meaning	This parameter enables a tracking output in the graphical display of a window, which shows information about the tracking efficiency, the tracking fake rate, the tracking ghost rate and the tracking clones rate in accordance to the absolute value of the momentum in z-direction accumulated for all events processed by the Hough tracking algorithm.
Standard	"false"
Range	{ "false"; "true" } or { 0; 1 }
Format	string or decimal
See	analysisInitMomentumEFGCEventPzEFGC, analysisInitMomentumTotal, analysisInitMomentumDisplay, analysisInitMomentumToRoot

Table A.75: Specification of the analysis parameter 'analysisInitMomentumEFGCTotalPzEFGC'

Name	**analysisInitMomentumEFGCEventPtEFGC**
Meaning	This parameter enables a tracking output in the graphical display of a window, which shows information about the tracking efficiency, the tracking fake rate, the tracking ghost rate and the tracking clones rate in accordance to the absolute value of the transversal momentum in xy-direction for each separate event processed by the Hough tracking algorithm.
Standard	"false"
Range	{ "false"; "true" } or { 0; 1 }
Format	string or decimal
See	analysisInitMomentumEFGCTotalPtEFGC, analysisInitMomentumEvent, analysisInitMomentumToRoot

Table A.76: Specification of the analysis parameter 'analysisInitMomentumEFGCEventPtEFGC'

Name	**analysisInitMomentumEFGCTotalPtEFGC**
Meaning	This parameter enables a tracking output in the graphical display of a window, which shows information about the tracking efficiency, the tracking fake rate, the tracking ghost rate and the tracking clones rate in accordance to the absolute value of the transversal momentum in xy-direction accumulated for all events processed by the Hough tracking algorithm.
Standard	"false"
Range	{ "false"; "true" } or { 0; 1 }
Format	string or decimal
See	analysisInitMomentumEFGCEventPtEFGC, analysisInitMomentumTotal, analysisInitMomentumDisplay, analysisInitMomentumToRoot

Table A.77: Specification of the analysis parameter 'analysisInitMomentumEFGCTotalPtEFGC'

Name	**analysisInitTrackPropagationEventPoint**
Meaning	This parameter enables an analysis, which evaluates the quality of the simulated input data for each event separately. For this purpose, the GEANE recomputed detector interaction coordinates of each propagated MC Track are compared to the original supplied MC Points by a distance distribution display.
Standard	"false"
Range	{ "false"; "true" } or { 0; 1 }
Format	string or decimal
See	analysisInitTrackPropagationEventHit, analysisInitTrackPropagationTotalPoint, analysisInitTrackPropagationTotalHit, analysisInitTrackPropagationToRoot

Table A.78: Specification of the analysis parameter 'analysisInitTrackPropagationEventPoint'

Name	**analysisInitTrackPropagationEventHit**
Meaning	This parameter enables an analysis, which evaluates the quality of the simulated input data for each event separately. For this purpose, the GEANE recomputed detector interaction coordinates of each propagated MC Track are compared to the original supplied Hits by a distance distribution display.
Standard	"false"
Range	{ "false"; "true" } or { 0; 1 }
Format	string or decimal
See	analysisInitTrackPropagationEventPoint, analysisInitTrackPropagationTotalPoint, analysisInitTrackPropagationTotalHit, analysisInitTrackPropagationToRoot

Table A.79: Specification of the analysis parameter 'analysisInitTrackPropagationEventHit'

Name	**analysisInitTrackPropagationTotalPoint**
Meaning	This parameter enables an analysis, which evaluates the quality of the simulated input data accumulated for all events. For this purpose, the GEANE recomputed detector interaction coordinates of each propagated MC Track are compared to the original supplied MC Points by a distance distribution display.
Standard	"false"
Range	{ "false"; "true" } or { 0; 1 }
Format	string or decimal
See	analysisInitTrackPropagationEventPoint, analysisInitTrackPropagationEventHit, analysisInitTrackPropagationTotalHit, analysisInitTrackPropagationDisplay, analysisInitTrackPropagationToRoot, analysisTrackPropagationPointDisplayMask, analysisTrackPropagationHitDisplayMask

Table A.80: Specification of the analysis parameter 'analysisInitTrackPropagationTotalPoint'

Name	**analysisInitTrackPropagationTotalHit**
Meaning	This parameter enables an analysis, which evaluates the quality of the simulated input data accumulated for all events. For this purpose, the GEANE recomputed detector interaction coordinates of each propagated MC Track are compared to the original supplied Hits by a distance distribution display.
Standard	"false"
Range	{ "false"; "true" } or { 0; 1 }
Format	string or decimal
See	analysisInitTrackPropagationEventPoint, analysisInitTrackPropagationEventHit, analysisInitTrackPropagationTotalPoint, analysisInitTrackPropagationDisplay, analysisInitTrackPropagationToRoot, analysisTrackPropagationPointDisplayMask, analysisTrackPropagationHitDisplayMask

Table A.81: Specification of the analysis parameter 'analysisInitTrackPropagationTotalHit'

Name	**analysisInitTrackPropagationDisplay**
Meaning	This parameter initializes a display, which shows all enabled accumulated track propagation analysis with an update after each event. So the progress of the accumulation can be seen. However it is important that just one of all displays can be enabled at the same time.
Standard	"false"
Range	{ "false"; "true" } or { 0; 1 }
Format	string or decimal
See	analysisInitTrackPropagationTotalPoint, analysisInitTrack-PropagationTotalHit, analysisInitTrackPropagationToRoot, analysisTrackPropagationPointDisplayMask, analysisTrack-PropagationHitDisplayMask, analysisInitMomentumDisplay, analysisInitProjectionDisplay, analysisInitMagnetfieldDisplay, analysisInitMagnetfieldConstantDisplay, analysisInitPrelu-tRangeDisplay

Table A.82: Specification of the analysis parameter 'analysisInitTrackPropagationDisplay'

Name	**analysisInitTrackPropagationToRoot**
Meaning	This parameter enables all graphical displays for the track propagation analysis to be written after each event into a file. Concerning the writing, there is no difference between the event analysis and the accumulated total analysis.
Standard	"false"
Range	{ "false"; "true" } or { 0; 1 }
Format	string or decimal
See	analysisInitTrackPropagationEventPoint, analysisInitTrack-PropagationEventHit, analysisInitTrackPropagationTotal-Point, analysisInitTrackPropagationTotalHit

Table A.83: Specification of the analysis parameter 'analysisInitTrackPropagationToRoot'

Name	**analysisTrackPropagationPointDisplayMask**
Meaning	This parameter can prevent each track propagation point distance distribution from displaying by setting the corresponding bit in the value, which is connected to the distribution by its position. Moreover if there are more distributions than bit positions, the assumed value for the missing bits is evidently always zero.
Standard	0
Range	$[\ 0\ ;\ 2^{maxDimSH} - 1\ ]$ (see appendix A.3 on page A-79)
Format	radix format (see appendix B on page B-1)
See	analysisInitTrackPropagationDisplay, analysisTrackPropagationHitDisplayMask

Table A.84: Specification of the analysis parameter 'analysisTrackPropagationPointDisplayMask'

Name	**analysisTrackPropagationHitDisplayMask**
Meaning	This parameter can prevent each track propagation hit distance distribution from displaying by setting the corresponding bit in the value, which is connected to the distribution by its position. Moreover if there are more distributions than bit positions, the assumed value for the missing bits is evidently always zero.
Standard	0
Range	$[\ 0\ ;\ 2^{maxDimSH} - 1\ ]$ (see appendix A.3 on page A-79)
Format	radix format (see appendix B on page B-1)
See	analysisInitTrackPropagationDisplay, analysisTrackPropagationPointDisplayMask

Table A.85: Specification of the analysis parameter 'analysisTrackPropagationHitDisplayMask'

Name	**analysisInitProjectionEFGCNEvent12EFGCN**
Meaning	This parameter enables an output in the graphical display of a window, which shows separately for each event the Hough space occupancy projection of the tracking efficiency, the tracking fake rate, the tracking ghost rate, the tracking clones rate and the tracking not-found track rate into the plane formed by dimension one and two.
Standard	"false"
Range	{ "false"; "true" } or { 0; 1 }
Format	string or decimal
See	analysisInitProjectionEFGCNTotal12EFGCN, analysisInitProjectionEFGCNEvent13EFGCN, analysisInitProjectionEFGCNEvent32EFGCN, analysisInitProjectionEvent, analysisInitProjectionToRoot

Table A.86: Specification of the analysis parameter 'analysisInitProjectionEFGCNEvent12EFGCN'

Name	**analysisInitProjectionEFGCNTotal12EFGCN**
Meaning	This parameter enables an output in the graphical display of a window, which shows accumulated for all events the Hough space occupancy projection of the tracking efficiency, the tracking fake rate, the tracking ghost rate, the tracking clones rate and the tracking not-found track rate into the plane formed by dimension one and two.
Standard	"false"
Range	{ "false"; "true" } or { 0; 1 }
Format	string or decimal
See	analysisInitProjectionEFGCNEvent12EFGCN, analysisInitProjectionEFGCNTotal13EFGCN, analysisInitProjectionEFGCNTotal32EFGCN, analysisInitProjectionTotal, analysisInitProjectionDisplay, analysisInitProjectionToRoot

Table A.87: Specification of the analysis parameter 'analysisInitProjectionEFGCNTotal12EFGCN'

Name	**analysisInitProjectionEFGCNEvent13EFGCN**
Meaning	This parameter enables an output in the graphical display of a window, which shows separately for each event the Hough space occupancy projection of the tracking efficiency, the tracking fake rate, the tracking ghost rate, the tracking clones rate and the tracking not-found track rate into the plane formed by dimension one and three.
Standard	"false"
Range	{ "false"; "true" } or { 0; 1 }
Format	string or decimal
See	analysisInitProjectionEFGCNEvent12EFGCN, analysisInitProjectionEFGCNTotal13EFGCN, analysisInitProjectionEFGCNEvent32EFGCN, analysisInitProjectionEvent, analysisInitProjectionToRoot

Table A.88: Specification of the analysis parameter 'analysisInitProjectionEFGCNEvent13EFGCN'

Name	**analysisInitProjectionEFGCNTotal13EFGCN**
Meaning	This parameter enables an output in the graphical display of a window, which shows accumulated for all events the Hough space occupancy projection of the tracking efficiency, the tracking fake rate, the tracking ghost rate, the tracking clones rate and the tracking not-found track rate into the plane formed by dimension one and three.
Standard	"false"
Range	{ "false"; "true" } or { 0; 1 }
Format	string or decimal
See	analysisInitProjectionEFGCNTotal12EFGCN, analysisInitProjectionEFGCNEvent13EFGCN, analysisInitProjectionEFGCNTotal32EFGCN, analysisInitProjectionTotal, analysisInitProjectionDisplay, analysisInitProjectionToRoot

Table A.89: Specification of the analysis parameter 'analysisInitProjectionEFGCNTotal13EFGCN'

Name	**analysisInitProjectionEFGCNEvent32EFGCN**
Meaning	This parameter enables an output in the graphical display of a window, which shows separately for each event the Hough space occupancy projection of the tracking efficiency, the tracking fake rate, the tracking ghost rate, the tracking clones rate and the tracking not-found track rate into the plane formed by dimension three and two.
Standard	"false"
Range	{ "false"; "true" } or { 0; 1 }
Format	string or decimal
See	analysisInitProjectionEFGCNEvent12EFGCN, analysisInitProjectionEFGCNEvent13EFGCN, analysisInitProjectionEFGCNTotal32EFGCN, analysisInitProjectionEvent, analysisInitProjectionToRoot

Table A.90: Specification of the analysis parameter 'analysisInitProjectionEFGCNEvent32EFGCN'

Name	**analysisInitProjectionEFGCNTotal32EFGCN**
Meaning	This parameter enables an output in the graphical display of a window, which shows accumulated for all events the Hough space occupancy projection of the tracking efficiency, the tracking fake rate, the tracking ghost rate, the tracking clones rate and the tracking not-found track rate into the plane formed by dimension three and two.
Standard	"false"
Range	{ "false"; "true" } or { 0; 1 }
Format	string or decimal
See	analysisInitProjectionEFGCNTotal12EFGCN, analysisInitProjectionEFGCNTotal13EFGCN, analysisInitProjectionEFGCNEvent32EFGCN, analysisInitProjectionTotal, analysisInitProjectionDisplay, analysisInitProjectionToRoot

Table A.91: Specification of the analysis parameter 'analysisInitProjectionEFGCNTotal32EFGCN'

Name	**analysisInitMomentumEvent**
Meaning	This parameter represents a global flag to simply evaluate in the source code whether the analysis for the momentum is enabled for each event separately. So this parameter is not used inside the source code to enable or disable something, but it must be set, if this momentum analysis is turned on.
Standard	"false"
Range	{ "false"; "true" } or { 0; 1 }
Format	string or decimal
See	analysisInitMomentumEFGCEventPzEFGC, analysisInitMomentumEFGCEventPtEFGC, analysisInitMomentumToRoot

Table A.92: Specification of the analysis parameter 'analysisInitMomentumEvent'

Name	**analysisInitMomentumTotal**
Meaning	This parameter represents a global flag to simply evaluate in the source code whether the analysis for the momentum is enabled accumulated for all events. So this parameter is not used inside the source code to enable or disable something, but it must be set, if this momentum analysis is turned on.
Standard	"false"
Range	{ "false"; "true" } or { 0; 1 }
Format	string or decimal
See	analysisInitMomentumEFGCTotalPzEFGC, analysisInitMomentumEFGCTotalPtEFGC, analysisInitMomentumDisplay, analysisInitMomentumToRoot

Table A.93: Specification of the analysis parameter 'analysisInitMomentumTotal'

Name	**analysisInitMomentumDisplay**
Meaning	This parameter initializes a display, which shows all enabled accumulated momentum analysis with an update after each event. So the progress of the accumulation can be seen. However it is important that just one of all displays can be enabled at the same time.
Standard	"false"
Range	{ "false"; "true" } or { 0; 1 }
Format	string or decimal
See	analysisInitMomentumEFGCTotalPzEFGC, analysisInitMomentumEFGCTotalPtEFGC, analysisInitMomentumTotal, analysisInitMomentumToRoot, analysisInitProjectionDisplay, analysisInitMagnetfieldDisplay, analysisInitMagnetfieldConstantDisplay, analysisInitPrelutRangeDisplay

Table A.94: Specification of the analysis parameter 'analysisInitMomentumDisplay'

Name	**analysisInitMomentumToRoot**
Meaning	This parameter enables all graphical displays for the momentum analysis to be written after each event into a file. Concerning the writing, there is no difference between the event analysis and the accumulated total analysis.
Standard	"false"
Range	{ "false"; "true" } or { 0; 1 }
Format	string or decimal
See	analysisInitMomentumEFGCEventPzEFGC, analysisInitMomentumEFGCTotalPzEFGC, analysisInitMomentumEFGCEventPtEFGC, analysisInitMomentumEFGCTotalPtEFGC, analysisInitMomentumEvent, analysisInitMomentumTotal, analysisInitMomentumDisplay

Table A.95: Specification of the analysis parameter 'analysisInitMomentumToRoot'

Name	**analysisInitProjectionEvent**
Meaning	This parameter represents a global flag to simply evaluate in the source code whether the analysis for the projection is enabled for each event separately. So this parameter is not used inside the source code to enable or disable something, but it must be set, if this projection analysis is turned on.
Standard	"false"
Range	{ "false"; "true" } or { 0; 1 }
Format	string or decimal
See	analysisInitProjectionEFGCNEvent12EFGCN, analysisInitProjectionEFGCNEvent13EFGCN, analysisInitProjectionEFGCNEvent32EFGCN, analysisInitProjectionToRoot

Table A.96: Specification of the analysis parameter 'analysisInitProjectionEvent'

Name	**analysisInitProjectionDisplay**
Meaning	This parameter initializes a display, which shows all enabled accumulated projection analysis with an update after each event. So the progress of the accumulation can be seen. However it is important that just one of all displays can be enabled at the same time.
Standard	"false"
Range	{ "false"; "true" } or { 0; 1 }
Format	string or decimal
See	analysisInitMomentumDisplay, analysisInitProjectionEFGCNTotal12EFGCN, analysisInitProjectionEFGCNTotal13EFGCN, analysisInitProjectionEFGCNTotal32EFGCN, analysisInitProjectionTotal, analysisInitProjectionToRoot, analysisInitMagnetfieldDisplay, analysisInitMagnetfieldConstantDisplay, analysisInitPrelutRangeDisplay

Table A.97: Specification of the analysis parameter 'analysisInitProjectionDisplay'

Name	analysisInitProjectionTotal
Meaning	This parameter represents a global flag to simply evaluate in the source code whether the analysis for the projection is enabled accumulated for all events. So this parameter is not used inside the source code to enable or disable something, but it must be set, if this projection analysis is turned on.
Standard	"false"
Range	{ "false"; "true" } or { 0; 1 }
Format	string or decimal
See	analysisInitProjectionEFGCNTotal12EFGCN, analysisInitProjectionEFGCNTotal13EFGCN, analysisInitProjectionEFGCNTotal32EFGCN, analysisInitProjectionDisplay, analysisInitProjectionToRoot

Table A.98: Specification of the analysis parameter 'analysisInitProjectionTotal'

Name	analysisInitProjectionToRoot
Meaning	This parameter enables all graphical displays for the projection analysis to be written after each event into a file. Concerning the writing, there is no difference between the event analysis and the accumulated total analysis.
Standard	"false"
Range	{ "false"; "true" } or { 0; 1 }
Format	string or decimal
See	analysisInitProjectionEFGCNEvent12EFGCN, analysisInitProjectionEFGCNTotal12EFGCN, analysisInitProjectionEFGCNEvent13EFGCN, analysisInitProjectionEFGCNTotal13EFGCN, analysisInitProjectionEFGCNEvent32EFGCN, analysisInitProjectionEFGCNTotal32EFGCN, analysisInitProjectionEvent, analysisInitProjectionTotal, analysisInitProjectionDisplay

Table A.99: Specification of the analysis parameter 'analysisInitProjectionToRoot'

Name	**analysisInitMagnetfieldX**
Meaning	This parameter enables an output in the graphical display of a window, which shows the actual used magnetic field. As furthermore the magnetic field has three components (B_x, B_y, B_z) based on a three dimensional coordinate (x, y, z), this display shows each component just on a varying x coordinate. Contrarily the y and z coordinates are fixed with standard values in the source code header file AnalysisLIB/include/analysis.h.
Standard	"false"
Range	{ "false"; "true" } or { 0; 1 }
Format	string or decimal
See	analysisInitMagnetfieldY, analysisInitMagnetfieldZ, analysisInitMagnetfieldDisplay, analysisInitMagnetfieldToRoot

Table A.100: Specification of the analysis parameter 'analysisInitMagnetfieldX'

Name	**analysisInitMagnetfieldY**
Meaning	This parameter enables an output in the graphical display of a window, which shows the actual used magnetic field. As furthermore the magnetic field has three components (B_x, B_y, B_z) based on a three dimensional coordinate (x, y, z), this display shows each component just on a varying y coordinate. Contrarily the x and z coordinates are fixed with standard values in the source code header file AnalysisLIB/include/analysis.h.
Standard	"false"
Range	{ "false"; "true" } or { 0; 1 }
Format	string or decimal
See	analysisInitMagnetfieldX, analysisInitMagnetfieldZ, analysisInitMagnetfieldDisplay, analysisInitMagnetfieldToRoot

Table A.101: Specification of the analysis parameter 'analysisInitMagnetfieldY'

Name	**analysisInitMagnetfieldZ**
Meaning	This parameter enables an output in the graphical display of a window, which shows the actual used magnetic field. As furthermore the magnetic field has three components (B_x, B_y, B_z) based on a three dimensional coordinate (x, y, z), this display shows each component just on a varying z coordinate. Contrarily the x and y coordinates are fixed with standard values in the source code header file AnalysisLIB/include/analysis.h.
Standard	"false"
Range	{ "false"; "true" } or { 0; 1 }
Format	string or decimal
See	analysisInitMagnetfieldX, analysisInitMagnetfieldY, analysisInitMagnetfieldDisplay, analysisInitMagnetfieldToRoot

Table A.102: Specification of the analysis parameter 'analysisInitMagnetfieldZ'

Name	**analysisInitMagnetfieldDisplay**
Meaning	This parameter initializes a display, which shows the actual magnetic field with all enabled features. However it is important that just one of all displays can be enabled at the same time.
Standard	"false"
Range	{ "false"; "true" } or { 0; 1 }
Format	string or decimal
See	analysisInitMomentumDisplay, analysisInitProjectionDisplay, analysisInitMagnetfieldX, analysisInitMagnetfieldY, analysisInitMagnetfieldZ, analysisInitMagnetfieldToRoot, analysisInitMagnetfieldConstantDisplay, analysisInitPrelutRangeDisplay

Table A.103: Specification of the analysis parameter 'analysisInitMagnetfieldDisplay'

Name	**analysisInitMagnetfieldConstantForEachEvent**
Meaning	This parameter enables the event by event evaluation of an optimal factor representing the magnetic field constant in the analytic formula for each detector plane. Further on, the same factors can be used for all events by defining them inside the source code (see appendix A.3 on page A-79).
Standard	"false"
Range	{ "false"; "true" } or { 0; 1 }
Format	string or decimal
See	analysisInitWeightedMagnetfieldConstant

Table A.104: Specification of the analysis parameter 'analysisInitMagnetfieldConstantForEachEvent'

Name	**analysisMagnetfieldConstantDisplayMask**
Meaning	This parameter can prevent the magnetic field constant distribution of a detector plane from displaying by setting the corresponding bit in the value, which is connected to the distribution by its position. Moreover if there are more distributions than bit positions, the assumed value for the missing bits is evidently always zero.
Standard	0
Range	[0 ; $2^{maxDimSH} - 1$] (see appendix A.3 on page A-79)
Format	radix format (see appendix B on page B-1)
See	analysisInitMagnetfieldConstantDisplay

Table A.105: Specification of the analysis parameter 'analysisMagnetfieldConstantDisplayMask'

Name	**analysisInitMagnetfieldToRoot**
Meaning	This parameter initializes all graphical displays for the magnetic field to be written into a file.
Standard	"false"
Range	{ "false"; "true" } or { 0; 1 }
Format	string or decimal
See	analysisInitMagnetfieldX, analysisInitMagnetfieldY, analysisInitMagnetfieldZ, analysisInitMagnetfieldDisplay

Table A.106: Specification of the analysis parameter 'analysisInitMagnetfieldToRoot'

Name	**analysisInitTime**
Meaning	This parameter enables a summary output on the standard output screen, which displays information about the actual time consumption of the Hough tracking algorithm.
Standard	"false"
Range	{ "false"; "true" } or { 0; 1 }
Format	string or decimal

Table A.107: Specification of the analysis parameter 'analysisInitTime'

Name	**analysisInitWeightedMagnetfieldConstant**
Meaning	This parameter enables the combination of the optimal factors representing the magnetic field constant in the analytic formula for each detector plane in the way that they are summed up and averaged event by event.
Standard	"false"
Range	{ "false"; "true" } or { 0; 1 }
Format	string or decimal
See	analysisInitMagnetfieldConstantForEachEvent

Table A.108: Specification of the analysis parameter 'analysisInitWeightedMagnetfieldConstant'

Name	**analysisInitMagnetfieldConstantDisplay**
Meaning	This parameter initializes a display, which shows the distribution of the possible magnetic field constants and their quality for all unmasked detector planes. However it is important that just one of all displays can be enabled at the same time.
Standard	"false"
Range	{ "false"; "true" } or { 0; 1 }
Format	string or decimal
See	analysisInitMomentumDisplay, analysisInitProjectionDisplay, analysisInitMagnetfieldDisplay, analysisInitMagnetfieldConstantForEachEvent, analysisInitWeightedMagnetfieldConstant, analysisMagnetfieldConstantDisplayMask, analysisInitMagnetfieldConstantToRoot, analysisInitPrelutRangeDisplay

Table A.109: Specification of the analysis parameter 'analysisInitMagnetfieldConstantDisplay'

Name	**analysisInitMagnetfieldConstantToRoot**
Meaning	This parameter initializes all graphical displays for the magnetic field constants to be written into a file.
Standard	"false"
Range	{ "false"; "true" } or { 0; 1 }
Format	string or decimal
See	analysisInitMagnetfieldConstantForEachEvent, analysisInitWeightedMagnetfieldConstant, analysisMagnetfieldConstantDisplay

Table A.110: Specification of the analysis parameter 'analysisInitMagnetfieldConstantToRoot'

Name	**analysisInitMagnetfieldVSConstants**
Meaning	This parameter enables three special displays. The first one shows the dependency of the magnetic field from the z position for a constant x and y position. The second display shows the corresponding dependency for the actual magnetic field constants. And the third display shows the integrated averaged magnetic field dependent from the z position.
Standard	"false"
Range	{ "false"; "true" } or { 0; 1 }
Format	string or decimal

Table A.111: Specification of the analysis parameter 'analysisInitMagnetfieldVSConstants'

Name	**analysisInitPrelutGoodness**
Meaning	This parameter enables the whole analysis package for the goodness of the first LUT, which is called 'prelut'. Further on, all results of this package can be displayed graphically or with absolute numbers on the standard output screen by using the parameters 'analysisInitAnalysisResultWarnings' and 'analysisInitAnalysisResultDisplays' in the configuration file (see table A.155 of appendix A.2 on page A-76, table A.156 of appendix A.2 on page A-76 and appendix A.3 on page A-79).
Standard	"false"
Range	{ "false"; "true" } or { 0; 1 }
Format	string or decimal
See	analysisInitAnalysisResultWarnings, analysisInitAnalysisResultDisplays

Table A.112: Specification of the analysis parameter 'analysisInitPrelutGoodness'

Name	**analysisInitLutGoodness**
Meaning	This parameter enables the whole analysis package for the goodness of the second LUT, which is called 'lut'. All results of this package can be displayed graphically or with absolute numbers on the standard output screen by using the parameters 'analysisInitAnalysisResultWarnings' and 'analysisInitAnalysisResultDisplays' in the configuration file (see table A.155 of appendix A.2 on page A-76, table A.156 of appendix A.2 on page A-76 and appendix A.3 on page A-79).
Standard	"false"
Range	{ "false"; "true" } or { 0; 1 }
Format	string or decimal
See	analysisInitAnalysisResultWarnings, analysisInitAnalysisResultDisplays

Table A.113: Specification of the analysis parameter 'analysisInitLutGoodness'

Name	**analysisInitHoughTransformGoodness**
Meaning	This parameter enables the whole analysis package for the goodness of the Hough transform using both LUT. All results of this package can be displayed graphically or with absolute numbers on the standard output screen by using the parameters 'analysisInitAnalysisResultWarnings' and 'analysisInitAnalysisResultDisplays' in the configuration file (see table A.155 of appendix A.2 on page A-76, table A.156 of appendix A.2 on page A-76 and appendix A.3 on page A-79).
Standard	"false"
Range	{ "false"; "true" } or { 0; 1 }
Format	string or decimal
See	analysisInitAnalysisResultWarnings, analysisInitAnalysisResultDisplays

Table A.114: Specification of the analysis parameter 'analysisInitHoughTransformGoodness'

Name	**analysisInitQuantizationGoodness**
Meaning	This parameter enables the whole analysis package for the goodness of the Hough space quantization. All results of this package can be displayed graphically or with absolute numbers on the standard output screen by using the parameters 'analysisInitAnalysisResultWarnings' and 'analysisInitAnalysisResultDisplays' in the configuration file (see table A.155 of appendix A.2 on page A-76, table A.156 of appendix A.2 on page A-76 and appendix A.3 on page A-79).
Standard	"false"
Range	{ "false"; "true" } or { 0; 1 }
Format	string or decimal
See	analysisInitAnalysisResultWarnings, analysisInitAnalysisResultDisplays

Table A.115: Specification of the analysis parameter 'analysisInitQuantizationGoodness'

Name	**analysisInitCreatedHistogramToRoot**
Meaning	This parameter enables the graphical display of the histogram to be written after the Histogramming step of the algorithm into a file. Further on, it is important that the parameter 'analysisInitJustOneCreatedHistogramToRoot' (see table A.119 of appendix A.2 on page A-59) controls whether one or all layers are written.
Standard	"false"
Range	{ "false"; "true" } or { 0; 1 }
Format	string or decimal
See	analysisInitEncodedHistogramToRoot, analysisInitFilteredHistogramToRoot, analysisInitJustOneCreatedHistogramToRoot, analysisInitCreatedHistogramToShow

Table A.116: Specification of the analysis parameter 'analysisInitCreatedHistogramToRoot'

Name	**analysisInitEncodedHistogramToRoot**
Meaning	This parameter enables the graphical display of the histogram to be written after the Encoding step of the algorithm into a file. Further on, it is important that the parameter 'analysisInitJustOneEncodedHistogramToRoot' (see table A.120 of appendix A.2 on page A-59) controls whether one or all layers are written.
Standard	"false"
Range	{ "false"; "true" } or { 0; 1 }
Format	string or decimal
See	analysisInitCreatedHistogramToRoot, analysisInitFilteredHistogramToRoot, analysisInitJustOneEncodedHistogramToRoot, analysisInitEncodedHistogramToShow

Table A.117: Specification of the analysis parameter 'analysisInitEncodedHistogramToRoot'

Name	**analysisInitFilteredHistogramToRoot**
Meaning	This parameter enables the graphical display of the histogram to be written after the 2D Peak-finding step of the algorithm into a file. Further on, it is important that the parameter 'analysisInitJustOneFilteredHistogramToRoot' (see table A.121 of appendix A.2 on page A-60) controls whether one or all layers are written.
Standard	"false"
Range	{ "false"; "true" } or { 0; 1 }
Format	string or decimal
See	analysisInitCreatedHistogramToRoot, analysisInitEncodedHistogramToRoot, analysisInitJustOneFilteredHistogramToRoot, analysisInitFilteredHistogramToShow

Table A.118: Specification of the analysis parameter 'analysisInitFilteredHistogramToRoot'

Name	**analysisInitJustOneCreatedHistogramToRoot**
Meaning	This parameter is able to prevent all histogram layers except the one, which is identified by the index 'analysisInitHistogramLayer' (see table A.129 of appendix A.2 on page A-64), from writing into a file after the Histogramming step of the algorithm.
Standard	"false"
Range	{ "false"; "true" } or { 0; 1 }
Format	string or decimal
See	analysisInitCreatedHistogramToRoot, analysisInitJustOneEncodedHistogramToRoot, analysisInitJustOneFilteredHistogramToRoot, analysisInitCreatedHistogramToShow, analysisInitHistogramLayer

Table A.119: Specification of the analysis parameter 'analysisInitJustOneCreatedHistogramToRoot'

Name	**analysisInitJustOneEncodedHistogramToRoot**
Meaning	This parameter is able to prevent all histogram layers except the one, which is identified by the index 'analysisInitHistogramLayer' (see table A.129 of appendix A.2 on page A-64), from writing into a file after the Encoding step of the algorithm.
Standard	"false"
Range	{ "false"; "true" } or { 0; 1 }
Format	string or decimal
See	analysisInitEncodedHistogramToRoot, analysisInitJustOneCreatedHistogramToRoot, analysisInitJustOneFilteredHistogramToRoot, analysisInitEncodedHistogramToShow, analysisInitHistogramLayer

Table A.120: Specification of the analysis parameter 'analysisInitJustOneEncodedHistogramToRoot'

Name	**analysisInitJustOneFilteredHistogramToRoot**
Meaning	This parameter is able to prevent all histogram layers except the one, which is identified by the index 'analysisInitHistogramLayer' (see table A.129 of appendix A.2 on page A-64), from writing into a file after the 2D Peak-finding step.
Standard	"false"
Range	{ "false"; "true" } or { 0; 1 }
Format	string or decimal
See	analysisInitFilteredHistogramToRoot, analysisInitJustOneCreatedHistogramToRoot, analysisInitJustOneEncodedHistogramToRoot, analysisInitFilteredHistogramToShow, analysisInitHistogramLayer

Table A.121: Specification of the analysis parameter 'analysisInitJustOneFilteredHistogramToRoot'

Name	**analysisInitCreatedHistogramToShow**
Meaning	This parameter initializes a display, which depicts the histogram layer identified by the index 'analysisInitHistogramLayer' (see table A.129 of appendix A.2 on page A-64) after the Histogramming step of the algorithm. However it is important that this functionality requires this one or all layers to be also written into file.
Standard	"false"
Range	{ "false"; "true" } or { 0; 1 }
Format	string or decimal
See	analysisInitCreatedHistogramToRoot, analysisInitJustOneCreatedHistogramToRoot, analysisInitEncodedHistogramToShow, analysisInitFilteredHistogramToShow, analysisInitHistogramLayer

Table A.122: Specification of the analysis parameter 'analysisInitCreatedHistogramToShow'

Name	**analysisInitEncodedHistogramToShow**
Meaning	This parameter initializes a display, which depicts the histogram layer identified by the index 'analysisInitHistogramLayer' (see table A.129 of appendix A.2 on page A-64) after the Encoding step of the algorithm. However it is important that this functionality requires this one or all layers to be also written into file.
Standard	"false"
Range	{ "false"; "true" } or { 0; 1 }
Format	string or decimal
See	analysisInitEncodedHistogramToRoot, analysisInitJustOneEncodedHistogramToRoot, analysisInitCreatedHistogramToShow, analysisInitFilteredHistogramToShow, analysisInitHistogramLayer

Table A.123: Specification of the analysis parameter 'analysisInitEncodedHistogramToShow'

Name	**analysisInitFilteredHistogramToShow**
Meaning	This parameter initializes a display, which depicts the histogram layer identified by the index 'analysisInitHistogramLayer' (see table A.129 of appendix A.2 on page A-64) after the 2D Peakfinding step of the algorithm. However it is important that this functionality requires this one or all layers to be also written into file.
Standard	"false"
Range	{ "false"; "true" } or { 0; 1 }
Format	string or decimal
See	analysisInitFilteredHistogramToRoot, analysisInitJustOnefilteredHistogramToRoot, analysisInitCreatedHistogramToShow, analysisInitEncodedHistogramToShow, analysisInitHistogramLayer

Table A.124: Specification of the analysis parameter 'analysisInitFilteredHistogramToShow'

Name	**analysisInitTotalAnalysis**
Meaning	This parameter enables a special output for analyses results, which are accumulated for multiple events of the Hough tracking. Furthermore it is clear that this possibility is applied only for analyses, which does not support such a functionality directly. Moreover all available results can be displayed graphically or with absolute numbers on the standard output screen by using the parameters 'analysisInitAnalysisResultWarnings', 'analysisInitAnalysisMoreResultWarnings', 'analysisInitAnalysisResultDisplays' and 'analysisInitAnalysisMoreResultDisplays' in the configuration file (see table A.155 of appendix A.2 on page A-76, table A.157 of appendix A.2 on page A-76, table A.156 of appendix A.2 on page A-76, table A.158 of appendix A.2 on page A-77 and appendix A.3 on page A-79).
Standard	"false"
Range	{ "false"; "true" } or { 0; 1 }
Format	string or decimal
See	analysisInitAnalysisResultWarnings, analysisInitAnalysisMoreResultWarnings, analysisInitAnalysisResultDisplays, analysisInitAnalysisMoreResultDisplays

Table A.125: Specification of the analysis parameter 'analysisInitTotalAnalysis'

Name	**analysisInitNumberOfTracksPerColumn**
Meaning	This parameter enables the evaluation of the number of found tracks occurring in one column of all histogram layers. Further on, the result of this analysis can be displayed graphically or with absolute numbers on the standard output screen by using the parameters 'analysisInitAnalysisResultWarnings' and 'analysisInitAnalysisResultDisplays' in the configuration file (see table A.155 of appendix A.2 on page A-76, table A.156 of appendix A.2 on page A-76 and appendix A.3 on page A-79).
Standard	"false"
Range	{ "false"; "true" } or { 0; 1 }
Format	string or decimal
See	analysisInitAnalysisResultWarnings, analysisInitAnalysisResultDisplays

Table A.126: Specification of the analysis parameter 'analysisInitNumberOfTracksPerColumn'

Name	**analysisInitNumberOfTracksPerRow**
Meaning	This parameter enables the evaluation of the number of found tracks occurring in one row of all histogram layers. Further on, the result of this analysis can be displayed graphically or with absolute numbers on the standard output screen by using the parameters 'analysisInitAnalysisResultWarnings' and 'analysisInitAnalysisResultDisplays' in the configuration file (see table A.155 of appendix A.2 on page A-76, table A.156 of appendix A.2 on page A-76 and appendix A.3 on page A-79).
Standard	"false"
Range	{ "false"; "true" } or { 0; 1 }
Format	string or decimal
See	analysisInitAnalysisResultWarnings, analysisInitAnalysisResultDisplays

Table A.127: Specification of the analysis parameter 'analysisInitNumberOfTracksPerRow'

Name	**analysisInitNumberOfTracksPerLayer**
Meaning	This parameter enables the evaluation of the number of found tracks occurring in one histogram layer. Further on, the result of this analysis can be displayed graphically or with absolute numbers on the standard output screen by using the parameters 'analysisInitAnalysisResultWarnings' and 'analysisInitAnalysisResultDisplays' in the configuration file (see table A.155 of appendix A.2 on page A-76, table A.156 of appendix A.2 on page A-76 and appendix A.3 on page A-79).
Standard	"false"
Range	{ "false"; "true" } or { 0; 1 }
Format	string or decimal
See	analysisInitAnalysisResultWarnings, analysisInitAnalysisResultDisplays

Table A.128: Specification of the analysis parameter 'analysisInitNumberOfTracksPerLayer'

Name	**analysisInitHistogramLayer**
Meaning	This parameter defines the index of the histogram layer, which is used in the analyses for the histogram visualization or the histogram file writing after a dedicated step of the algorithm.
Standard	100
Range	$[\ 0\ ;\ +2^{16}-1\ ]$ **More limiting is the size of the third dimension of the Hough space (see parameter 'trackfinderGammaStep' in table A.37 of appendix A.2 on page A-20)**
Format	unsigned decimal number
See	analysisInitJustOneCreatedHistogramToRoot, analysisInitJustOneEncodedHistogramToRoot, analysisInitJustOneFilteredHistogramToRoot, analysisInitCreatedHistogramToShow, analysisInitEncodedHistogramToShow, analysisInitFilteredHistogramToShow

Table A.129: Specification of the analysis parameter 'analysisInitHistogramLayer'

Name	**analysisInitHitReadoutDistribution**
Meaning	This parameter enables the evaluation of a statistical summary about the hit to histogram layer ratio, the mean hit processing ratio and an approximation of the time consumption for the histogram processing in a FPGA implementation. Further on, the result of this analysis can be displayed graphically or with absolute numbers on the standard output screen by using the parameters 'analysisInitAnalysisResultWarnings' and 'analysisInitAnalysisResultDisplays' in the configuration file (see table A.155 of appendix A.2 on page A-76, table A.156 of appendix A.2 on page A-76 and appendix A.3 on page A-79).
Standard	"false"
Range	{ "false"; "true" } or { 0; 1 }
Format	string or decimal
See	analysisInitReadoutColumnsInParallel, analysisInitInstantiateAllLayersInParallel, analysisInitHistogramReadoutToLayerRatio, analysisInitAnalysisResultWarnings, analysisInitAnalysisResultDisplays

Table A.130: Specification of the analysis parameter 'analysisInitHitReadoutDistribution'

Name	**analysisInitReadoutColumnsInParallel**
Meaning	This parameter enables the computation of the time consumption for the histogram processing in a FPGA implementation with a parallel read-out column-wise. Otherwise the parallel read-out is assumed to be realized row-wise.
Standard	"false"
Range	{ "false"; "true" } or { 0; 1 }
Format	string or decimal
See	analysisInitHitReadoutDistribution, analysisInitInstantiateAllLayersInParallel, analysisInitHistogramReadoutToLayerRatio

Table A.131: Specification of the analysis parameter 'analysisInitReadoutColumnsInParallel'

Name	**analysisInitInstantiateAllLayersInParallel**
Meaning	This parameter enables the computation of the time consumption for the histogram processing in a FPGA implementation with all parallel histogram layer instantiated. Otherwise a group of parallel instantiated histogram layers has to be used serially.
Standard	"false"
Range	{ "false"; "true" } or { 0; 1 }
Format	string or decimal
See	analysisInitHitReadoutDistribution, analysisInitReadoutColumnsInParallel, analysisInitHistogramReadoutToLayerRatio

Table A.132: Specification of the analysis parameter 'analysisInitInstantiateAllLayersInParallel'

Name	**analysisInitHistogramReadoutToLayerRatio**
Meaning	This parameter sets the ratio between the number of parallel instantiated histogram read-out chains and the parallel instantiated histogram layers. So the value one means that each histogram has its own corresponding read-out chain. Otherwise the read-out chains have to be used serially until the amount of parallel histogram layers is processed.
Standard	1
Range	[1; 'trackfinderGammaStep' (see table A.37 of appendix A.2 on page A-20)]
Format	unsigned decimal number
See	analysisInitHitReadoutDistribution, analysisInitReadoutColumnsInParallel, analysisInitInstantiateAllLayersInParallel

Table A.133: Specification of the analysis parameter 'analysisInitHistogramReadoutToLayerRatio'

Name	**analysisInitPrelutRangeForEachEvent**
Meaning	This parameter enables the event by event evaluation of an optimal prelut range for the equation in the analytic formula, which is used to compute the first LUT. Further on, the same range can be used for all events by defining them with the parameters 'trackfinderPrelutRadiusMin' (see table A.44 of appendix A.2 on page A-22) and 'trackfinderPrelutRadiusMax' (see table A.45 of appendix A.2 on page A-23).
Standard	"false"
Range	{ "false"; "true" } or { 0; 1 }
Format	string or decimal
See	trackfinderPrelutRadiusMin, trackfinderPrelutRadiusMax, analysisInitWeightedPrelutRange

Table A.134: Specification of the analysis parameter 'analysisInitPrelutRangeForEachEvent'

Name	**analysisInitPrelutRangeDisplay**
Meaning	This parameter initializes a display, which shows all enabled results for the prelut range analysis with an update after each event. However it is important that just one of all displays can be enabled at the same time.
Standard	"false"
Range	{ "false"; "true" } or { 0; 1 }
Format	string or decimal
See	analysisInitMomentumDisplay, analysisInitProjectionDisplay, analysisInitMagnetfieldDisplay, analysisInitMagnetfieldConstantDisplay, analysisInitPrelutRangeForEachEvent, analysisInitWeightedPrelutRange, analysisInitPrelutRangeDisplayMode, analysisPrelutRangeStationDisplayMask, analysisPrelutRangeStationSumDisplayMask, analysisPrelutRangeConstraintDisplayMask, analysisPrelutRangeConstraintSumDisplayMask, analysisPrelutRangeRelativeDisplayMask, analysisInitPrelutRangeToRoot

Table A.135: Specification of the analysis parameter 'analysisInitPrelutRangeDisplay'

Name	**analysisInitWeightedPrelutRange**
Meaning	This parameter enables the evaluation of the prelut range for the analytic formula, which is used to compute the first LUT, to be updated by the computation of the optimal prelut range for each event.
Standard	"false"
Range	{ "false"; "true" } or { 0; 1 }
Format	string or decimal
See	trackfinderPrelutRadiusMin, trackfinderPrelutRadiusMax, analysisInitPrelutRangeForEachEvent

Table A.136: Specification of the analysis parameter 'analysisInitWeighted-PrelutRange'

Name	**analysisInitPrelutRangeDisplayMode**
Meaning	This parameter sets the mode of the displays, which would be drawn for the prelut range analysis, if it is enabled.
Standard	MAINRELATIVEDISPLAYMODE
Range	{ CORRECTRELATIVEDISPLAYMODE, CUTCORRECTRELATIVEDISPLAYMODE, NORMALCORRECTDISPLAYMODE, MAINRELATIVEDISPLAYMODE, CUTMAINRELATIVEDISPLAYMODE, NORMALMAINDISPLAYMODE }
Format	unsigned decimal number
See	analysisInitPrelutRangeForEachEvent, analysisInitWeightedPrelutRange, analysisInitPrelutRangeDisplay, analysisPrelutRangeStationDisplayMask, analysisPrelutRangeStationSumDisplayMask, analysisPrelutRangeConstraintDisplayMask, analysisPrelutRangeConstraintSumDisplayMask, analysisPrelutRangeRelativeDisplayMask

Table A.137: Specification of the analysis parameter 'analysisInitPrelutRangeDisplayMode'

Name	**analysisPrelutRangeStationDisplayMask**
Meaning	This parameter can prevent the separate results for each detector station of the prelut range analysis from displaying by setting the corresponding bit in the value, which is connected to the distribution by its position. Moreover if there are more distributions than bit positions, the assumed value for the missing bits is evidently always zero.
Standard	0
Range	[0 ; $2^{maxDimSH} - 1$] (see appendix A.3 on page A-79)
Format	radix format (see appendix B on page B-1)
See	analysisInitPrelutRangeDisplay

Table A.138: Specification of the analysis parameter 'analysisPrelutRangeStationDisplayMask'

Name	**analysisPrelutRangeStationSumDisplayMask**
Meaning	This parameter can prevent the accumulated result for all detector stations of the prelut range analysis from displaying.
Standard	"false"
Range	{ "false"; "true" } or { 0; 1 }
Format	string or decimal
See	analysisInitPrelutRangeDisplay

Table A.139: Specification of the analysis parameter 'analysisPrelutRangeStationSumDisplayMask'

Name	**analysisPrelutRangeConstraintDisplayMask**
Meaning	This parameter can prevent the separate constraints for each detector station of the prelut range analysis from displaying by setting the corresponding bit in the value, which is connected to the distribution by its position. Moreover if there are more distributions than bit positions, the assumed value for the missing bits is evidently always zero.
Standard	0
Range	$[\,0\,;\,2^{maxDimSH}-1\,]$ (see appendix A.3 on page A-79)
Format	radix format (see appendix B on page B-1)
See	analysisInitPrelutRangeDisplay

Table A.140: Specification of the analysis parameter 'analysisPrelutRangeConstraintDisplayMask'

Name	**analysisInitPercentageOfHitsForPrelutRange**
Meaning	This parameter sets the percentage for the number of hits, which must be found in the accurate histogram layer corresponding to the MC Tracks. Hence this parameter is used to evaluate a good range for the analytic formula, which computes the first LUT.
Standard	95
Range	[0 ; 100]
Format	unsigned decimal number
See	analysisInitPrelutRangeForEachEvent, analysisInitWeightedPrelutRange

Table A.141: Specification of the analysis parameter 'analysisInitPercentageOfHitsForPrelutRange'

Name	**analysisPrelutRangeConstraintSumDisplayMask**
Meaning	This parameter can prevent the accumulated constraint for all detector stations of the prelut range analysis from displaying.
Standard	"false"
Range	{ "false"; "true" } or { 0; 1 }
Format	string or decimal
See	analysisInitPrelutRangeDisplay

Table A.142: Specification of the analysis parameter 'analysisPrelutRange-ConstraintSumDisplayMask'

Name	**analysisPrelutRangeRelativeDisplayMask**
Meaning	This parameter can prevent the relative results for all detector stations of the prelut range analysis from displaying.
Standard	"false"
Range	{ "false"; "true" } or { 0; 1 }
Format	string or decimal
See	analysisInitPrelutRangeDisplay

Table A.143: Specification of the analysis parameter 'analysisPrelutRan-geRelativeDisplayMask'

Name	**analysisInitPrelutRangeToRoot**
Meaning	This parameter initializes all graphical displays for the prelut range analysis to be written into a file after each event.
Standard	"false"
Range	{ "false"; "true" } or { 0; 1 }
Format	string or decimal
See	analysisInitPrelutRangeForEachEvent, analysisInitWeighted-PrelutRange, analysisInitPrelutRangeDisplay

Table A.144: Specification of the analysis parameter 'analysisInitPrelu-tRangeToRoot'

Name	**analysisChooseConstraintPrelutRange**
Meaning	This parameter selects the evaluated prelut range result of the MC Tracks, which feature the minimum constraint of at least a certain number of hits in combination with the parameter 'analysisChooseAccuratePrelutRange' (see table A.154 of appendix A.2 on page A-75), for usage. If this parameter is unset, the prelut range result, which possess the maximum number of hits, is taken without any constraints.
Standard	"false"
Range	{ "false"; "true" } or { 0; 1 }
Format	string or decimal
See	analysisInitPrelutRangeForEachEvent, analysisInitWeightedPrelutRange, analysisInitPercentageOfHitsForPrelutRange, analysisChooseMainPrelutRange

Table A.145: Specification of the analysis parameter 'analysisChooseConstraintPrelutRange'

Name	**analysisInitPercentageOfHitsInSignature**
Meaning	This parameter sets the percentage of hits, which must be included in a signature to accept it. Moreover it has to be set carefully, because it is used to analyze the goodness of both LUTs as well as to generate the table files on-line with the parameters 'inputCodingTableMode' (see table A.15 of appendix A.2 on page A-12), 'inputGradingPTableMode' (see table A.17 of appendix A.2 on page A-12) and 'inputGradingRTableMode' (see table A.19 of appendix A.2 on page A-13) in the configuration file.
Standard	70
Range	[0; 100]
Format	unsigned decimal number
See	inputCodingTableMode, inputGradingPTableMode, inputGradingRTableMode

Table A.146: Specification of the analysis parameter 'analysisInitPercentageOfHitsInSignature'

Name	**analysisInitPercentageOfTracksForSignature**
Meaning	This parameter sets the percentage of MC Tracks, which must be found by the group of accepted signatures. Moreover it has to be set carefully, because it is used to analyze the goodness of both LUTs as well as to generate the table files based on the LUT results with the parameters 'inputCodingTableMode' (see table A.15 of appendix A.2 on page A-12), 'inputGradingPTableMode' (see table A.17 of appendix A.2 on page A-12) and 'inputGradingRTableMode' (see table A.19 of appendix A.2 on page A-13) in the configuration file.
Standard	75
Range	[0; 100]
Format	unsigned decimal number
See	inputCodingTableMode, inputGradingPTableMode, inputGradingRTableMode

Table A.147: Specification of the analysis parameter 'analysisInitPercentageOfTracksForSignature'

Name	**analysisPrelutRangeMinStart**
Meaning	This parameter defines the start value of the computation range for the parameter 'trackfinderPrelutRadiusMin' (see table A.44 of appendix A.2 on page A-22).
Standard	-0.1
Range	[$-1.7 \cdot 10^{308}$; $+1.7 \cdot 10^{308}$]
Format	signed decimal number
See	analysisPrelutRangeMinStop, analysisPrelutRangeMinSteps

Table A.148: Specification of the analysis parameter 'analysisPrelutRangeMinStart'

Name	**analysisPrelutRangeMinStop**
Meaning	This parameter defines the stop value of the computation range for the parameter 'trackfinderPrelutRadiusMin' (see table A.44 of appendix A.2 on page A-22).
Standard	+0.1
Range	$[-1.7 \cdot 10^{308} ; +1.7 \cdot 10^{308}]$
Format	signed decimal number
See	analysisPrelutRangeMinStart, analysisPrelutRangeMinSteps

Table A.149: Specification of the analysis parameter 'analysisPrelutRangeMinStop'

Name	**analysisPrelutRangeMinSteps**
Meaning	This parameter sets the number of quantization steps and thus the resolution for the computation range of the parameter 'trackfinderPrelutRadiusMin' (see table A.44 of appendix A.2 on page A-22).
Standard	21
Range	$[0 ; +2^{16} - 1]$
Format	unsigned decimal number
See	analysisPrelutRangeMinStart, analysisPrelutRangeMinStop

Table A.150: Specification of the analysis parameter 'analysisPrelutRangeMinSteps'

Name	**analysisPrelutRangeMaxStart**
Meaning	This parameter defines the start value of the computation range for the parameter 'trackfinderPrelutRadiusMax' (see table A.45 of appendix A.2 on page A-23).
Standard	−0.1
Range	$[-1.7 \cdot 10^{308} ; +1.7 \cdot 10^{308}]$
Format	signed decimal number
See	analysisPrelutRangeMaxStop, analysisPrelutRangeMaxSteps

Table A.151: Specification of the analysis parameter 'analysisPrelutRangeMaxStart'

Name	**analysisPrelutRangeMaxStop**
Meaning	This parameter defines the stop value of the computation range for the parameter 'trackfinderPrelutRadiusMax' (see table A.45 of appendix A.2 on page A-23).
Standard	+0.1
Range	$[\,-1.7 \cdot 10^{308}\,;\,+1.7 \cdot 10^{308}\,]$
Format	signed decimal number
See	analysisPrelutRangeMaxStart, analysisPrelutRangeMaxSteps

Table A.152: Specification of the analysis parameter 'analysisPrelutRange-MaxStop'

Name	**analysisPrelutRangeMaxSteps**
Meaning	This parameter sets the number of quantization steps and thus the resolution for the computation range of the parameter 'trackfinderPrelutRadiusMax' (see table A.45 of appendix A.2 on page A-23).
Standard	21
Range	$[\,0;\,+2^{16}-1\,]$
Format	unsigned decimal number
See	analysisPrelutRangeMaxStart, analysisPrelutRangeMaxStop

Table A.153: Specification of the analysis parameter 'analysisPrelutRange-MaxSteps'

Name	**analysisChooseAccuratePrelutRange**
Meaning	This parameter selects the evaluated prelut range result of the MC Tracks, which are found either in the accurate histogram layer or in the correct one, for usage.
Standard	"true"
Range	{ "false"; "true" } or { 0; 1 }
Format	string or decimal
See	analysisInitPrelutRangeForEachEvent, analysisInitWeighted-PrelutRange, analysisChooseConstraintPrelutRange

Table A.154: Specification of the analysis parameter 'analysisChooseAccu-ratePrelutRange'

Name	**analysisInitAnalysisResultWarnings**
Meaning	This parameter sets which result of an analysis should be printed on the standard output screen by a warning message.
Standard	0 (see appendix A.3 on page A-79)
Range	[(NUMBEROFTRACKSWHICHCANNOTBEFOUND); $+2^{32}-1$]
Format	unsigned decimal number
See	analysisInitAnalysisResultDisplays , analysisInitAnalysisMoreResultWarnings, analysisInitAnalysisMoreResultDisplays

Table A.155: Specification of the analysis parameter 'analysisInitAnalysisResultWarnings'

Name	**analysisInitAnalysisResultDisplays**
Meaning	This parameter sets which result of an analysis should be displayed with a graphical output in a window.
Standard	0 (see appendix A.3 on page A-79)
Range	[(TRACKWITHMOMENTAERRORDISTRIBUTION | TRACKSPERCOLUMNDISTRIBUTION | TRACKSPERLAYERDISTRIBUTION); $+2^{32}-1$]
Format	unsigned decimal number
See	analysisInitAnalysisResultWarnings , analysisInitAnalysisMoreResultWarnings, analysisInitAnalysisMoreResultDisplays

Table A.156: Specification of the analysis parameter 'analysisInitAnalysisResultDisplays'

Name	**analysisInitAnalysisMoreResultWarnings**
Meaning	This parameter sets which result of an analysis should be printed on the standard output screen by a warning message.
Standard	0 (see appendix A.3 on page A-79)
Range	[0 ; $+2^{32}-1$]
Format	unsigned decimal number
See	analysisInitAnalysisResultWarnings , analysisInitAnalysisResultDisplays, analysisInitAnalysisMoreResultDisplays

Table A.157: Specification of the analysis parameter 'analysisInitAnalysisMoreResultWarnings'

Name	**analysisInitAnalysisMoreResultDisplays**
Meaning	This parameter sets which result of an analysis should be displayed with a graphical output in a window.
Standard	0 (see appendix A.3 on page A-79)
Range	[0; $+2^{32}-1$]
Format	unsigned decimal number
See	analysisInitAnalysisResultWarnings , analysisInitAnalysisResultDisplays, analysisInitAnalysisMoreResultWarnings

Table A.158: Specification of the analysis parameter 'analysisInitAnalysisMoreResultDisplays'

Name	**analysisWriteCellFiles**
Meaning	This parameter enables the production of the necessary files to run the Cell BE implementation.
Standard	"false"
Range	{ "false"; "true" } or { 0; 1 }
Format	string or decimal
See	analysisHitCellFileName, analysisPrelutCellFileName, analysisLutCellFileName

Table A.159: Specification of the analysis parameter 'analysisWriteCellFiles'

Name	**analysisHitCellFileName**
Meaning	This parameter holds the name of the file containing the hits in a format, which is used for the Cell BE implementation.
See	analysisWriteCellFiles, analysisPrelutCellFileName, analysisLutCellFileName

Table A.160: Specification of the analysis parameter 'analysisHitCellFileName'

Name	**analysisPrelutCellFileName**
Meaning	This parameter holds the name of the file containing the prelut in a format, which is used for the Cell BE implementation.
See	analysisWriteCellFiles, analysisHitCellFileName, analysisLutCellFileName

Table A.161: Specification of the analysis parameter 'analysisPrelutCellFileName'

Name	**analysisLutCellFileName**
Meaning	This parameter holds the name of the file containing the lut in a format, which is used for the Cell BE implementation.
See	analysisWriteCellFiles, analysisHitCellFileName, analysisPrelutCellFileName

Table A.162: Specification of the analysis parameter 'analysisLutCellFileName'

Name	**initStatus**
Meaning	This parameter enables a processing status for different parts of the Hough tracking algorithm, which is displayed on the standard output screen.
Standard	"false"
Range	{ "false"; "true" } or { 0; 1 }
Format	string or decimal

Table A.163: Specification of the analysis parameter 'initStatus'

As all parameters related to the analyses are successfully illustrated now, it is obvious that the cumulative list of general parameters is terminated. However it is clear that additional parameters exist, which can not be changed so easily by any user. But this is not a big disadvantage, because these ones require a deeper knowledge about details of the algorithm, the software implementation as well as this special kind of Hough tracking. Nevertheless the following appendix A.3 on page A-79 lists some of them in combination with their location, which is means certainly more precisely the naming of the source code file containing the related define statement.

A.3 Definitions

This section presents general parameters, which are not accessible via the configuration file, because they are a lot more sensible to the algorithm and their configuration require a deeper knowledge. Nevertheless an exhaustive list of all these parameters including their locations inside the source code files is shown, before a short description of the major important ones follows.

- MiscLIB/include/memoryDef.h
  - CALLOC_HISTOGRAM_RUNTIME
  - maxDimTheta
  - maxDimRadius
  - maxDimGamma
  - maxDimSH
- DataRootObjectLIB/src/trackfinderInputMagnetidField.cxx
  - numberOfDefaultMagnetfieldFactors
  - defaultMagnetfieldFactors
  - INTEGRATEDMAGNETICFIELD
- RootFrameworkLIB/src/hitProducer.cxx
  - binningXResolution
  - binningYResolution
  - smearingXSigma
  - smearingYSigma
- MiscLIB/include/defs.h
  - LUTVERSION
- InputLIB/include/inputRoot.h
  - STSPoint
  - STSHit
  - MCTrack

- OutputResultLIB/include/outputTrack.h
  - StsTrack
- InputLIB/include/inputData.h
  - DETECTORINFO
  - TRACKINFO
  - HITINFO
- DataRootObjectLIB/include/trackfinderInputData.h
  - NOTEXISTINGTRACKIDWARNING
  - TRACKWITHNOPOINTSWARNING
  - TRACKWITHNOHITSWARNING
  - HITWITHNOPOINTWARNING
  - HITWITHNOTRACKWARNING
- DataRootObjectLIB/src/peakfindingGeometry.cxx
  - RELATIVERANGE
- HoughTransformation/src/houghTransformation.cxx
  - PRINTBORDERSTOFILE
  - PRINTCREATEDHISTOGRAMLAYERSTOFILE
- HistogramTransformation/include/histogramTransformation.h
  - PRINTORIGINALHISTOGRAMLAYERSTOFILE
  - PRINTENCODEDHISTOGRAMLAYERSTOFILE
  - PRINTDIAGONALIZEDHISTOGRAMLAYERSTOFILE
  - PRINTFILTEREDHISTOGRAMLAYERSTOFILE
  - PRINTFINALIZEDHISTOGRAMLAYERSTOFILE
  - PRINTPREFILTEREDTRACKLAYERSTOFILE
  - PRINTFILTEREDTRACKLAYERSTOFILE
- HistogramTransformationLIB/src/autoFinder.cxx
  - SPEEDUPPEAKFINDING
  - DEBUGPEAKFINDINGGEOMETRY

- FileioLIB/include/fileio.h
  - fileCmdSeparator
  - fileCommentSeparator
  - fileDisableCmdSeparator
- MiscLIB/include/coordinateSystem.h
  - DIM1, DIM2 and DIM3
  - HRADIUS, HTHETA and HGAMMA
  - PX, PY and PZ
  - MRADIUS, MTHETA and MGAMMA
- MiscLIB/include/hitArray.h
  - hitArray
- MiscLIB/include/bitArray.h
  - bitArray
- DataRootObjectLIB/include/inputHitSpecialList.h
  - minimalArraySize
- DataObjectLIB/src/filterGeometry.cxx
  - RELATIVERANGE
- DataRootObjectLIB/include/inputHitSpecialArray.h
  - minimalArraySize
- MiscLIB/include/defs.h
  - ...
- AnalysisLIB/include/analysisDef.h
  - ...
- AnalysisLIB/include/projectionEFGCNAnalysisDef.h
  - ...

- AnalysisLIB/include/momentumEFGCAnalysisDef.h
  - ...
- AnalysisLIB/include/magnetfieldFactorAnalysisDef.h
  - ...
- AnalysisLIB/include/magnetfieldAnalysisDef.h
  - ...
- AnalysisLIB/include/histogramAnalysisDef.h
  - ...
- AnalysisLIB/include/analysis.h
  - FACTORS
  - FACTORMIN
  - FACTORMAX
- AnalysisLIB/src/analysis.cxx
  - NUMBEROFDIFFERENTMOMENTA
  - NUMBEROFDIFFERENTCORRECTHITS
  - STARTSLOPE
  - INCRSLOPE
- {*Name*}LIB/include/{*name*}WarningMsg.h
  - NO{*NAME*}WARNINGMESSAGE
- MiscLIB/include/errorHandling.h
  - NOERRORMESSAGE
- MiscLIB/include/projects
  - ...

Before starting now with the description of some selected major important parameters, it has to be noticed that the three dots inside this list represent always an amount of parameters, which is too huge to exhibit. Furthermore almost all of these parameters define just names, which occur for instance at the axis of a graphical display. Thence it is naturally not that essential to present them here separately by name.

Coming now to the point, the element of the Hough tracking algorithm causing the major significant parameter is of course determined by the memory, which is used by the histogram due to the huge dimensions of the Hough space. Furthermore it is clear that even if the algorithm is restructured to be able to compute each histogram layer of the Hough space serially, the amount of necessary memory would remain anyhow big.
For that reason, the dynamic allocation of memory for such a single layer defines obviously still a problem, because it takes at least a lot of runtime. As the opposite behavior is further on commonly identified by the static allocation of memory, a first parameter is evidently required, which is able to switch between these two possibilities. For this purpose, the source code file MiscLIB/include/memoryDef.h contains the parameter 'CALLOC_HISTOGRAM_RUNTIME', which features exactly this functionality. So if this parameter is defined, the dynamic allocation of the total amount of needed memory during runtime would be enabled. Otherwise, it is clear that a maximum of static memory is allocated, which can be not exceedingly used during runtime.
As however the maximal available size for each dimension of the Hough space must be then of course fixed even at the source code compilation, it is obvious that the additional parameters 'maxDimTheta', 'maxDimRadius' and 'maxDimGamma', which correspond to the dimensions, are essential.
Besides this, the maximal available amount of memory, which is consumed by the signature inside a single cell of the histogram, has to be contrarily defined independently always by the parameter 'maxDimSH'. The reason for this special treatment is identified by the implementation, which is realized either by an integrated data type or by the bitset template of the STL.

Further on, the next important parameter group concerns the factors, which represent the magnetic field for each detector station in the analytic formula, because even if these factors can be derived either by the given magnetic field map or by given MC data, it would be good to fix the result somewhere in a source code file to avoid the rerun of the analysis for each run.
For this purpose, that functionality is exactly offered by the source code file DataRootObjectLIB/src/trackfinderInputMagnetidField.cxx exhibiting the parameter array 'defaultMagnetfieldFactors', which has to contain the factors,

and the parameter 'numberOfDefaultMagnetfieldFactors', which has to define the number of factors. Moreover these factors determine naturally also a backup solution, that would be applied, if the analysis for the constant factors dependent on MC data is disabled and the magnetic field map deactivated.
Besides this, it has to be additionally noticed that the parameter 'INTEGRATEDMAGNETICFIELD' features the possibility to evaluate the value of the magnetic field map instead of the averaged integrated one.

Besides this, the implemented simple form of a hit producer, which is actually deprecated but has featured the generation of hits based on MC Points in former times completely independent of the CBMROOT framework, requires unerringly some control parameters Hence the parameters 'binningXResolution', 'binningYResolution', 'smearingXSigma', 'smearingYSigma', which are located in the source code file RootFrameworkLIB/src/hitProducer.cxx, realize configuration parameters for the available binning and smearing algorithm, which can be selected by the parameter 'trackfinderHitProducer' in the configuration file (see table A.24 of appendix A.2 on page A-15).

Coming now to a more hardware related point of view, it has to be reminded that a prerequisite of the systolic histogram processing is determined by the fact that the histogram curves have to possess always a slope bigger than 45 degrees, which leads to the following rule for the position of a digital sample point with index i in such a curve:

$$\begin{aligned} positionDim1[i] &= positionDim1[i+1] + x \quad , x \in \{0;1\} \\ positionDim2[i] &= positionDim2[i+1] + 1 \end{aligned}$$

As however the computation of such a Hough curve bases obviously just on the evaluation an equation and especially on the ratio of the histogram layer dimension sizes, this rule has to be obviously enforced afterwards. In addition to this, it is clear that this rule can be certainly softened. But this circumstance will result evidently in an exhaustive hardware overhead. Thus the parameter 'LUTVERSION', which is located in the source code file file MiscLIB/include/defs.h, is able to enable or disable this rule in the software to offer the possibility to analyze the effect on the performance of the algorithm. Maybe the price for the additional hardware might be acceptable to improve the performance.

Beyond that, the automatic generation of the Peak-finding geometry requires the next mentionable parameter, because the join patches for different relative peak clusters, which are apparently defined by the histogram cells containing each correct peak, imply a crucial problem. This means more precisely that the applied algorithm forms such a geometry by matching the occurring peak clusters statistically for some sample MC data, while the histogram cell, which contains the correct peak, must be of course not invariably covered, but is at all times nearby. As such geometries should be however also accepted to build statistics for the relative Peak-finding geometry, the parameter 'RELATIVERANGE' located in the source code file DataObjectLIB/src/filterGeometry.cxx realizes simply a security region concerning the number of cells, which are allowed to lay in between the cluster geometry and the join patch.

Further on, another important group of parameters consists of identifier names, which enable the access of the STS related data in the ROOT based file. So caused by this circumstance, it is clear that these names are defined by 'STSPoint', 'STSHit', 'MCTrack' and 'STSTrack'. Moreover this group is collaterally separated into the data, which should be read out of the file, and the data, which should be written into the file due to the result of the Hough tracking algorithm. Hence the parameters for the input can be found in the source code file InputLIB/include/inputRoot.h, while the ones for the output are located in the source code file OutputResultLIB/include/outputTrack.h. In this connection, it has to be additionally noticed that the identifiers for the analyses, which can be written into file, are placed in the corresponding analysis definition file. Finally it has to be mentioned here that more information on this topic is given later in the actual section.

Beyond that, another also fundamental group of parameters belongs to the consistency crosscheck of the input data. This means more precisely that warning messages would be produced, if strange conditions are found during the application of different checks to the input data. So caused by this circumstance, the most useful parameters, which are located in the source code file DataRootObjectLIB/include/trackfinderInputData.h, are determined by 'NOTEXISTINGTRACKIDWARNING', 'TRACKWITHNOPOINTSWARNING', 'TRACKWITHNOHITSWARNING', 'HITWITHNOPOINTWARNING' and 'HITWITHNOTRACKWARNING'.
In this context, the first parameter checks the data for the number of hits, which have the attribute 'trackID' set to a value representing no real existing MC Track. Although it is not that obvious at a first glance, this information is very important, because it enables the identification of hits, which result from an unknown source.
Moreover the subsequent parameter checks the input data for the number of MC Tracks, which exhibit not any point in the STS. This information is also very important, because such a track can be obviously not found due to no available hit for the Hough transform process.

In addition to this, the next parameter checks the data for the number of MC Tracks, which have no hits in the STS. So in analogousness, such a track can be surely not found too, but the huge difference is that the missing hits can occur due to the modeled detector inefficiency in the hit producers.
Further on, the successive parameter checks the input data for the number of hits, which does not belong to a point. In early versions of the CBM-ROOT framework, such a situation describes evidently a problem, but since the introduction of the micro-strip detector stations, this number represents simply the created fake hits by the corresponding hit producer.
Supplementary the last parameter checks the data for the number of hits, which can not be assigned to a MC Track. So this number is obviously composed of the fake hits in combination with the hits, which have a value set for the attribute 'id' representing no real existing MC Track.

In addition to all that, there are also some other input debugging parameters, which are located in the source code file InputLIB/include/inputData.h. These parameters enable simply a printout of some information about the detector stations as well as an excerpt of the given input data for the MC Tracks and the STS Hits on the standard output screen. For this purpose, the corresponding parameters 'DETECTORINFO', 'TRACKINFO' and 'HITINFO' feature an imagination about the input data of the Hough tracking algorithm.

Besides this, the parameters, which are located in the source code files HoughTransformation/src/houghTransformation.cxx and HistogramTransformation/include/histogramTransformation.h, offer the deepest available insight in the operation mode of each step of the Hough tracking algorithm, because they feature the memory dump for all interesting objects like the histogram for instance into separate files before and/or after all major modifications. Since such a memory dump is furthermore very useful for debugging or to gain simply an imagination about the content, the actual version of the software is able to visualize the histogram layer memory graphically by enabling the corresponding functionality in the configuration file (see table A.122 of appendix A.2 on page A-60, table A.123 of appendix A.2 on page A-61, and table A.124 of appendix A.2 on page A-61).

Further on, the two parameters 'SPEEDUPPEAKFINDING' and 'DEBUGPEAKFINDINGGEOMETRY', which are located in the source code file HistogramTransformationLIB/src/autoFinder.cxx, should be not forget, even if they are of a negligible importance, because the first one enables a timing speedup by skipping the Peak-finding process for histogram cells containing the value zero, while the second one simply enables a printout of some debug information to the standard output screen.

Beyond that, it should be also noticed that the meta tags, which are used in all ASCII formatted files to separate different objectives, can be changed in the source code file FileioLIB/include/fileio.h by adjusting the parameters 'fileCmdSeparator', 'fileCommentSeparator' and 'fileDisableCmdSeparator'.

Moreover, the next bigger group of parameters, which is located in the source code file MiscLIB/include/coordinateSystem.h, belongs to the different available coordinate systems of the Hough space and the momentum. Thus only advanced users should change their value, just if another solution for a problem is impossible, because there are many dependencies of the coordinate systems, which are not that obvious like these ones caused by the split of the 3D Hough space into 2D layers.
Nevertheless the Hough space system requires actually the parameters 'DIM1', 'DIM2' and 'DIM3' as well as the formula parameters 'HRADIUS', 'HTHETA' and 'HGAMMA', which are simply used to imply the transformation of the coordinate systems into each other. Combining these ones now with the parameters 'PX', 'PY' and 'PZ' for the momentum p_x, p_y and p_z and especially the analog formula parameters 'MRADIUS', 'MTHETA' and 'MGAMMA', such a transformation looks like the easy function call *momentum.set(hough.get(HGAMMA), MGAMMA)*.
By the way, it is needless to say that the transformation is not as totally simple as shown here, but the principle of hiding the system orientation can be envisaged. Finally it has to be mentioned that more information about the different coordinate systems can be found in the comment section of the source code file.

Coming now to a very interesting topic, the files MiscLIB/include/bitArray.h and MiscLIB/include/hitArray.h contain respectively just a single parameter, which belong both to the implementation of the histogram cells with regard to the memory usage strategy. Furthermore the class 'bitArray' specifies obviously the one, which has to store the bits of the signature in a histogram cell, while the class 'hitArray' determines evidently the one, which has to store the hits responsible for an enabled bit in the 'bitArray'.
Moreover it is clear that two different representations have to be considered for the 'bitArray', which features either optimal bit access or high-performance arithmetic computation, because these two topics determine the major used operations and thus the essentials for optimization. But as the bit access is heavily used during the Histogramming and the arithmetic computation for the Peak-finding, the optimal rated representation is naturally constituted by the number of operations in each task. Since the Histogramming is further on dependent of the number of hits, while the Peak-finding depends only on the histogram dimensions, it is certainly impossible to pre-

fer a representation just because of its dependencies. Nevertheless it is clear that such a decision has to be made. So for all that reasons, only the measurement of the needed time for both representations in the simulation of an event is naturally able to offer aid.
Besides this, the evaluation of the best rated representation for the 'hitArray' is even more complex, because nobody knows the amount of required memory, which additionally varies potentially dramatically. Hence two different representations are also considered. Within this context, the first one is simply characterized by a number of statically allocated arrays, which would be widened dynamically, if they are not big enough. In contrast to this, the second approach uses just the 'std::list' template of the STL for static and dynamic memory. Further on, it is self-evident that the size of all statically allocated memory, which has to be predefined in the source code files DataRootObjectLIB/include/inputHitSpecialList.h and DataRootObjectLIB/include/inputHitSpecialArray.h, has to be collaterally optimized, because the dynamic allocation takes a lot more runtime and the dependencies are comparable to the ones of the signature memory. So regardless of that, the easiest way to receive the best representation in combination with the best static memory size is again determined by the measurement of the needed time for both representations in the simulation of an event.

Finally the last but not least group of parameters is characterized by their global field of application. So in this connection, the source code file MiscLIB/include/defs.h contains the global parameters, which are almost needed for compatibility or debugging modes. In addition to this, the second source code file AnalysisLIB/include/analysisDef.h consists of global parameters for the analyses like exemplary the bit positions for the configuration file parameters 'analysisInitAnalysisResultWarnings' (see table A.155 of appendix A.2 on page A-76), 'analysisInitAnalysisResultDisplays' (see table A.156 of appendix A.2 on page A-76), 'analysisInitAnalysisMoreResultWarnings' (see table A.157 of appendix A.2 on page A-76) and 'analysisInitAnalysisMoreResultDisplays' (see table A.158 of appendix A.2 on page A-77), which enable or disable the displaying of analyses results. As this circumstance is surely of major importance, the following tables give a short summary about the bit positions for the parameters 'analysisInitAnalysisResultWarnings' and 'analysisInitAnalysisResultDisplays'.

Name	TRACKWITHHITINCORRECTLAYERDISTRIBUTION
Meaning	Benignity results for the first LUT with regard to the correct histogram layer
Bit	0

Table A.164: Specification of the bit position TRACKWITHHITIN-CORRECTLAYERDISTRIBUTION

Name	TRACKWITHHITINBESTLAYERDISTRIBUTION
Meaning	Benignity results for the first LUT with regard to the best histogram layer
Bit	1

Table A.165: Specification of the bit position TRACKWITHHITINBEST-LAYERDISTRIBUTION

Name	DISTANCEOFBESTANDCORRECTLAYER
Meaning	Benignity results for the first LUT with regard to the distance of the best and the correct histogram layer
Bit	2

Table A.166: Specification of the bit position DISTANCEOFBESTAND-CORRECTLAYER

Name	TRACKWITHHITWITHCORRECTSLOPEDISTRIBUTION
Meaning	Benignity results for the first LUT with regard to the correct slopes
Bit	3

Table A.167: Specification of the bit position TRACKWITHHITWITH-CORRECTSLOPEDISTRIBUTION

Name	TRACKWITHYZPROJECTION
Meaning	Benignity results for the first LUT with regard to the yz-projection of a single MC Track
Bit	4

Table A.168: Specification of the bit position TRACKWITHYZ-PROJECTION

Name	TRACKWITHHITINCORRECTCELLDISTRIBUTION
Meaning	Benignity results for the second LUT with regard to the correct histogram cell
Bit	5

Table A.169: Specification of the bit position TRACKWITHHITIN-CORRECTCELLDISTRIBUTION

Name	TRACKWITHHITINBESTCELLDISTRIBUTION
Meaning	Benignity results for the second LUT with regard to the histogram cell
Bit	6

Table A.170: Specification of the bit position TRACKWITHHITINBEST-CELLDISTRIBUTION

Name	DISTANCEOFBESTANDCORRECTCELL
Meaning	Benignity results for the second LUT with regard to the distance of the best and the correct histogram cell
Bit	7

Table A.171: Specification of the bit position DISTANCEOFBESTAND-CORRECTCELL

Name	TRACKWITHHITINCORRECTHISTOGRAM-DISTRIBUTION
Meaning	Benignity results for the combination of both LUTs with regard to the correct histogram cell in the correct histogram layer
Bit	8

Table A.172: Specification of the bit position TRACKWITHHITIN-CORRECTHISTOGRAMDISTRIBUTION

Name	TRACKWITHHITINBESTHISTOGRAMDISTRIBUTION
Meaning	Benignity results for the combination of both LUTs with regard to the same histogram cell in the same histogram layer
Bit	9

Table A.173: Specification of the bit position TRACKWITHHITINBEST-HISTOGRAMDISTRIBUTION

Name	DISTANCEOFBESTANDCORRECTHISTOGRAM
Meaning	Benignity results for the combination of both LUTs with regard to the distance of the best and the correct histogram cell
Bit	10

Table A.174: Specification of the bit position DISTANCEOFBESTAND-CORRECTHISTOGRAM

Name	TRACKWITHSIGNATUREDISTRIBUTION
Meaning	Benignity results for the combination of both LUTs with regard to the MC Tracks' hit signatures
Bit	11

Table A.175: Specification of the bit position TRACKWITHSIGNATURE-DISTRIBUTION

Name	TRACKWITHMOMENTUMERRORDISTRIBUTION
Meaning	Benignity results for the combination of both LUTs with regard to the momentum error of the MC Tracks and their corresponding found track
Bit	12

Table A.176: Specification of the bit position TRACKWITHMOMENTUM-ERRORDISTRIBUTION

Name	QUANTIZEDHOUGHSPACEDIMDISTRIBUTION
Meaning	Quantization benignity results separate for each dimension of the Hough space
Bit	13

Table A.177: Specification of the bit position QUANTIZEDHOUGHSPACE-DIMDISTRIBUTION

Name	QUANTIZEDMOMENTADIMDISTRIBUTION
Meaning	Quantization benignity results separate for each dimension of the momenta
Bit	14

Table A.178: Specification of the bit position QUANTIZEDMOMENTA-DIMDISTRIBUTION

Name	QUANTIZEDHOUGHSPACEDISTRIBUTION
Meaning	Quantization benignity results for the combined dimensions of the Hough space
Bit	15

Table A.179: Specification of the bit position QUANTIZEDHOUGHSPACE-DISTRIBUTION

Name	QUANTIZEDMOMENTADISTRIBUTION
Meaning	Quantization benignity results for the combined dimensions of the momenta
Bit	16

Table A.180: Specification of the bit position QUANTIZEDMOMENTA-DISTRIBUTION

Name	SAMEHOUGHSPACECELLDISTRIBUTION
Meaning	Quantization benignity results for the histogram cells with more than one track
Bit	17

Table A.181: Specification of the bit position SAMEHOUGHSPACECELL-DISTRIBUTION

Name	NUMBEROFTRACKSWHICHCANNOTBEFOUND
Meaning	Number of tracks, which cannot be found because of the quantization benignity
Bit	18

Table A.182: Specification of the bit position NUMBEROFTRACKS-WHICHCANNOTBEFOUND

Name	PEAKDISTANCEDIMDISTRIBUTION
Meaning	MC Tracks's peak distance distribution separate for each dimension of the Hough space
Bit	19

Table A.183: Specification of the bit position PEAKDISTANCEDIM-DISTRIBUTION

Name	PEAKDISTANCEDISTRIBUTION
Meaning	MC Track' peak distance distribution for the combined dimensions of the Hough space
Bit	20

Table A.184: Specification of the bit position PEAKDISTANCE-DISTRIBUTION

Name	AVERAGEPEAKDISTANCE
Meaning	Average MC Track' peak distance
Bit	21

Table A.185: Specification of the bit position AVERAGEPEAKDISTANCE

Name	MINIMALTRACKSDISTRIBUTION
Meaning	Distribution of the minimal number of found track candidates
Bit	22

Table A.186: Specification of the bit position MINIMALTRACKS-DISTRIBUTION

Name	MAXIMALTRACKSDISTRIBUTION
Meaning	Distribution of the maximal number of found track candidates
Bit	23

Table A.187: Specification of the bit position MAXIMALTRACKS-DISTRIBUTION

Name	AVERAGETRACKSDISTRIBUTION
Meaning	Distribution of the average number of found track candidates
Bit	24

Table A.188: Specification of the bit position AVERAGETRACKS-DISTRIBUTION

Name	FIFOSDISTRIBUTION
Meaning	Distribution of the FIFO sizes for the found track candidates
Bit	25

Table A.189: Specification of the bit position FIFOSDISTRIBUTION

Name	HITREADOUTDISTRIBUTION
Meaning	Distribution of the number of hits with an entry in multiple histogram layers
Bit	26

Table A.190: Specification of the bit position HITREADOUT-DISTRIBUTION

Name	HITREADOUTAVERAGEDISTRIBUTION
Meaning	Distribution of the average number of histogram layers for one hit with different parallel instantiated layers
Bit	27

Table A.191: Specification of the bit position HITREADOUTAVERAGE-DISTRIBUTION

Name	FPGAHISTOGRAMPROCESSINGTIMING
Meaning	Time consumption for the FPGA histogram implementations with multiple parallel implemented histogram layers
Bit	28

Table A.192: Specification of the bit position FPGAHISTOGRAM-PROCESSINGTIMING

Name	MAGNETFIELDFACTORINFORMATION
Meaning	Information about the evaluated magnetic field factors
Bit	29

Table A.193: Specification of the bit position MAGNETFIELDFACTOR-INFORMATION

Name	PRELUTRANGEINFORMATION
Meaning	Information about the evaluated acceptance range for the first LUT
Bit	30

Table A.194: Specification of the bit position PRELUTRANGE-INFORMATION

In addition to these ones, the next tables give a summary about the bit positions for the parameters 'analysisInitAnalysisMoreResultWarnings' and 'analysisInitAnalysisMoreResultDisplays'.

Name	TRACKSPERLAYERDISTRIBUTION
Meaning	Distribution of the number of found tracks per histogram layer
Bit	0

Table A.195: Specification of the bit position TRACKSPERLAYER-DISTRIBUTION

Name	TRACKDENSITYPERLAYERDISTRIBUTION
Meaning	Distribution of the found track densities per consecutive histogram layers
Bit	1

Table A.196: Specification of the bit position TRACKDENSITYPERLAYERDISTRIBUTION

Name	FPGASERIALIZATIONPROCESSINGTIME
Meaning	Time consumption for the FPGA Serialization implementations with multiple parallel implemented histogram layers
Bit	2

Table A.197: Specification of the bit position FPGASERIALIZATIONPROCESSINGTIME

Name	MCTRACKVISUALIZATION
Meaning	Information about the MC Tracks
Bit	3

Table A.198: Specification of the bit position MCTRACKVISUALIZATION

Name	FOUNDTRACKVISUALIZATION
Meaning	Information about the found tracks
Bit	4

Table A.199: Specification of the bit position FOUNDTRACKVISUALIZATION

Name	PEAKFINDINGGEOMETRYVISUALIZATION
Meaning	Information about the filter geometry which is used in the Peak-finding process
Bit	5

Table A.200: Specification of the bit position PEAKFINDINGGEOMETRY-VISUALIZATION

Name	PROJECTEDPEAKFINDINGGEOMETRY-VISUALIZATION
Meaning	Information about the projected filter geometry which is used in the Peak-finding process
Bit	6

Table A.201: Specification of the bit position PROJECTED-PEAKFINDINGGEOMETRYVISUALIZATION

Name	COVEREDPEAKFINDINGGEOMETRYVISUALIZATION
Meaning	Information about the covered filter geometry which is used in the Peak-finding process
Bit	7

Table A.202: Specification of the bit position COVEREDPEAKFINDING-GEOMETRYVISUALIZATION

Name	COVEREDPROJECTEDPEAKFINDINGGEOMETRY-VISUALIZATION
Meaning	Information about the covered and projected filter geometry which is used in the Peak-finding process
Bit	8

Table A.203: Specification of the bit position COVEREDPROJECTED-PEAKFINDINGGEOMETRYVISUALIZATION

Name	LUTCORRECTIONCOUNTER
Meaning	Number of corrections in the second LUT implementing the math formula
Bit	9

Table A.204: Specification of the bit position LUTCORRECTION-COUNTER

Name	GOODSIGNATURES
Meaning	Information about the number of good signatures in both LUTs
Bit	10

Table A.205: Specification of the bit position GOODSIGNATURES

Name	GOODSIGNATURETABLES
Meaning	Information about the good signatures in both LUTs
Bit	11

Table A.206: Specification of the bit position GOODSIGNATURETABLES

Name	USEDSIGNATURETABLES
Meaning	Information about the used signature tables
Bit	12

Table A.207: Specification of the bit position USEDSIGNATURETABLES

Name	HOUGHPICTURES
Meaning	Draw static pictures for Hough transform
Bit	13

Table A.208: Specification of the bit position HOUGHPICTURES

Since all these parameters can only enable or disable the output of analyses, another group of parameters is doubtless required, which determines the look of the corresponding graphical display. For this purpose, a special source

code file exists for each analysis, which contains all strings for the graphical displays like the axis title, the display title, the display name and the folder name inside the ROOT based file whether saving is possible.
Furthermore the names of these correspondent files are AnalysisLIB/include/projectionEFGCNAnalysisDef.h, AnalysisLIB/include/momentumEFGCAnalysisDef.h, AnalysisLIB/include/magnetfieldFactorAnalysisDef.h, AnalysisLIB/include/magnetfieldAnalysisDef.h and AnalysisLIB/include/histogramAnalysisDef.h.
In addition to this, it has to be also noticed that two supplementary files named AnalysisLIB/include/analysis.h and AnalysisLIB/src/analysis.cxx offer some standard parameters for special analyses, which evaluate the optimal constant magnetic field factors (see section 4.4.2 on page 225), the quality of the first LUT (see section 4.5.1 on page 230) and the quality of both combined LUTs (see section 4.5.3 on page 236).

Finally the last but not less important parameter group belongs to the extensive warning and error message concepts, which are essential in the development phase of the software, but might be negligible for the usage. Hence a separate parameter, which is located in the source code files {<*Library name*>}LIB/include/{<*Library name*>}WarningMsg.h in addition to the relevant classes, is able to enable or disable all warning messages originating from the corresponding library. However just a unique parameter, which is located in the source code file MiscLIB/include/errorHandling.h, is applied to enable or disable all error messages for all libraries, because all these messages are classified to be fundamental. Further on, it is self-evident, but should be explicitly noticed, that this concept does not support different severity levels than the warning messages.
Nevertheless both concepts apply an identifying string, which belongs to the name of the library, because it is of course really significant to know the originating one. The only difference in that case is determined by the function call, because the error messages require the string to be committed, while the warning messages realize this functionality internal. At this juncture, it is additionally clear that these strings are also parameters, which are defined in the source code file MiscLIB/include/projects.h.

A.4 Table Files

This section presents all necessary details about the table files, which form important parts of the implemented Hough tracking software. For this purpose, it has to be at first noticed here that a table file realizes simply a file, which consists of a table in a special format. So it is clear that the word table file is also often used synonymously for table and vice versa. Further on, three different table files are actually responsible for duties in diverse positions of the Hough tracking algorithm. Therefore figure A.1 depicts their corresponding field of operation.

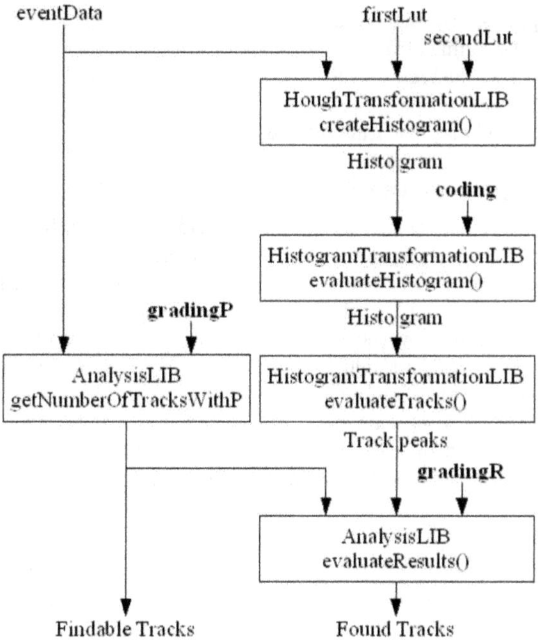

Figure A.1: Place of usage in the algorithm for the table files

A closer look to this figure shows that the first table, which is called 'gradingP', can be found at the left side related to the function *getNumberOfTracksWithP(...)* of the library *AnalysisLIB*. Furthermore the task of this function is obviously determined by the computation of the number of MC

Tracks, which have a higher momentum than a predefined cut. However the usage of this table file features an additional constraint for the signature of these MC Tracks, because only these ones are counted, whose signature is included in this table with a priority class higher than zero. But as the priority class of a signature is of course not really used here, it is nevertheless an important part of this file to build a uniform structure with the other table files, because such a unity offers the possibility to use each file for all purposes. Anyhow it is self-evident that an applicable table file is based somehow on the data of the actual event, because it is used to determine the internal quality of the Hough tracking.

Besides this, the second table, which is just called 'coding', can be found at the right upper side of the figure. As that name already tells, this table is just used for the Encoding of the histogram, which means more precisely that the signature of each histogram cell is simply substituted for the corresponding priority class. For this purpose, the related priority class-signature pair inside the table is obviously identified by its signature. In this connection, it is additionally important to realize that missing pairs are generated by the fiction of the priority class zero.

Beyond that, the last table, which is called 'gradingR', can be found at the right lower side of the figure related to the function *evaluateResults(...)* of the library *AnalysisLIB*. As this function is furthermore obviously used to evaluate the actual quality of the Hough tracking, the task of this table is simply determined by the identification of the track candidates, which represent found peaks in the histogram, by inspecting the hit contribution of the corresponding MC Tracks. So this means more precisely that each track candidate exhibits a separate signature for each MC Tracks contributing at least a single hit, which has to be contained in this table to assume the actual peak to be possibly a found track representing this MC Tracks.
Nevertheless there are evidently additional analyses, which determine the exact character of a track candidate to be a found real track, a clone track, a ghost track or a fake track.

Having now the separate duties clarified, the focus is set to more abstract information, which counts for all table files. For this purpose, it has to be at first mentioned that such a file is obviously implemented by using the common ASCII format, because this circumstance enables all files to be modifiable by almost every available editor and to modify their content without the requirement of a source code recompilation.
Coming now to the point, the layout of such a file is naturally of major interest. So such a file starts as usual with a header, which exposes information about the content, followed by the data. Further on, such a header is divided

into two sections, whose entries are characterized by being either processing overhead or human understanding improvements.
Based on this concept, the processing overhead, which is required for a successful processing by the source code and thus given at first, contains a name, a usage information and the number of entries. Furthermore it is immediately clear that the number of entries is the most important fact, because the value defines the number of objects, which are read afterwards from the data section and must thus be available.
In contrast to this, the human understanding improvements, which are predefined in the source code file DataObjectLIB/src/tableFile.cxx and thus not read from the table file, consists of the 'structureSeparator', the 'blockSeparator', the content and the format of the data section. Further on, it is obvious that the major important items are the 'structureSeparator' and the 'blockSeparator', because the 'structureSeparator' separates properties of an object, while the 'blockSeparator' separates complete objects in the data section.
If the header information is then finished, the file would continue with the data.
A more impressive imagination about the structure of such a table can be gained by the following example in addition to the explanation.

```
/* FILE HEADER STARTS
/* Name of the look up table
name := codingTable
/* Usage of the look up table
usage := Table for coding the histogram signatures
/* Number of entries in the look up table
numberOfEntries := 3
/* The separator for the members of the data structure
/**/ structureSeparator := TABULATOR
/* The separator for the blocks of the data
/**/ blockSeparator := NEWLINE
/* The content of the file
/**/ content := Signature => Classification
/* The format for an entry in the file
/**/ format := [Radix]x[Number][TABULATOR] \\
              [Radix]x[Number][NEWLINE]
/* FILE HEADER STOPS
/*
/*
```

/ FILE DATA STARTS*
2x011111 10x2
2x101111 10x2
2x111111 10x3
/ FILE DATA STOPS*

With regard to this example, it is self-evident that the data comprises three objects, which are separated by the 'blockSeparator' End-Of-Line (EOL). Further on, it is also patent that each object contains two properties or member elements, which are separated by the 'structureSeparator' tabulator. Since this information is apparently enough to be able to realize such a file access, the example is also used to explain some additional circumstances with regard to the table. So as furthermore the content information of the header specifies an object to possess a signature followed by a priority class, the interpretation of the two member elements is clarified. Remembering now the general duty of such a table, it is clear that a test signature has to completely match one of these table signatures to receive the correspondingly defined priority class. Within this context, it is further on clear that the assumed priority class would be zero, if the test signature is not found in the table.

However this is not the total story, because an additional solution for the problem of unequal bit sizes of the test signature and the table signatures must obviously exist too. Hence two easy special cases are considered.
So if the table signatures exhibits more bits than the test signature, the highest bit positions of the table signatures would be simply cut off until the number of bits are equal.
And if otherwise the test signature features more bits than the table signatures, each existing table signature would be extended to the corresponding size by adding enough bit positions in front containing the value 'x'. Within this context, it is furthermore clear that the value 'x' determines an allowed meta tag, which is simply used to represent the two signatures owning a set and an unset bit at that position respectively. So such a meta tag can be easily eliminated by doubling this signature and replacing the 'x' once by a zero and once an one. In this connection, it is also important to realize that the corresponding priority class remains untouched in this process, which means that it stays equal for both new objects.

Besides this, the last important information in the header represents the format of both values, which is accepted. So returning to the given example, it is obvious that the standard is applied for both numbers, which means that they have to start with a radix followed by a 'x' and the number. The reason for this is quite simple, because commonly the signature is set in binary or hexadecimal mode, while the priority class is set in decimal mode. Further on, more information about this format can be found in appendix B on page B-1.

Since the format of the table files is explained in detail, the focus is set in the following to meaningful content. So the earlier mentioned circumstance, that an applicable table file bases on the data of the events, introduces obviously an essential challenge, because it is of course really difficult to generate all required table files containing the tables with an ASCII editor from scratch. Hence diverse algorithms are developed, which offer the possibility to produce more or less good table files automatically. However it is clear that not only a single parameter can offer this functionality, but rather a combination of the parameters 'inputCodingTableMode', 'inputGradingPTableMode', 'inputGradingRTableMode', 'inputCodingTableFileName', 'inputGradingPTableFileName', 'inputGradingRTableFileName', 'inputCodingTableWrite', 'inputGradingPTableWrite' and 'inputGradingRTableWrite' is needed. As the details of all these parameters can be found in appendix A.2 on page A-4, the focus is set only to these ones, which define the mode and respectively the available algorithm.

Besides this, it is more than obvious that all automatically produced tables are suboptimal and can just be generated by some simple algorithms, which are at least able to avoid the problem of unequal bit sizes, because the number of actual used detector stations defines the bit size of the actual required signatures. Moreover the meta tag 'x' in a signature, which is produced automatically, represents the interesting fact that the detector station at this corresponding position delivers no useful information to such a signature. Furthermore it is clear that if all signatures contain this meta tag at the same position, the corresponding detector station could be removed completely, because it would not contribute any useful information at all. As such a circumstance is apparently very interesting, the software analyzes the generated signatures with regard to this situation by the common Quine-McCluskey method, which is able to find such possible meta tag bit positions.

Coming now to the point, the following paragraphs explain each algorithm, which is implemented to generate a table. But at first a short list summarizes all available ones.

- NOTABLE
- CHECKSUMTABLE (recommendation for 'gradingR')
- ONLINETABLE (recommendation for 'gradingP')
- FILETABLE
- LUTGOODNESSTABLE (recommendation for 'coding')

With regard to this listing, the first algorithm, which is identified by the parameter value 'NOTABLE', is obviously also the easiest one, because it simply inserts the value of the signature for the priority class of the same signature. So if such a table is applied for the Encoding step of the algorithm, nothing would be changed in the histogram, because the signature and the corresponding priority class is equal.

Further on, the second algorithm, which is tagged by the parameter value 'CHECKSUMTABLE', inserts the sum of digits of ones in the signature for the priority class of the same signature. In this connection, it has to be additionally noticed that it has been shown that this algorithm is very good to generate the 'gradingR' table in combination with the parameter 'trackfinderMinClassGradingR' of the configuration file (see appendix A.2 on page A-4). However it is important to realize that this supplementary parameter has to be adapted to a value, which is related to the number of used detector stations. So actually the standard value of this parameter is set to 30% of the number of stations, which is evidently evaluated to the value three.

Afterwards the third algorithm, which is selected by the parameter value 'ONLINETABLE', is quite more complex and can not be characterized as easy as the previous ones. Hence a closer look has to be taken at first on the signatures, before the correspondent priority classes can be defined.
So while doing this, it can be seen that such a table consists only of signatures, which have at least a certain percentage of bit positions in each signature set to one. Furthermore this circumstance is not surprising, because the parameter 'analysisInitPercentageOfHitsInSignature' in the configuration file (see appendix A.2 on page A-4) determines this percentage and controls thus the amount of bit positions. Supplementary the parameter 'CONSECUTIVEONLINELAYERS', which is located in the source code

file DataRootObjectLIB/src/tables.cxx, offers the possibility to force these bit positions to be additionally consecutive.
Having now detailed information about the signatures of such a table, the corresponding priority classes are quite simple to evaluate, because each one is computed to the number of bits in the signature set to one minus the minimum number of ones, which is implicitly defined by the percentage, plus one.
In this connection, it has to be also noticed that it has been shown that this algorithm is very good to generate the 'gradingP' table in combination with the parameter 'trackfinderMinClassGradingP' of the configuration file (see appendix A.2 on page A-4). However it is important to realize that this additional parameter can be set constant to one in contrast to the parameter 'trackfinderMinClassGradingR', which is used in the algorithm identified by the value 'CHECKSUMTABLE'.

Subsequently the fourth algorithm, which is picked by the parameter value 'FILETABLE', constitutes a special case, because it does not enable an automatic production algorithm. However as such a table is read out of a file with a given name, it determines nevertheless an automatic generation somehow.

Finally, the last algorithm, which can be chosen by the parameter value 'LUTGOODNESSTABLE', represents the most complex and the only MC data dependent one, because it takes the analysis for the goodness of the first and the second LUT into account. This means more precisely that every MC Track of an event (see event simulation) is transformed into the Hough space by using the histogram layer information of both LUTs, while the resulting signature of the cell, which contains the corresponding found track, is used for the table. So within this context, it is obvious that the corresponding priority class is at a first glance set to one. However if this algorithm processes more than just a single event, these priority classes would be apparently not only set to one, but to the frequency of occurrence for each signature above the used events.
In this connection, it has to be additionally noticed that it has been shown that this algorithm is very good to generate the 'coding' table in combination with the parameter 'trackfinderMinClassCoding' of the configuration file (see appendix A.2 on page A-4). However it is important to realize that this supplementary parameter can be also set constant to one in contrast to the parameter 'trackfinderMinClassGradingR', which is used in the algorithm identified by the value 'CHECKSUMTABLE'.

Since all available algorithms, which offer the possibility to produce the table files automatically, are exhaustively explained, it must be also mentioned that the quality of the Hough tracking can be easily enhanced by improving

these tables manually, because no algorithm suits apparently perfectly. For instance, even if the parameter values 'LUTGOODNESSTABLE', 'ONLINE-TABLE' and 'CROSSFOOTTABLE' constitute the best available solutions for an automatic production of the 'coding', the 'gradingP' and the 'gradingR' table, the huge amount of fake hits, which occur due to the usage of the micro-strip detector stations, disturb the perfect world of tracking.
So improving the 'coding' table means in that case commonly the reduction of the amount of signatures, which exhibit a small frequency of occurrence, because the consequence is determined by less peaks surviving the Encoding step of the algorithm, while the tracking efficiency hopefully remains. However it is immediately clear that the removed signatures concern peaks, which are caused by fake hits as well as peaks representing MC Tracks. So for that reason, it is apparently impractical to configure a good trade-off automatically and the cut must be employed manually. Furthermore a degradation of 2 % or 3 % for the tracking efficiency might be acceptable to receive a degradation of 10 % to 15 % for the amount of fake tracks.
Besides this, improving the 'gradingP' table means commonly the enhancement of the signatures representing these tracks, which should be accepted and found, because they form the standard for the performance measurement. So if just tracks, which exhibit the best signature described by all bit positions set to one, occurs for the search, the Hough tracking performance would be obviously much better than compared to the search of signatures including at least three of seven actually available bit positions set to one.
Finally it should be additionally mentioned that improving the 'gradingR' table means commonly just to amend the parameter 'trackfinderMinClass-GradingR', which is setup in the configuration file (see appendix A.2 on page A-4, because it defines the required number of hits for an accepted signature.

A.5 Scripts

The following script illustrates a simulation script, which produces new data by using UrQMD input and GEANT. Afterwards this data can be obviously used for input to the Hough tracking algorithm in the second script, which demonstrates a reconstruction script. Further on, both scripts are apparently located in the folder `macro/htrack`.

However it has to be finally noticed that the shown simulation script, whose file name is runProduceFile.C, is standard and can be thus used basically for data production.

```
//
// Macro for standard transport simulation using UrQMD
   input and GEANT3
// CBM setup with STS only
//
// C. Steinle    23/01/2007
//

{

  // Adjust this part according to your requirements

  // Output folder for the files
  TString folder      = "files";

  // Collision system
  TString system      = "auau";

  // Beam momentum
  TString beam        = "25gev";

  // Trigger (centrality)
  TString trigger     = "centr";
  // Number of events
  Int_t   nEvents     = 1;

  // Output file name
  TString outFile     = folder + "/" + system + "." +
     beam + "." + trigger + ".mc.root";

  // Parameter file name
  TString parFile     = folder + "/" + system + "." +
     beam + "." + trigger + ".params.root";

  // Cave geometry
  TString caveGeom    = "cave.geo";

```

```
// Target geometry
TString targetGeom   = "target_au_250mu.geo";

// Beam pipe geometry
TString pipeGeom     = "pipe_standard.geo";

// Magnet geometry and field map
TString magnetGeom   = "magnet_standard.geo";
TString fieldMap     = "FieldActive";
Double_t fieldZ      = 50.;      // z position of field
    centre
Double_t fieldScale  = 1.;       // field scaling
    factor

// STS geometry
TString stsGeom      = "sts_Standard_s3055AAFK5.SecD.
    geo";


// In general, the following parts need not be
    touched

// Input file name
TString inFile       = "/d/cbm03/urqmd/" + system + "/
    " + beam + "/"
                     + trigger + "/urqmd." + system +
                         "." + beam + "."
                     + trigger + ".0000.ftn14";

// Debug option
gDebug = 0;

// Timer
TStopwatch timer;
timer.Start();

// Load libraries
gROOT->LoadMacro("$VMCWORKDIR/gconfig/basiclibs.C");
basiclibs();
gSystem->Load("libGeoBase");
gSystem->Load("libParBase");
```

```
gSystem->Load("libBase");
gSystem->Load("libCbmBase");
gSystem->Load("libCbmData");
gSystem->Load("libField");
gSystem->Load("libGen");
gSystem->Load("libPassive");
gSystem->Load("libSts");

// Create detectors and passive volumes
CbmModule* cave    = new CbmCave("CAVE");
cave->SetGeometryFileName(caveGeom);

CbmModule* pipe    = new CbmPipe("PIPE");
pipe->SetGeometryFileName(pipeGeom);

CbmModule* target = new CbmTarget("Target");
target->SetGeometryFileName(targetGeom);

CbmModule* magnet = new CbmMagnet("MAGNET");
magnet->SetGeometryFileName(magnetGeom);

CbmDetector* sts   = new CbmSts("STS", kTRUE);
sts->SetGeometryFileName(stsGeom);

// Create magnetic field
if ( fieldMap == "FieldActive" || fieldMap == "FieldIron")
  CbmFieldMapSym3* magField = new CbmFieldMapSym3(fieldMap);
else if ( fieldMap == "FieldAlligator" )
  CbmFieldMapSym2* magField = new CbmFieldMapSym2(fieldMap);
else {
  CbmField*           magField = NULL;
  cout << "===> ERROR: Field map " << fieldMap << " unknown! " << endl;
  exit;
}
if (magField != NULL) {
  magField->SetPosition(0., 0., fieldZ);
  magField->SetScale(fieldScale);
```

```
}

// Create PrimaryGenerator
CbmPrimaryGenerator* primGen = new CbmPrimaryGenerator();
CbmUrqmdGenerator*  urqmdGen = new CbmUrqmdGenerator(inFile);
primGen->AddGenerator(urqmdGen);

// Create simulation run
CbmRunSim* fRun = new CbmRunSim();
fRun->SetName("TGeant3");                    // Transport engine
fRun->SetOutputFile(outFile);                // Output file
fRun->SetGenerator(primGen);                 // PrimaryGenerator
fRun->SetMaterials("media.geo");             // Materials
fRun->AddModule(cave);
fRun->AddModule(pipe);
fRun->AddModule(target);
fRun->AddModule(magnet);
fRun->AddModule(sts);
fRun->SetField(magField);
fRun->Init();

// Fill parameter containers
CbmRuntimeDb* rtdb          = fRun->GetRuntimeDb();
if (magField != NULL) {
  CbmFieldPar* fieldPar  = (CbmFieldPar*) rtdb->getContainer("CbmFieldPar");
  fieldPar->SetParameters(magField);
  fieldPar->setChanged();
}
Bool_t kParameterMerged  = kTRUE;
CbmParRootFileIo* parOut = new CbmParRootFileIo(kParameterMerged);
parOut->open(parFile.Data());
rtdb->setOutput(parOut);
rtdb->saveOutput();
rtdb->print();

```

```
  // Start run
  fRun->Run(nEvents);

  // Finish
  timer.Stop();
  Double_t rtime = timer.RealTime();
  Double_t ctime = timer.CpuTime();
  cout << endl << endl;
  cout << "Macro finished succesfully." << endl;
  cout << "Output file is "    << outFile << endl;
  cout << "Parameter file is " << parFile << endl;
  cout << "Real time " << rtime << " s, CPU time " << ctime
     << "s" << endl << endl;

}
```

If the data production is now finished and the simulation script terminates, the next step would be obviously determined by running the following reconstruction script. In this connection, it should be mentioned that the file name of this script, which implements a standard Hough tracking without modifying any parameter, is runHoughStsTrackfinder.C. Nevertheless if parameters have to be adapted, appendix A.1 on page A-1, appendix A.2 on page A-4, appendix A.3 on page A-79 and appendix A.4 on page A-101 would feature detailed information.

```
//
// Macro for reconstruction in STS from MC data
//
// Tasks:  CbmStsDigitise
//         CbmStsFindHits
//         CbmStsMatchHits
//         CbmStsFindTracks
//
// C. Steinle    23/01/2007
//

{

  // Adjust this part according to your requirements

```

```
// Input folder for the files
TString inputFolder  = "files";

// Input folder for the files
TString outputFolder = "files";

// Collision system
TString system       = "auau";

// Beam momentum
TString beam         = "25gev";

// Trigger (centrality)
TString trigger      = "centr";

// Input file (MC events)
TString inFile       = inputFolder  + "/" + system +
    "." + beam + "." + trigger + ".mc.root";

// Output file
TString outFile      = outputFolder + "/" + system +
    "." + beam + "." + trigger + ".reco.root";

// Parameter file
TString parFile      = inputFolder  + "/" + system +
    "." + beam + "." + trigger + ".params.root";

// Analysis file
TString anaFile      = outputFolder + "/" + system +
    "." + beam + "." + trigger + ".ana.root";

// Digitisation file
TString digiFile     = "sts_Standard_s3055AAFK5.SecD.
    digi.par";

// Number of events to process
Int_t nEvents        = 1;

// Verbosity level (0=quiet, 1=event level, 2=track
    level, 3=debug)
Int_t iVerbose       = 1;
```

```


// In general, the following parts need not be touched

// Load libraries
gROOT->LoadMacro("$VMCWORKDIR/gconfig/basiclibs.C");
basiclibs();
gSystem->Load("libGeoBase");
gSystem->Load("libParBase");
gSystem->Load("libBase");
gSystem->Load("libCbmBase");
gSystem->Load("libCbmData");
gSystem->Load("libField");
gSystem->Load("libPassive");
gSystem->Load("libSts");
gSystem->Load("libTrkBase");
gSystem->Load("libGeane");
gSystem->Load("libHTrack");

// Timer
TStopwatch timer;
timer.Start();

// Reconstruction run
CbmRunAna *fRun                         = new CbmRunAna()
    ;
fRun->SetInputFile(inFile);
fRun->SetOutputFile(outFile);

// GEANE run
CbmGeane* Geane                         = new CbmGeane(
    inFile);

// STS digitizer
CbmStsDigitize* stsDigitize             = new
    CbmStsDigitize("STSDigitize", iVerbose);
fRun->AddTask(stsDigitize);

// STS hit finding
CbmStsFindHits* findHits                = new
```

```
    CbmStsFindHits("STS␣Hit␣Finder", iVerbose);
fRun->AddTask(findHits);

// STS hit matching
CbmStsMatchHits* matchHits              = new
    CbmStsMatchHits("STS␣Hit␣Matcher", iVerbose);
fRun->AddTask(matchHits);

// STS Track finding
CbmStsFindTracks* findTracks            = new
    CbmStsFindTracks(iVerbose, NULL, kFALSE, "STS␣
    Track␣Finder");
fRun->AddTask(findTracks);

// STS TrackFinder
CbmHoughStsTrackFinder* trackFinder = new
    CbmHoughStsTrackFinder();
trackFinder->SetOutputFile(anaFile);
findTracks->UseFinder(trackFinder);

// Parameter database
TString stsDigiFile                     = gSystem->Getenv
    ("VMCWORKDIR");
stsDigiFile                            += "/parameters/
    sts/";
stsDigiFile                            += digiFile;
CbmRuntimeDb* rtdb                      = fRun->
    GetRuntimeDb();
CbmParRootFileIo*  parIo1               = new
    CbmParRootFileIo();
CbmParAsciiFileIo* parIo2               = new
    CbmParAsciiFileIo();
parIo1->open(parFile.Data());
parIo2->open(stsDigiFile.Data(),"in");
rtdb->setFirstInput(parIo1);
rtdb->setSecondInput(parIo2);
rtdb->setOutput(parIo1);
rtdb->saveOutput();
fRun->LoadGeometry();

// Intialise and run
```

```
    fRun->Init();
    Geane->SetField(fRun->GetField());
    fRun->Run(0,nEvents);

    // Finish
    timer.Stop();
    Double_t rtime = timer.RealTime();
    Double_t ctime = timer.CpuTime();
    cout << endl << endl;
    cout << "Macro finished succesfully." << endl;
    cout << "Output file is "    << outFile << endl;
    cout << "Parameter file is " << parFile << endl;
    cout << "Real time " << rtime << " s, CPU time " <<
       ctime << " s" << endl;
    cout << endl;

}
```

As mentioned earlier, these two scripts would form now everything, which is needed to use the Hough tracking algorithm for the STS in the CBMROOT framework, if the environment for the framework is already setup correctly.

Appendix B

Conversion Routines

Since the library called *fileio* (see section 3.2.2 on page 125) offers the functionality to write and read standard files consisting of a header and a data section, there is obviously a need for a general process, which is able to serialize and deserialize objects, because such a representation is of course easily written or read. For that purpose, this section contains a brief description of the available conversion routines for integrated data types representing numbers like int or double into strings and vice versa, because they define the basic routines to implement such a general process for each available more complex object.

Furthermore it is clear that these conversion routines are also very useful in diverse situations. For that reason, a global point of implementation, which is formed by the file MiscLIB/include/conversionRoutines.h, has to be chosen. Further on, the content of this file is surely determined by a function for the conversion of each integrated data type into a string and vice versa.
Moreover the naming convention is illustrated easily with an example. So the function *int itos(int value, char* string, int radix = 10, int digits = intConversionDigits)* converts a signed integer into a string and the function *int stoi(char* value, int radix = 10)* does it vice versa.
A closer look to the first function shows now that the required parameters contain commonly a buffer to place the output, the length of the buffer, the value of the number and the radix for the string representation. In addition to this, the returning value informs apparently about the number of counted characters in the resulting string for a successful conversion or about the error otherwise.
So if a problem occurs, which might be for example a too small the buffer, the returning value would be minus one and the buffer would consist of a special character, which implements a flag representing the reason of the problem.

In contrast to this, the parameters for the second function are just the string to convert and the radix, while the returning value is obviously the number.

However it is important to notice that the data type boolean supports two different string representations, which are identified by a bit representation with zero and one as well as the adequate strings false and true.

Besides this, there is evidently also the need for a special function, which implements the generation of the format identifier based on the radix of the value for the integrated data type in the converted string. This function is apparently called *int addRadix(int radix, char* value)* and has the task to add the radix of the counting system to interpret the subsequent number correctly.
In addition to this, there is obviously the requirement for the reverse function, which extracts the radix for the conversion of the string into an integrated data type. So in analogousness, the name of this function is *int extractRadix(int* radix, char* value)*.

Coming now to the details, the format of such a string is called radix format and is defined as [sign][radix][number] or [radix][sign][number]. The allowed values for the radix identifier are '2x' till '36x' with the number as base and the 'x' as separator to the value. Additionally the common radixes like 'b', 'B' for binary, 'o', 'O' for octal, 'd', 'D' for decimal and 'h', 'H' for hexadecimal are also supported.

So based on this concept, it is now very easy to handle the file access for complex objects, which base on integrated data types, because the serialization into a string representation and the deserialization into the object can be simply implemented by the combination of these basic functions and a derived class using the library *fileio*.

Appendix C

Synthesis Extras

Before presenting additional information for the FPGA implementation, this chapter has to clarify at first the meaning of the word synthesis, because it is commonly used in many different fields to represent usually a process, which combines two or more pre-existing elements to form something new. But within this thesis, the word synthesis refers to the process of converting a high-level Electronic Design Automation (EDA) design into a low-level FPGA behavior specification.
For this purpose, the startup is determined by the creation of a chip independent behavior description by using either a HDL like VHDL and Verilog or a schematic designer. Further on, this description is afterwards committed to an electronic design automation tool, which generates a technology-based netlist. This netlist is then fitted to the actual FPGA device by using a process called mapping. Subsequently such a mapped netlist, which realizes a gate level description, is further used to evaluate a good positioning for the elements on the device by the place-and-route process. By the way, it has to be noticed that these two processes are commonly performed by the FPGA company's proprietary map and place-and-route software. Later on, the map as well as the place-and-route results have to be validated via timing analyses, simulation and other verification methodologies. Once the design and validation process is completed, the binary file, which is generated by using the FPGA company's proprietary software once again, is used to configure or reconfigure the FPGA device.

Besides this short summary, detailed information about the whole synthesis flow can be found in [Flo], while exact information about the configuration can be found in [Vira].
Moreover, it has to be also noticed here that the design of complex FPGA descriptions is simplified by the existence of supporting libraries consisting

of predefined complex functions and circuits, that have been tested and optimized to speed up the design process. Commonly these predefined circuits are called IP cores and are available from FPGA vendors and third-party IP suppliers. However these cores are rarely free and typically released under proprietary licenses. Other predefined circuits are also available from developer communities such as OpenCores and other sources. But in contrast to the previous ones, these are typically free and released under the GNU General Public License (GPL), Berkeley Software distribution license (BSD) or similar license.

Coming now to the point, the following sections present detailed information about the synthesis configuration, which is used to produce the results exhibited in section 5.5 on page 327, in combination with report files and selected absolute results for the single-chip and the multi-chip implementation of the Hough tracking algorithm.

C.1 Main Batch Files

This section presents general information about the main batch files, which are applied to reduce the complexity of the synthesis process. So the main folder of each functional unit contains such a batch file, which implements apparently a complete synthesis run. In this connection, it is obvious that the synthesis parameters (see [Syn00]), which are setup in the HDL source files, are not modified by these batch files. Hence a change belonging to them must be done manually and the new results must be obtained by a rerun of the batch file afterwards. Moreover a typical batch file, which bases on this concept, looks like:

```
@echo off
cd <library>\ise\<implementation>
REM xst.exe -f xst.xmd
REM ngdbuild.exe -f ngdbuild.xmd
synplify_pro.exe -batch <unit>.spr
ngdbuild.exe -f ngdbuild.smd
map.exe -f map.cmd
par.exe -f par.cmd
trce.exe -f trce.cmd
cd..
cd..
cd..
```

Within such a file, the first interesting circumstance can be found in line three and four, because they are not executed due to the REM command. But the reason for this circumstance is obvious, because it is simply identified by the fact that the synthesis startup is realized by the tool SYNPLIFY PRO in line five and six instead of the tool XST. So moving the REM from line three and four to five and six easily switches between these two tools.
Further on, the following two commands implement the mapping and the place-and-route processes, while the final command traces the result with regard to the timing.
Besides this, it is important to realize that each command needs special information, which is supplied by separate additional configuration files specified by the option '-f'. For that reason, the next few sections introduce detailed information about each separate program in combination with its configuration file.

C.2 Synthesis Tool XST

The synthesis tool XST, which is part of the vendor specific software suite ISE 8.2.03i Application Version I.34 (see [ISE]) from the company Xilinx (see [Xila]), offers the possibility to produce netlist files by the usage of an option file typically looking like:

```
-intstyle ise
-ifn <unit>.xst
-ofn <unit>.syr
```

A closer look to the details of such a file shows that it contains just a definition for the style, the input file name and the output file name. However it is important at this juncture that this input file specifies another option file, which looks usually like:

```
set -xsthdpdir ./xst


run
-ifn <unit>.xpr
-ifmt mixed
-ofn <unit>
-ofmt NGC
-top <unit>
-p xc5vlx110-ff1760-3
-opt_mode Speed
```

```
-opt_level 1
-fsm_encoding Auto
-rtlview NO
-iuc NO
-resource_sharing YES
-safe_implementation NO
-bus_delimiter <>
-write_timing_constraints YES
-read_cores YES
-sd ./cores
-decoder_extract YES
-shreg_extract YES
-priority_extract YES
-shift_extract YES
-fsm_extract YES
-fsm_style lut
-ram_extract YES
-ram_style Auto
-rom_extract YES
-rom_style Auto
-mux_extract YES
-mux_style Auto
-xor_collapse YES
-cross_clock_analysis NO
-iob Auto
-iobuf YES
-keep_hierarchy YES
-optimize_primitives NO
-move_first_stage YES
-move_last_stage YES
-register_duplication YES
-slice_packing YES
-slice_utilization_ratio 100
-slice_utilization_ratio_maxmargin 5
-glob_opt allclocknets
-register_balancing No
```

While looking at this content, it is clear that this secondary file consists of all device and mapping options in combination with some special definitions. Furthermore the two major important special definitions are determined by the output format, which is set by the parameter '-ofmt' to a Native Generic

Circuit (NGC) file, and the parameter '-ifn', which specifies an additional input file submitting all required HDL source files. In this connection, it is interesting that such another input file generally looks like:

```
vhdl  hbuffer_lib  <source file 1>
vhdl  hbuffer_lib  <source file 2>
<and so on>
```

So based on this concept, the synthesis tool XST requires three different option files for a successful processing, while the generated report file is distinguished by the file extension SYnthesis Report (SYR).

C.3 Synthesis Tool SYNPLIFY PRO

The synthesis tool SYNPLIFY PRO Version 9.0.1 (see [Synb]) from the company Synplicity (see [Syna]) offers also the possibility to produce netlist files by the usage of an option file, which customarily looks like:

```
#add_file options
add_file -vhdl -lib work "<source␣file␣1>"
add_file -vhdl -lib work "<source␣file␣2>"
<and so on>

#implementation: "rev_1"
impl -add rev_1 -type fpga

#device options
set_option -technology VIRTEX5
set_option -part XC5VLX110
set_option -package FF1760
set_option -speed_grade -3

#compilation/mapping options
set_option -default_enum_encoding default
set_option -resource_sharing 1
set_option -use_fsm_explorer 0
set_option -top_module "<unit>"

#map options
set_option -frequency auto
set_option -run_prop_extract 1
```

```
set_option -fanout_limit 10000
set_option -disable_io_insertion 0
set_option -pipe 1
set_option -update_models_cp 0
set_option -verification_mode 0
set_option -retiming 0
set_option -no_sequential_opt 0
set_option -fixgatedclocks 3
set_option -fixgeneratedclocks 3

#sequential_optimizations options
set_option -symbolic_fsm_compiler 1

#simulation options
set_option -write_verilog 0
set_option -write_vhdl 0

#VIF options
set_option -write_vif 1

#automatic place and route (vendor) options
set_option -write_apr_constraint 1

#set result format/file last
project -result_file "rev_1/<unit>.edf"

#implementation attributes
set_option -vlog_std v2001
set_option -synthesis_onoff_pragma 0
set_option -project_relative_includes 1
impl -active "rev_1"
```

A more detailed look to this content shows now that this synthesis tool applies in contrast to the tool XST just a single option file, which consists of all HDL source files, device and mapping options as well as some special definitions like the output file format set by the parameter 'project -result_file' to Electronic Data interchange Format file (EDF). Besides this, the produced report file is branded also contrarily by the file extension Synthesis Results and Reports (SRR).

C.4 Synthesis Tool NGDBUILD

The synthesis tool NGDBUILD, which is part of the vendor specific software suite ISE 8.2.03i Application Version I.34 (see [ISE]) from the company Xilinx (see [Xila]), performs all necessary steps to create a file in the Native Generic Database (NGD) format based on a netlist. Such a created file contains then naturally a logical description of the design, which is reduced to Xilinx NGD primitives, as well as a description in terms of the original hierarchy, which is expressed in the input netlist. But as the initial netlist can be produced by the application of two different tools, the option file used here must be apparently adapted. For this purpose, the following lines show the typical option file corresponding to the tool XST:

```
-intstyle ise
-dd _ngo
-nt timestamp
-p xc5vlx110-ff1760-3
-uc <unit>.ucf
<unit>.ngc
<unit>.ngd
```

Besides this, the next few lines inform about a typical option file for the tool SYNPLIFY PRO.

```
-intstyle ise
-dd _ngo
-nt timestamp
-p xc5vlx110-ff1760-3
-uc <unit>.ucf
./rev_1/<unit>.edf
<unit>.ngd
```

With regard to these two structures, the only difference is obviously determined by the netlist format of the input file, because the first one expects a XST evaluated NGC file, while the second one anticipates an EDF file, which is produced by the synthesis tool SYNPLIFY PRO. Beyond that, a secondary but not less interesting input file is identified by the parameter '-uc', because it contains user constraints in the User Constraints File (UCF) format. As such a file contains furthermore just timing requirements, its organization is very simple and looks typically like:

```
NET "clk" TNM_NET = clk;
TIMESPEC TS_clk = PERIOD "clk" 7.3 ns \
    INPUT_JITTER 0 ps PRIORITY 1;
```

Finally it has to be noticed here that the generated report file is earmarked by the file extension ngdBuiLD (BLD).

C.5 Synthesis Tool MAP

The synthesis tool MAP, which is part of the vendor specific software suite ISE 8.2.03i Application Version I.34 (see [ISE]) from the company Xilinx (see [Xila]), performs all needed steps to map the NGD netlist to the actual chosen FPGA device by the application of an option file usually looking like:

```
-intstyle ise
-p xc5vlx110-ff1760-3
-w
-logic_opt off
-ol high
-t 1
-cm area
-detail
-pr b
-k 6
-c 100
-tx off
-o <unit>_map.ncd <unit>.ngd <unit>.pcf
```

So based on that options, this tool generates two separate files, which are identified by the Native Circuit Description (NCD) format containing the physical description of the design and the Physical Constraint File (PCF) format consisting of the physical constraints. Besides this, the produced report file is characterized by the file extension Map RePort (MRP).

C.6 Synthesis Tool PAR

The synthesis tool PAR, which is part of the vendor specific software suite ISE 8.2.03i Application Version I.34 (see [ISE]) from the company Xilinx (see [Xila]), evaluates a good positioning for the mapped FPGA device elements by the place-and-route process, which is configured by an option file customarily looking like:

```
-intstyle ise
-w
-ol high
-t 1
<unit>_map.ncd <unit>.ncd <unit>.pcf
```

This tool generates with regard to the options a single file in the NCD format, which contains the physical description of the design in combination with the place-and-route information. Besides this, the resulting report file, whose exemplary content can be found in appendix C.8.1 on page C-10, is marked by the file extension PAR.

C.7 Optional Tool TRCE

The optional tool TRCE, which is part of the vendor specific software suite ISE 8.2.03i Application Version I.34 (see [ISE]) from the company Xilinx (see [Xila]), creates timing simulation data after the place-and-route process by the usage of an option file generally looking like:

```
-intstyle ise
-e 3
-l 3
-s 3
-xml <unit> <unit>.ncd
-o <unit>.twr <unit>.pcf
-ucf <unit>.ucf
```

It has to be noticed here that the timing data, which is produced by this tool, is of course of major importance, because it could be used to determine, if the timing requirements and the functionality of the design have been met. For this purpose, the output file, which can be exemplary found in appendix C.8.2 on page C-15, contains the timing analysis result and is characterized by the file extension Timing Wizard Report (TWR).

C.8 Example Files

As the previous sections introduce diverse types of report files, the two most important ones, which are obviously determined by the file extension PAR and TWR, are exemplarily shown in the following to receive an imagination and a deeper insight. In contrast to them, the other report files are not presented here for completeness, because they are almost too big and not that important.

C.8.1 Place-and-route Report Report File

```
Release 8.2.03i par I.34
Copyright (c) 1995-2006 Xilinx, Inc.  All rights reserved.

MP-LAP35::  Thu Jul 17 09:09:30 2008

par -intstyle ise -w -ol high -t 1 hbuffer_map.ncd hbuffer.ncd hbuffer.pcf

Constraints file: hbuffer.pcf.
Loading device for application Rf_Device from file '5vlx110.nph' in environment C:\Programme\Xilinx.
   "hbuffer" is an NCD, version 3.1, device xc5vlx110, package ff1760, speed -3

Initializing temperature to 85.000 Celsius. (default - Range: 0.000 to 85.000 Celsius)
Initializing voltage to 0.950 Volts. (default - Range: 0.950 to 1.050 Volts)

Device speed data version:  "ADVANCED 1.48 2006-09-05".

INFO:Par:253 - The Map -timing placement will be retained since it is likely to achieve better performance.

Device Utilization Summary:

```

```
    Number of BUFGs                          1 out of 32      3%
    Number of External IOBs                 71 out of 800     8%
       Number of LOCed IOBs                  0 out of 71      0%

    Number of RAMB36_EXPs                   43 out of 128    33%
    Number of Slice Flip Flops            4054 out of 69120   5%
    Number of Slice LUTS                  7306 out of 69120  10%
    Number of Slice LUT-Flip Flop pairs 7371 out of 69120  10%

Overall effort level (-ol):   High
Router effort level (-rl):    High

Starting initial Timing Analysis.  REAL time: 59 secs
Finished initial Timing Analysis.  REAL time: 1 mins

Starting Router

Phase 1: 46171 unrouted;       REAL time: 1 mins 2 secs

Phase 2: 42949 unrouted;       REAL time: 1 mins 5 secs

Phase 3: 31877 unrouted;       REAL time: 1 mins 26 secs

Phase 4: 31877 unrouted; (3403895)      REAL time: 1 mins 28 secs

Phase 5: 31834 unrouted; (53996)      REAL time: 8 mins 4 secs

Phase 6: 31895 unrouted; (0)      REAL time: 9 mins 12 secs

```

```
Phase 7: 0 unrouted; (0)      REAL time: 11 mins 36 secs

Phase 8: 0 unrouted; (0)      REAL time: 11 mins 48 secs

Phase 9: 0 unrouted; (0)      REAL time: 12 mins 8 secs

Total REAL time to Router completion: 12 mins 8 secs
Total CPU time to Router completion: 12 mins 2 secs

Partition Implementation Status
-------------------------------

  No Partitions were found in this design.

-------------------------------

Generating "PAR" statistics.

**************************
Generating Clock Report
**************************

|        Clock Net       |     Resource    |Locked|Fanout|Net Skew(ns)|Max Delay(ns)|
|                  clk_c | BUFGCTRL_X0Y0| No   | 2075 |  0.474      | 1.399       |

* Net Skew is the difference between the minimum and maximum routing
only delays for the net. Note this is different from Clock Skew which
is reported in TRCE timing report. Clock Skew is the difference between
the minimum and maximum path delays which includes logic delays.

   The Delay Summary Report

```

```
The NUMBER OF SIGNALS NOT COMPLETELY ROUTED for this
   design is: 0

   The AVERAGE CONNECTION DELAY for this design is:
              1.433
   The MAXIMUM PIN DELAY IS:
                                   6.108
   The AVERAGE CONNECTION DELAY on the 10 WORST NETS is
      :    4.817

   Listing Pin Delays by value: (nsec)

    d < 1.00    < d < 2.00  < d < 3.00  < d < 4.00   < d
      < 7.00  d >= 7.00
   ----------  ----------  ----------  ----------
      ----------  ----------
        16856       17831        6899         2177
                 919            0

Timing Score: 0

Asterisk (*) preceding a constraint indicates it was
   not met.
   This may be due to a setup or hold violation.

  Constraint                                  | Requested
       | Actual      | Logic   | Absolute     |Number of
                                              |
                                                 |
                                                 | Levels
                                                  | Slack
                                                        |
                                                 errors
  TS_clk = PERIOD TIMEGRP "clk" 7.3 ns HIGH | 7.300ns
       | 7.300ns     | 8        | 0.000ns      | 0
   50% INPUT_JITTER 0 ns PRIORITY 1           |
                    |           |       |           |

All constraints were met.
```

```

Generating Pad Report.

All signals are completely routed.

Total REAL time to PAR completion: 12 mins 18 secs
Total CPU time to PAR completion: 12 mins 12 secs

Peak Memory Usage:  476 MB

Placer: Placement generated during map.
Routing: Completed - No errors found.
Timing: Completed - No errors found.

Number of error messages: 0
Number of warning messages: 0
Number of info messages: 1

Writing design to file hbuffer.ncd

PAR done!
```

C.8.2 Timing Wizard Report Report File

```
Release 8.2.03i Trace
Copyright (c) 1995-2006 Xilinx, Inc.  All rights reserved.

trce.exe -intstyle ise -e 3 -l 3 -s 3 -xml hbuffer hbuffer.ncd -o hbuffer.twr
hbuffer.pcf -ucf hbuffer.ucf

Design file:              hbuffer.ncd
Physical constraint file: hbuffer.pcf
Device,speed:             xc5vlx110,-3 (ADVANCED 1.48 2006-09-05)
Report level:             error report

Environment Variable      Effect
--------------------      ------
NONE                      No environment variables were set

INFO:Timing:2752 - To get complete path coverage, use the unconstrained paths
   option. All paths that are not constrained will be reported in the
   unconstrained paths section(s) of the report.

Timing constraint: TS_clk = PERIOD TIMEGRP "clk" 7.3 ns HIGH 50% INPUT_JITTER 0 ns PRIORITY 1;

 1373949 items analyzed, 0 timing errors detected. (0 setup errors, 0 hold errors)
 Minimum period is   7.300ns.

All constraints were met.

Data Sheet report:
-----------------
All values displayed in nanoseconds (ns)

Clock to Setup on destination clock clk
```

```
                  | Src:Rise| Src:Fall| Src:Rise| Src:Fall|
Source Clock      |Dest:Rise|Dest:Rise|Dest:Fall|Dest:Fall|
clk               |    7.300|         |         |         |

Timing summary:
---------------

Timing errors: 0  Score: 0

Constraints cover 1373949 paths, 0 nets, and 40306 connections

Design statistics:
   Minimum period:   7.300ns   (Maximum frequency: 136.986MHz)

Analysis completed Thu Jul 17 09:22:19 2008

Trace Settings:
-------------------------
Trace Settings

Peak Memory Usage: 399 MB
```

C.9 Absolute Values for the FPGA Occupancy

This section presents the absolute values for the FPGA resource occupancy of different implementations, which realize the Hough transform for the first level tracking purpose in the CBM experiment. Furthermore this detailed information is additionally split into two sections related to the versions, which means single-chip and multi-chip.

C.9.1 Single-Chip Version

This section contains the absolute values for the results of two different Hough transform implementations, which fit into a single FPGA. Furthermore these implementations differ only in the realization strategy of the histogram read-out. So it is clear that the two chosen implementations identify the best available one for each strategy. For that reason, the building blocks, which are selected for each functional unit, are summarized in table C.1.

Read-out	Prelut	HBuffer	Lut	Generator	Histogram	Encoder	Diagonalizer	Peakfinder2d	Serializer	LBuffer	Peakfinder3d
↓	-	2	-	2	1	4	1	1	1	-	-
→	-	2	-	12	27	10	2	2	2	-	-

Table C.1: Summary of the selected building blocks for the Hough transform fitting into a single FPGA

Coming now to the point, it is obvious that the absolute synthesis results of each building block, which are presented in section 5.5 on page 327, enable the calculation of the total hardware resource consumption for each implementation. Hence table C.2 presents this computation for the row-wise histogram read-out at the bottom, while table C.3 exhibits the calculation for the column-wise histogram read-out at right.

Unit	Slice Logic Utilization				Slice Logic Distribution		Specific Feature Utilization (BlockRAM)
	Flip-Flops	LUTs			Either LUT or Flip-Flop	Both LUT and Flip-Flop	
		Logic	Memory	Route Thrus			
Network In	2× 350	2× 350	2× 0	2× 0	2× 600	2× 200	2× 72
Prelut	1× 2,500	1× 5,000	1× 0	1× 0	1× 5,000	1× 2,500	1× 0
HBuffer	1× 4,054	1× 7,304	1× 0	1× 2	1× 7,371	1× 3,989	1× 1,548
Lut	1× 2,500	1× 5,000	1× 0	1× 0	1× 5,000	1× 2,500	1× 0
Generator	1× 61	1× 101	1× 30	1× 0	1× 189	1× 3	1× 0
Histogram	1×24,309	1×26,330	1× 0	1× 0	1×27,171	1×23,468	1× 0
Encoder	1× 0	1× 363	1× 0	1× 0	1× 363	1× 0	1× 1,710
Diagonalizer	1× 568	1× 188	1× 0	1× 0	1× 568	1× 188	1× 0
Peakfinder2d	1× 769	1× 494	1× 0	1× 0	1× 998	1× 265	1× 0
Serializer	1× 572	1× 1,031	1× 517	1× 194	1× 1,789	1× 525	1× 72
LBuffer	-	-	-	-	-	-	-
Peakfinder3d	-	-	-	-	-	-	-
Network Out	2× 350	2× 350	2× 0	2× 0	2× 600	2× 200	2× 72
Sum	36,733	47,211	547	196	50,849	34,238	3,618
xc5vlx110	÷ 69,120	÷ 69,120	÷ 17,920	÷ 69,120	÷ 69,120	÷ 69,120	÷ 4,608
Occupancy	53.14 %	68.30 %	3.05 %	0.28 %	73.57 %	49.53 %	78.52 %

Table C.2: Occupancy of the FPGA for the Hough transform realization with a single-chip and a row-wise histogram read-out at the bottom

Unit	Slice Logic Utilization				Slice Logic Distribution		Specific Feature Utilization (BlockRAM)
	Flip-Flops	LUTs			Either LUT or Flip-Flop	Both LUT and Flip-Flop	
		Logic	Memory	Route Thrus			
Network In	2× 350	2× 350	2× 0	2× 0	2× 600	2× 200	2× 72
Prelut	1× 2,500	1× 5,000	1× 0	1× 0	1× 5,000	1× 2,500	1× 0
HBuffer	1× 4,054	1× 7,304	1× 0	1× 2	1× 7,371	1× 3,989	1× 1,548
Lut	1× 2,500	1× 5,000	1× 0	1× 0	1× 5,000	1× 2,500	1× 0
Generator	1× 133	1× 56	1× 26	1× 0	1× 180	1× 35	1× 0
Histogram	1× 1,176	1× 798	1× 868	1× 1	1× 2,176	1× 667	1× 0
Encoder	1× 0	1× 134	1× 0	1× 0	1× 134	1× 0	1× 558
Diagonalizer	1× 184	1× 60	1× 0	1× 0	1× 184	1× 60	1× 0
Peakfinder2d	1× 312	1× 132	1× 0	1× 0	1× 324	1× 120	1× 0
Serializer	1× 363	1× 553	1× 538	1× 2	1× 1,114	1× 342	1× 72
LBuffer	-	-	-	-	-	-	-
Peakfinder3d	-	-	-	-	-	-	-
Network Out	2× 350	2× 350	2× 0	2× 0	2× 600	2× 200	2× 72
Sum	12,622	20,437	1,432	5	23,883	11,013	2,466
xc5vlx110	÷ 69,120	÷ 69,120	÷ 17,920	÷ 69,120	÷ 69,120	÷ 69,120	÷ 4,608
Occupancy	18.26 %	29.57 %	7.99 %	0.01 %	34.55 %	15.93 %	53.52 %

Table C.3: Occupancy of the FPGA for the Hough transform realization with a single-chip and a column-wise histogram read-out at right

C.9.2 Multi-Chip Version

This section contains the absolute values for the results of three different Hough transform implementations, which do not fit into a single FPGA, but can be realized with a multi-chip platform. Furthermore these implementations differ only in the parallelism level of the histogram layers and the parallelism level of the histogram read-out. So it is clear that the three chosen implementations identify the best available one for each level combination. For that reason, the building blocks, which are selected for each functional unit, are summarized in table C.4.

Id	**Layer parallelism**	**Read-out parallelism**	**Prelut**	**HBuffer**	**Lut**	**Generator**	**Histogram1**	**Histogram2**	**Encoder**	**Diagonalizer**	**Peakfinder2d**	**Serializer**	**LBuffer**	**Peakfinder3d**
Input interface FPGA														
1	9	9	-	-	-	10	27	26	10	2	2	2	-	-
2	0	0	-	-	-	x	x	x	x	x	x	x	-	-
3	11	3	-	-	-	10	27	26	10	2	2	2	-	-
Processor FPGA														
1	14	14	-	-	-	10	27	26	10	2	2	2	-	-
2	24	1	-	-	-	10	27	26	10	2	2	2	-	-
3	20	5	-	-	-	10	27	26	10	2	2	2	-	-

Table C.4: Summary of the selected building blocks for a multi-chip platform realizing the Hough transform

Coming now to the point, it is obvious that the absolute synthesis results of each building block, which are presented in section 5.5 on page 327, enable the calculation of the total hardware resource consumption for each implementation. Hence the tables C.5 and C.6 exhibit this computation for both FPGA types of id one. In addition to this, the tables C.7 and C.8 present naturally the calculation for the FPGA types of id two, while the tables C.9 and C.10 show finally the analog evaluation for the FPGA types of id three.

Unit	Slice Logic Utilization				Slice Logic Distribution		Specific Feature Utilization (BlockRAM)
	Flip-Flops	LUTs			Either LUT or Flip-Flop	Both LUT and Flip-Flop	
		Logic	Memory	Route Thrus			
Network In	3× 350	3× 350	3× 0	3× 0	3× 600	3× 200	3× 72
Prelut	1× 2,500	1× 5,000	1× 0	1× 0	1× 5,000	1× 5,000	1× 0
HBuffer	0× 4,054	0× 7,304	0× 0	0× 2	0× 7,371	0× 3,989	0× 1,548
Lut	1× 2,500	1× 5,000	1× 0	1× 0	1× 5,000	1× 5,000	1× 0
Generator	1× 507	1× 30	1× 0	1× 0	1× 524	1× 13	1× 0
Histogram1	6× 1,176	6× 798	6× 868	6× 1	6× 2,176	6× 667	6× 0
Histogram2	3× 936	3× 768	3× 0	3× 1	3× 1,299	3× 406	3× 558
Encoder	9× 237	9× 93	9× 0	9× 0	9× 250	9× 80	9× 0
Diagonalizer	9× 184	9× 60	9× 0	9× 0	9× 184	9× 60	9× 0
Peakfinder2d	9× 312	9× 132	9× 0	9× 0	9× 324	9× 120	9× 0
Serializer	9× 363	9× 553	9× 538	9× 2	9× 1,114	9× 342	9× 72
Network Out	6× 350	6× 350	6× 0	6× 0	6× 600	6× 200	6× 72
Sum	28,385	27,814	10,050	18	49,725	22,451	2,970
xc5vlx110	÷ 69,120	÷ 69,120	÷ 17,920	÷ 69,120	÷ 69,120	÷ 69,120	÷ 4,608
Occupancy	41.07 %	40.24 %	56.08 %	0.03 %	71.94 %	32.48 %	64.45 %

Table C.5: Occupancy of the input interface FPGA which is used for the multi-chip platform with id one realizing the Hough transform

Unit	Slice Logic Utilization				Slice Logic Distribution		Specific Feature Utilization
	Flip-Flops	LUTs			Either LUT or Flip-Flop	Both LUT and Flip-Flop	(BlockRAM)
		Logic	Memory	Route Thrus			
Network In	4× 350	4× 350	4× 0	4× 0	4× 600	4× 200	4× 72
Generator	1× 507	1× 30	1× 0	1× 0	1× 524	1× 13	1× 0
Histogram1	10× 1,176	10× 798	10× 868	10× 1	10× 2,176	10× 667	10× 0
Histogram2	4× 936	4× 768	4× 0	4× 1	4× 1,299	4× 406	4× 558
Encoder	14× 237	14× 93	14× 0	14× 0	14× 250	14× 80	14× 0
Diagonalizer	14× 184	14× 60	14× 0	14× 0	14× 184	14× 60	14× 0
Peakfinder2d	14× 312	14× 132	14× 0	14× 0	14× 324	14× 120	14× 0
Serializer	14× 363	14× 553	14× 538	14× 2	14× 1,114	14× 342	14× 72
Network Out	2× 350	2× 350	2× 0	2× 0	2× 600	2× 200	2× 72
Sum	33,455	24,914	16,212	42	57,288	17,935	3,672
xc5vlx110	÷ 69,120	÷ 69,120	÷ 17,920	÷ 69,120	÷ 69,120	÷ 69,120	÷ 4,608
Occupancy	48.40 %	36.04 %	90.47 %	0.06 %	82.88 %	25.95 %	79.69 %

Table C.6: Occupancy of the input interface FPGA which is used for the multi-chip platform with id one realizing the Hough transform

Unit	Slice Logic Utilization				Slice Logic Distribution		Specific
	Flip-Flops	LUTs			Either LUT	Both LUT	Feature
		Logic	Memory	Route Thrus	or Flip-Flop	and Flip-Flop	Utilization (BlockRAM)
Network In	3× 350	3× 350	3× 0	3× 0	3× 600	3× 200	3× 72
Prelut	1× 2,500	1× 5,000	1× 0	1× 0	1× 5,000	1× 5,000	1× 0
HBuffer	0× 4,054	0× 7,304	0× 0	0× 2	0× 7,371	0× 3,989	0× 1,548
Lut	1× 2,500	1× 5,000	1× 0	1× 0	1× 5,000	1× 5,000	1× 0
Generator	0× 507	0× 30	0× 0	0× 0	0× 524	0× 13	0× 0
Histogram1	0× 1,176	0× 798	0× 868	0× 1	0× 2,176	0× 667	0× 0
Histogram2	0× 936	0× 768	0× 0	0× 1	0× 1,299	0× 406	0× 558
Encoder	0× 237	0× 93	0× 0	0× 0	0× 250	0× 80	0× 0
Diagonalizer	0× 184	0× 60	0× 0	0× 0	0× 184	0× 60	0× 0
Peakfinder2d	0× 312	0× 132	0× 0	0× 0	0× 324	0× 120	0× 0
Serializer	0× 363	0× 553	0× 538	0× 2	0× 1,114	0× 342	0× 72
Network Out	4× 350	4× 350	4× 0	4× 0	4× 600	4× 200	4× 72
Sum	7,450	12,450	0	0	14,200	11,400	504
xc5vlx110	÷ 69,120	÷ 69,120	÷ 17,920	÷ 69,120	÷ 69,120	÷ 69,120	÷ 4,608
Occupancy	10.78 %	18.01 %	0.00 %	0.00 %	20.54 %	16.49 %	10.94 %

Table C.7: Occupancy of the input interface FPGA which is used for the multi-chip platform with id two realizing the Hough transform

Unit	Slice Logic Utilization				Slice Logic Distribution		Specific Feature Utilization (BlockRAM)
	Flip-Flops	LUTs			Either LUT or Flip-Flop	Both LUT and Flip-Flop	
		Logic	Memory	Route Thrus			
Network In	4× 350	4× 350	4× 0	4× 0	4× 600	4× 200	4× 72
Generator	1× 507	1× 30	1× 0	1× 0	1× 524	1× 13	1× 0
Histogram1	18× 1,176	18× 798	18× 868	18× 1	18× 2,176	18× 667	18× 0
Histogram2	6× 936	6× 768	6× 0	6× 1	6× 1,299	6× 406	6× 558
Encoder	1× 237	1× 93	1× 0	1× 0	1× 250	1× 80	1× 0
Diagonalizer	1× 184	1× 60	1× 0	1× 0	1× 184	1× 60	1× 0
Peakfinder2d	1× 312	1× 132	1× 0	1× 0	1× 324	1× 120	1× 0
Serializer	1× 363	1× 553	1× 538	1× 2	1× 1,114	1× 342	1× 72
Network Out	2× 350	2× 350	2× 0	2× 0	2× 600	2× 200	2× 72
Sum	30,487	21,940	16,162	26	52,958	16,257	3,852
xc5vlx110	÷ 69,120	÷ 69,120	÷ 17,920	÷ 69,120	÷ 69,120	÷ 69,120	÷ 4,608
Occupancy	44.11 %	31.74 %	90.19 %	0.04 %	76.62 %	23.52 %	83.59 %

Table C.8: Occupancy of the input interface FPGA which is used for the multi-chip platform with id two realizing the Hough transform

Unit	Slice Logic Utilization				Slice Logic Distribution		Specific Feature Utilization (BlockRAM)
	Flip-Flops	LUTs			Either LUT or Flip-Flop	Both LUT and Flip-Flop	
		Logic	Memory	Route Thrus			
Network In	3× 350	3× 350	3× 0	3× 0	3× 600	3× 200	3× 72
Prelut	1× 2,500	1× 5,000	1× 0	1× 0	1× 5,000	1× 5,000	1× 0
HBuffer	0× 4,054	0× 7,304	0× 0	0× 2	0× 7,371	0× 3,989	0× 1,548
Lut	1× 2,500	1× 5,000	1× 0	1× 0	1× 5,000	1× 5,000	1× 0
Generator	1× 507	1× 30	1× 0	1× 0	1× 524	1× 13	1× 0
Histogram1	7× 1,176	7× 798	7× 868	7× 1	7× 2,176	7× 667	7× 0
Histogram2	4× 936	4× 768	4× 0	4× 1	4× 1,299	4× 406	4× 558
Encoder	3× 237	3× 93	3× 0	3× 0	3× 250	3× 80	3× 0
Diagonalizer	3× 184	3× 60	3× 0	3× 0	3× 184	3× 60	3× 0
Peakfinder2d	3× 312	3× 132	3× 0	3× 0	3× 324	3× 120	3× 0
Serializer	3× 363	3× 553	3× 538	3× 2	3× 1,114	3× 342	3× 72
Network Out	6× 350	6× 350	6× 0	6× 0	6× 600	6× 200	6× 72
Sum	23,921	24,352	7,690	17	41,968	19,912	3,096
xc5vlx110	÷ 69,120	÷ 69,120	÷ 17,920	÷ 69,120	÷ 69,120	÷ 69,120	÷ 4,608
Occupancy	34.61 %	35.23 %	42.91 %	0.02 %	60.72 %	28.81 %	67.19 %

Table C.9: Occupancy of the input interface FPGA which is used for the multi-chip platform with id three realizing the Hough transform

Unit	Slice Logic Utilization				Slice Logic Distribution		Specific Feature Utilization (BlockRAM)
	Flip-Flops	LUTs: Logic	LUTs: Memory	LUTs: Route Thrus	Either LUT or Flip-Flop	Both LUT and Flip-Flop	
Network In	4× 350	4× 350	4× 0	4× 0	4× 600	4× 200	4× 72
Generator	1× 507	1× 30	1× 0	1× 0	1× 524	1× 13	1× 0
Histogram1	14× 1,176	14× 798	14× 868	14× 1	14× 2,176	14× 667	14× 0
Histogram2	6× 936	6× 768	6× 0	6× 1	6× 1,299	6× 406	6× 558
Encoder	5× 237	5× 93	5× 0	5× 0	5× 250	5× 80	5× 0
Diagonalizer	5× 184	5× 60	5× 0	5× 0	5× 184	5× 60	5× 0
Peakfinder2d	5× 312	5× 132	5× 0	5× 0	5× 324	5× 120	5× 0
Serializer	5× 363	5× 553	5× 538	5× 2	5× 1,114	5× 342	5× 72
Network Out	2× 350	2× 350	2× 0	2× 0	2× 600	2× 200	2× 72
Sum	30,167	22,100	14,842	30	51,742	15,997	4,140
xc5vlx110	÷ 69,120	÷ 69,120	÷ 17,920	÷ 69,120	÷ 69,120	÷ 69,120	÷ 4,608
Occupancy	43.64 %	31.97 %	82.82 %	0.04 %	74.86 %	23.14 %	89.84 %

Table C.10: Occupancy of the input interface FPGA which is used for the multi-chip platform with id three realizing the Hough transform

Appendix D

Sony Playstation III Extras

This chapter presents additional information for the Sony Playstation III implementation, which includes the technical system specification, the used makefile and the detailed results for the fastest algorithm configuration.

D.1 Technical System Specification

This section summarizes details about the Sony Playstation III system, which runs the Yellow Dog Linux (see [Ter]). For this purpose, the command 'df -h' is used at first to receive the successive information about the file system:

Filesystem	Size	Used	Available	Use	Mountpoint
/dev/ps3da1	99 MB	18 MB	77 MB	19 %	/boot
/dev/ps3da2	43 GB	4.2 GB	36 GB	11 %	/
tmpfs	106 MB	0 MB	106 MB	0 %	/dev/shm

Table D.1: Filesystem information of the Playstation III system

Afterwards, the command 'cat /proc/meminfo' is used to get the following information about the memory:

Information	Value
MemTotal	215,672 kB
MemFree	11,728 kB
Buffers	57,568 kB
Cached	87,748 kB
SwapCached	0 kB
Active	86,776 kB
Inactive	71,708 kB
SwapTotal	425,712 kB
SwapFree	425,652 kB
Dirty	4 kB
Writeback	0 kB
AnonPages	13,220 kB
Mapped	8,588 kB
Slab	38,744 kB
SReclaimable	28,800 kB
SUnreclaim	9,944 kB
PageTables	1,040 kB
NFS_Unstable	0 kB
Bounce	0 kB
CommitLimit	533,548 kB
Committed_AS	34,588 kB
VmallocTotal	8,589,934,592 kB
VmallocUsed	2,072 kB
VmallocChunk	8,589,931,916 kB
HugePages_Total	0
HugePages_Free	0
HugePages_Rsvd	0
Hugepagesize	16,384 kB

Table D.2: Memory information of the Playstation III system

By using finally the command 'cat /proc/cpuinfo', the subsequent platform information is produced:

Information	Value
Processor	0
CPU	Cell Broadband Engine, AltiVec supported
Clock	3.192 GHz
Revision	5.1 (pvr 0070 0501)
Processor	1
CPU	Cell Broadband Engine, AltiVec supported
Clock	3.192 GHz
Revision	5.1 (pvr 0070 0501)
Timebase	79,800,000
Platform	Playstation III

Table D.3: CPU information of the Playstation III system

D.2 Makefile

This section presents the makefile for the Sony Playstation III implementation, which is used to generate the executable program and enables thus the production of the results exhibited in section 5.6 on page 370.

```
# DIRECTORIES
PARDIR          = ../
RELEASE         = $(PARDIR)Release/
OBJ             = $(RELEASE)Obj/
H               = include/
CPP             = src/
DEPFILE         = global.dep
VECFILE         = vectorTrackingFunctions
SCALFILE        = scalarTrackingFunctions
PPUEXEFILE      = ppuTrackfinder
PPUSPUEXEFILE = spuTrackfinder
MISCLIB = MiscLIB/
FILEIOLIB = FileioLIB/
DATAOBJECTLIB = DataObjectLIB/
DATAROOTOBJECTLIB = DataRootObjectLIB/
```

```
LUTGENERATORLIB = LutGeneratorLIB/
PPUCODE = PpuCode/
SPUCODE = SpuCode/

# COMPILATION
PPUCOMPILER     = ppu-g++
PPULFLAGS       = -lspe2 -lpthread -m64 -mcpu=cell -mtune=cell -O3
PPUCFLAGS       = -c -maltivec -m64 -mcpu=cell -mtune=cell -O3 -I /usr/lib/gcc/ppu/4.1.1/include
PPUDFLAGS       = -E -MM -maltivec -I /usr/lib/gcc/ppu/4.1.1/include
EMBEDDER        = embedspu
EMBEDDERFLAGS = -m64
SPUCOMPILER     = spu-g++
SPULFLAGS       = -mtune=cell -O3
SPUCFLAGS       = -c -mtune=cell -O3
SPUDFLAGS       = -E -MM

# EXTENSIONS
OF              = .o
HF              = .h
CF              = .cxx
PPUOF           = .ppo
PPUDF           = .ppd
SPUOF           = .spo
SPUDF           = .spd

# OBJECTS
MISCLIBOBJ = $(patsubst $(PARDIR)$(MISCLIB)$(CPP)%$(CF),$(OBJ)$(MISCLIB)%$(OF),$(wildcard $(PARDIR)$(MISCLIB)$(CPP)*$(CF)))
FILEIOLIBOBJ = $(patsubst $(PARDIR)$(FILEIOLIB)$(CPP)%$(CF),$(OBJ)$(FILEIOLIB)%$(OF),$(wildcard $(PARDIR)$(FILEIOLIB)$(CPP)*$(CF)))
DATAOBJECTLIBOBJ = $(patsubst $(PARDIR)$(DATAOBJECTLIB)$(CPP)%$(CF),$(OBJ)$(DATAOBJECTLIB)%$(OF),$(wildcard $(PARDIR)$(DATAOBJECTLIB)$(CPP)*$(CF)))
DATAROOTOBJECTLIBOBJ = $(patsubst $(PARDIR)$(DATAROOTOBJECTLIB)$(CPP)%$(CF),$(OBJ)$(DATAROOTOBJECTLIB)%$(OF),$(wildcard $(PARDIR)$(
```

```
DATAROOTOBJECTLIB)$(CPP)*$(CF)))
LUTGENERATORLIBOBJ = $(patsubst $(PARDIR)$(LUTGENERATORLIB)$(CPP)%$(CF),$(OBJ)$(LUTGENERATORLIB)%$(OF),$(wildcard $(PARDIR)$(LUTGENERATORLIB)$(CPP)*$(CF)))
PPUCODEOBJ = $(patsubst $(PARDIR)$(PPUCODE)$(CPP)%$(CF),$(OBJ)$(PPUCODE)%$(OF),$(wildcard $(PARDIR)$(PPUCODE)$(CPP)*$(CF)))
SPUCODEOBJ = $(patsubst $(PARDIR)$(SPUCODE)$(CPP)%$(CF),$(OBJ)$(SPUCODE)%$(OF),$(wildcard $(PARDIR)$(SPUCODE)$(CPP)*$(CF)))

PPUOBJECTS = ${PPUCODEOBJ} ${MISCLIBOBJ} ${FILEIOLIBOBJ} ${DATAOBJECTLIBOBJ} ${DATAROOTOBJECTLIBOBJ} ${LUTGENERATORLIBOBJ}
SPUOBJECTS = ${SPUCODEOBJ}

VECOBJECTS          = $(foreach dir,$(VECFILE),$(filter %$(dir)$(OF),$(PPUOBJECTS) $(SPUOBJECTS)))
SCALOBJECTS         = $(foreach dir,$(SCALFILE),$(filter %$(dir)$(OF),$(PPUOBJECTS) $(SPUOBJECTS)))

PPUEXEOBJECTS       = $(foreach dir,$(PPUEXEFILE),$(filter %$(dir)$(OF),$(PPUOBJECTS)))
PPUSPUEXEOBJECTS    = $(foreach dir,$(PPUSPUEXEFILE),$(filter %$(dir)$(OF),$(PPUOBJECTS)))

PPUGENERALOBJECTS = $(filter-out $(SCALOBJECTS) $(VECOBJECTS) $(PPUEXEOBJECTS) $(PPUSPUEXEOBJECTS),$(PPUOBJECTS))
SPUGENERALOBJECTS = $(filter-out $(SCALOBJECTS) $(VECOBJECTS),$(SPUOBJECTS))

PPUOBJ              =
SPUOBJ              =

ifeq ($(strip $(MAKECMDGOALS)),)
PPUOBJ              = $(patsubst %$(OF),%$(PPUOF),$(PPUGENERALOBJECTS) $(PPUSPUEXEOBJECTS))
SPUOBJ              = $(patsubst %$(OF),%$(SPUOF),$(SPUGENERALOBJECTS) $(VECOBJECTS))
```

```
endif

ifeq ($(strip $(MAKECMDGOALS)),spuvec)
PPUOBJ                  = $(patsubst %$(OF),%$(PPUOF),$(PPUGENERALOBJECTS) $(PPUSPUEXEOBJECTS))
SPUOBJ                  = $(patsubst %$(OF),%$(SPUOF),$(SPUGENERALOBJECTS) $(VECOBJECTS))
endif

ifeq ($(strip $(MAKECMDGOALS)),spuscal)
PPUOBJ                  = $(patsubst %$(OF),%$(PPUOF),$(PPUGENERALOBJECTS) $(PPUSPUEXEOBJECTS))
SPUOBJ                  = $(patsubst %$(OF),%$(SPUOF),$(SPUGENERALOBJECTS) $(SCALOBJECTS))
endif

ifeq ($(strip $(MAKECMDGOALS)),ppuvec)
PPUOBJ                  = $(patsubst %$(OF),%$(PPUOF),$(PPUGENERALOBJECTS) $(VECOBJECTS) $(PPUEXEOBJECTS))
SPUOBJ                  = $(patsubst %$(OF),%$(SPUOF),$(SPUGENERALOBJECTS))
endif

ifeq ($(strip $(MAKECMDGOALS)),ppuscal)
PPUOBJ                  = $(patsubst %$(OF),%$(PPUOF),$(PPUGENERALOBJECTS) $(SCALOBJECTS) $(PPUEXEOBJECTS))
SPUOBJ                  = $(patsubst %$(OF),%$(SPUOF),$(SPUGENERALOBJECTS))
endif

# PROGRAM WITH VECTOR ARITHMETIC ON THE SPU
spuvec: spu ppu
        @$(SPUCOMPILER) $(SPULFLAGS) -o $(RELEASE)htrack.spu $(SPUOBJ)
        @awk '{commendMode = 0; while ((getline lineInFile) > 0) { lineComment = match(lineInFile,/[/][/]+/); commentStart = match(lineInFile,/[/]+[*]+/); commentStop = match(lineInFile,/[*]+[/]+/); lineStartIndex = match(lineInFile,/[[:space:]]*extern[[:space:]]*spe_program_handle_t/); lineStopIndex =
```

```
            match(lineInFile,/[;]/); if ((lineStartIndex
             > commentStart) && (commentStart > 0)) {
            commentMode = 1; }; if ((lineStartIndex >
            commentStop) && (commentStop > 0)) {
            commentMode = 0; }; if ((commentMode == 0)
            && ((lineComment > lineStopIndex) || (
            lineComment == 0)) && (lineStopIndex >
            lineStartIndex) && (lineStartIndex > 0)) {
            handlePart = substr(lineInFile,
            lineStartIndex,lineStopIndex-lineStartIndex)
            ; numberOfHandleNames = split(handlePart,
            handleName,/[[:space:]]*/); system("$(
            EMBEDDER)␣$(EMBEDDERFLAGS)␣" handleName[
            numberOfHandleNames] "␣$(RELEASE)htrack.spu␣
            $(RELEASE)htrack.spu_csf.o");}}}' $(wildcard
             $(PARDIR)$(PPUCODE)$(CPP)*$(CF))
        @$(PPUCOMPILER) $(PPULFLAGS) -o $(RELEASE)
            htrack $(PPUOBJ) $(RELEASE)htrack.spu_csf.o

# PROGRAM WITH SCALAR ARITHMETIC ON THE SPU
spuscal: spu ppu
        @$(SPUCOMPILER) $(SPULFLAGS) -o $(RELEASE)
            htrack.spu $(SPUOBJ)
        @awk '{commendMode = 0; while ((getline
            lineInFile) > 0) { lineComment = match(
            lineInFile,/[/][/]+/); commentStart = match(
            lineInFile,/[/]+[*]+/); commentStop = match(
            lineInFile,/[*]+[/]+/); lineStartIndex =
            match(lineInFile,/[[:space:]]*extern[[:space
            :]]*spe_program_handle_t/); lineStopIndex =
            match(lineInFile,/[;]/); if ((lineStartIndex
             > commentStart) && (commentStart > 0)) {
            commentMode = 1; }; if ((lineStartIndex >
            commentStop) && (commentStop > 0)) {
            commentMode = 0; }; if ((commentMode == 0)
            && ((lineComment > lineStopIndex) || (
            lineComment == 0)) && (lineStopIndex >
            lineStartIndex) && (lineStartIndex > 0)) {
            handlePart = substr(lineInFile,
            lineStartIndex,lineStopIndex-lineStartIndex)
            ; numberOfHandleNames = split(handlePart,
```

```
handleName,/[[:space:]]*/); system("$(
EMBEDDER)␣$(EMBEDDERFLAGS)␣" handleName[
numberOfHandleNames] "␣$(RELEASE)htrack.spu␣
$(RELEASE)htrack.spu_csf.o");}}}' $(wildcard
 $(PARDIR)$(PPUCODE)$(CPP)*$(CF))
	@$(PPUCOMPILER) $(PPULFLAGS) -o $(RELEASE)
	    htrack $(PPUOBJ) $(RELEASE)htrack.spu_csf.o

# PROGRAM WITH VECTOR ARITHMETIC ON THE PPU
ppuvec: ppu
	@$(PPUCOMPILER) $(PPULFLAGS) -o $(RELEASE)
	    htrack $(PPUOBJ)

# PROGRAM WITH SCALAR ARITHMETIC ON THE PPU
ppuscal: ppu
	@$(PPUCOMPILER) $(PPULFLAGS) -o $(RELEASE)
	    htrack $(PPUOBJ)

# BUILDING PPU
ppu: $(PPUOBJ)

# BUILDING SPU
spu: $(SPUOBJ)

# BUILDING OBJECTS
%$(PPUOF):
	@$(PPUCOMPILER) $(PPUCFLAGS) -o $@ $(dir $(
	    patsubst $(OBJ)%$(PPUOF),$(PARDIR)%$(CF),$@)
	    )$(CPP)$(notdir $(patsubst $(OBJ)%$(PPUOF),$
	    (PARDIR)%$(CF),$@))

%$(SPUOF):
	@$(SPUCOMPILER) $(SPUCFLAGS) -o $@ $(dir $(
	    patsubst $(OBJ)%$(SPUOF),$(PARDIR)%$(CF),$@)
	    )$(CPP)$(notdir $(patsubst $(OBJ)%$(SPUOF),$
	    (PARDIR)%$(CF),$@))

# BUILDING DEPENDENCIES
%$(PPUDF):
	@mkdir -p $(@D)
	@$(PPUCOMPILER) $(PPUDFLAGS) -MQ $(patsubst %$(
```

```
            PPUDF),%$(PPUOF),$@) -MF $@ $(dir $(patsubst
              $(OBJ)%$(PPUDF),$(PARDIR)%$(CF),$@))$(CPP)$
              (notdir $(patsubst $(OBJ)%$(PPUDF),$(PARDIR)
              %$(CF),$@))

%$(SPUDF):
        @mkdir -p $(@D)
        @$(SPUCOMPILER) $(SPUDFLAGS) -MQ $(patsubst %$(
            SPUDF),%$(SPUOF),$@) -MF $@ $(dir $(patsubst
              $(OBJ)%$(SPUDF),$(PARDIR)%$(CF),$@))$(CPP)$
              (notdir $(patsubst $(OBJ)%$(SPUDF),$(PARDIR)
              %$(CF),$@))

%$(DEPFILE):
        @mkdir -p $(@D)
        @echo > $(OBJ)$(DEPFILE)
        @for dir in $(PPUDEP); do( \
                echo $$dir | awk '{numFileSep = split(
                    $$dir,fileSep,"/"); if (numFileSep >
                    0) { directory = "$(PARDIR)"
                    fileSep[numFileSep - 1] "/"; } else
                    { directory = "";}; file = fileSep[
                    numFileSep]; gsub(/\$(PPUDF)/,"$(CF)
                    ",file); printf "%s: %s$(CPP)%s\n",
                    $$dir, directory, file}' >> $(OBJ)$(
                    DEPFILE) \
        );done;:

        @for dir in $(SPUDEP); do( \
                echo $$dir | awk '{numFileSep = split(
                    $$dir,fileSep,"/"); if (numFileSep >
                    0) { directory = "$(PARDIR)"
                    fileSep[numFileSep - 1] "/"; } else
                    { directory = "";}; file = fileSep[
                    numFileSep]; gsub(/\$(SPUDF)/,"$(CF)
                    ",file); printf "%s: %s$(CPP)%s\n",
                    $$dir, directory, file}' >> $(OBJ)$(
                    DEPFILE) \
        );done;:

# CLEANING
```

```
clean:
	@rm -rf $(RELEASE)*

# DEFINING DEPENDENCIES
PPUDEP = $(patsubst %$(PPUOF),%$(PPUDF),$(PPUOBJ))
SPUDEP = $(patsubst %$(SPUOF),%$(SPUDF),$(SPUOBJ))

# INCLUDING DEPENDENCIES
ifneq ($(MAKECMDGOALS),clean)
-include $(OBJ)$(DEPFILE)
endif

ifneq ($(MAKECMDGOALS),spu)
ifneq ($(MAKECMDGOALS),clean)
-include $(PPUDEP)
endif
endif

ifneq ($(MAKECMDGOALS),ppuvec)
ifneq ($(MAKECMDGOALS),ppuscal)
ifneq ($(MAKECMDGOALS),ppu)
ifneq ($(MAKECMDGOALS),clean)
-include $(SPUDEP)
endif
endif
endif
endif
```

D.3 Details for the Fastest Algorithm Configuration

This section presents the detailed result for the fastest available algorithm configuration, which runs on the Sony Playstation III system. However it is clear that the corresponding configuration has to be evidently shown before. For this purpose, table D.4 summarizes at first the parameters, which are set via define statements in the source code.

Parameter	Value
CODEVERSION	3
MEMORYVERSION	1
OPTIMIZEJOBCREATIONBYSAVEHIT	false
OPTIMIZEJOBCREATIONBYSAVETRANSFORM	false
DISABLESECURITYCHECKS	true
MAXIMUMNUMBEROFHITS	10,000
PERCENTAGEOFMEMORYSECURITY	4
MAXIMUMUSEDSPUS	0
MAXIMUMMEMORYPARALLELISMLEVEL	16
MAXIMUMDATAPARALLELISMLEVEL	16
TIMEBASE	79,800,000
WATCHDOGTIMERLIMIT	-1
DIAGONALIZATIONSUMMAND	1
DEBUGCONFIG	true
DEBUGTIME	true
DEBUGTHREADTIME	false
DEBUGMEMORY	true
DEBUGOUTPUT	true
DEBUGTHREADS	false
DEBUGTHREADSTATES	false
DEBUGMFCCOMMUNICATION	false
DISABLEPEAKFINDING2D	false
DISABLEPEAKFINDING3D	false
DEBUGINPUTFILES	false
DEBUGALGORITHM	0

Table D.4: Configuration parameters of the fastest algorithm running on the Sony Playstation III supplied via define statement

Afterwards the next table D.5 summarizes in addition the configuration parameters of the algorithm, which are set by a configuration file:

Parameter	Value
inputFileNameHits	../../Data/cellHits.txt
inputFileNamePrelut	../../Data/cellPrelut.txt
inputFileNameLut	../../Data/cellLut.txt
inputFileNameCodingTable	../../Data/codingTable.txt
inputFileNamePeakfindingGeometry	../../Data/filterGeometry.txt
outputFileNameTracks	../../Data/cellTracks.txt
trackfinderGammaMin	−0.467,64
trackfinderGammaMax	+0.467,64
trackfinderGammaStep	191
trackfinderThetaMin	−0.467,64
trackfinderThetaMax	+0.467,64
trackfinderThetaStep	95
trackfinderPrelutRadiusMin	−0.035
trackfinderPrelutRadiusMax	+0.035
trackfinderLutRadiusMin	−1.111,11
trackfinderLutRadiusMax	+1.111,11
trackfinderLutRadiusStep	31

Table D.5: Configuration parameters of the fastest algorithm running on the Sony Playstation III supplied via configuration file

So based on this configuration, the resulting interface memory is illustrated in the subsequent table D.6.

Information	Memory [kB]	
Limit for the jobs	157.00	
Hit file		244.67
Prelut file		30.60
Lut file		122.35
Input data	397.62	397.62
Output data	2.55	

Table D.6: Detailed result for the interface memory of the fastest Sony Playstation III implementation

With regard to this table, it is important to realize that the values, which are located among each other without a dividing line, are added to receive the sum, which is afterwards additionally located in the row after the final summand and one column left. Further on, this kind of displaying information is selected here, because the supplementary benefit is determined by the possibility to give precise extra information for interesting characteristics. In this connection, it is also clear that this concept is not special to the interface memory, but generally used for all following information.

Besides this, the detailed timing result of the fastest Sony Playstation III implementation is exhibited in table D.7.

Algorithm step	**Time** [ms]		
Initialization	921.54		
Setup threads	37.47		
Configuration	0.69		
Job prerequisites		1.25	
Job creation			6.11
Job processing		18.27	
3D peakfinding		3.21	
Event	22.74	22.74	
Finish threads	35.40		
Total for 6 threads	1,017.84		

Table D.7: Detailed result for the timing of the fastest Sony Playstation III implementation

Having the time and the interface memory clarified up to here, all following tables contain the detailed job memory results for the fastest Sony Playstation III implementation.

Information for job 0	Value		
Start index	0		
Memory parallelism	16		
Data parallelism	16		
Number of input data	1,582		
Number of output data	4		
CodingTable [kB]			0.25
GeometryTable [kB]			0.02
Input [kB]			12.36
Histogram [kB]			46.02
Input data [kB]		58.64	58.64
Output data [kB]		0.02	
Total data [kB]	58.66	58.66	

Table D.8: Detailed memory result for the first job of the fastest Sony Playstation III implementation

Information for job 1	Value		
Start index	16		
Memory parallelism	16		
Data parallelism	16		
Number of input data	2,754		
Number of output data	4		
CodingTable [kB]			0.25
GeometryTable [kB]			0.02
Input [kB]			21.52
Histogram [kB]			46.02
Input data [kB]		67.80	67.80
Output data [kB]		0.02	
Total data [kB]	67.82	67.82	

Table D.9: Detailed memory result for the second job of the fastest Sony Playstation III implementation

Information for job 2	Value		
Start index	32		
Memory parallelism	16		
Data parallelism	16		
Number of input data	3,484		
Number of output data	20		
CodingTable [kB]			0.25
GeometryTable [kB]			0.02
Input [kB]			27.22
Histogram [kB]			46.02
Input data [kB]		73.50	73.50
Output data [kB]		0.08	
Total data [kB]	73.58	73.58	

Table D.10: Detailed memory result for the third job of the fastest Sony Playstation III implementation

Information for job 3	Value		
Start index	48		
Memory parallelism	16		
Data parallelism	16		
Number of input data	6,644		
Number of output data	36		
CodingTable [kB]			0.25
GeometryTable [kB]			0.02
Input [kB]			51.91
Histogram [kB]			46.02
Input data [kB]		98.19	98.19
Output data [kB]		0.14	
Total data [kB]	98.33	98.33	

Table D.11: Detailed memory result for the fourth job of the fastest Sony Playstation III implementation

Information for job 4	Value		
Start index	64		
Memory parallelism	16		
Data parallelism	16		
Number of input data	9,320		
Number of output data	132		
CodingTable [kB]			0.25
GeometryTable [kB]			0.02
Input [kB]			72.81
Histogram [kB]			46.02
Input data [kB]		119.10	119.10
Output data [kB]		0.52	
Total data [kB]	119.61	119.61	

Table D.12: Detailed memory result for the fifth job of the fastest Sony Playstation III implementation

Information for job 5	Value		
Start index	80		
Memory parallelism	16		
Data parallelism	9		
Number of input data	9,502		
Number of output data	168		
CodingTable [kB]			0.25
GeometryTable [kB]			0.02
Input [kB]			74.24
Histogram [kB]			46.02
Input data [kB]		120.52	120.52
Output data [kB]		0.66	
Total data [kB]	121.18	121.18	

Table D.13: Detailed memory result for the sixth job of the fastest Sony Playstation III implementation

Information for job 6	Value		
Start index	89		
Memory parallelism	16		
Data parallelism	9		
Number of input data	9,212		
Number of output data	104		
CodingTable [kB]			0.25
GeometryTable [kB]			0.02
Input [kB]			71.97
Histogram [kB]			46.02
Input data [kB]		118.25	118.25
Output data [kB]		0.41	
Total data [kB]	118.66	118.66	

Table D.14: Detailed memory result for the seventh job of the fastest Sony Playstation III implementation

Information for job 7	Value		
Start index	98		
Memory parallelism	16		
Data parallelism	8		
Number of input data	9,482		
Number of output data	112		
CodingTable [kB]			0.25
GeometryTable [kB]			0.02
Input [kB]			74.08
Histogram [kB]			46.02
Input data [kB]		120.36	120.36
Output data [kB]		0.44	
Total data [kB]	120.80	120.80	

Table D.15: Detailed memory result for the eighth job of the fastest Sony Playstation III implementation

Information for job 8	Value		
Start index	106		
Memory parallelism	16		
Data parallelism	14		
Number of input data	9,622		
Number of output data	132		
CodingTable [kB]			0.25
GeometryTable [kB]			0.02
Input [kB]			75.17
Histogram [kB]			46.02
Input data [kB]		121.46	121.46
Output data [kB]		0.52	
Total data [kB]	121.97	121.97	

Table D.16: Detailed memory result for the ninth job of the fastest Sony Playstation III implementation

Information for job 9	Value		
Start index	120		
Memory parallelism	16		
Data parallelism	16		
Number of input data	7,880		
Number of output data	76		
CodingTable [kB]			0.25
GeometryTable [kB]			0.02
Input [kB]			61.56
Histogram [kB]			46.02
Input data [kB]		107.85	107.85
Output data [kB]		0.30	
Total data [kB]	108.14	108.14	

Table D.17: Detailed memory result for the tenth job of the fastest Sony Playstation III implementation

Information for job 10	Value		
Start index	136		
Memory parallelism	16		
Data parallelism	16		
Number of input data	4,542		
Number of output data	16		
CodingTable [kB]			0.25
GeometryTable [kB]			0.02
Input [kB]			35.49
Histogram [kB]			46.02
Input data [kB]		81.77	81.77
Output data [kB]		0.06	
Total data [kB]	81.83	81.83	

Table D.18: Detailed memory result for the eleventh job of the fastest Sony Playstation III implementation

Information for job 11	Value		
Start index	152		
Memory parallelism	16		
Data parallelism	16		
Number of input data	2,296		
Number of output data	8		
CodingTable [kB]			0.25
GeometryTable [kB]			0.02
Input [kB]			17.94
Histogram [kB]			46.02
Input data [kB]		64.22	64.22
Output data [kB]		0.03	
Total data [kB]	64.25	64.25	

Table D.19: Detailed memory result for the twelfth job of the fastest Sony Playstation III implementation

Information for job 12	Value		
Start index	168		
Memory parallelism	16		
Data parallelism	16		
Number of input data	1,870		
Number of output data	4		
CodingTable [kB]			0.25
GeometryTable [kB]			0.02
Input [kB]			14.61
Histogram [kB]			46.02
Input data [kB]		60.89	60.89
Output data [kB]		0.02	
Total data [kB]	60.91	60.91	

Table D.20: Detailed memory result for the thirteenth job of the fastest Sony Playstation III implementation

Information for job 13	Value		
Start index	184		
Memory parallelism	16		
Data parallelism	7		
Number of input data	174		
Number of output data	0		
CodingTable [kB]			0.25
GeometryTable [kB]			0.02
Input [kB]			1.36
Histogram [kB]			46.02
Input data [kB]		47.64	47.64
Output data [kB]		0.00	
Total data [kB]	47.64	47.64	

Table D.21: Detailed memory result for the fourteenth job of the fastest Sony Playstation III implementation

Glossary

1000BASE-T

1000BASE-T, which is also known as IEEE 802.3ab, is a network specification standard for gigabit Ethernet over copper wiring. It defines for instance the maximum length of such a network segment to be 100 meters when utilizing Category 5 cable at a minimum. In a departure from both 10BASE-T and 100BASE-TX, 1000BASE-T uses all four cable pairs for simultaneous transmission in both directions through the use of echo cancellation and a 5-level pulse amplitude modulation (PAM-5) technique. In addition to this, the symbol rate is identical to that of 100BASE-TX, which is 125 Mbaud. Further on, the noise immunity of the 5-level signaling is also identical to that of the 3-level signaling in 100BASE-TX, since 1000BASE-T uses 4-dimensional Trellis Coded Modulation (TCM) to achieve a 6 dB coding gain across the 4 pairs. 190

accumulator

In the common Hough transform, an accumulator means a discrete n-dimensional array field, which implements the discrete form of the Hough space, with one dimension for each feature variable. So each cell of this array field contains the accumulation of the number of Hough curves, which intersect for the corresponding feature parameter combination represented by the cell. 42, 51, 53, 54, 96, 97, 101, 102, 138, 141, 225, 259, 279, 297

AltiVec

AltiVec is a floating point and integer SIMD instruction set, which is designed and owned by Apple, IBM and Freescale Semiconductor. Commonly it is implemented on various versions of the PowerPC. Moreover AltiVec is a standard part of the new Power ISA v.2.03 specification. 186, D-3

antiproton

The antiproton references the antiparticle of the proton. It is generally stable but typically short-lived, because any collision with a proton will cause both particles to be annihilated in a burst of energy. Further on, this particle consists of two anti-up quarks and one anti-down quark. 7

baryon

Baryons name the family of composite particles, which are made of three quarks. They are further on opposed to the mesons, which term the family of composite particles made of one quark and one antiquark. However baryons and mesons are part of the larger particle family comprising all particles made of quarks, which are called the hadrons. 6, 8, 9

Bevalac

The Bevalac references the combination of the SuperHILAC linear accelerator, a diverting tube and the Bevatron. It is located at the Berkeley Radiation Laboratory and operates from about 1970 till 1993. The Bevalac is a linear accelerator followed by the so-called 'Race track' to be able to accelerate any and all sufficiently-stable nuclei. Further on, the outcome of its experiments was defined by the observation of compressed nuclear matter and depositing ions in tumors in cancer research. 106

Bevatron

The Bevatron is located at the Berkeley Radiation Laboratory and operates from 1954 till about 1970. It's shape is determined by the so-called 'Race track'. Moreover protons can be accelerated to a kinetic energy of 6.2 GeV. Further on, the made experiments were about strange particles with the discovering of antiprotons, antineutrons and resonances. 4, 106

bubble chamber

A bubble chamber is a vessel filled with a transparent liquid, most often liquid hydrogen, heated to just below its boiling point. As particles enter the chamber, a piston suddenly decreases its pressure and the liquid enters into a superheated, metastable phase. So charged particles create an ionization track, around which the liquid vaporises, forming microscopic bubbles. The bubble density around a track is proportional to a particle's energy loss. Further on the bubbles grow in size as the chamber expands until they are large enough to be seen or photographed. Commonly several cameras are mounted around the vessel allowing a three-dimensional image of an event to be captured. While bubble chambers were extensively used in the past, they have now mostly been supplanted by wire chambers and spark chambers. 2, 6, 35

c

C is a general-purpose computer programming language, which is developed in 1972 by Dennis Ritchie at the Bell Telephone Laboratories for use with the Unix operating system. Although C was designed for implementing system software, it is also widely used for developing portable application software. It is also one of the most popular programming languages of all time and only very few computer architectures are not supported by a C compiler. Moreover C has greatly influenced many other popular programming languages most notably C++, which began as an extension to C. 189

c++

C++ is a statically typed, free-form, multi-paradigm, compiled and general-purpose programming language. It is regarded as a middle-level language as it comprises a combination of both high-level and low-level language features. It was developed by Bjarne Stroustrup starting in 1979 at Bell Labs as an enhancement to the C programming language and originally named 'C with Classes'. Afterwards it was renamed C++ in 1983. 95, 106, 107, 109, 111, 113, 189

cartesian coordinate system

A Cartesian coordinate system specifies each point uniquely in a plane by a set of numerical coordinates, which are defined by the signed distances from the point to all available fixed perpendicular directed lines measured in the same unit of length. Moreover each reference line is called coordinate axis or just axis of the system, and the point where they meet determines its origin. Besides that, the coordinates can be also defined via the positions of the perpendicular projections of the point onto the axes expressed as a signed distances from the origin. 46

Cell BE

The Cell Broadband Engine combines a general-purpose Power Architecture core of modest performance with streamlined co-processing elements, which greatly accelerate multimedia and vector processing applications as well as many other forms of dedicated computation. It is a microprocessor designed to bridge the gap between conventional desktop processors (such as the well known Pentium and PowerPC families) and more specialized high-performance processors such as NVIDIA and ATI graphics-processors (GPUs). The first major commercial application of a Cell BE processor is defined by Sony's Playstation III game console or the dual Cell blade server of Mercury Computer Systems. iii, xii, xiii, h, j, 41, 152, 158, 184–189, 191–201, 205, A-77, A-78

Cell BE

Die Cell Broadband Engine vereinigt eine universelle Power Architekturkern bescheidener Leistung mit angeschlossenen Koprozessorelementen, die Multimedia und Vektor Verarbeitungsprogramme genauso wie viele andere Arten geeigneter Berechnugen außerordentlich beschleunigen. Es ist ein Mikroprozessor, der die Lücke zwischen klassischen Desktop-Prozessoren (wie zum Beispiel die namhaften Pentium und PowerPC Familien) und spezialisierteren Hochleistungsprozessoren wie beispielsweise NVIDIA und ATI Grafikprozessoren (GPUs) schließen soll. Die erste kommerzielle Verwertung eines Cell BE Prozessors ist durch Sony's Playstation III Spielekonsole oder durch den Zwei-Cell-Bladeserver von Mercury Computer Systems bestimmt. c, d

charm

The charm is the third most massive of all quarks. Like all quarks, the charm is an elementary fermion with spin minus one half and experiences all four fundamental interactions, which are defined by gravitation, electromagnetism, weak interaction and strong interaction. 8, 9

clone track

A clone track is a track, which already exists. This means more precisely that there is at least one other track, which corresponds to the same MC Track related to their hits. 251–254, 258, 261, 263, 264, 312, A-102

Cockcroft-Walton generator

The Cockcroft-Walton generator is a circuit design, which is used by John Douglas Cockcroft and Ernest Thomas Sinton Walton in 1932, to power the particle accelerator performing the first artificial nuclear disintegration in history. It is made up of a voltage multiplier ladder network of capacitors and diodes to convert AC or pulsing DC electrical power from a low voltage level to a higher DC voltage level. Using only capacitors and diodes, these voltage multipliers can step up relatively low voltages to extremely high values, while at the same time being far lighter and cheaper than transformers. 4

constructor

In object-oriented programming, a constructor (ctor) in a class is a special block of statements, which is called when an object is created, declared or dynamically constructed on the heap through the keyword 'new'. Further on, a constructor is similar to a class method, but it differs in the fact that it never has an explicit return type, it is not inherited and usually has different rules for scope modifiers. Constructors are often distinguished by having the same name as the declaring class. Their responsibility is to initialize the object's data members and to establish the invariant of the class. Thus it is immediately clear that it would fail, if the invariant is not valid. Moreover a properly written constructor will leave the object in a 'valid' state. Therefore immutable objects must be initialized in a constructor. 215, A-3, A-4

Conway's Game of Life

Conway's Game of Life is a cellular automaton defining a zero-player game, which means that its evolution is determined by its initial state requiring no further input from humans. One interacts with the Game of Life by creating an initial configuration and observing how it evolves. Despite its simplicity, the system achieves an impressive diversity of behavior, which fluctuates between apparent randomness and order. Besides this, one of the most obvious features of the Game of Life is the frequent occurrence of gliders, which mean arrangements of cells that essentially move themselves across the grid. These cells are furthermore really interesting, because it is possible to arrange the automaton so that the gliders interact to perform computations. So it has been shown after much effort that the Game of Life can emulate a universal Turing machine. 40

Core 2

Core 2 terms a brand encompassing a range of INTEL's consumer x86-64 single-, dual- and quad-core CPUs, which base on the INTEL Core microarchitecture. Compared to the preceding NetBurst microarchitecture, the Core 2 returns to lower clock rates while improving the usage of power and both available clock cycles. Further on, it provides more efficient decoding stages, execution units, caches and buses. Besides this, common Core-based processors do not have Hyper-Threading Technology and L3 cache. 41

Core i7

Core i7 names a family of several INTEL desktop and laptop x86-64 processors, which include the first ones released using the INTEL Nehalem micro-architecture. As this family defines furthermore the successor to the INTEL Core 2 family, it is interesting that a L3 cache and Hyper-threading is present. Besides this, it is also noticeable that the upcoming Core i7-6xx 'Arrandale' mobile processors exhibit only two cores, although all current models possess quad-core processors. 41

Cygwin

Cygwin terms a collection of free software tools, which are originally developed by Cygnus Solutions to allow various versions of Microsoft Windows to act similar to a Unix system. These tools aim mainly at porting software to enable POSIX system software basing on Linux, BSD or another Unix derivative to run on Windows with just a little more costs than a recompilation. 111

Diagonalization

In the context of this thesis, the process of Diagonalization defines a special fragment of the Hough transform, which is essential to work around the cluster problem in the histogram. xii, xiii, xviii, xxiii–xxv, 100, 101, 118, 120, 126, 144, 147, 156, 174–180, 192, 207, 209, 210, 326, 356, 361–364, 367, 383, 385, 397, 400, 409, 410

Don't-care

In the boolean algebra, a Don't-care represents an input to a boolean function, which has no effect on the function's result. So this means more precisely that a change of this input from zero to one or otherwise will not result in any change to the result. 147

DØ

DØ names one of two major experiments, which are located at the Tevatron Collider at the Fermi National Accelerator Laboratory in Batavia, Illinois. The research of this experiment is focused on precise studies of interactions of protons and antiprotons at the highest available energies. Therefore it involves an intense search for subatomic clues that reveal the character of the building blocks of the universe. 5

Encoding

In the context of this thesis, the process of Encoding defines a special fragment of the Hough transform, which evaluates the priority classes based on the signatures via a table look-up for each cell. xi–xiii, xv, xxii–xxiv, j, 99, 100, 118, 120, 126, 141, 142, 144–146, 156, 172, 174–176, 192, 203, 204, 207, 208, 255, 256, 268, 271, 307, 308, 317, 318, 321, 326, 356, 358–361, 364, 367, 383, 384, 397, 400, 409, 410, 433, A-12, A-14, A-58, A-59, A-61, A-102, A-106, A-108

endianness
In computing, endianness denotes the ordering of individually addressable subunits within a longer data word stored in memory. The most typical cases are the ordering of Bytes within a 16-, 32- or 64-bit word, where endianness is often simply referred to as byte order. The most common concepts are the most and the least significant byte first, which is called big-endian and little-endian respectively. But moreover mixed forms are also possible. For example the ordering of Bytes within a 16-bit word may be different from the ordering of 16-bit words within a 32-bit word. Although being rare, such cases are sometimes collectively referred to as mixed-endian or middle-endian. 433

Enkodierung
Im Rahmen dieser Arbeit bezeichnet der Prozess der Enkodierung einen speziellen Teil der Hough-Transformation, der die Prioritätsklassen anhand der Signaturen mittels Tabellensuche für jede Zelle ermittelt. d

ethernet
Ethernet terms a family of frame-based computer networking technologies for local area networks (LANs). Moreover the name comes generally from the physical concept of the ether. It defines a number of wiring and signaling standards for the Physical Layer of the OSI networking model through means of network access at the Media Access Control (MAC) /Data Link Layer and a common addressing format. Ethernet is standardized as IEEE 802.3. Besides this, the combination of the twisted pair versions of Ethernet for connecting end systems to the network along with the fiber optic versions for site backbones determines the most widespread wired LAN technology. It has been in use from around 1980 to the present, largely replacing competing LAN standards such as token ring, FDDI and ARCNET. 190, 393, 407, 420, 431

event simulation
In event simulation, the operation of a system is commonly represented as a chronological sequence of events. Furthermore each event occurs naturally at an instant in time and marks a change of state in the

system. In particle physics, an event describes generally one set of particle interactions, which occur in a brief span of time and are typically recorded together. At modern particle accelerators, this refers obviously to the interactions that occur as a result of one beam crossing inside a detector. 102, 110, 113, 117, 119, 121–123, A-2, A-7, A-107

fake hit
A fake hit names a hit, which corresponds not to a real hit from the physics experiment. So in the case of using MAPS, a fake hit is simply caused by detector noise. Moreover in the case of using STRIPS, a fake hit represents a 'true' combination of a front and a back side strip with either a displaced coordinate or a bad strip overlapping. ix, 19, 23, 24, 44, 219, 259, 261, 320, 439, A-86, A-108

fake track
A fake track terms a track, which just exists because of fake hits. So if the fake hits are removed, this track would not exist anymore. 251–254, 259, 263, 264, 312, 319, 321, A-102, A-108

feature extraction
In pattern recognition and image processing, feature extraction determines generally a special form of dimensionality reduction. So if the actual data set is considered to be huge and almost redundant, it could be obviously simplified by a transformation into a reduced representation set of features (also named feature vector). Therefore this transforming process is then called feature extraction. If the features are further on carefully chosen, it is expected that the feature set will extract the relevant information from the data set. h, 45, 46, 52, 54, 56, 58, 98, 101

first level tracking
In high energy physics, first level tracking names the track reconstruction process, which is applied at the first level. Thence this process has to run with full speed on all produced data, and the input is almost always directly delivered by the front end read-out electronics of the detectors. So commonly the major aim of the first level tracking is characterized by an enormous data reduction. 34, 44, 95, 392, 394, 400, 407, 408, 421, 433, 434, C-17

flip-flop
In digital circuits, a flip-flop determines a kind of bistable multivibrator possessing two stable states, which offer thereby the capability

of serving as one bit of memory. Further on, simple flip-flops can be built by two cross-coupled inverting elements of vacuum tubes, bipolar transistors, field effective transistors, inverters or inverting logic gates perhaps augmented by some enable/disable gating mechanism. Moreover clocked non-transparent flip-flops are typically implemented as master-slave devices where two basic flip-flops plus some additional logic collaborate to make it insensitive to spikes and noise between the short clock transitions. Nevertheless they often include also asynchronous clear or set inputs, which may be used to change the current output independent of the clock. 97, 153, 240, 332, 333, 343, 345, 348, 350, 352, 355, 361, 400, 410

FLUKA
The FLUKA package is a tool for particle transport and interactions with matter (see particle transport engine). 108

g++
G++ is a C++ compiler, which is distributed freely with the GNU Compiler Collection (see GCC). 111

GEANE
GEANE names a software package, which offers the possibility to compute the average motion trajectory of particles through detector stations. So it is not really surprising this package is directly derived from the GEANT package (see GEANT). 109, 110, 220, A-19, A-39, A-40

ghost track
A ghost track names a track, which combines not enough hits from a single MC Track to represent such a MC Track. Furthermore a ghost track occurs commonly due to the overlapping of enough entries for hits, which belong to different MC Tracks. 251–254, 259, 263, 264, 312, 319, 321, 323, A-102

gluons
Gluons are elementary expressions of quark interaction, which are indirectly involved with the binding of protons and neutrons together in atomic nuclei. In technical terms, they are vector gauge bosons that mediate strong color charge interactions of quarks in quantum chromodynamics. Unlike the electrically neutral photon of quantum electrodynamics, gluons themselves carry color charge and therefore participate in the strong interaction in addition to mediating it, which

makes quantum chromodynamics significantly harder to analyze than quantum electrodynamics. 2, 7

GNU

The GNU Project is a free software and mass collaboration project, which initiated the GNU operating system. Its name is a recursive acronym for 'GNU's Not Unix!'. This name was chosen, because GNU's design is Unix-like, but differs from Unix by being free software and containing no Unix code. The founding goal of the project was in the words of its initial announcement to develop 'a sufficient body of free software [...] to get along without any software that is not free.'. 95, 111

GTX 280

The GeForce 200 Series determines the tenth generation of nVIDIA's GeForce graphics processing units, which represents also the continuation of the company's unified shader architecture introduced with the GeForce 8 Series and the GeForce 9 Series. Further on, the GeForce GTX 280 and GTX 260 are based on the same processor core, which is separated by quality tests during the manufacturing process. Moreover the GT 200 series defines actually the largest commercial GPU ever constructed. It consists of 1.4 billion transistors covering a 576 square millimeter die surface area built on a 65 nm process. The theoretical shader performance of the GTX 280 is 933 GFLOPS, while the thermal design power reaches 236 Watts. Typical clock frequencies are about 602 MHz for the core, 1296 MHz for the shader and 2214 MHz for the GDDR3 memory, which is able to deliver a bandwidth of 141.7 GiB/s. 41

H1

H1 names one of the four experiments located at HERA in DESY, Hamburg, Germany. While H1 contains a general purpose detection system based on different subdetectors, the main design feature is an asymmetric construction to cope with the boosted center of mass in the laboratory frame due to the large energy imbalance of the colliding beams. The most interesting physics topics treated there include the diffraction, the production of heavy quarks, tests of QCD in jet and particle production, the cross section measurements of reactions with charged and neutral electroweak currents, the studies of proton structure, the determination of quark and gluon parton distribution functions as well as the search for physics beyond the Standard Model. 5

hadrons

In particle physics, a hadron defines a particle, which is made of quarks and held together by the strong force. Furthermore hadrons are either mesons or baryons. Other combinations such as tetraquarks and pentaquarks may be possible, but no evidence conclusively suggests their existence. Moreover the best known mesons are pions and kaons, while the best known baryons are protons and neutrons. 2, 5, 8

HERMES

The HERMES experiment is conducted using the HERA particle accelerator, which is located at DESY in Hamburg, Germany. Its goal is to investigate the quark-gluon structure of matter by examining how a nucleon's constituents affect its spin. 5

histogram

see accumulator. ii–v, x–xix, xxi–xxiv, j, 42, 53, 58, 60, 61, 63, 64, 72, 79, 80, 96–101, 115, 117–120, 124, 126–128, 131, 132, 134, 136–141, 143–145, 147–151, 156–161, 163–174, 176–184, 191, 192, 194, 200, 202–210, 216, 225, 230–235, 237, 238, 240, 242, 244–247, 249–257, 265–272, 274, 275, 277, 279, 283–294, 296–304, 309, 315–317, 320, 323, 325, 326, 328, 332, 334–357, 359, 362, 363, 365, 366, 368, 371, 375–384, 388, 391, 393, 394, 397, 399–404, 408–411, 413, 415–417, 421–423, 426, 427, 437, A-2, A-12, A-14, A-20–A-22, A-57–A-61, A-63–A-66, A-70, A-75, A-83–A-87, A-89–A-91, A-93, A-95–A-97, A-102, A-106, A-107, C-17–C-20

Histogramming

In the context of this thesis, the process of Histogramming defines a special fragment of the Hough transform, which generates the initial histogram by inserting entries for the occurring hits based on the computed borders of the look-up-tables. xv, xviii, xxiv, 53, 99, 100, 120, 128, 137, 141, 156, 173, 192, 202, 203, 255, 256, 287, 317, 325, 376, 380, 381, 385, 386, 397, 409, 410, 413, A-57, A-59, A-60, A-87

Hit

In dieser Arbeit bezeichnet ein Hit im Allgemeinen den Interaktionspunkt eines Teilchens mit einer Detektorstation. Darüber hinaus beinhaltet solch ein Hit im Gegensatz zu einem Point alle maßgeblichen Informationen bezüglich der Detektorebene wie Verwischung, Mehrfachstreuung oder etwas anderes. d, b

hit

In this thesis, a hit defines generally the interaction point of a particle with a detector station. Furthermore such a hit contains in contrast to a point all relevant information based on the detector plane, like smearing, multiple scattering or something else. iv, ix, xv–xix, xxii, 13, 17–24, 26, 30, 31, h, 34, j, 36, 37, 39, 40, 42, 44, 55, 56, 66–69, 71, 79–84, 96, 98–100, 105, 113–116, 118, 124, 128–131, 133–137, 140, 142, 143, 156, 159, 160, 165, 168, 190, 191, 194, 196, 200–206, 210, 216–220, 225, 230–238, 257, 259, 261, 262, 268–277, 279, 283–287, 296, 304–309, 317–319, 321–325, 334, 342, 371, 375–377, 392, 393, 402–404, 406, 408, 410, 413, 415, 417, 419–423, 426, 432, 435–439, A-1, A-6, A-11, A-15–A-17, A-26, A-39, A-40, A-42, A-65, A-70, A-72, A-77, A-84–A-87, A-91, A-95, A-102, A-108

hybrid pixel

Hybrid pixel detectors are composed of sensor cells of 50x50 μm^2 yielding a spatial resolution of about 15 μm. Furthermore the sensors and the matching read-out electronics are placed on two separate chips, which are interconnected pixel by pixel through microscopic solder balls. Though hybrid pixel detectors are relatively thick and require presumably active cooling in the acceptance, the stations contribute in the current STS concept with unambiguous space points to the track finding where the track densities are high. However GSI do not carry out R&D on such hybrid pixel detectors at the moment. i, ix, 7, 14, 15, 17, 18, 115, 318, 319, A-17

Hybrid-Pixel

Hybrid-Pixel Detektoren bestehen aus 50x50 μm^2 Sensorzellen, die eine räumliche Auflösung von ungefähr 15 μm erbringen. Darüber hinaus sind die Sensoren und die entsprechende Ausleseelektronik auf zwei getrennten Chips plaziert, die pixelweise durch mikroskopisch kleine Lötpunkte verbunden sind. Obwohl Hybrid-Pixel Detektoren verhältnismäßig dick sind und im Einsatzbereich voraussichtlich aktive Kühlung benötigen, tragen die Stationen im gegenwärtigen STS-Konzept mit eindeutigen Raumpunkten zur Spurensuche bei hohen Spurdichten bei. Das GSI führt jedoch derzeit keine Forschung und Entwicklung für solche Hybrid-Pixel Detektoren durch. d

hyperon

In particle physics, a hyperon terms any baryon, which contains one or more strange quarks, but no charm quarks or bottom quarks. 8

ion

An ion is characterized to be an atom or molecule where the total number of electrons is not equal to the total number of protons, which implies a net positive or negative electrical charge. 3, 5–9

Kalman filter

The Kalman filter is an efficient recursive filter that estimates the state of a linear dynamic system from a series of noisy measurements. It is used in a wide range of engineering applications from radar to high energy physics and constitutes an important topic in control theory and control systems engineering. Besides this, an easy example for a typical application is given by the need for accurate and continuously updated information about the position and velocity of an object, which is determined by only a sequence of observations about its position including some error. So the Kalman filter is now able to exploit the trusted model of the dynamics of the target, which describes the kind of movement possible by the target, to remove the effects of the noise due to a good estimate of the location of the target at the present time, at a future time or at a time in the past. 35, 36, 435, 439

L3

The main objective of L3 is to understand some of the fundamental laws of particle physics. Therefore electrons and positrons are accelerated in the ring of LEP in opposite directions. The enforced collision right in the middle of the L3 detector can then be used to study the annihilation of two original particles and the generation of another pair of particle and anti-particle by conserving the energy and providing that their mass is not too big. 5

lepton

Leptons are a family of elementary particles alongside quarks and gauge bosons. Further on, leptons are fermions and are subject to the electromagnetic force, the gravitational force and the weak interaction like quarks. But in contrast to this, leptons do unlike quarks not participate in the strong interaction. 8, 9, 34

Lorentz Force

The Lorentz force $F = q \cdot (E + v \times B)$ names the force, which takes effect on a point charge due to electromagnetic fields. Furthermore the general well-known rule of this effect is determined by the circumstance that a positively charged particle will be accelerated in the same linear orientation as the E field, but will curve perpendicularly to the instantaneous velocity vector v and the B field according to the right-hand rule. Besides this, the term $q \cdot E$ is additionally called electric force, while the term $q \cdot (v \times B)$ is called magnetic force. 75, 84, 87

Make

In software development, Make is a utility, which features the automation of the build process for large applications. For this purpose, special files called Makefiles are required, which specify instructions. Further on, Make is an expert system, which tracks the files that have changed since the last time the project was built and invokes the compiler on only those source code files and their dependencies. Moreover Make is most commonly used in C/C++ projects, but in principle it can be used with almost any software project. 95, 111, 372

Merkmalsextraktion

In der Mustererkennung und Bildverarbeitung bedingt die Merkmalsextraktion im Allgemeinen eine spezielle Form der Dimensionalitätsreduktion. Wenn also der eigentliche Datensatz für zu zu groß und nahezu redundant erachtet wird, kann er offensichtlich durch eine Transformation in eine verringerte Merkmaldarstellung (auch Merkmalvektor gennant) vereinfacht werden. Folglich wird dieser Transformationsprozess dann Merkmalsextraktion genannt. Wenn die Merkmale weiterhin sorgfältig ausgewählt werden, wird erwartet, daß die Merkmalzusammenstellung alle maßgeblichen Informationen des Datensatzes herausziehen wird. b

mesons

In particle physics, mesons are subatomic particles, which are composed of one quark and one antiquark. Further on, these particles are part of the hadron particle family. The main difference between mesons and baryons is determined by the fact that mesons are bosons, while baryons are fermions. 8, 9

micro-strip

In this thesis, micro-strip stands for a low-mass Silicon micro-strip detector, which is used for the track point measurement. GSI is actually conducting R&D on thin double-sided silicon micro-strip detectors with 50 μm strip pitch and a moderately large stereo angle, that can serve as building blocks of detector modules for the tracking stations with read-out electronics outside of the STS acceptance. i, ix, xxii, 7, 14, 15, 19–24, 40, 115, 219, 318–320, A-17, A-86, A-108

Microsoft Visual Studio
Microsoft Visual Studio centers on an integrated development environment, which lets programmers create standalone applications, web sites, web applications and web services, that run on any platforms supported by Microsoft's .NET Framework (for all versions after Visual Studio 6). Supported platforms include Microsoft Windows servers and workstations, PocketPC, Smartphones as well as World Wide Web browsers. Moreover Visual Studio includes compilers for Basic, C++, C#, J# and ASP. Some versions include also an edition of Microsoft SQL Server. 110, 111

Mikro-Strip
In dieser Arbeit steht Mikro-Strip für einen massearmen Silizium-Mikro-Streifen Detektor, der für die Spurpunktmessung verwendet wird. Das GSI betreibt derzeitig Forschung und Entwicklung für dünne doppelseitige Silizium-Mikro-Streifen Detektoren mit 50 μm Streifenbreite und angemessen großem Stereowinkel, die als Baustein für Detektormodule der Trackingstationen mit Ausleseelektronik außerhalb der STS-Bereichs dienen können. d

multiple scattering
Scattering is a general physical process, which forces some forms of radiation like moving particles to deviate from a common trajectory by one or more localized non-uniformities in the medium through which it passes. As it is further on very common to group multiple non-uniformities together, it is clear that the radiation may scatter many times, which is then evidently called multiple scattering. 70, 101, 110

multiplexer
In electronics, a multiplexer or mux terms a device, that performs multiplexing. For this purpose, such a device simply selects one of many analog or digital input signals and forwards the selected input into a single output line. So within this context, it is clear that a multiplexer with 2^n inputs requires n select lines to temporarily interconnected a single input to the output. Based on this functionality, an electronic multiplexer features thus the sharing of a single device or resource for several clients like for example a communication line for multiple senders instead of having one device per sender. 97, 98, 166, 168, 170, 171, 182

muon
The muon is an elementary particle similar to the electron with negative

electric charge and a spin of one half. Further on, it is classified to be a lepton as well as the electron, the tauon and the three neutrinos. Moreover this particle constitutes the unstable subatomic particle with the second longest mean lifetime behind the neutron. Besides this, the muon has obviously a corresponding antiparticle like all elementary particles of opposite charge but equal mass and spin, which is called antimuon. 9

Mustererkennung
Die Mustererkennung verfolgt allgemein das Ziel Datenmuster zu klassifizieren, die üblicherweise durch Messwertgruppen oder Beobachtungspunkte in einem mehrdimensionalen Raum bestimmt werden. Weiterhin besteht ein vollständiges System zur Mustererkennung normalerweise aus einem Sensor, der Beobachtungen erfasst, einem Mechanismus für die Merkmalsextraktion, der numerische oder symbolische Informationen für die Beobachtungen berechnet und einem Ordnungssystem, um die Beobachtungen hinsichtlich der extrahierten Merkmale zu klassifizieren. b

not-found track
A not-found track names a MC Track which exists, but is not found by the Hough transform. 251–254

nVIDIA
nVIDIA is an American multinational corporation, which is specialized in the manufacturing of graphics-processor technologies for workstations, desktop computers and handheld devices. Moreover this company, which is headquartered in Santa Clara, California, has become a major supplier of integrated circuits used for personal-computer motherboard chip sets, graphics processing units and gaming consoles. Notable product lines include the GeForce series for gaming and the Quadro series for graphics processing on professional workstations as well as the nForce series of integrated motherboard chip sets. 41, 188

particle event generator
In this thesis, particle event generators refer to software libraries, that generate simulated high-energy particle physics events (see also event simulation). 104, 106, 107

particle transport engine
In this thesis, a particle transport engine is generally able to describe

the passage of elementary particles through matter (see GEANT). 104, 107–109

pattern recognition

Universally pattern recognition aims to classify data patterns, which are usually determined by groups of measurements or observations defining points in an appropriate multidimensional space. Further on, a complete pattern recognition system consists naturally of a sensor that gathers the observations, a feature extraction mechanism that computes numeric or symbolic information from the observations and a classification scheme to classify the observations with regard to the extracted features. h, 34, 35, 37, 38, 45, 46, 52

peak

In this thesis, a peak names a local maximum in the histogram. So commonly these local maxima are called peaks before the filtering process (encoding and peak finding) is applied. Afterwards the surviving ones are called track candidates. Besides this, it should be mentioned that there are almost no exact peaks, because of the clustering problem. iii, xv, xvii, xviii, 53, 54, 58, 60, 61, 69, 72, 97–100, 118, 145, 147, 148, 216, 230–234, 237, 238, 240, 242, 244, 247–251, 255–262, 268, 279, 298, 302–304, 306–310, 312, 314, 315, 317, 321, 323, 436, 437, 439, A-84, A-85, A-93, A-94, A-102, A-108

Peak-finding

In the context of this thesis, the process of Peak-finding defines a special fragment of the Hough transform, which filters the filled histogram in the way that the background noise is removed. So at the best, just the peaks, which represent a real track, will survive this process. iv, xi–xiii, xv–xvii, xix, xxiii, xxv, 53, 54, 97, 99–101, 118–120, 126, 127, 141, 144, 147–152, 156–158, 174, 176, 179–183, 192, 203, 204, 207, 210, 211, 223, 248, 255–257, 265, 279, 280, 294, 302, 309, 310, 318, 322, 323, 326, 356, 364–368, 370, 371, 383, 386–389, 393, 397, 398, 400, 408–410, A-25–A-29, A-31, A-32, A-58, A-60, A-61, A-84–A-87, A-98

Phobos

The Phobos experiment names one of the five experiments, which are located at the Relativistic Heavy Ion Collider at the Brookhaven National Laboratory in Long Island, New York. It examines gold collisions with accelerated ions to center of mass energies of $\sqrt{s_{NN}} = 56$ GeV, 130 GeV and 200 GeV, because such high energies are suggested to create nuclear matter at a sufficient high density and temperature to allow a phase transition to a new state of matter, which is called quark-gluon plasma. Moreover the aim of first measurements, which are performed by the PHOBOS collaboration, was to investigate the properties of particle production in these new energy regimes and verify, if the extreme conditions needed for a phase transition in nuclear matter have actually been achieved. 5

PLUTO
PLUTO is a software package, which implements a customized event generator for hadronic and electromagnetic decays (see particle event generator). 104, 107

point
In this thesis, a point stands for the interaction coordinate of a particle with a detector station. However it is important to realize in this connection that this generated point does generally not include the correct detector interaction, but represents rather just a GEANT measurement of the tracked particle at the positions where detector stations should be. 12, 15, 24, 44, 104, 113, 114, 131, 219, 220, 318, A-6, A-15, A-16, A-32, A-39, A-40, A-42, A-84–A-86

pointer
In computer science, a pointer is a programming language data type, whose value refers directly to or 'points to' another value stored elsewhere in the computer memory using its address. For high-level programming languages, pointers effectively take the place of general purpose registers in low-level languages such as assembly language or machine code, but may be in available memory. So a pointer references a location in memory more or less restrictive to obtain the value at this location. Such an access method is commonly known as dereferencing the pointer. Further on, pointers significantly improve performance for repetitive operations such as traversing strings, lookup tables, control tables and tree structures. In particular, it is often much cheaper in time and space to copy and dereference pointers than to copy and access the data to which the pointers point. 113–116, 118, 136, 137, 140, 143, 148

polyamide
A polyamide is a polymer containing monomers of amides, which are joined by peptide bonds. Furthermore this polymer can occur both naturally and artificially. Examples are given by proteins such as wool

as well as silk, which can be made artificially through step-growth polymerization. More examples are cited by nylons, aramids and sodium poly(aspartate). Besides this, polyamides are commonly used in textiles, automotives, carpet and sportswear due to their extreme durability and strength. 24

POPOP
POPOP or 1.4-bis(5-phenyloxazol-2-yl) benzene is a scintillator, which can be used as wavelength shifter, because it converts shorter wavelength light to longer wavelength light. Furthermore its output spectrum peaks at 410nm, which is violet. Generally POPOP is used in both solid and liquid organic scintillators. 31

Power Architecture
Power Architecture is a broad term, which describes similar RISC instruction sets for microprocessors developed and manufactured by such companies as IBM, Freescale, AMCC, Tundra and P.A. Semi. Further on, the governing body is Power.org comprising over 40 companies and organizations. 186

pp2pp
The pp2pp experiment is one of the five experiments, which are studied at the Relativistic Heavy Ion Collider at the Brookhaven National Laboratory in Long Island, New York. It is designed to measure the elastic scattering of protons from $\sqrt{s} = 60$ GeV to $\sqrt{s} = 500$ GeV in two kinematic regions. Therefore the detector consists of silicon strip detectors, which are mounted in Roman Pots and installed in the RHIC ring 60 m from the interaction region. 5

priority class
In this thesis, a priority class represents the probability of a signature to correspond to a real track. Moreover priority classes are taken into account for the filtering process, which should remove the background noise in the histogram. iii, xiii, j, 99–101, 118, 145–148, 150, 151, 172, 174–176, 178, 179, 207, 209, 214, 217, 307–309, 321–323, 371, 383, 388, 393, A-23, A-24, A-34, A-102, A-104–A-107

Prioritätsklasse
In dieser Arbeit repräsentiert eine Prioritätsklasse die Wahrscheinlichkeit für die Zugehörigkeit einer Signatur zu einer echten Spur. Ferner werden Prioritätsklasse beim Filterprozess, der das Hintergrundrauschen im Histogramm beseitigen soll, mit einbezogen. d

PROOF

PROOF names the parallel implementation of ROOT. 106

proton

The proton is a subatomic particle with an electric charge of plus the elementary charge. Further on, it is found in the nucleus of each atom along with neutrons, but is also stable by itself and has a second identity as the hydrogen ion. Moreover a proton is composed of three fundamental particles, which are identified by two up quarks and one down quark. 7, 9

quarks

A quark is an elementary particle and a fundamental constituent of matter. Furthermore quarks combine to form composite particles called hadrons, which define the components of atomic nuclei. Due to a phenomenon known as color confinement, quarks are never found in isolation, but rather within hadrons. For this reason, much of what is known about quarks has been drawn from observations of the hadrons themselves. 2, 7, 8

queue

A queue is a particular kind of collection, which keeps their entities in order. For this purpose, the principal operations of such a collection are the addition of entities to the rear terminal position and the removal of entities from the front terminal position. So the first element added to the queue will be the first one to be removed. Further on, a queue is an example of a linear FIFO data structure, which provides services in computer science, transport and operations research where various entities such as data, objects, persons or events are stored to be processed later. So generally the queue performs the function of a buffer. 342, 346, 347, 352, 355

Quine-McCluskey

The Quine-McCluskey algorithm identifies a method to minimize boolean functions. It is functionally identical to the Karnaugh mapping, but the tabular form makes it more efficient for use in computer algorithms. Moreover this method involves two steps, which are defined by finding all prime implicants and further removing the unnecessary ones. 147, A-105

Reality Synthesizer
The RSX graphics processing unit is a graphics chip design, which is co-developed by NVIDIA and Sony for the Playstation III computer console. In this connection, staff at Sony were quoted in the Playstation Magazine saying that the RSX shares a lot of inner workings with the NVIDIA 7800, which is based on the G70 architecture. Hence the RSX was expected to feature also 24 parallel pixel and eight vertex shader pipelines, which are capable of carrying out 136 shader operations per clock cycle. Moreover NVIDIA CEO Jen-Hsun Huang stated during Sony's pre-show press conference at E3 2005, that the RSX would be more powerful than two GeForce 6800 Ultra video cards combined. 188

ROOT
ROOT is an object-oriented software framework, which was originally designed for particle physics data analyses. For this purpose, the developers at CERN add evidently several features specific to this field. Nevertheless this framework is also common in many other fields such as astronomy and data mining. 102, 106–108, 110, 111, 114, 123, 124, 128–131, A-8, A-10, A-85, A-100

Serialization
In the context of this thesis, the process of Serialization defines a special fragment of the Hough transform, which sorts the finally surviving track candidates inside the histogram into a serial stream of data. Therefore the common three dimensional Hough transform applies this process for the final stage after the Peak finding to write the results in a buffer. However it has to be explicitly noticed that the algorithm, which uses the decomposed histogram, requires the Serialization process naturally between the two and the three dimensional Peak finding step. xii, xiii, xvii, xix, xxiii, xxv, 97, 101, 118, 126, 144, 148, 150, 151, 156, 174, 179, 182, 183, 192, 207, 210, 284, 288, 290, 291, 293, 294, 326, 356, 364, 368, 369, 383, 386, 387, 397, 400, 403, 409, 410, 417, 426, 427, A-97

shift register
In digital circuits, a shift register is a cascade of flip-flops, which share the same clock and have the output of any one except the last flip-flop connected to the input of the next one in the chain. Hence such a circuit shifts each bit of the flip-flops in every clock cycle to the next one, while the data present at its input is shifted in and the data present at the output, which is determined by the last flip-flop in the chain, is shifted out. 98, 167, 172, 173, 176, 177

Signatur

In dieser Arbeit bezeichnet eine Signatur die Hit-Kombination in einer einzelnen Histogrammzelle. Genauer ausgedrückt bedeutet dies, daß eine Zelle des Akkumulators nicht wie gewöhnlich die Anzahl der vorhandenen Hits zählt, sondern vielmehr ein einzelnes Bit in der Signatur für jede Detektorstation, in der ein Hit aufgetreten ist, setzt. c, d

signature

In this thesis, a signature identifies the hit-combination of a single histogram cell. This means more precisely that one cell of the accumulator does not count the number of hits falling into as usual, but sets rather a single bit in the signature for each detector station where a hit has occured. i, iv, x, xiv, xvi, xvii, xxi, xxii, j, 99, 100, 115, 117, 118, 124, 137, 139–143, 145–147, 160, 166–172, 174, 176, 205, 207–209, 216, 234, 236, 237, 240, 261, 262, 265, 268–270, 272–279, 304–309, 317, 318, 321, 325, 334, 383, 438, A-2, A-72, A-73, A-83, A-87, A-88, A-91, A-99, A-102, A-104–A-108

spark chamber

Spark chamber detectors are generally less accurate than bubble chamber detectors, but can be made highly selective with the help of auxiliary detectors, which enable them to be useful in the search for very rare events. Further on, a spark chamber consists of metal plates, which are placed in a sealed box that is filled with a gas such as helium, neon or a mixture of both. So as a charged particle travels through such a detector, the gas will be ionized between the plates. Furthermore a trigger system is used to apply a high voltage to the plates to create an electric field immediately after the particle goes through the chamber, which produces sparks on its exact trajectory. 2

stoichiometry

Stoichiometry names a field in natural science, which addresses the calculation of quantitative measurable relationships of reactants and products in a balanced chemical reaction. It can be for instance used to calculate quantities such as the amount of products, that can be produced with the given reactants and percent yield. 1

task

Commonly a task defines a set of actions, which accomplish a job, problem or assignment. In computing, multitasking is further on a method to share common processing resources such as a CPU for multiple tasks. But in general, only one task is said to be running at any point in time, which means that the CPU is actively executing instructions for that task. Hence the task concept schedules the tasks at any given time to be running or waiting. 34, 35, 42, 102, 110, 113, 115, 120, 123, 155, A-1, A-2

Tevatron
The Tevatron names a circular particle accelerator at the Fermi National Accelerator Laboratory in Batavia, Illinois, which is the second highest energy particle collider in the world after the Large Hadron Collider at CERN. Furthermore it is a synchrotron, that accelerates protons and antiprotons in a 6.28 km ring to energies of up to 1 TeV. 5

Toshiba corporation
Toshiba is a multinational manufacturing company, which is headquartered in Tokyo, Japan. The company's businesses are in high technology, electrical engineering and many electronic fields. For example, Toshiba-made semiconductors are among the worldwide Top 20 semiconductor sales leaders. 185

track
In this thesis, a track references a grouped number of measured particle interactions with detector stations. In addition to this, the characteristic of such a group is determined by the possible motion trajectory of one particle (see tracking, see also peak). i, ii, iv, ix, x, xiv–xviii, xxi, 7, 13, 15, 20, 21, 24, 25, h, 34–36, 39–44, 55, 58, 59, 61, 63, 64, 66–69, 80, 83, 84, 88, 98, 99, 101, 104, 109, 113, 114, 118, 122, 124, 130, 134, 135, 152, 194, 216–220, 230–240, 242–247, 251–254, 258–265, 269, 273, 274, 279–283, 296, 297, 300–306, 309, 310, 312–315, 317–320, 323, 391, 408, 438, 439, A-1, A-2, A-7, A-12–A-14, A-18, A-26, A-32–A-34, A-39–A-42, A-63, A-64, A-70, A-72, A-73, A-75, A-85, A-86, A-90–A-94, A-96, A-97, A-102, A-107, A-108

track candidate
In this thesis, a track candidate names an encoded peak in the histogram, which represents thus a priority class instead of a signature. Further on, it is just a candidate, because it is not sure that the local maximum corresponds to a real Monte Carlo Track. Moreover this name is kept up to the end to have the possibility to append another improving algorithm without getting confusion, which is related to the naming of the concerned objects. iv, xv, xvii, xix, 34–37, 39–41, 100,

101, 118, 119, 126, 142, 148–151, 179, 181–183, 203, 211, 251–255, 257, 258, 260–262, 274, 282, 284, 288–294, 388, 397, 403, 404, 408, 415, 417, 420, 421, 426–428, 432, 435–439, A-30, A-31, A-94, A-95, A-102

tracking
In high energy physics, tracking or track reconstruction terms the process, which groups particle interactions with detector stations to possible motion trajectories of particles. Moreover such groups of detector interaction points are then obviously called tracks. ii, iv, x, xii, xv, xvii, xviii, xxi, xxii, 9, 10, 12, 14–16, 25–27, 34–43, 52, 85, 96, 102, 105, 110, 113–115, 119–121, 123, 125, 131, 136, 138, 140, 142–144, 151, 156, 184, 214–218, 221, 222, 240, 258–265, 268, 272, 281, 295, 296, 299, 300, 307, 309–313, 315–323, 370, 371, 391, 435, 436, 438, A-1–A-7, A-9, A-11, A-15, A-32–A-38, A-53, A-62, A-78, A-83, A-85, A-86, A-101, A-102, A-107, A-108, A-113, A-117, C-2

tracking clones rate
The tracking clones rate is defined by the quotient of the number of track candidates, which correspond to an already found MC Track, and the number of existing track candidates. Thence this quotient gives evidently information about the multiplicity of the MC Tracks related to the data set. 122, 260, 262, 268, 313, 314, 318, 320, 322, 323, A-35–A-38, A-43–A-45

tracking efficiency
The tracking efficiency, which is used here, is defined by the quotient of the number of track candidates, which can be identified to correspond to a findable MC Track, and the number of all findable MC Tracks. Important is that track candidates, which correspond to the same MC Track, are just counted once, while the others are consequently counted in the clones rate. Further on, the number of findable MC Tracks is obviously defined by the MC Tracks, which exhibit a targeted momentum cut and special predefined hit signatures. 122, 260–263, 313, 314, 318–324, 391, A-35–A-38, A-43–A-45, A-108

tracking fake rate
The tracking fake rate is defined by the quotient of the number of track candidates, which correspond not to a findable MC Track, and the number of existing track candidates. Important is that the track candidates, which are counted here, would not exist in contrast to the ones accounted for the ghost rate, if the fake hits are removed from the data input to the tracking. So the number of track candidates, which

does not correspond to a findable MC Track, is evidently split in the ghost and fake rate. 122, 261, 262, 268, 313, 314, 319–321, A-35–A-38, A-43–A-45

tracking ghost rate
The tracking ghost rate is defined by the quotient of the number of track candidates, which correspond not to a findable MC Track, and the number of existing track candidates. Important is that in contrast to the fake rate these track candidates would also exist, if the fake hits are removed from the data input to the tracking. So the number of track candidates which does not correspond to a findable MC Track is split into the ghost and fake rate. 72, 122, 261, 262, 268, 313, 314, 318–323, A-35–A-38, A-43–A-45

tracking identification rate
The tracking identification rate is defined by the quotient of the number of track candidates, which can be identified to correspond to a findable MC Track, and the number of all track candidates. Important is that track candidates, which correspond to the same MC Track, are just counted once here. The others are counted in the clones rate. i, 261, 262, 268, 321, 391, A-35, A-36

tracking not-found track rate
The tracking not-found track rate is defined by the quotient of the number of findable MC Tracks, which are not identified by a track candidate, and the number of all findable MC Tracks. 122, A-43–A-45

tracking reduction rate
The tracking reduction rate is defined by the quotient of the number of hits minus the number of peaks and the number of hits. So this quotient gives information about the data reduction, because the number of data output is correlated to the number of data input. i, 261, 391, A-35, A-36

Trackingidentifizierungsrate
Die Trackingidentifizierungsrate ist durch den Quotient der Anzahl an Spurkandidaten, die einer auffindbaren MC Spur zugeordent werden können, und der Anzahl aller Spurkandidaten bestimmt. Wichtig ist, daß Spurkandidaten, die der gleichen MC Spur entsprechen, hier nur einmal mitgezählt werden. Die anderen werden in der Klonrate mitgezählt. d

Trackingreduktionsrate

Die Trackingreduktionsrate ist durch den Quotient der Anzahl an Hits minus der Anzahl an Peaks und der Anzahl an Hits bestimmt. Demnach gibt dieser Quotient Auskunft über die Reduzierung der Daten, denn die Anzahl an Ausgangsdaten wird zu der Anzahl an Eingangsdaten in Bezug gesetzt. d

Trigger

In der Teilchenphysik versteht man unter einem Trigger ein System, das einfache Kriterien verwendet, um schnell zu entscheiden welche Ereignisse in einem Teilchendetektor aufbewahrt werden müssen wenn nur eine kleiner Teil aufgezeichnet werden kann. Trigger-Systeme sind notwendig aufgrund der in der Realität existierenden Einschränkungen für die Speicherkapazität und Datenraten. Da Experimente normalerweise nach "interessanten" Ereignissen forschen (wie zum Beispiel den Zerfällen seltener Teilchen), die mit einer ziemlich geringen Rate auftreten, werden Trigger-Systeme verwendet, um die Ereignisse zu ermitteln, die für nachträgliche Untersuchungen aufgezeichnet werden sollen. a, b, e

trigger

In particle physics, a trigger is a system that uses simple criteria to rapidly decide which events in a particle detector to keep when only a small fraction of the total can be recorded. Trigger systems are necessary due to real-world limitations in data storage capacity and rates. Since experiments are typically searching for "interesting" events (such as decays of rare particles) that occur at a relatively low rate, trigger systems are used to identify the events that should be recorded for later analyses. ix, 1, g, 33, 34, j, 37, 156, 190

tyvek

Tyvek is a brand of flashspun high-density polyethylene fibers which are nonwoven and consists of spunbond olefin. The name is a registered trademark of DuPont. 31, 32

vertex

Here the vertex is the point of origin (creation of the particle) for a reconstructed particle track. The process to find such a vertex for all reconstructed and fitted tracks is called vertex determination or vertex finding. 7, 10, 12–14, 16, 34–40, 56, 94

vertexing

In particle physics vertexing or vertex reconstruction means the process of finding the starting point of a particle motion trajectory. Such a starting point or vertex is defined by the interaction point where the particle collides. For fixed target experiments, such an interaction region is given by the area where the beam and the target interacts. Here the vertex is called primary. In contrast to this, a vertex would be called secondary, if the starting point is not in this special area but afterwards inside the detector. 9, 10, 12, 34, 37

von Neumann

The von Neumann architecture is a design model for a digital computer, that uses a central processing unit and a single separate storage structure to hold both instructions and data. It is named after the mathematician and early computer scientist John von Neumann. Such computers implement a universal Turing machine and have a sequential architecture. 204

watchdog timer

A watchdog timer or computer operating properly timer is a computer hardware or software timer that could trigger a system reset or other corrective action, if the main program neglects to regularly service the watchdog by writing a 'service pulse', which is also referred to as 'kicking the dog', 'petting the dog', 'feeding the watchdog' or 'waking the watchdog', due to some fault condition like a hang for instance. Thus the intention of such a timer is to bring the system back from the unresponsive state into normal operation. The most common use is in embedded systems, where it is often realized as built-in unit of a microcontroller. Watchdog timers may also trigger fail-safe control systems to move into a safety state, such as turning off motors, high-voltage electrical outputs or other potentially dangerous subsystems until the fault is cleared. 194

ZEUS

The ZEUS experiment studies the internal structure of the proton through measurements of deep inelastic scattering by colliding leptons (electrons or positrons) with protons in the interaction point. These measurements are also used to test and study the Standard Model of particle physics, as well as searching for particles beyond this model.

Therefore the ZEUS detector operates on the HERA particle accelerator at DESY and consists of tracking components, a calorimeter and muon detectors. 5, 35

Acronyms

AGS: Alternating Gradient Synchrotron

The location is at the Brookhaven National Laboratory in New York, USA. The years of operation were from 1960 till now. It can accelerate protons to an energy of 33 GeV. The discoveries made are the subatomic particle J/ψ, muon neutrinos and the CP violation in kaons. 4, 5, 9, 106

ALEPH: Apparatus for LEP PHysics

The goal of the ALEPH experiment is a general study of the Standard Model of particle physics and the search for characteristics of unknown and new physics beyond the Standard Model. 5

ALICE: A Large Ion Collider Experiment

ALICE is one of the six detector experiments being constructed at the LHC at CERN. Pb-Pb nuclei collisions will be studied at a center of mass energy of 5.5 TeV per nucleon. The resulting temperature and energy density are expected to be large enough to generate a quark-gluon plasma, which represents a state of matter wherein quarks and gluons are deconfined. 5, 27, 28, 37

ALU: Arithmetic Logic Unit

In computing, an ALU is a digital circuit that performs arithmetic and logical operations. Further on, the ALU is generally a fundamental building block of a CPU. Even the simplest microprocessors contain one for purposes such as maintaining timers. Moreover modern CPUs and GPUs exhibit very powerful and very complex ALUs. 203, 205, 206, 380, 381, 383

APD: Avalanche PhotoDiode

An APD is a photodetector, which can be regarded as the semiconductor analog to the photomultiplier tube (PMT). By applying a high reverse bias voltage, APDs show an internal current gain effect (around

100) due to the impact ionization (avalanche effect). However, some silicon APDs employ alternative doping and beveling techniques compared to traditional APDs. This allows greater voltage to be applied before the breakdown is reached, which results hence in a greater operating gain (> 1000). In general, the higher the reverse voltage is, the higher is the gain. 31

API: Application Programming Interface
An API is an interface implemented by a software program to enable interaction with other software. It is quite similar to the way a user interface facilitates interaction between humans and computers. APIs are implemented by applications, libraries and operating systems to determine the vocabulary and calling conventions the programmer should employ to use their services. It may include specifications for routines, data structures, object classes and protocols used to communicate between the consumer and implementer of the API. 102, 120, 123, A-6, A-10, A-15

ARES: Analyser of Rare EventS
The ARES experiment studies rare decays of muons and pions as well as other rare nuclear processes. For the purpose of investigating such rare decays and rare nuclear processes, a spectrometric facility is designed and built at the Laboratory of Nuclear Problems of JINR in Dubna. The spectrometer consists of cylindrical proportional chambers with gas supply and gas leakage systems, cylindrical scintillation hodoscopes, a magnet, electronics with power supplies and a data acquisition system. 43

ASCII: American Standard Code for Information Interchange
ASCII is a character-encoding scheme based on the ordering of the English alphabet. ASCII codes represent text in computers, communications equipment, and other devices that use text. Most modern character-encoding schemes, which support many more characters than the original, are based on ASCII. Historically, ASCII is developed from telegraphic codes. Its first commercial use was as a seven-bit teleprinter code promoted by Bell data services. 117, 125, 128, 257, A-5, A-8, A-10, A-87, A-102, A-105

ASIC: Application Specific Integrated Circuit
An ASIC is an integrated circuit (IC), which is customized for a particular use rather than intended for general-purpose. 153, 155

ATLAS: A Toroidal LHC ApparatuS
ATLAS is one of the six particle detector experiments (ALICE, ATLAS, CMS, TOTEM, LHCb, and LHCf) constructed at the LHC at CERN. The experiment is designed to observe phenomena that involve highly massive particles, which were not observable using earlier lower-energy accelerators and might shed light on new theories of particle physics beyond the Standard Model. Moreover ATLAS is especially designed to detect the so-called Higgs boson. 5, 27, 33, 37

BLD: ngdBuiLD
The BLD file contains information about the NGDBuild run and its subprocesses. C-8

BMBF: Bundesministerium für Bildung und Forschung
The BMBF is a ministry in the German Cabinet headquartered in Bonn including an office in Berlin. The BMBF provides funding for research projects and institutions and regulates general educational policy. However, a large part of educational policy in Germany is decided at the state level, strongly limiting the influence of the ministry. E

BRAHMS: Broad RAnge Hadron Magnetic Spectrometers
The Brahms experiment is one of the five experiments, which are located at the Relativistic Heavy Ion Collider at the Brookhaven National Laboratory in Long Island, New York. It is designed to measure charged hadrons over a wide range of rapidity and transverse momentum to study the reaction mechanisms of the relativistic heavy ion reactions and the properties of the highly excited nuclear matter formed in these reactions. 5

BSD: Berkeley Software distribution license
BSD licenses represent a family of permissive free software licenses. The original was used for the Berkeley Software Distribution, a Unix-like operating system for which the license is named. The first version of the license was revised, and the resulting licenses are more properly called modified BSD licenses. Permissive licenses, sometimes with important differences pertaining to license compatibility, are referred to as 'BSD-style licenses'. Several BSD-like licenses, including the New BSD license, have been vetted by the Open Source Initiative as meeting their definition of open source. The licenses have few restrictions compared to other free software licenses such as the GNU General Public License or even the default restrictions provided by copyright, putting it relatively closer to the public domain. C-2

BTeV: B physics experiment at 2-TeV proton-antiproton accelerator

The BTeV experiment was designed to challenge the Standard Model explanation of the CP violation, mixing and rare decays of bottom and charm quark states. The Standard Model has been the baseline particle physics theory for several decades and BTeV aimed to find out what lies beyond the Standard Model. In this connection, the BTeV results could have contributed to shed light on phenomena associated with the early universe such as why the universe is made up of matter and not anti-matter. 25

CBM: Compressed Baryonic Matter
CBM is a physics experiment at GSI which creates highest baryon densities in nucleus-nucleus collisions to explore the properties of superdense nuclear matter by searching for in-medium modifications of hadrons like the transition from dense hadronic matter to quark-gluon matter and for the critical endpoint in the phase diagram of strongly interacting matter. i, v, ix, 1, 6–12, 14, 15, 28–31, g, h, 34, j, 36–40, 42–45, 52, 55–57, 72, 74, 77, 86, 95, 104, 120, 152, 295, 307, 315, 320, 322, 391–394, 399, 400, 406–408, 413, 419, 421, 422, 431, 433–435, C-17

CBM: Komprimierte Baryonische Materie
CBM ist ein Physikexperiment am GSI, das höchste Baryondichten in Nukleon-Nukleon Kollisionen erzeugt, um die Eigenschaften von superdichter Kernmaterie durch Suchen nach gegenstandsinternen Hadronenänderungen wie den Übergang von dichter Hadronenmaterie zu Quark-Gluon Materie und nach dem kritischen Endpunkt im Phasendiagramm der stark wechselwirkenden Materie zu erforschen. c, d, b, e

CDF: Collider Detector at Fermilab
The CDF experimental collaboration studies high energy particle collisions at the Tevatron, which is located at the Fermi National Accelerator Laboratory in Batavia, Illinois. The goal is to discover the identity and properties of the particles that make up the universe and to understand the forces and interactions between those particles. 5

CERN: Conseil Européen pour la Recherche Nucléaire
The CERN is the world's largest particle physics laboratory situated just northwest of Geneva on the border between France and Switzerland. Its main function is to provide the particle accelerators and other

infrastructure needed for high-energy physics research. Numerous experiments have been constructed at CERN by international collaborations to make use of them. 5, 33, 35, 106

CLB: Configurable Logic Block
The CLB is the basic logic unit in a FPGA. Exact numbers and features vary from device to device, but every CLB consists of a configurable switch matrix with 4 or 6 inputs, some selection circuitry (MUX, etc) and flip-flops. Furthermore the switch matrix is highly flexible and can be configured to handle combinatorial logic, shift registers or RAM. 153, 154, 332

CMOS: Complementary Metal Oxide Semiconductor
CMOS is a major technology class of integrated circuits. It is used in digital logic circuits as well as in a wide variety of analog circuits. Two important characteristics of CMOS devices are high noise immunity and low static power consumption. Significant power is only drawn when the transistors in the CMOS device are switching between on and off states. Consequently CMOS devices do not produce as much waste heat as other forms of logic like transistor-transistor logic (TTL) or NMOS logic, which uses all n-channel devices without p-channel devices, for instance. The phrase 'metal-oxide-semiconductor' is a reference to the physical structure of certain field-effect transistors having a metal gate electrode placed on top of an oxide insulator, which in turn is on top of a semiconductor material. 15, 16

CMS: Compact Muon Solenoid
The CMS experiment is one of two large general-purpose particle physics detectors built on the Large Hadron Collider (LHC) accelerator at CERN. The main goals of this experiment are the physics exploration at the TeV scale, the discovery of the Higgs boson, the searching for an evidence of physics beyond the standard model such as supersymmetry or extra dimensions and finally the study of different aspects of heavy ion collisions. 5, 37

CPLD: Complex Programmable Logic Device
A CPLD is a programmable logic device with complexity between that of PALs and FPGAs and the architectural features of both. The building block of a CPLD is a macro cell, which contains logic implementing disjunctive normal form expressions and more specialized logic operations. 153

CPU: Central Processing Unit

A CPU or sometimes just called processor represents a class of logic machines that can execute computer programs, which are commonly described by a series of instructions that are kept in some kind of computer memory. So the fundamental operation of most CPUs, which is regardless of the physical form they take, is to execute a sequence of stored instructions called program. Such logic machines conform usually to the von Neumann architecture. 39, 188, 189, 222

CRT: Cathode Ray Tube

The CRT is a vacuum tube containing a source of electrons and a fluorescent screen, which is used by internally accelerating and deflecting the electron beam to create images in the form of light. These images, which are emitted from the fluorescent screen, commonly represent electrical waveforms in oscilloscopes, pictures in televisions or computer monitors, radar targets and many others. The earliest version of the CRT was invented by the German physicist Ferdinand Braun in 1897 and is also known as the Braun tube. Today other display technologies, such as flat plasma displays, liquid crystal displays, DLP, OLED displays have replaced CRTs in many applications and become increasingly popular as costs decline. 4

CVDD: Chemical Vapor Deposition of Diamond

CVDD is a method of producing synthetic diamond by creating the circumstances necessary for carbon atoms in a gas to settle on a substrate in crystalline form. This method offers the production of diamonds, which were previously too expensive or too difficult to build economically. The diamond growth typically occurs under low pressure. Varying amounts of energized gases, which contain a carbon source often in combination with hydrogen, has to be fed into a chamber that fulfills conditions for diamond growth on the substrate. The amount of used gas types vary immense depending on the type of diamond which should be grown. 30

DAQ: Data Acquisition

The DAQ system is used to transport the experimental data from a source to a destination via a network. In this thesis, the source is defined by the front-end electronic of each detector station, which produces the data, while the destination is determined by the trigger system. But as there is not just one trigger system, which needs all data, the network must combine and deliver a configured part of the experimental data to all special trigger system. 34, 156

DELPHI: DEtector with Lepton, Photon and Hadron Identification

DELPHI is one of the four main detectors of the Large Electron-Positron Collider (LEP) at CERN. It weights 3500 tons and has the shape of a cylinder over 10 meters in length and diameter. In normal operation, electrons and positrons from the accelerator go through a pipe to the center of the cylinder to collide in the middle of the detector. Further on, a large number of subdetectors, which are designed to identify the nature and trajectories of the particles produced by such a collision, analyze then the resulting collision products traveling outwards from the pipe. 5, 35

DEPFET: DEpleted P-channel Field Effective Transistor

The DEPFET is a field effective transistor, which can be used as an image photon sensor. It is usually formed in a fully-depleted substrate and acts as a sensor, amplifier and memory node at the same time. Further on the common field effective transistor relies on an electric field to control the shape and hence the conductivity of a channel of one type of charge carrier in a semiconductor material. E

DESY: Deutsches Elektron SYnchrotron

The DESY is the biggest German research center for particle physics with sites in Hamburg and Zeuthen. It is financed by the public authorities and is a member of the Helmholtz Association of National Research Centres. The main purposes are fundamental research in particle physics and research with synchrotron radiation. For these subjects, the DESY develops and runs several particle accelerators. 5, 35, 36

DFG: Deutsche ForschungsGemeinschaft

The DFG is an important German research funding organization. The DFG supports research in science, engineering and the humanities through a large variety of grant programs, prizes and by funding infrastructure. The self-governed organization is based in Bonn and financed by the German states and the federal government. E

DMA: Direct Memory Access

DMA is a feature of modern computers and microprocessors, that allows certain hardware subsystems within the computer to access system memory for reading and/or writing independently of the CPU allowing computation and data transfer concurrency. Many hardware systems use DMA including disk drive controllers, graphics cards, network cards and sound cards. DMA is also used for intra-chip data transfer in multi-core processors, especially in multiprocessor system-on-chips, where its processing element is equipped with a local memory and DMA is used for transferring data between the local memory and the main memory. 194, 202, 210, 211, 377, 388

DPRAM: Dual-Ported Random Access Memory

DP-RAM is a type of random access memory that allows two parallel accesses at the same time, which can be any combination of read and write access. It has to be however mentioned that problems could occur , if the access is at the same address. Moreover this type of RAM determines obviously an upgrade of the common single-ported RAM, which does not allow any parallel access. 161, 163, 164, 171, 172, 174, 331, 332, 346, 347, 352, 355, 359

DRAM: Dynamic Random Access Memory

DRAM is a type of memory that stores each bit of data in a separate capacitor within an integrated circuit. Since real capacitors leak charge, the information eventually fades unless the capacitor charge is refreshed periodically. Due to this refresh requirement, it is a dynamic memory as opposed to SRAM and other static memory. The advantage of DRAM is its structural simplicity, because only one transistor and one capacitor is required per bit, which is much less compared to the six transistors in the SRAM technology. This allows DRAM to reach very high density. Like SRAM, it is in the class of volatile memory devices, since it loses its data when the power supply is removed. Unlike SRAM however, data may still be recovered for a short time after power-off. 159

DSP: Digital Signal Processing

DSP algorithms have traditionally run on specialized processors called digital signal processors (DSPs). Moreover algorithms, which require more performance than DSPs could provide, are typically implemented using application-specific integrated circuit (ASICs). Today however there are a number of technologies used for digital signal processing. These include more powerful general purpose microprocessors, field-programmable gate arrays (FPGAs), digital signal controllers (mostly for industrial applications such as motor control) and stream processors among others. 154

ECAL: Electromagnetic CALorimeter

In particle physics, an ECAL is an experimental apparatus that measures the energy of particles, which interact primarily via the electromagnetic interaction. Generally most particles enter the calorimeter and initiate a particle shower. So its energy is deposited in the calorimeter, collected and measured. This energy may be further on sampled or measured in its entirety requiring total containment of the particle shower. Typically calorimeters are segmented transversely to provide information about the direction of the particle or particles as well as the energy deposited. Besides this, the longitudinal segmentation can provide information about the identity of the particle based on the shape of the shower as it develops. Finally it has to be mentioned that the calorimetry design is an active area of research in particle physics. i, 9, 11, 12, 27, 30, 31

EDA: Electronic Design Automation

EDA is the category of tools for designing and producing electronic systems ranging from printed circuit boards to integrated circuits. C-1

EDF: Electronic Data interchange Format file

An industry standard file format for specifying a design netlist. C-6, C-7

EEPROM: Electrically Erasable Programmable Read-Only Memory

EEPROM (also written E^2PROM) is a type of non-volatile memory, which is used in computers and other electronic devices to store small amounts of data that must be saved when power is removed. Examples are determined by calibration tables or a device configuration. 153

EIB: Element Interconnect Bus

EIB names the internal communication bus of the Cell BE processor, which connects the various on-chip system elements. These elements are further defined by the PPE processor, the memory controller (MIC), the eight SPE co-processors and two off-chip I/O interfaces, which are in total 12 participants. For this purpose, the EIB has to include obviously an arbitration unit which functions as a set of traffic lights. 186, 187

EOL: End-Of-Line

In computing, an end-of-line defines a special character or sequence of characters signifying the end of a line of text. The actual codes representing an end-of-line vary across operating systems, which can be evidently a problem when exchanging data between systems with different representations. A-104

FAIR: Facility for Antiprotons and Ions Research
The proposed project FAIR is an international accelerator facility of the next generation, which incorporates new technological concepts to technological developments and experience already made at the existing GSI facility in Darmstadt. 6–8, g, 435

FAIR: Einrichtung für Antiprotonen- und Ionenforschung
Das beantragte Projekt FAIR beinhaltet eine internationale Beschleunigereinrichtung der nächsten Generation, welche neue technologische Konzepte bis zu Technologie-Entwicklungen und Erfahrungen, die bereits an der bestehenden GSI Einrichtung in Darmstadt gemacht wurden, umfasst. b

FEE: Front-End Electronic
The FEE is used to convert the detector response of a particle interaction into a digital representation, which can be then transported via network. Commonly the FEE produces a coordinate, which represents the position of the interaction and often a value to mark different interactions. 24, 26, 33, 156

FIFO: First In First Out
The FIFO strategy describes the principle of a queue processing technique, which implements the preference schema that is well-known as first-come, first-served (FCFS). This means in other words: What comes in first, is handled first, and what comes in next waits until the first is finished, etc. So this technique defines an abstraction in the ways of organizing and manipulating data relative to time and preference. 182, 288, 290, 397, 425, A-95

FLOPS: Floating Point Operations Per Second
In computing, FLOPS is a measure for computer performance, especially in fields of scientific calculations that make heavy use of floating point calculations. Since computing devices exhibit an enormous range of performance levels in floating-point applications, it makes sense to introduce larger units than FLOPS. For this purpose, the standard SI prefixes can be used to receive the units GFLOPS (one billion or 10^9 FLOPS), TFLOPS (one trillion or 10^{12} FLOPS) and PFLOPS (one quadrillion or 10^{15} FLOPS). 189

FOPI: Four Pi

As suggested by the name, FOPI stands for 4π, which is a synonym for the entire solid angle. The corresponding experiment is expected to produce interesting information on the character of the interaction in hot and compressed nuclear matter. Therefore the used detector detects, identifies and determines the momentum of all charged particles, which are emitted in a heavy-ion reaction. To this the detector is potentially able to identify neutral particles indirectly because of their decays. 30

FPGA: Field Programmable Gate Array

A FPGA is a semiconductor device containing programmable logic components called 'logic blocks' and programmable interconnects. Logic blocks can be programmed to perform the function of basic logic gates such as AND and XOR or even more complex combinational functions such as decoders or simple mathematical functions. In most FPGAs, the logic blocks also include memory elements, which may be simple flip-flops or more complex blocks of memories like 'blockram'. iii, v, vi, xi, xii, xvi, xvii, xix, xxii, h, j, 44, 95, 152–160, 162, 165–167, 169, 171, 173, 175, 177, 179, 180, 182–184, 191, 221, 287, 293, 294, 297, 327, 328, 332, 333, 343, 345, 348, 350, 352, 355, 364, 367, 370, 392, 394, 396, 399–402, 406–416, 419–424, 426–428, 431–433, 435, A-65, A-66, A-96, A-97, C-1, C-2, C-8, C-9, C-17–C-26

FPGA: Feld programmierbare (Logik-)Gatter-Anordnung

Ein FPGA ist ein Halbleiterbauelement, welches programmierbare Logikkomponenten, die sogenannten Logik-Blöcke, und programmierbare Verbindungsnetzwerke enthält. Logik-Blöcke können programmiert werden, um die Funktion grundlegender Logikgatter wie zum Beispiel UND und EXKLUSIVES ODER oder auch komplizierteren Schaltungsfunktionen wie etwa Dekodierer oder einfache mathematische Funktionen zu verrichten. In den meisten FPGAs schließen die Logik-Blöcke auch Speicherelemente ein, was einfache Flip-flops oder kompliziertere Speicherblöcke wie 'Blockram' sein können. c, d, e

FTP: File Transfer Protocol

FTP is a standard network protocol, which is used to copy a file from one host to another over a TCP/IP-based network such as the Internet. Further on, it is built on a client-server architecture and utilizes separate control and data connections between the client and server applications, which solve the problem of different end host configurations like the operating systems or file names. FTP can be used with user-based password authentication or with anonymous user access. Applications were originally interactive command-line tools with a standardized command syntax, but graphical user interfaces have been developed for all desktop operating systems in use today. 106

GCC: GNU Compiler Collection; originally: GNU C Compiler
The GNU Compiler Collection determines a set of programming language compilers produced by the GNU Project. It defines a key component of the GNU toolchain and is freely distributed by the Free Software Foundation (FSF) under the GNU General Public License (GNU GPL) and GNU Lesser General Public License (GNU LGPL). Further on, it is the standard compiler for the free software Unix-like operating systems and Apple Mac OS X. 111, 189

GDDR3: Graphics Double Data Rate 3
GDDR3 is a graphics card-specific memory technology, which is designed by ATI Technologies and the collaboration of JEDEC. It has almost the same technological base as DDR2, but the power and heat dispersal requirements have been reduced somewhat, which allows for higher-speed memory modules and simplified cooling systems. Unlike the DDR2 used on graphics cards, GDDR3 is unrelated to the JEDEC DDR3 specification. This memory uses internal terminators enabling it to better handle certain graphics demands. Further on, GDDR3 memory transfers 4 bits of data per pin in 2 clock cycles to improve bandwidth. 188

GEANT: GEometry ANd Tracking
GEANT is the name of a series of simulation software, which is designed to describe the passage of elementary particles through matter by the usage of Monte Carlo methods (see particle transport engine). Originally developed at CERN for high energy physics experiments, today GEANT has uses in many other fields. 107–110, 216, 220, A-108

GEM: Gas Electron Multiplier
The GEM is a type of gaseous ionization detector, which is used in nuclear and particle physics to detect radiation. It is able to collect the electrons, which are released by ionizing radiation, guiding them to a region with a large electric field and thereby initiating an electron avalanche. The avalanche is then able to produce enough electrons to create a current or charge, which is large enough to be detected by electronics. Moreover GEMs create the large electric field in small

holes in a thin polymer sheet while the avalanche occurs inside of these holes. The resulting electrons are ejected from the sheet. Afterwards a separate system must be used to collect the electrons and guide them towards the read-out. 26

GPL: GNU General Public License
The GPL is a widely used free software license, originally written by Richard Stallman for the GNU project. The GPL is the most popular and well-known example of the type of strong copyleft license that requires derived works to be available under the same copyleft. Under this philosophy, the GPL is said to grant the recipients of a computer program the rights of the free software definition and uses copyleft to ensure the freedoms are preserved, even when the work is changed or added to. This is in distinction to permissive free software licenses, of which the BSD licenses are the standard examples. C-2

GPU: Graphics Processing Unit
A GPU is a dedicated graphics rendering device for a personal computer, workstation or game console. Modern GPUs are very efficient for manipulating and displaying computer graphics, while their highly parallel structure makes them more effective than general-purpose CPUs for a range of complex algorithms. A GPU can sit on top of a video card or can be integrated directly into the motherboard. 188, 189

GSI: Gesellschaft für Schwerionenforschung
Das Ziel der wissenschaftlichen Forschung, die von der GSI in Darmstadt, Deutschland durchgeführt wird, besteht darin, die Struktur und das Verhalten der Welt, die uns umgibt, zu verstehen. Das GSI betreibt eine große Beschleunigereinrichtung für Schwerionenstrahlen. Forscher nutzen die Einrichtung für Experimente, die helfen den Weg zu neuen und faszinierenden Entdeckungen in der Grundlagenforschung zu weisen. Das Programm am GSI deckt eine breite Palette von Aktivitäten ab, die sich von Kern- und Atomphysik über Plasma- und Materialforschung bis zu Biophysik und Krebstherapie erstrecken. Vermutlich sind die bekanntesten Ergebnisse die Entdeckung sechs neuer chemischer Elemente und die Entwicklung einer neuartigen Tumortherapie, die Ionenstrahlen verwendet. b

GSI: Gesellschaft für Schwerionenforschung
The goal of the scientific research conducted at GSI in Darmstadt, Germany is to understand the structure and behavior of the world that surrounds us. GSI operates a large accelerator facility for heavy-ion

beams. Researchers use the facility for experiments that help point the way to new and fascinating discoveries in basic research. The program at GSI covers a broad range of activities extending from nuclear and atomic physics to plasma and materials research to biophysics and cancer therapy. Probably the best-known results are the discovery of six new chemical elements and the development of a new type of tumor therapy using ion beams. 6, h, 110, 111, E

HADES: High Acceptance Di-Electron Spectrometer
The HADES experiment contains a detector system for di-electron pair spectroscopy. The physics program of HADES is broad and includes the study of electron-positron pair emission in relativistic heavy ion collisions, di-electron production in elementary reactions and experiments aimed to study the structure of hadrons. With the available beam energies of 4.5 GeV for protons and 1-2 AGeV for heavy ions, the interest is focused on the pair invariant masses up to 1 $\frac{GeV}{c^2}$. 25

HDL: Hardware Description Language
In contrast to programming languages, which are inherently procedural (single-threaded) with limited syntactical and semantic support to handle concurrency, HDLs can model multiple parallel processes that automatically execute independent of each other. Both programming languages and HDLs are processed by a compiler, but with different goals. For HDLs compiler refers to the synthesis, which is a process of transforming the HDL source code into a device configuring bit-file. 153, 169, 368, C-1, C-2, C-5, C-6

HERA: Hadron Elektron Ring Anlage
HERA is a particle accelerator, which is on the energy frontier in certain regions of the kinematic range. It is able to collide electrons or positrons with protons at a center of mass energy of 318 GeV. Additionally it is the only lepton-proton collider in the world. HERA is located under the DESY site nearby the Volkspark in Hamburg, Germany. It is around 15 to 30 m underground and has a circumference of 6.3 km. Moreover four interaction regions are used by the experiments H1, ZEUS, HERMES and HERA-B. 5

HERAB: Hadron Electron Ring Accelerator B hadrons
HERAB was an innovative particle physics detector of the HERA accelerator at DESY. Its primary aim was determined by the measurement of the CP violation in the decays of heavy B-mesons in the late 1990s,

which was several years ahead of the Large Hadron Collider and the BaBar programs. Unlike most particle physics detectors, the particles are produced not by colliding two circulating beams head-on nor by slamming the beam into a stationary target, but by moving a thin wire target directly into the waste 'halo' of the circulating proton beam of the HERA accelerator. Hence the beam is unaffected by this 'scraping', but the produced collision rate could be made extremely high, which means around 5 to 10 million interactions per second (5-10 MHz). 5, 30, 31, 36, 43

IBM: International Business Machines corporation
IBM is a multinational computer technology and consulting corporation headquartered in New York, USA. The company is one of the few information technology companies with a continuous history dating back to the 19th century. IBM manufactures and sells computer hardware and software, offers infrastructure services, hosting services and consulting services in areas ranging from mainframe computers to nanotechnology. 157, 158, 185

IDE: Integrated Development Environment
An IDE, which is also known as integrated design environment or integrated debugging environment, is a software application providing comprehensive facilities to computer programmers for software development. For this purpose, an IDE consists ordinary of a source code editor, a compiler and/or an interpreter, build automation tools and a debugger. Sometimes a version control system and various tools are integrated to simplify the construction of a GUI. Many modern IDEs include also a class browser, an object inspector and a class hierarchy diagram, which is profitable in object-oriented software development. 153

INTEL: INTegrated ELectronics corporation
INTEL is one of the world's largest semiconductor chip manufacturer, which is headquartered in Santa Clara, California, USA. The company is for example the inventor of the x86 series of microprocessors, which are found in most personal computers. INTEL produces also motherboard chip sets, network interface controllers, integrated circuits, flash memory, graphic chips, embedded processors and other devices related to communications and computing. 41

JTAG: Joint Test Action Group

JTAG is the usual name for the IEEE 1149.1 standard, which entitles the Standard Test Access Port and Boundary-Scan Architecture for test access ports. It is commonly used for testing printed circuit boards using boundary scan. Nowadays JTAG is the common interface for transferring the software generated bit-file, which bases on a HDL description, to a FPGA to configure the logic device. 153

KOPIO: K zero Pi zero: $K_\perp^0 \longrightarrow \pi^0\nu\bar{\nu}$

The KOPIO experiment is looking for a very rare reaction with the potential to explain the observed lack of symmetry between matter and anti-matter in the universe. Using the AGS proton accelerator at Brookhaven National Laboratory, the KOPIO experiment tries to create an intense beam of kaons to study special very rare decays. The discovery and observation of these rare reactions will incisively probe charge parity (CP) symmetry violation, which is a fundamental evidence that a mirror-image anti-universe would look slightly different from our own. 30

LEP: Large Electron-positron Collider

The LEP is a circular collider, which has a circumference of 27 kilometers and is built in a tunnel straddling the border of Switzerland and France. It accelerates the electrons and positrons to a total energy of 45 GeV each to enable production of the Z boson, which has a mass of approximately 91 GeV. The accelerator was upgraded to enable the production of a pair of W bosons weighting respectively approximately 80 GeV. The LEP collider energy eventually topped at 209 GeV at the end in the year 2000, because it is shut down and then dismantled in order to make room in the tunnel for the construction of the Large Hadron Collider. 5

LHC: Large Hadron Collider

The LHC is the world's largest and highest-energy particle accelerator, which is intended to collide opposing beams of protons or lead ions moving each approximately close to the speed of light. It is built by the European Organization for Nuclear Research (CERN) with the intention for testing various predictions of high-energy physics including the existence of the hypothesized Higgs boson and the large family of new particles predicted by the supersymmetry. The accelerator possesses a circumference of 27 kilometers and lies underneath the Franco-Swiss border between the Jura Mountains and the Alps near Geneva,

Switzerland. It is funded by and built in collaboration with over 10,000 scientists and engineers from over 100 countries as well as hundreds of universities and laboratories. 5, 9, 37, 109

LHCb: Large Hadron Collider beauty
The LHCb experiment is one of six particle physics detector experiments built on the Large Hadron Collider (LHC) accelerator at CERN. LHCb is a specialized b-physics experiment particularly aimed at measuring the parameters of the CP violation in the interactions of b-hadrons, which are heavy particles containing a bottom quark. 5, 30, 31, 37, 43

LQCD: Lattice Quantum ChromoDynamics
In physics, LQCD is a theory of quarks and gluons, which is formulated on a space-time lattice. So this theory defines a lattice model of quantum chromodynamics, which is a special case of a lattice gauge theory or lattice field theory. At the moment, this is the most well established non-perturbative approach to solve the theory of Quantum Chromodynamics. 7

LUT: Look-Up-Tabelle
Look-Up-Tabellen werden oft verwendet, um sehr komplizierte Berechnungen in der Informatik und insbesonders in der Hardware-Entwicklung bis auf die LUT-Zugriffszeit zu beschleunigen, denn die Look-Up-Tabellen sind normalerweise in der Lage die entsprechend gespeicherten Ergebnisse, die zu Anfang off-line berechnet werden, auszuliefern. Somit werden die erforderlichen Berechnungen aus dem kritischen Pfad, der zur zeitlichen Einschränkung gehört, herausgezogen. Nichtsdestotrotz ist ein Nachteil offensichtlich durch den Umstand bestimmt, daß die zeitbezogene Beschleunigung Hand in Hand mit den Kosten für den notwendigen Speicher, der die Ergebnisse in der LUT speichert, geht. c

LUT: Look-up-table
LUTs are often used to speed up very complex calculations in computer science and especially hardware development to just the LUT access time, because the LUTs are naturally able to deliver the correspondingly stored results, which are calculated initially off-line. Hence the required calculations are moved out of the critical path belonging to the timing constraints. Nevertheless a disadvantage is obviously determined by the circumstance that the speed up in time goes hand in hand with the costs for the necessary memory to store the results in

the LUT. i–iii, v, xi, xiii, xiv, xvi–xviii, xxi, h, 52, 73, 78, 79, 82, 84, 96, 99–101, 116–118, 123–125, 128–130, 132–137, 145, 153, 156, 158–160, 165, 176, 190, 192, 201, 202, 225, 227, 229–238, 265, 268, 273–277, 298, 304–307, 324, 325, 328, 332, 342, 343, 345, 350, 355–358, 361, 371, 376, 383, 384, 392, 397, 400, 408, 410, 415, 435–437, A-18, A-19, A-22, A-23, A-55, A-56, A-67, A-68, A-70, A-72, A-73, A-89–A-92, A-96, A-99, A-100, A-107

MAPD: Micropixel Avalanche PhotoDiode
A MAPD is a micropixel avalanche photodiode, which works in limited Geiger mode with the internal gain up to 10^6. Moreover it has no nuclear counting effect due to the pixel structure. 32, 33

MAPMT: Multiple Anode PhotoMultiplier Tube
The MAPMT can detect the temporal and spatial distribution of signal intensity equivalent to having multiple PMTs located in a single housing. Further on, the PMT is well-known as a single-photon-sensitive detector that measures the intensity as a function of time into the gigahertz range. However the PMT can not measure the spatial distribution of the intensity. 25

MAPS: Monolithische-Aktive-Pixel-Sensor
Ein MAPS ist ein sehr dünner Pixelsensor, der widerstandsfähig gegenüber Strahlung ist, und eine hohe räumliche Auflösung aufweist. Ferner vereint diese Technologie den Sensor und die einzelnen Vorverstärkern auf dem gleichen CMOS Chip. Im CBM Experiment wird dieser Typ verwendet, um den MVD aufzubauen. d

MAPS: Monolithic Active Pixel Sensor
A MAPS is a very thin pixel sensor, which is radiation hard and features a high spatial resolution. Moreover this technology combines the sensor and the individual preamplifiers on the same CMOS chip. In the CBM experiment this type is used to build the MVD. i, xviii, xxi, 15–17, 115, 316, 317, A-17

MC: Monte Carlo
Monte Carlo methods represent a widely used class of computational algorithms for simulating the behavior of various physical and mathematical systems. In general, these methods are used to solve various problems by generating suitable random numbers and observing that fraction of the numbers obeying some properties. Hence they are useful to obtain numerical solutions for problems, which are too complicated

to solve analytically or to model phenomenas with significant uncertainty in inputs (see also GEANT). ix, xiv–xviii, 39, 43, 44, 58, 61, 72, 79, 80, 104, 105, 107, 108, 110, 113, 121, 122, 131, 216–220, 222, 229–240, 242–247, 251–254, 258–265, 269, 273, 274, 279–283, 296, 300–309, 312–315, 318–320, 323, A-6, A-7, A-12–A-15, A-26, A-32–A-34, A-39, A-40, A-70, A-72, A-73, A-75, A-83–A-86, A-90–A-94, A-97, A-101, A-102, A-107, A-108

MGT: RocketIO Multi-Gigabit Tranceiver
The MGT is a Xilinx technology, which implements a serial optical high-speed network with maximal transfer rates above 1 Gbit per second. 156, 406, 420, 421, 431, 433

MIC: Memory Interface Controller
The MIC is a controller, which interfaces the memory in the Cell BE architecture, to enable the XDRRAM memory to be available for all other components in the system. 186, 187

MRP: Map RePort
The MRP file contains information about the Map run and its subprocesses. C-8

MUCH: Muon Chamber
The MUCH detector is designed to identify muons by detecting them outside most of the iron of the hadron calorimeter, because all other charged particles are likely to have been absorbed within the calorimeter. Since muons are furthermore minimum ionizing particles with a specific energy loss in the iron, any particles with momenta above 2-3 GeV/c can penetrate the well-chosen length of iron. i, 9, 12, 25, 26

MVD: Micro Vertex Detector
Between the target and the STS, a MVD enables to distinguish particle decay vertices from the event vertex. i, iv, ix, xxi, 11–17, 316, 317, 320, A-17

MWPC: Multiwire Proportional Chamber
A MWPC is a detector for particles of ionizing radiation, which defines an advancement to the concept of the Geiger counter and the proportional counter. Moreover a proportional counter uses a Geiger-Müller tube, which contains a wire under high voltage running down the length of a metal tube whose walls are held at ground potential. Further on, this tube is filled with carefully chosen gas, such that any ionizing particle that passes through the tube will ionize surrounding

gaseous atoms. The resulting ions and electrons are accelerated by the potential on the wire causing a cascade of ionization, which is collected on the wire and results in an electric current. This circumstance offers then the possibility to count particles and determine their energy in the case of the proportional counter. 27, 28

NCD: Native Circuit Description
The NCD file contains a physical description of the design in terms of the components in the target Xilinx device. Such a file can contain either mapping information and/or place-and-route information. C-8, C-9

NEMO: Neutrino Ettore Majorana Observatory
The NEMO experiment is an international collaboration of scientists searching for neutrinoless double beta decays, which would be the evidence that neutrinos are Majorana particles. The observation of such a process would also provide fundamental informations about neutrinos like their absolute mass scale, their inner nature concerning the matter/antimatter asymmetry or maybe the evidence for supersymmetry. The experiment has a cylindrical shape with 20 sectors that contain different isotopes in the form of thin foils with a total surface of about 20 square meter. 43

NGC: Native Generic Circuit
The NGC file is a Xilinx netlist file that contains both logical design data and constraints. C-4, C-7

NGD: Native Generic Database
A NGD file describes the logical design reduced to Xilinx primitives. C-7, C-8

OPAL: Omni-Purpose Apparatus for Lep
The main goals of the OPAL experiment are the exact determination of the Z^0 mass, the generation of W-boson pairs including their decay products for testing of the electroweak interaction and the search for hints of the Higgs mechanism. 5

PAR: Place-and-route Report
The PAR file includes summary information for all placement and routing iterations. vi, 348, 350, C-9, C-10

PCF: Physical Constraint File

The PCF file is an ASCII text file containing the constraints specified during design entry expressed in terms of physical elements. C-8

PDG: Particle Data Group

The PDG is an international collaboration of particle physicists that compiles and reanalyzes published results related to the properties of particles and fundamental interactions. Moreover this group publishes also reviews of theoretical results that are phenomenologically relevant including those in related fields such as cosmology. Further on, the PDG maintains the standard numbering scheme or code for particles in event generators in association with the event generator authors. A-19

PHENIX: Pioneering High Energy Nuclear Interactions eXperiment

PHENIX is one of the five experiments at the Relativistic Heavy Ion Collider located at the Brookhaven National Laboratory. Its research goals are the discovery and the examination of the quark gluon plasma, which defines a state of matter where the strong interaction is dominant and the quarks and gluons are not bound in hadrons, and the analysis of the spin structure of the proton. PHENIX consists further of several detector types that are designed to detect photonic, leptonic and hadronic signals. 5, 30

PLB: Processor Local Bus

The PLB is a high-speed and low-latency general data bus, which belongs to the CoreConnect microprocessor bus architecture from IBM and is established for system-on-a-chip designs. Furthermore it exhibits a standard interface between processor cores and bus controllers, which applies separate read and write data buses supporting 32, 64 and 128 bit concurrent access. 157

PLL: Phase Locked Loop

A PLL is a control system that generates a signal that has a fixed relation to the phase of a reference signal. So a PLL circuit responds to the frequency and the phase of the input signals. For this purpose, such a circuit automatically raises or lowers the frequency of a controlled oscillator until it is matched to the reference in frequency and phase. 154

PMT: PhotoMultiplier Tube

A PMT is a vacuum phototube, which is extremely sensitive to light in the ultraviolet, the visible and the near-infrared ranges of the electromagnetic spectrum. Therefore such a detector multiplies the signal, which is produced by incident light, about a 100 million times. For that reason, single photons could be detected individually, if the incident flux of light is very low. 25, 31

PPC: PowerPC
A PPC is a RISC microprocessor architecture, which is created by the Apple-IBM-Motorola alliance known as AIM in 1991. Originally intended for personal computers, the PPC processors have become popular in embedded and high-performance computing as well. 155, 157, 158, 183, 188, 370

PPE: Power Processing Element
The PPE is the power architecture based two-way multi-threaded core, which acts as controller for the eight SPEs of a Cell BE microprocessor to handle most of the computational workload. This element will work with conventional operating systems due to its similarity to other 64-bit PowerPC processors. In addition to this, IBM has also included an AltiVec unit, which is fully pipelined for single precision floating point. However this Altivec unit does not support double precision floating-point vectors. xii, xxiv, xxv, 186, 188, 189, 193, 195–199, 201, 202, 210, 211, 372, 375, 377–387, 389

PSD: Projectile Spectator Detector
The PSD can be used for the precise determination of the centrality of nuclear collisions in combination with the number of participants as well as the reconstruction of the reaction plane and the beam intensity monitor by detecting the electromagnetic dissociated neutrons. i, 32

pTP: para-TerPhenyl
pTP is a combustible toxic liquid, which is boiling at 405 degree Celsius. Moreover crystals of this liquid are commonly used for scintillation counters, while the polymerization with styrene results in plastic phosphor. 31

QCD: Quantum ChromoDynamics
QCD is a theory explaining the strong interaction, which defines a fundamental force describing the interactions of the quarks and gluons found in hadrons. These particles are generally made of quarks or gluons such as the proton, neutron or pion. Further on, the QCD determines an important part of the Standard Model of particle physics

and stands for a quantum field theory of a special kind, which is called a non-abelian gauge theory. 2, 6, 7

QGP: Quark-Gluon Plasma
QGP is a phase of quantum chromodynamics, which exists at extremely high temperature and/or density. This phase consists of almost free quarks and gluons, which are the basic building blocks of matter. 6, 8, 9

RAM: Random Access Memory
Today RAM determines a kind of storage in the form of integrated circuits that allow the stored data to be accessed in any unspecified order. In this context, the word random refers thus to the fact that any piece of data can be returned in a constant time regardless of its physical location and whether or not it is related to the previous piece of data. So this contrasts obviously with storage mechanisms such as tapes, magnetic discs and optical discs, which rely on the physical movement of the recording medium or a reading head. In these devices, the movement takes longer than the data transfer, and the retrieval time varies depending on the physical location of the next item. xii, xviii, xxiii, 159, 165, 171, 173, 175, 188, 327, 328, 332, 336, 345–347, 350–355, 357, 361, 410

RHIC: Relativistic Heavy Ion Collider
The years of operation for the RHIC last from today back to the year 2000, in which it was successfully established with a dimension of 3800 meters for the ring at the Brookhaven National Laboratory in New York, USA. It can accelerate Au-Au, Cu-Cu, d-Au and polarized pp proton-proton pairs to an energy of 100 GeV per nucleon. Experiments using this accelerator are STAR, PHENIX, Brahms, Phobos and pp2pp. 5, 9, 106

RICH: Ring Imaging CHerenkov
A RICH detector is a particle detector that can determine the velocity of a charged fundamental particle. The concept used to realized this characteristic is determined by an indirect measurement of the Cherenkov angle, which means for example the angle between the emitted Cherenkov radiation and the particle path. i, 9, 11, 12, 24, 25, 27, 30

RISC: Reduced Instruction Set Computer
The acronym RISC represents a CPU design strategy emphasizing the

insight that simplified instructions that 'do less' may still provide higher performance than CISCs (Complex Instruction Set Computers), because this simplicity can be utilized to make instructions execute very quick. Even if many proposals for a 'precise' definition have been attempted, the term would be slowly replaced by the more descriptive load-store architecture due to its simplicity. Well known RISC families include Alpha, ARC, ARM, AVR, MIPS, PA-RISC, Power Architecture including PowerPC, SuperH and SPARC. 157, 158, 186

RPC: Resistive Plate Chamber
A RPC consists of a stack of planar electrode plates of high resistivity, which are kept by spacers at fixed distances between 0.2 and 0.35 mm typically. Further on, an anode is normally placed in the center of the stack, while two cathodes enclose it at its outer surfaces. So operated in the right gas mixture under a uniform electric field of about 10 kV/mm between the electrodes, the counters deliver fast avalanche signals, which can be derived as single pulses from the anode or between the anode and the cathodes in differential mode. 9, 28, 29

RTL: Register Transfer Level
In integrated circuit design, the RTL description is a way of describing the operation of a synchronous digital circuit. In RTL design, a circuit's behavior is defined in terms of the flow of signals or transfer of data between hardware registers and the logical operations performed on those signals. RTL abstraction is used in HDLs like Verilog or VHDL to create high-level representations of a circuit, from which lower-level representations and ultimately actual wiring can be derived. 183, 332

SATA: Serial Advanced Technology Attachment
SATA defines a computer bus interface for connecting host bus adapters to mass storage devices such as hard disk drives and optical drives. It was designed to replace the older ATA standard, which is also known as EIDE. SATA is able to use the same low level commands, but host-adapters and devices communicate via a high-speed serial cable over two pairs of conductors. In contrast, PATA, which renames the legacy ATA specification, uses 16 data conductors with each operating at a much lower speed. SATA offers further several compelling advantages over the older PATA interface like hot swapping, reduced cable-bulk from 80 wires to seven and thus costs as well as faster and more efficient data transfers. 188

SCEI: Sony Computer Entertainment Incorporated

The SCEI is a multinational conglomerate corporation, which is one of the world's largest media conglomerates headquartered in Tokyo, Japan. This corporation is one of the leading manufacturers in electronics, video, communications, video game consoles and information technology products for the consumer and professional markets.Additionally SCEI handles the research and development, production and sales of hardware and software for their high-selling Playstation line of handheld and video game consoles. The corporation acts also as developer and publisher of video games for their systems. 185, 188

SIMD: Single Instruction Multiple Data

In computing, SIMD is a technique, which is employed to achieve data level parallelism in a vector processor. First made popular in large-scale supercomputers (contrary to MIMD parallelization), smaller-scale SIMD operations have now become widespread in personal computer hardware. So today the term is associated almost entirely with these smaller units. 186, 205, 372

SiPM: Silicon PhotoMultiplier

A SiPM is a silicon single photon sensitive device, which is built from an avalanche photodiode (APD) array on common Si substrate. The idea behind this device is determined by the detection of single photon events in sequentially connected Si APDs. So every APD in a SiPM operates in Geiger mode and is coupled with the others by a polysilicon quenching resistor. Although the device works in digital/switching mode, the SiPM is an analog device, because all the microcells are read in parallel, which enables the generation of signals within a dynamic range from a single photon to 1000 photons for just a single square millimeter area device. Besides this, the supply voltage depends obviously on the used APD technology. 31

SIS: Schwer-Ionen-Synchrotron

The SIS is a heavy ion synchrotron, which is located at the GSI in Darmstadt, Germany and operates since 1989. It is able to accelerate ions to an energy from 50 to 1000 Mev/u by applying a ring shape with a diameter of 70 meters. 6, 7, 9, 106

SISD: Single Instruction Single Data

In computing, SISD is a term referring to a computer architecture, in which a uniprocessor executes a single instruction stream consecutively to operate on single data elements. SISD is one of the four main classifications as defined in Flynn's taxonomy. In this system, classifications are based upon the number of concurrent instructions and data streams that are present in the computer architecture. 205, 372

SM: Standard Model of particle physics

The SM is a theory, which describes the well-known elementary particles and their fundamental interactions with each other. It contains three different interactions, which are the strong interaction, the weak interaction and the electromagnetic interaction. Further on, the SM is a relativistic quantum field theory, which obeys the laws of the special theory of relativity. The fundamental objects are fields in space-time (field theory), which are only changed in discrete packets (quantum theory). By the way, these discrete packets correspond in an accurate representation to the observed particles. Although particle physics experiments nearly prove predictions of the SM, the gravity is however not included and some observations can not be explained. 2

SOI: Silicon On Insulator

The SOI technology refers to the use of a layered silicon-insulator-silicon substrate in place of conventional silicon substrates in semiconductor manufacturing to reduce parasitic device capacitance and thereby improving performance. SOI-based devices differ from conventional silicon-built devices in the way that the silicon junction is above an electrical insulator, which is typically made of silicon dioxide or sapphire. Furthermore this choice depends largely on the intended application. Therefore sapphire is used for radiation-sensitive applications and silicon dioxide preferred for improved performance and diminished short channel effects in microelectronics devices. Moreover the insulating layer and topmost silicon layer also vary widely with the application. 15

SPE: Synergistic Processing Element

A SPE is a RISC processor with SIMD organization for single and double precision instructions including an excellent vectorized floating point code execution. With the current generation of the Cell BE, each SPE contains 256kB embedded SRAM for instruction and data called local storage, which is visible to the PPE and can be addressed directly by software. Further on, the local store does not operate like a conventional CPU cache, as it is neither transparent to software nor does it contain hardware structures that predict which data to load.

xii, xxiv, xxv, 186, 188, 189, 192–194, 196, 202–205, 207, 210, 211, 371, 372, 375–390

SPRAM: Single-Ported Random Access Memory
SP-RAM is a synonym for the most common type of RAM. In contrast to the DP-RAM, the SP-RAM allows no concurrent access. 161, 358, 359

SPS: Super Proton Synchrotron
The years of operation for the SPS last from 1981 till 1984 at CERN. It's dimension reaches 6900 meters for the ring and can accelerate proton-antiproton pairs to an energy of 400 GeV. Experiments using this accelerator are UA1 and UA2. 5, 9, 10, 106

SRAM: Static Random Access Memory
SRAM is a type of semiconductor memory, which does not need to be periodically refreshed like dynamic RAM (DRAM), because SRAM uses bistable latching circuitry to store each bit. Further on, SRAM exhibits data remanence, but is still volatile in the conventional sense that data is eventually lost when the memory is not powered. 159, 160, 328

SRL: Shift Register Look-up-table
In digital circuits, a memory element, which follows the FIFO principle, is often implemented with sequential shift registers. But in contrast to this, a ring buffer can also be used to offer such a functionality. Vendor specific FPGAs, like the Xilinx Virtex V family, has often the capability to use dedicated LUTs in CLBs to realize such a ring buffer. As such a LUT has commonly 4 inputs, a 16 times shift can be realized. This leads then to the vendor specific name SRL16. 343, 352

SRR: Synthesis Results and Reports
The SRR file provides information on the Synplify Pro synthesis run, as well as area and timing reports. C-6

STAR: Solenoidal Tracker at RHIC
STAR is one of the five experiments at the Relativistic Heavy Ion Collider, which is located at the Brookhaven National Laboratory. The primary physics task of STAR is to study the formation and characteristics of the quark gluon plasma, which is a state of matter believed to exist at sufficiently high energy densities. Further on, detecting and

understanding the QGP enables the better understanding of the universe in the moments after the Big Bang, where the symmetries and lack of symmetries of our surroundings were put into motion. 5

STI: Sony-Toshiba-IBM joint venture
STI is an alliance of the Sony Computer Entertainment Incorporated, the Toshiba corporation and the IBM corporation. These companies work together in a joint venture to develop the Cell BE microprocessor architecture, which is the central processing unit of the Sony Playstation III gaming console. 41, 185

STL: Standard Template Library
The STL provides a ready-made set of common classes such as containers and associative arrays, which can be used with any type that supports some elementary operations such as copying and assignment. Moreover STL algorithms are independent of containers, which significantly reduces the complexity of the library. Further on, the STL achieves its results through the use of templates. This approach provides compile-time polymorphism, which is often more efficient than traditional run-time polymorphism. 136–138, 141, 143, 150, 151, A-83, A-88

STS: Silizium Tracking System
Das STS besteht aus Siliziumdetektorgeräten, die den Halbleiter Silizium verwenden, um durchkreuzende geladene Teilchen oder die Absorbierung von Photonen zu detektieren. Darüber hinaus dienen diese Geräte zur Spur- und Impulsmessung aller geladener Teilchen, die am "Target" (Zeilscheibe) erzeugt werden. Weiterhin werden diese Geräte in dieser Arbeit auch Detektorstationen oder Detektorebenen genannt. b

STS: Silicon Tracking System
The STS consists of silicon detector devices, which uses the semiconductor silicon to detect traversing charged particles or the absorption of photons. Moreover these devices serve for track and momentum measurement of all charged particles, which are produced at the target. Further on, the devices are also called detector stations or detector planes in this thesis. i, iv, ix, 7, 9–15, 25, 28, h, 34, 40, 95, 105, 156, 281, 282, 298, 317, 392, A-1, A-2, A-10, A-15–A-18, A-32, A-85, A-86, A-117

SuperHILAC: Super Heavy Ion Linear Accelerator
The HILAC was initially built in 1957 and afterwards continually im-

proved over the years until 1972, as it was upgraded to the SuperHilac, which is a more powerful machine. Ions accelerated at the SuperHilac are passed through a transfer line where they are accelerated to nearly the speed of light. With the later modifications in the early 1980's, the renamed Bevalac became the only machine in the world, which is capable of accelerating all of the elements in the periodic table up to and including Uranium to relativistic energies. 106

SYR: SYnthesis Report

The SYR file consists of three parts. The first one contains the synthesis options, the second one contains messages, which are generated during the synthesis run and the third part gives a final report. C-5

THGEM: THick Gas Electron Multiplier

THGEM is an augmented version of a GEM detector, which exhibits mechanically drilled holes on thick FR4 plates having a thickness of 500μ and more. Furthermore these holes are obviously larger and operate thus at higher voltages, which leads naturally also to an easier manufacturing process. 26

TOF: Time Of Flight

A TOF detector is a particle detector, which can discriminate between a lighter and a heavier elementary particle of same momentum using their time of flight information measured between two scintillators. i, 9, 11, 12, 25, 28–30

TOTEM: TOTal Elastic and diffractive cross section Measurement

The TOTEM experiment is one of the six detector experiments, which are constructed at the Large Hadron Collider at CERN. The goal of this experiment is the detection of the size of a proton in an all-time exactness and the observation of the luminosity of the particle accelerator. Therefor the detector aims the measurement of total cross section, elastic scattering and diffractive processes. 5

TR: Transition Radiation

TR is produced by relativistic charged particles when they cross the interface of two media of different dielectric constants. Further on, the emitted radiation is determined by the homogeneous difference between the two inhomogeneous solutions of Maxwell's equations for the electric and magnetic fields of the moving particle in each medium separately. 27

TRD: Transition Radiation Detector
A TRD is a particle detector, which utilizes the gamma-dependent threshold of the transition radiation in a stratified material. Such a detector contains naturally many layers of materials with different indices of refraction. So at each interface between materials, the probability of transition radiation increases with the relativistic gamma factor. Thus particles with large gamma give off many photons, while small gamma results in few. This circumstance allows then obviously for a given energy to discriminate between lighter particles, which have a high gamma and therefore radiates, and heavier particles, which have a low gamma and radiates thus much less. i, xxi, 9, 11, 12, 25, 27, 28, 30

TWR: Timing Wizard Report
A TWR file is a special report file, which is dedicated to the timing analysis after the place-and-route processed NCD file. vi, C-9, C-10, C-15

UCF: User Constraints File
he User Constraints File is an ASCII file that you create with a text editor or the Xilinx Constraints Editor. The file contains timing and layout constraints that affect how the logical design is implemented in the target device. The constraints in the file are added to the information in the output NGD file. C-8

UNILAC: UNIversal Linear ACcelerator
The UNILAC is a heavy ion linear accelerator, which is located at the GSI in Darmstadt, Germany. It can provide beams of accelerated ions for elements from proton to Uranium with energies of up to 11.4 MeV/u. Further on, the UNILAC is used to send beams of heavy ions to experiments as well as to load the SIS with high-energy ions. In the past 20 years, experiments using beams from UNILAC have produced the elements 107 to 112 of the periodic table of elements. 6

UrQMD: Ultrarelativistic Quantum Molecular Dynamics
The UrQMD stands for a specialized software, which offers the possibility to simulate (ultra)relativistic heavy ion collisions. Thence this software is often used for event generation (see particle event generator). 104, 106, 107, A-108

VMC: Virtual Monte Carlo
see Monte Carlo. 102, 104, 107, 108, 110

WLSF: Wavelength Shifting Fiber

A WLSF is a photofluorescent fiber that absorbs higher frequency photons and emits lower frequency photons. In most cases, the used fiber absorbs one photon and emits multiple lower-energy photons. Further on, properties of WLSF are transmission attenuation and transverse absorption. 25, 30–32

XDRDRAM: Extreme Data Rate Dynamic Random Access Memory

The XDRDRAM is a high-performance RAM interface, which defines the successor to the Rambus RDRAM. Therefore this type competes with the DDR2 SDRAM and GDDR4 technology. For this purpose, the XDRDRAM is designed to be effective in small and high-bandwidth consumer systems, high-performance memory applications and high-end GPUs. Moreover the XDRDRAM eliminates the unusually high latency problems that plagued early forms of RDRAM. In addition to this, it emphasizes heavily on the bandwidth per pin, which can benefit further the cost control on the PCB production. This fact is obviously determined by th circumstance that fewer lanes are needed for the same amount of bandwidth. By the way, Rambus owns the rights to the technology, which is used by Sony in the Playstation III console. 187, 188, 201, 202, 211, 371, 376, 379, 384

XMB: XrossMediaBar

The XMB developed by Sony is a graphical user interface, which features icons that are spread horizontally across the screen. Furthermore these icons organize the options, which are available to the user, in categories, while implementing also the navigation via the movement instead of the common application of a cursor. So if an icon is selected on the horizontal bar, several more would appear vertically, above and/or below, which are selectable by the up and down directions on a directional pad. Even if originally used on the PSX, the XrossMediaBar is generally used as default interface on both the Playstation Portable and the Playstation III. 188

Bibliography

[AAA+06] J. Allison, K. Amako, J. Apostolakis, H. Araujo, P. Arce Dubois, M. Asai, G. Barrand, R. Capra, S. Chauvie, R. Chytracek, G. Cirrone, G. Cooperman, G. Cosmo, G. Cuttone, G. Daquino, M. Donszelmann, M. Dressel, G. Folger, F. Foppiano, J. Generowicz, V. Grichine, S. Guatelli, P. Gumplinger, A. Heikkinen, I. Hrivnacova, A. Howard, S. Incerti, V. Ivanchenko, T. Johnson, F. Jones, T. Koi, R. Kokoulin, M. Kossov, H. Kurashige, V. Lara, S. Larsson, F. Lei, O. Link, F. Longo, M. Maire, A. Mantero, B. Mascialino, I. McLaren, P. Mendez Lorenzo, K. Minamimoto, K. Murakami, P. Nieminen, L. Pandola, S. Parlati, L. Peralta, J. Perl, A. Pfeiffer, M. Pia, A. Ribon, P. Rodrigues, G. Russo, S. Sadilov, G. Santin, T. Sasaki, D. Smith, N. Starkov, S. Tanaka, E. Tcherniaev, B. Tomé, A. Trindade, P. Truscott, L. Urban, M. Verderi, A. Walkden, J. Wellisch, D. Williams, D. Wright, and H. Yoshida, "Geant4 Developments and Applications," *IEEE Transactions on Nuclear Science*, vol. 53, no. 1, pp. 270–278, February 2006.

[AAB+06] P. Akishin, E. Akishina, S. Baginyan, V. Ivanov, V. Ivanov, I. Kisel, B. Kostenko, E. Litvinenko, G. Ososkov, A. Raportirenko, A. Soloviev, P. Zrelov, and V. Uzhinsky, "Methods for event reconstruction in the CBM experiment," JINR Communication, E10-2006-48, Dubna, Russia, December 2006, `http://www.gsi.de/documents/DOC-2008-Mar-139_e.html`.

[AAFN96] M. Albanesi, A. Antola, M. Ferretti, and R. Negrini, "A chipset for the Generalized Hough Transform," *The Journal of VLSI Signal Processing*, vol. 12, no. 2, pp. 115–134, 1996.

[AATB+04] J. Adamczewski, M. Al-Turany, D. Bertini, H. Essel, and S. Linev, "Analysis Organization: Go4 Analysis Steps and TTask," ROOT workshop, February 2004.

[Act] "Actel," http://www.actel.com/products/devices.aspx.

[Add] "Adding a sub-package to the CBMROOT framework," http://cbmroot.gsi.de/General/adding_sub.htm.

[AEGK02a] I. Abt, D. Emeliyanov, I. Gorbounov, and I. Kisel, "Cellular automaton and Kalman filter based track search in the HERA-B pattern tracker," *Nuclear Instruments and Methods in Physics Research Section A: Accelerators, Spectrometers, Detectors and Associated Equipment*, vol. 490, no. 3, pp. 546–558, September 2002, http://www.sciencedirect.-com/science?_ob=ArticleURL&_udi=B6TJM-461XNJ6-3-&_user=2717328&_coverDate=09%2F11%2F2002&_alid=-1136326726&_rdoc=1&_fmt=high&_orig=search&_cdi=-5314&_docanchor=&view=c&_ct=1&_acct=C000056831-&_version=1&_urlVersion=0&_userid=2717328&md5=-ea8ac5e54b58711a3f0b23636092cac8.

[AEGK02b] I. Abt, D. Emeliyanov, I. Gorbunov, and I. Kisel, "Cellular automaton and Kalman filter based track search in the HERA-B pattern tracker," *Nuclear Instruments and Methods in Physics Research*, vol. 490, no. 3, pp. 546–558, 2002.

[AEKM02] I. Abt, D. Emeliyanov, I. Kisel, and S. Masciocchi, "CATS: a cellular automaton for tracking in silicon for the HERA-B vertex detector ," *Nuclear Instruments and Methods in Physics Research Section A: Accelerators, Spectrometers, Detectors and Associated Equipment*, vol. 489, no. 1-3, pp. 389–405, August 2002, http://www.sciencedirect.-com/science?_ob=ArticleURL&_udi=B6TJM-45TTPV5-15-&_user=2717328&_coverDate=08%2F21%2F2002&_alid=-1136300036&_rdoc=3&_fmt=high&_orig=search&_cdi=5314-&_sort=r&_docanchor=&view=c&_ct=6&_acct=C000056831-&_version=1&_urlVersion=0&_userid=2717328&md5=-8571d0da4311d12be7e086392668cfcd.

[ALI] "ALICE - A Large Ion Collider Experiment," http://alice-info.cern.ch/Public/Welcome.html.

[Alta] "Altera," http://www.altera.com/products/devices/dev--index.jsp.

[Altb] "Alternating Gradient Synchrotron," http://www.bnl.gov/-bnlweb/facilities/AGS.asp.

[Ama] "Amazon: Playstation 3 40GB," http://www.amazon.com/-Sony-98007-Playstation-3-40GB/dp/B000XGJH10/ref=-pd_bbs_sr_1?ie=UTF8&s=videogames&qid=1212504953&sr=-8-1.

[Ame99] N. Amelin, "PHYSICS AND ALGORITHMS OF THE HADRONIC MONTE-CARLO EVENT GENERATORS. Notes for a developer." November 1999.

[And06] A. Andronic, "The TRD of the CBM experiment," *Nuclear Instruments and Methods in Physics Research - Section A*, vol. 563, no. 2, pp. 349–354, July 2006, http://www.sciencedirect.com/science?_ob=-ArticleURL&_udi=B6TJM-4JJ27XM-5&_user=2717328-&_coverDate=07%2F15%2F2006&_alid=1130425185-&_rdoc=1&_fmt=high&_orig=search&_cdi=5314&_sort=-r&_docanchor=&view=c&_ct=1&_acct=C000056831-&_version=1&_urlVersion=0&_userid=2717328&md5=-3a8711879fd4fa6fc94e7aa48b83ad13.

[App] "Application-specific integrated circuit," http://en.wiki-pedia.org/wiki/ASIC.

[ASIa] "ASIC Prototyping Using Off-the-Shelf FPGA Boards: How to Save Months of Verification Time and Tens of Thousands of Dollars," http://www.synplicity.com/literature/-whitepapers/pdf/proto_wp06.pdf.

[ASIb] "ASIC Prototyping with Stratix III FPGAs," http://-www.altera.com/products/devices/stratix2/features/-st2-asic_proto.html.

[AT06] M. Al-Turany, Ed., *CBM Simulation & Analysis Framework - Cbmroot 2*, GSI Darmstadt. GSI Darmstadt, February 2006.

[ATFG+07] M. Al-Turany, A. Fontana, P. Genova, L. Lavezzi, A. Panzarasa (Finuda), and A. Rotondi, "GEANE in FairRooT: status and perspectives," CBM Software Meeting, February 2007, http://cbm-wiki.gsi.de/cgi-bin/view/-CbmRoot/CbmSoft070215.

[ATL] "ATLAS," http://atlas.ch.

[Avn] "Avnet Electronics Marketing - Products," `http://em.avnet.-com/part/xlx/c/XC5VLX110-1FFG1760C`.

[AYBD+] S. Amar-Youcef, A. Besson, M. Deveaux, M. Dorokhov, W. Dulinski, I. Fröhlich, M. Goffe, D. Grandjean, F. Guilloux, S. Heini, J. Heuser, A. Himmi, C. Hu, K. Jaaskelainen, C. Müntz, M. Pellicioli, E. Scopelliti, A. Shabetei, J. Stroth, M. Szelezniak, I. Valinund, and M. Winter, "Strahlenhärte von Monolithic Active Pixel Sensors (MAPS) im Kontext des CBM-Experiments."

[Bal81] D. Ballard, "Generalizing the Hough transform to detect arbitrary shapes," *Pattern Recognition*, vol. 13, no. 2, pp. 111–122, 1981.

[BB91] H. Bässmann and W. Besslich, *Bildverstehen - Ad Oculus*. Springer Verlag, Berlin, 1991, pp. 101–121.

[BBF+97] R. Brun, N. Buncic, V. Fine, M. Goto, F. Rademakers, G. Roland, and A. Sandoval, "ROOT - An Interactive Object Oriented Framework and its Application to NA49 Data Analysis," 1997, `http://root.cern.ch/root/Publications.html`.

[BCC+] J. Berst, G. Claus, C. Colledani, G. Deptuch, W. Dulinski, U. Goerlach, Y. Gornushkin, D. Husson, Y. Hu, J. Le Normand, G. Orazi, J. Riester, R. Turchetta, and M. Winter, "Monolithic Active Pixel Sensors for High Resolution Vertex Detectors."

[BCHM] R. Brun, F. Carminati, I. Hřivnáčová, and A. Morsch, "Virtual Monte Carlo," `http://root.cern.ch/root/vmc/VirtualMC.-html`.

[Bes93] D. Best, "Tracking mit der Hough-Transformation für die zentrale Driftkammer des GSI-4π-Experiments," February 1993, gSI-93-11.

[Bil89] P. Billoir, "Progressive track recognition with a Kalman-like fitting procedure," *Computer Physics Communications*, vol. 57, no. 1-3, pp. 390–394, December 1989, `http://-www.sciencedirect.com/science?_ob=ArticleURL&_udi=-B6TJ5-46S6153-2J&_user=2717328&_rdoc=1&_fmt=&_orig=-search&_sort=d&_docanchor=&view=c&_acct=C000056831-&_version=1&_urlVersion=0&_userid=2717328&md5=-83e513bdf0dc02f9e6351e4b933e3cce`.

[BJ95] S. Ben Yacoub and J. Jolion, "Hierarchical line extraction," *IEE Proceedings - Vision, Image, and Signal Processing*, vol. 142, no. 1, pp. 7–14, 1995.

[Blaa] "Blade - Press Releases - Pricing and Availability," http://-www.bladenetwork.net/?pageid=834.

[BLAb] "BLADE NETWORK TECHNOLOGIES - RackSwitch G8124 - 10G SFP+ Low Latency Switch - Product Brief," http://-www.bladenetwork.net/RackSwitch-G8124.html.

[BQ90a] P. Billoir and S. Qian, "Further test for the simultaneous pattern recognition and track fitting by the Kalman filtering method," *Nuclear Instruments and Methods in Physics Research Section A: Accelerators, Spectrometers, Detectors and Associated Equipment*, vol. 295, no. 3, pp. 492–500, November 1990, http://www.sciencedirect.com/science?_ob=ArticleURL&_udi=B6TJM-473FTVD-12&_user=-2717328&_coverDate=11%2F01%2F1990&_fmt=abstract-&_orig=search&_cdi=5314&view=c&_acct=C000056831-&_version=1&_urlVersion=0&_userid=2717328&md5=-f249b91413b845aedcdc56ee762288f7&ref=full.

[BQ90b] P. Billoir and S. Qian, "Simultaneous pattern recognition and track fitting by the Kalman filtering method," *Nuclear Instruments and Methods in Physics Research Section A: Accelerators, Spectrometers, Detectors and Associated Equipment*, vol. 294, no. 1-2, pp. 219–228, September 1990, http://www.sciencedirect.com/science?_ob=ArticleURL&_udi=B6TJM-4729K35-J&_user=-2717328&_coverDate=09%2F01%2F1990&_fmt=abstract-&_orig=search&_cdi=5314&view=c&_acct=C000056831-&_version=1&_urlVersion=0&_userid=2717328&md5=-f5595fb513398562ad1e971203f1c261&ref=full.

[BRa] R. Brun and F. Rademakers, "Performance Comparison between ROOT, Objectivity/DB and LHC++ histOOgrams," http://-root.cern.ch/root/Publications.html.

[BRb] R. Brun and F. Rademakers, "ROOT - An Object Oriented Data Analysis Framework," http://root.cern.ch/-root/Publications.html.

[BSMM05] I. Bronstein, K. Semendjajew, G. Musiol, and H. Mühlig, *Taschenbuch der Mathematik*, 6th ed. Verlag Harri Deutsch, 2005.

[BTe] "BTeV Home Page," `http://www-btev.final.gov/`.

[BTS90] D. Ben-Tzvi and M. Sandler, "A combinatorial Hough transform," *Pattern Recognition Letters*, vol. 11, no. 3, pp. 167–174, 1990.

[CBM04] CBM Collaboration, "Letter of Intent for a Compressed Baryonic Matter Experiment at the Future Accelerator Facility in Darmstadt," January 2004.

[CBM05] CBM Collaboration, "Technical Status report CBM Experiment," January 2005, `http://www.gsi.de/documents/-QCD_CBM-report-2005-001.html`.

[CBM07] CBM Collaboration, "Cbm progress report 2007," 2007, `http://www.gsi.de/documents/DOC-2008-May-3-1.pdf`.

[CBM08] CBM Collaboration, "The Compressed Baryonic Matter experiment," October 2008, `http://www.gsi.de/documents/-DOC-2007-Apr-21-1.pdf`.

[Cela] "Cell Broadband Engine," `http://www-01.ibm.com/chips/-techlib/techlib.nsf/products/Cell_Broadband_Engine`.

[Celb] "Cell Broadband Engine Architecture," `http://www-01.-ibm.com/chips/techlib/techlib.nsf/techdocs/-1AEEE1270EA2776387257060006E61BA`.

[Celc] "Cell Broadband Engine Programming Handbook," `http://-www-01.ibm.com/chips/techlib/techlib.nsf/techdocs/-9F820A5FFA3ECE8C8725716A0062585F`.

[Celd] "Cell (microprocessor)," `http://en.wikipedia.org/wiki/-CELL`.

[CER94] *GEANT - Detector Description and Simulation Tool: GEANT User's Guide*, 3rd ed., April 1994.

[Cha08] S. Chattopadhyay, "Physics at high baryon density at FAIR," *Journal of Physics G: Nuclear and Particle Physics*, vol. 35, no. 10, October 2008, `http://www.gsi.de/documents/-DOC-2008-Jul-24-1.pdf`.

[Com] "Complex programmable logic device," `http://en.wikipedia.org/wiki/CPLD`.

[Coo06] D. Coombes, "PlayStation University: Introduction to PS3," Game Developers Conference, March 2006, `https://www.cmpevents.com/GD06/a.asp?option=C&V=11&SessID=2817`.

[Cos04] G. Cosmo, "The Geant4 geometry modeler," *IEEE Nuclear Science Symposium Conference Record*, vol. 4, pp. 2196–2198, Octobre 2004, 0-7803-8700-7.

[Cyg] "Cygwin," `http://en.wikipedia.org/wiki/Cygwin`.

[Dat] "Data Acquisition," `http://cbm-wiki.gsi.de/cgi-bin/view/DAQ/WebHome`.

[DBD+05] M. Deveaux, J. Berst, W. De Boer, M. Caccia, G. Claus, G. Deptuch, W. Dulinski, G. Gaycken, D. Grandjean, L. Jungermann, J. Riester, and M. Winter, "Charge collection properties of X-ray irradiated monolithic active pixel sensors," *Nuclear Instruments and Methods in Physics Research*, vol. 552, no. 1-2, pp. 118–123, 2005.

[DDZ94] D. Detlefs, A. Dosser, and B. Zorn, "Memory Allocation Costs in Large C and C++ Programs," *Software: Practice and Experience*, vol. 24, no. 6, pp. 527 – 601, 1994 1994, `http://www3.interscience.wiley.com/journal/113446193/issue`.

[Dea81] S. Deans, "Hough transform from Radon transform," *IEEE Transactions on pattern analysis and machine intelligence*, vol. 3, pp. 185–188, 1981.

[Det] "Detector Simulation of the Main Tracker (STS)," `http://cbm-wiki.gsi.de/cgi-bin/view/CbmRoot/CbmStsDigi`.

[dFBTS90] L. da Fontoura Costa, D. Ben-Tzvi, and M. Sandler, "Performance improvements to the Hough transform," *UK IT 1990 Conference*, pp. 98–103, 1990.

[DH72] R. Duda and P. Hart, "Use of the Hough transformation to detect lines and curves in pictures," *Communications of the Association for Computing Machinery (ACM)*, vol. 15, no. 1, pp. 11–15, 1972.

[Don06] J. Dongarra, "The Impact of Multicore on Math Software and Exploiting Single Precision in Obtaining Double Precision," Fall Creek Falls Conference, October 2006, `http://www.-ccs.ornl.gov/workshops/FallCreek06/presentations/-j_dongarra.pdf`.

[DST92] M. Dillencourt, H. Samet, and M. Tamminen, "A General Approach to Connected-Component Labeling for Arbitrary Image Representations," *Journal of the Association for Computing Machinery (JACM)*, vol. 39, no. 2, pp. 253–280, 1992.

[EEP] "EEPROM," `http://en.wikipedia.org/wiki/EEPROM`.

[Eis99] J. Eischen, "Bildanalytische und rheologische untersuchungen zum orientierungs- und strukturierungsverhalten von faserförmigen partikeln in laminaren scherströmungen," Ph.D. dissertation, Eidgenössische Technische Hochschule Zürich, 1999.

[Faz98] A. Fazekas, "New optical tracking methods for robot cars," *Acta Mathematica Academiae Paedagogicae Nyíregyháziensis*, vol. 14, pp. 63–69, 1998, the European Mathematical Information Service.

[FFG03] C. Finck, P. Fonte, and A. Gobbi, "Results concerning understanding and applications of timing GRPCs ," *Nuclear Instruments and Methods in Physics Research Section A: Accelerators, Spectrometers, Detectors and Associated Equipment*, vol. 508, no. 1-2, pp. 63–69, August 2003.

[FFR+03] A. Fassó, A. Ferrari, S. Roesler, P. Sala, G. Battistoni, F. Cerutti, E. Gadioli, M. Garzelli, F. Ballarini, A. Ottolenghi, A. Empl, and J. Ranft, Eds., *The physics models of FLUKA: status and recent developments.* Computing in High Energy and Nuclear Physics Conference (CHEP), March 2003, (paper MOMT005), eConf C0303241 (2003), arXiv:hep-ph/0306267.

[FFRS05] A. Fassó, A. Ferrari, J. Ranft, and P. Sala, "FLUKA: a multi-particle transport code," 2005, cERN-2005-10 (2005), INFN/TC_05/11, SLAC-R-773.

[FGI+03] V. Friese, J. Gläß, V. Ivanov, A. Jerusalimov, I. Kisel, V. Lindenstruth, W. Müller, V. Pechenov, O. Rogachevsky, and P. Senger, "Track reconstruction for the CBM experiment," May 2003, gSI Scientific Report 2003.

[FHC] FHCPA-FPGA High Performance Computing Alliance, http://www.fhpca.org/maxwell.html.

[Fie] "Field-programmable gate array," http://en.wikipedia.-org/wiki/FPGA.

[Flo] "Flow," http://toolbox.xilinx.com/docsan/xilinx4/-data/docs/xug/imp_fpga2.html.

[FLU] "FLUKA - official FLUKA site Milano- online manual," http://pcfluka.mi.infn.it/manual/Online.shtml.

[FRB+00] R. Frühwirth, M. Regler, R. Bock, H. Grote, and D. Notz, *Data Analysis Techniques for High-energy Physics*, 2nd ed. Cambridge University Press, August 2000, iSBN:0-5216-3548-9.

[Fri06a] V. Friese, "Strangeness and charm in the CBM experiment," *Journal of Physics G: Nuclear and Particle Physics*, vol. 32, no. 12, pp. 439–446, December 2006, http://www.gsi.de/-documents/DOC-2007-Apr-5-1.pdf.

[Fri06b] V. Friese, "Strangeness and charm in the CBM experiment," *Journal of Physics G: Nuclear and Particle Physics*, vol. 32, pp. 439–446, November 2006, http://www.iop.org/-EJ/abstract/0954-3899/32/12/S53/.

[Fri06c] V. Friese, "The CBM experiment at GSI/FAIR," *Nuclear Physics A*, vol. 774, pp. 377–386, August 2006, http://www.-gsi.de/documents/DOC-2006-Dec-91-1.pdf.

[Frö07] I. Fröhlich, "The Pluto++ Event Generator," XI International Workshop on Advanced Computing and Analysis Techniques in Physics Research (ACAT), April 2007.

[FS89] L. Fontoura Costa and M. Sandler, "Improving parameter space for Hough transform," *Electronics Letters*, vol. 25, no. 2, pp. 134–136, 1989.

[GBM90] J. Gläß, R. Baur, and R. Männer, "Programmable trigger for electron pairs in ring image Cherenkov counters," *IEEE Transactions on Nuclear Science*, vol. 37, no. 2, pp. 241–247, April 1990.

[GCC] "GCC, the GNU Compiler Collection," http://gcc.gnu.org/.

[GCC+05] A. Gay, G. Claus, C. Colledani, G. Deptuch, M. Deveaux, W. Dulinski, Y. Gornushkin, D. Grandjean, A. Himmi, C. Hu, I. Valin, and M. Winter, "High resolution CMOS sensors for a vertex detector at the linear collider," *Nuclear Instruments and Methods in Physics Research*, vol. 549, no. 1-3, pp. 99–102, 2005.

[Gea07a] Geant4 Collaboration, *Geant4 User's Guide for Application Developers*, 8th ed., May 2007.

[Gea07b] Geant4 Collaboration, *Geant4 User's Guide for Toolkit Developers*, 8th ed., May 2007.

[Gea07c] Geant4 Collaboration, *Introduction to Geant4*, 8th ed., May 2007.

[Gea07d] Geant4 Collaboration, *Physics Reference Manual*, 8th ed., May 2007.

[GK05] S. Gorbunov and I. Kisel, "An analytic formula for track extrapolation in an inhomogeneous magnetic field," CBM-SOFT-note-2005-001 and I3HP-FutureDAQ-note-2005-001, March 2005, `https://www.gsi.de/documents/DOC-2005-Jun-49-1.pdf`.

[GK08] S. Gorbunov and I. Kisel, "Simdized cellular automaton based track finder," IEEE Nuclear Science Symposium and Medical Imaging Conference, Dresden, Germany, October 2008, `http://www.gsi.de/documents/DOC-2008-Oct-94-1.pdf`.

[GKK+09] S. Gorbunov, I. Kisel, I. Kulakov, I. Rostovtseva, and I. Vassiliev, "First Level Event Selection Package of the CBM Experiment," 17th International Conference on Computing in High Energy and Nuclear Physics, Prague, Czech Republic, March 2009, `http://www.gsi.de/documents/DOC-2009-Mar-294-1.pdf`.

[GKKO93] A. Glazov, I. Kisel, E. Konotopskaya, and G. Ososkov, "Filtering tracks in discrete detectors using a cellular automaton," *Nuclear Instruments and Methods in Physics Research Section A: Accelerators, Spectrometers, Detectors and Associated Equipment*, vol. 329, no. 1-2, pp. 262–268, May 1993, `http://www.sciencedirect.com/science?-_ob=ArticleURL&_udi=B6TJM-4731HF1-21&_user=2717328-&_rdoc=1&_fmt=&_orig=search&_sort=d&_docanchor=-&view=c&_searchStrId=1136283687&_rerunOrigin=google-`

&_acct=C000056831&_version=1&_urlVersion=0&_userid=-2717328&md5=8379fa2c0d45ad4f669e58466a4e1489.

[GNUa] "GNU + Cygnus + Windows = cygwin," http://www.cygwin.-com/.

[GNUb] "GNU Make," http://www.gnu.org/software/make/.

[Grä08] M. Gräbner, "Wie die Rettung der Welt die Erde zerstört," *HEISE Online*, November 2008, http://www.heise.de/tp/-r4/artikel/29/29215/1.html.

[GSI] "GSI - Gesellschaft für Schwerionenforschung," http://www.-gsi.de/.

[GSI06] GSI Collaboration, "Cbm progress report 2006," GSI Darmstadt, Tech. Rep., 2006, http://www.gsi.de/documents/-DOC-2007-Mar-137-1.pdf.

[GSM05] J. Gläss, C. Steinle, and R. Männer, "Tracking in the Silicon Tracker System of the CBM-Experiment using Hough Transform," 14th IEEE-NPSS REAL TIME Conference, Stockholm, Sweden, June 2005.

[GSTB05] P. Giacon, S. Saggin, G. Tommasi, and M. Busti, "Implementing DSP Algorithms Using Spartan-3 FPGAs," *DSP magazine*, no. 1, pp. 16–19, October 2005, http://www.xilinx.com/-publications/magazines/dsp_01/index.htm.

[GW00] R. Gonzalez and R. Woods, *Digital Image Processing*. Addison-Wesley Publishing Company, 2000, pp. 432–438.

[HAB+03] I. Hřivnáčová, D. Adamová, V. Berejnoi, R. Brun, F. Carminati, A. Fassò, E. Futó, A. Gheata, and I. Morsch, A.and González Caballero, Eds., *The Virtual Monte Carlo*. Computing in High Energy and Nuclear Physics Conference (CHEP), March 2003.

[HAD] "HADES," http://www-hades.gsi.de/.

[Har] "Hardware description language," http://en.wikipedia.-org/wiki/Hardware_description_language.

[Har06] J. Harris, "How to Apply FPGAs to Robotic Control Systems," *Embedded Magazine*, no. 4, pp. 32–35, November 2006, http://www.xilinx.com/publications/magazines/-emb_04/index.htm.

[HBD+06] J. Heuser, K. Banicz, A. David, M. Floris, M. Keil, C. Lourenço, H. Ohnishi, E. Radermacher, R. Shahoyan, and G. Usai, "Experience with the NA60 silicon pixel vertex tracker in a harsh radiation environment," *Nuclear Instruments and Methods in Physics Research Section A: Accelerators, Spectrometers, Detectors and Associated Equipment*, vol. 560, no. 1, pp. 9–13, May 2006.

[HDD+08] C. Höhne, S. Das, M. Dürr, T. Galatyuk, P. Koczon, S. Lebedev, A. Maevskaya, and G. Ososkov, "Development of a RICH detector for electron identification in CBM," *Nuclear instruments and Methods in Physics Research. Section A, Accelerators, spectrometers, detectors and associated equipment*, vol. 595, no. 1, pp. 187–189, 2008, http://cat.inist.fr/?aModele=-afficheN&cpsidt=20770278.

[HDL] "HDL Designer," http://www.mentor.com/products/fpga_-pld/hdl_design/hdldesigner_series/index.cfm.

[HDMS06] J. Heuser, M. Deveaux, C. Müntz, and J. Stroth, "Requirements for the Silicon Tracking System of CBM," *Nuclear Instruments and Methods in Physics Research Section A: Accelerators, Spectrometers, Detectors and Associated Equipment*, vol. 568, no. 1, pp. 258–262, November 2006, http://www.sciencedirect.-com/science?_ob=ArticleURL&_udi=B6TJM-4K8R2VD-13-&_user=2717328&_coverDate=11%2F30%2F2006&_alid=-1130390288&_rdoc=3&_fmt=high&_orig=search&_cdi=5314-&_sort=r&_docanchor=&view=c&_ct=3&_acct=C000056831-&_version=1&_urlVersion=0&_userid=2717328&md5=-ca7e73b30db0c109abd8424efa16905e.

[Heu06] J. Heuser, "Development of a Silicon Tracking and Vertex Detection System for the CBM Experiment at FAIR," 15th International Workshop on Vertex Detectors, Perugia, Italy, September 29 2006, to be published in the journal Nuclear Instruments and Methods in Physics Research Section A: Accelerators, Spectrometers, Detectors and Associated Equipment.

[Heu07] J. Heuser, "Development of a silicon tracking and vertex detection system for the CBM experiment at FAIR," *Nuclear Instruments and Methods in Physics Research Section A: Accelerators, Spectrometers, Detectors and Associated Equipment*, vol.

582, no. 3, pp. 910–915, December 2007, http://www.gsi.de/-documents/DOC-2006-Dec-19.html.

[HKM+99] C. Hinkelbein, A. Kugel, R. Männer, M. Sessler, H. Simmler, H. Singpiel, J. Baines, R. Bock, and M. Smizanska, "Pattern Recognition in the TRT for the ATLAS B-Physics Trigger," CERN, Geneva, Tech. Rep., September 1999.

[HKP93] J. Han, L. Koczy, and T. Poston, "Fuzzy Hough transform," *Second IEEE International Conference on Fuzzy Systems*, vol. 2, pp. 803–808, 1993.

[HLSH00] M. Hani, T. Lin, and N. Shaikh-Husin, "FPGA implementation of RSA public-key cryptographic coprocessor," *Proceedings of the TENCON 2000*, vol. 3, pp. 6–11, September 2000, http://-ieeexplore.ieee.org/xpls/abs_all.jsp?arnumber=892209.

[HMS05a] J. Heuser, W. Müller, and P. Senger, "A high-performance silicon tracker for the Compressed Baryonic Matter experiment at FAIR," Quark Matter 2005, Budapest, Hungary, August 4-9 2005, czech. J. Phys. 55 (2005) 1649-1653.

[HMS+05b] J. Heuser, W. Müller, P. Senger, C. Müntz, and J. Stroth, "A High-Performance Silicon Tracker for the CBM Experiment at FAIR," Particles and Nuclei International Conference Conference (PANIC05), Santa Fe, New Mexico, USA, October 24-28 2005, aIP Conference Proceedings No. 842, Particles and Nuclei, 1073-1075.

[HMS+05c] J. Heuser, W. Müller, P. Senger, C. Müntz, and J. Stroth, "A high-performance silicon tracker for the Compressed Baryonic Matter experiment at FAIR," *Czechoslovak Journal of Physics*, vol. 55, no. 12, pp. 1649–1653, December 2005, http://www.-gsi.de/documents/DOC-2006-Nov-43-1.pdf.

[Höh07] C. Höhne, "The Cbm Experiment at Fair Exploring the QCD Phase Diagram at High Net Baryon Densities," *International Journal of Modern Physics E*, vol. 16, no. 7-8, pp. 2419–2424, 2007, http://www.gsi.de/documents/-DOC-2007-Mar-126-1.pdf.

[Hou59] P. Hough, "Machine Analysis of Bubble Chamber Pictures," International Conference on High Energy Accelerators and Instrumentation, CERN, Meyrin, Suiss, 1959.

[Hou62] P. Hough, "Method and means for recognizing complex patterns," United States Patent, Nr. 3069654, 1962.

[Ht08] J. Heuser and the CBM Collaboration, "The Compressed Baryonic Matter experiment at FAIR: physics of strangeness and charm, status of preparations," *Journal of Physics G: Nuclear and Particle Physics*, vol. 35, no. 4, pp. 44–49, April 2008, `http://www.gsi.de/documents/DOC-2007-Oct-120-1.pdf`.

[IBMa] "IBM," `http://en.wikipedia.org/wiki/IBM`.

[IBMb] "IBM United States," `http://www.ibm.com/us/`.

[IK87] J. Illingworth and J. Kittler, "The adaptive Hough transform," *IEEE Transactions on Pattern Analysis and Machine Intelligence*, vol. 9, no. 5, pp. 690–698, 1987.

[IK88] J. Illingworth and J. Kittler, "A survey of the hough transform," *Computer Vision, Graphics and Image Processing*, vol. 44, no. 1, pp. 87–116, 1988.

[Ila01] A. Ilachinski, *Cellular Automata: A Discrete Universe.* World Scientific Publishing Co. Pte. Ltd., August 2001, iSBN:981-02-4623-4.

[IMN91] V. Innocente, M. Maire, and E. Nagy, "GEANE: Average TRacking and Error Propagation Package," July 1991, cERN IT-ASD W5013-E GEANE.

[IND92] M. Ibrahim, E. Ngau, and M. Daemi, "Weighted Hough transform," *Proceedings of the International Society of Photo-Optical Instrumentation Engineers (SPIE)*, vol. 1607, pp. 237–241, 1992.

[Inn] Innovasic Semiconductor, "IA64250 - Data Sheet - Histogram/-Hough Transform Processor," `http://www.innovasic.com/-pdfs/IA64250DS.pdf`.

[Int] "Intel Core2Duo E8400 2048MB 80GB DVDROM onBoard Grafik (PC-Einsteiger) - Computer Shop - Hardware, Notebook & Software by Mindfactory.de," `http://www.-mindfactory.de/product_info.php/info/p432143_Core2-Duo-E8400-2048MB-80GB-DVDROM-onBoard-Grafik-PC-Ein-steiger-.html`.

[ISE] "ISE Foundation Software," http://www.xilinx.com/ise/-logic_design_prod/foundation.htm.

[Jäh05] B. Jähne, *Digital Image Processing*, 6th ed. Springer Verlag Berlin, Heidelberg, New York, 2005, pp. 436–439.

[Jer05] A. Jerusalimov, "Reconstruction of track parameters in nonuniform magnetic field," January 2005, http://www.gsi.de/-Documents/DOC-2005-Jan-58.html.

[Joi] "Joint Test Action Group," http://en.wikipedia.org/wiki/-JTAG.

[JS72] M. Jobes and H. Shaylor, "Data analysis techniques in high energy physics," *Reports on Progress in Physics*, vol. 35, no. 3, pp. 1077–1172, September 1972, http://www.iop.org/EJ/-article/0034-4885/35/3/303/rpv35i3p1077.pdf.

[KBS75] C. Kimme, D. Ballard, and J. Sklansky, "Finding circles by an array of accumulators," *Communications of the Association for Computing Machinery (ACM)*, vol. 18, no. 2, pp. 120–122, 1975, http://portal.acm.org/citation.cfm?id=360677.

[KDH+05] J. Kahle, M. Day, H. Hofstee, C. Johns, T. Maeurer, and D. Shippy, "Introduction to the Cell multiprocessor," *IBM Journal: Research and Development*, vol. 49, no. 4/5, pp. 589–604, July/September 2005, http://researchweb.watson.-ibm.com/journal/rd/494/kahle.pdf.

[KHXO95] H. Kälviäinen, P. Hirvonen, L. Xu, and E. Oja, "Probabilistic and Non-probabilistic Hough Transforms: Overview and Comparisons," *Image and Vision Computing*, vol. 13, no. 4, pp. 239–252, 1995.

[Kis03] I. Kisel, "Tracking in Magnetic Field Based in the Cellular Automaton Method," July 2003, https://www.gsi.de/-documents/DOC-2003-Nov-123-2.pdf.

[Kis04] I. Kisel, "Cellular Automaton Method for Track Finding (HERA-B, LHCb, CBM)," 2nd FutureDAQ Workshop, GSI, Darmstadt, germany, September 2004, https://www.gsi.de/-documents/DOC-2004-Sep-107-2.pdf.

[Kis06] I. Kisel, "Event reconstruction in the CBM experiment ," *Nuclear Instruments and Methods in Physics Research Section A: Accelerators, Spectrometers, Detectors and Associated Equipment*, vol. 566, no. 1, pp. 85–88, October 2006, http://www.sciencedirect.com/science?_ob=ArticleURL-&_udi=B6TJM-4K3CY3J-6&_user=2717328&_rdoc=1&_fmt=-&_orig=search&_sort=d&_docanchor=&view=c&_acct=-C000056831&_version=1&_urlVersion=0&_userid=2717328-&md5=6ae901db7cdd86b268db06f30001fc31.

[KOM95] F. Klefenz, M. Oberle, and R. Männer, "VLSI Implementierung eines parallelen Hough-Transformations-Prozessors mit dynamisch nachladbaren Mustern," Department for Mathematics and Computer Science, University of Mannheim, Tech. Rep., 1995, http://www.informatik.uni-mannheim.de/tb/html/-TR-95-022.html.

[KRM+06] R. Karanam, A. Ravindran, A. Mukherjee, C. Gibas, and A. Wilkinson, "Using FPGA-Based Hybrid Computers for Bioinformatics Applications," *Xcell Journal*, no. 58, pp. 80–83, 2006, http://www.xilinx.com/publications/-xcellonline/xcell_58/index.htm.

[Lap] "Laptops, LCD Televisions, Projectors, Medical Imaging & More - Toshiba America Inc," http://www.toshiba.com/tai/.

[Lat] "Lattice Semiconductor Corporation," http://www.latticesemi.com/products/fpga/index.cfm?source=topnav&-jsessionid=ba30e88ddbbbEC861$.

[Lea93] V. Leavers, "Which Hough transform?" *CVGIP: Image Understanding*, vol. 58, no. 2, pp. 250–264, 1993.

[LHC] "LHCb - Large Hadron Collider beauty experiment," http://-lhcb-public.web.cern.ch/lhcb-public/.

[Lina] "Linux development on the PlayStation 3, Part 1: More than a toy," http://www.ibm.com/developerworks/linux/-library/l-linux-ps3-1/.

[Linb] "Linux development on the PlayStation 3, Part 1: More than a toy," http://www.ibm.com/developerworks/linux/-library/l-linux-ps3-3/.

[Linc] "Linux for PlayStation 3," http://en.wikipedia.org/wiki/-Linux_for_PlayStation_3.

[Lin07] M. Linklater, "Optimizing Cell Core," *Game Developer Magazine*, pp. 15–18, April 2007.

[Lis] "List of accelerators in particle physics," http://en.wikipedia.org/wiki/ List_of_accelerators_in_particle_-physics.

[LSI] LSI Logic Corporation, "L64250 - Histogram/Hough Transform Processor."

[LT95] R. Lo and W. Tsai, "Gray-scale hough transform for thick line detection in gray-scale images," *Pattern Recognition*, vol. 28, no. 5, pp. 647–661, 1995.

[LYK+07] S. Lu, P. Yiannacouras, R. Kassa, M. Konow, and T. Suh, "An FPGA-based Pentium in a complete desktop system," *Proceedings of the 2007 ACM/SIGDA 15th international symposium on Field programmable gate arrays*, pp. 53–59, February 2007, http://portal.acm.org/citation.cfm?-doid=1216919.1216927.

[Mai85] H. Maitre, "Un panorama de la transformation de hough," *Traitemente du Signal*, vol. 2, no. 4, pp. 305–317, 1985.

[mak] "make (software)," http://en.wikipedia.org/wiki/Make_-%28software%29.

[Man97] R. Mankel, "A concurrent track evolution algorithm for pattern recognition in the HERA-B main tracking system ," *Nuclear Instruments and Methods in Physics Research Section A: Accelerators, Spectrometers, Detectors and Associated Equipment*, vol. 395, no. 2, pp. 169–184, August 1997, http://www.sciencedirect.com/science?_ob=-ArticleURL&_udi=B6TJM-3SPTFJX-D0&_user=2717328-&_coverDate=08%2F11%2F1997&_rdoc=3&_fmt=high&_-orig=browse&_srch=doc-info(%23toc%235314%231997%-23996049997%2310729%23FLP%23display%23Volume)&_cdi=-5314&_sort=d&_docanchor=&_ct=19&_acct=C000056831-&_version=1&_urlVersion=0&_userid=2717328&md5=-713767dfd3868ec343620c51b6a97f79.

[Men] "Mentor Graphics," http://www.mentor.com/.

[Mer80] M. Mermikides, "Data Analysis for Bubble Chambers and Hybrid Systems," 6th CERN School of Computing, Vraona-Attiki, Greece, September 1980, http://cdsweb.cern.ch/record/-1049896?ln=sv.

[Mic] "Microsoft Visual Studio," http://en.wikipedia.org/wiki/-Microsoft_Visual_Studio.

[Mod] "ModelSim SE," http://www.mentor.com/products/fpga_-pld/simulation/modelsim_se/index.cfm.

[MON] "MONTE CARLO PARTICLE NUMBERING SCHEME," http://pdg.lbl.gov/2008/reviews/rpp2008-rev-monte--carlo-numbering.pdf.

[MQR02] S. Melnikoff, S. Quigley, and M. Russell, "Implementing a Simple Continuous Speech Recognition System on an FPGA," *Proceedings of the 10th Annual IEEE Symposium on Field-Programmable Custom Computing Machines*, pp. 275–276, 2002, http://portal.acm.org/citation.cfm?coll=GUIDE&-dl=GUIDE&id=795961.

[Mul05] B. Mullins, "Xilinx Teams with Optos," *Embedded magazine*, no. 1, pp. 18–20, March 2005, http://www.xilinx.com/-publications/magazines/emb_01/index.htm.

[NETa] "NETGEAR 48-Port Gigabit Advanced Smart Switch GS748AT," http://www.netgear.com/Products/Swit-ches/AdvancedSmartSwitches/GS748AT.aspx?detail=Spe-cifications.

[NETb] "NETGEAR Smart Managed Gigabit Switch Advanced mit 4 SFP GBIC slots - Computer Shop - Hardware, Notebook & Software by Mindfactory.de," http://www.mindfactory.-de/product_info.php/info/p497563_NETGEAR-Smart--Managed-Gigabit-Switch-Advanced-mit-4-SFP-GBIC--slots.html.

[NKK+97] NEMO Collaboration, I. Kisel, V. Kovalenko, F. Laplanche, R. Arnold, C. Augier, A. Barabash, D. Blum, V. Brudanin, J. Campagne, D. Dassié, V. Egorov, R. Eschbach, J. Guyonnet, F. Hubert, P. Hubert, S. Jullian, O. Kochetov, V. Kornoukov,

D. Lalanne, F. Leccia, I. Link, C. Longuemare, F. Mauger, P. Mennrath, H. Nicholson, A. Nozdrin, F. Piquemal, O. Purtov, J. Reyss, F. Scheibling, J. Suhonen, C. Sutton, G. Szklarz, V. Tretyak, V. Umatov, I. Vanushin, A. Vareille, Y. Vasilyev, T. Vylov, and V. Zerkin, "Cellular automaton and elastic net for event reconstruction in the NEMO-2 experiment," *Nuclear Instruments and Methods in Physics Research Section A: Accelerators, Spectrometers, Detectors and Associated Equipment*, vol. 387, no. 3, pp. 433–442, March 1997, http://www.sciencedirect.com/science?_ob=ArticleURL&_udi=B6TJM-3SPGW62-42&_user=2717328&_coverDate=03%2F11%2F1997&_alid=1136348354&_rdoc=3&_fmt=high&_orig=search&_cdi=5314&_docanchor=&view=c&_ct=5&_acct=C000056831&_version=1&_urlVersion=0&_userid=2717328&md5=54711ba26a64e3e5b07704eef97e617c.

[NO86] Y. Noguchi and A. Ono, "Global pattern recognition in layered track chambers," *Nuclear Instruments and Methods in Physics Research Section A: Accelerators, Spectrometers, Detectors and Associated Equipment*, vol. 253, no. 1, pp. 27–37, 1986.

[OC76] F. O'Gorman and M. Clowes, "Finding Picture Edges Through Collinearity of Feature Points," *Transactions on Computers*, vol. C-25, no. 4, pp. 449–456, 1976, http://ieeexplore.ieee.org/xpls/abs_all.jsp?arnumber=1674627.

[OPP02] G. Ososkov, A. Polanski, and I. Puzynin, "Current Methods of Processing Experimental Data in High Energy Physics," *Physics of Particles and Nuclei*, vol. 33, no. 3, pp. 347–382, 2002.

[Para] "Partial re-configuration," http://en.wikipedia.org/wiki/Partial_re-configuration.

[Parb] "Particle Data Group," http://en.wikipedia.org/wiki/Particle_Data_Group.

[PFTV92] W. Press, B. Flannery, S. Teukolsky, and W. Vetterling, *Numerical Recipes in C: The Art of Scientific Computing*, 2nd ed. Cambridge University Press, 1992, ch. 16.1.

[PHE] "PHENIX front page," http://www.phenix.bnl.gov/.

[PIK89] J. Princen, J. Illingworth, and J. Kittler, "A hierarchical approach to line extraction," *Proceedings on the Computer Society Conference on Computer Vision and Pattern Recognition (CVPR)*, pp. 92–97, 1989.

[Plaa] "PlayStation 3," http://en.wikipedia.org/wiki/PlayStation_3.

[Plab] "PlayStation 3 accessories," http://en.wikipedia.org/wiki/PlayStation_3_accessories#AV_Cables.

[Plac] "Playstation 3: Sparsameres 40-GByte-Modell weiterhin mit 90-nm-Chips," http://www.heise.de/newsticker/Playstation-3-Sparsameres-40-GByte-Modell-weiterhin-mit--90-nm-Chips-/meldung/98380.

[Plad] "PlayStation.com - PLAYSTATION 3," http://www.us.playstation.com/PS3.

[PLC] J. Park, C. Looney, and H. Chen, "Fast Connected Component Labeling Alogrithm Using A Divide And Conquer Technique," http://cs.ua.edu/research/TechnicalReports/TR-2000-04.pdf.

[Plu] "Pluto++: A Monte Carlo simulation tool for hadronic physics," http://www-hades.gsi.de/computing/pluto/html/pluto.html.

[PPU] "PPU & SPU C/C++ Language Extension Specification," http://www-01.ibm.com/chips/techlib/techlib.nsf/techdocs/30B3520C93F437AB87257060006FFE5E.

[PRK88] F. Pühlhofer, D. Röhrich, and R. Keidel, "Track recognition in digitized streamer chamber pictures," *Nuclear Instruments and Methods in Physics Research Section A: Accelerators, Spectrometers, Detectors and Associated Equipment*, vol. 263, no. 2-3, pp. 360–367, 1988.

[Proa] "Programming high-performance applications on the Cell BE processor, Part 1: An introduction to Linux on the PLAYSTATION 3," http://www.ibm.com/developerworks/power/library/pa-linuxps3-1/.

[Prob] "Programming high-performance applications on the Cell BE processor, Part 5: Programming the SPU in C/C++," http://www.ibm.com/developerworks/power/library/-pa-linuxps3-5/.

[PSSW86] J. Perl, A. Schwarz, A. Seiden, and A. Weinstein, "Track finding with the Mark II/SLC drift chamber," *Nuclear Instruments and Methods in Physics Research Section A: Accelerators, Spectrometers, Detectors and Associated Equipment*, vol. 252, no. 2-3, pp. 616–620, 1986.

[PYIK89] J. Princen, H. Yuen, J. Illingworth, and J. Kittler, "A comparison of Hough transform methods," *Third International Conference on Image Processing and its Applications*, pp. 73–77, 1989.

[Rad] D. Rademakers, "The Power of Object-Oriented Frameworks," http://root.cern.ch/root/Publications.html.

[Reca] "Reconfigurable computing," http://en.wikipedia.org/-wiki/Reconfigurable_computing.

[Recb] "Reconstruction and Analysis in cbmroot," http://cbm-wiki.-gsi.de/cgi-bin/viewauth/CbmRoot/CbmReconstruction.

[RF05] L. Roman and B. Fayette, "Emulate 8051 Microprocessor in PicoBlaze IP Core," *Embedded magazine*, no. 1, pp. 44–47, March 2005, http://www.xilinx.com/publications/-magazines/emb_01/index.htm.

[RHI] "RHIC- relativistic heavy ion collider," http://www.bnl.gov/-RHIC/.

[RJ07] O. Rogachevsky and A. Jerusalimov, "Status of the LHE track finder/fitter for STS," 9th Collaboration Meeting, GSI, Germany, Darmstadt, February - March 2007, http://www.gsi.-de/documents/DOC-2007-Mar-45-1.pdf.

[ROO] "ROOT and Visual C++.NET 2003 (7.1)," http://www.slac.-stanford.edu/ gentit/cas/XPDebug.html.

[Ros69] A. Rosenfeld, "Picture Processing by Computer," *Association for Computing Machinery (ACM) Computing Surveys (CSUR)*, vol. 1, no. 3, pp. 147–176, 1969.

[RP66] A. Rosenfeld and J. Pfaltz, "Sequential Operations in Digital Picture Processing," *Journal of the Association for Computing Machinery (JACM)*, vol. 13, no. 4, pp. 471–494, 1966, http://portal.acm.org/citation.cfm?id=321357&dl=-ACM&coll=GUIDE.

[SC05] P. Schumacher and W. Chung, "FPGA-Based MPEG-4 Codec," *DSP magazine*, no. 1, pp. 8–9, October 2005, http://www.-xilinx.com/publications/magazines/dsp_01/index.htm.

[Sch] "Schlanke Endgeräte reduzieren den Energieverbrauch massiv," http://www.computerzeitung.de/articles/schlanke_-endgeraete_reduzieren_den_energieverbrauch_massiv:/-2008009/ 31405144_ha_CZ.html?thes=.

[Sch07] M. Schiller, "Standalone track reconstruction for the Outer Tracker of the LHCb experiment using a cellular automaton," Diploma thesis in Physics, University of Heidelberg, July 2007, http://www-linux.gsi.de/ ikisel/reco/LHCb/-diploma_manuel_schiller.pdf.

[SE05] R. Smith and D. Ellgen, "A Weapon Detection System Built with Xilinx FPGAs," *DSP magazine*, no. 1, pp. 68–71, October 2005, http://www.xilinx.com/publications/-magazines/dsp_01/index.htm.

[Sea] "Search For Quark-Gluon Plasma Began At Berkeley in 1984," http://www.lbl.gov/Science-Articles/Archive/-quark-gluon-berkeley.html.

[Sen02] P. Senger, Ed., *The nucleus-nucleus collision research program at the future facility at GSI.* Proceedings of XXX International Workshop on Ultrarelativistic Heavy-Ion Collisions, Hirschegg, Austria, January 2002, http://www.gsi.de/-documents/DOC-2003-Dec-12-1.pdf.

[Sen06] P. Senger, "The CBM experiment at FAIR," *Journal of Physics: Conference Series*, vol. 50, pp. 357–360, 2006, http://www.-iop.org/EJ/abstract/1742-6596/50/1/048.

[Sen07] P. Senger, "Strange Particles and Neutron Stars - Experiments at GSI," *International Journal of Modern Physics E*, vol. 16, no. 4, pp. 1135–1147, 2007, http://www.gsi.de/documents/-DOC-2006-Nov-49-1.pdf.

[Sie05] K. Sienski, "Virtex-4 FPGAs for Software Defined Radio," *DSP magazine*, no. 1, pp. 44–45, October 2005, http://www.xilinx.com/publications/magazines/dsp_01/index.htm.

[Sil] "Silicon Tracking System and Micro-Vertex Detector," http://cbm-wiki.gsi.de/cgi-bin/view/Public/PublicSts.

[Sona] "Sony," http://en.wikipedia.org/wiki/Sony.

[Sonb] "Sony Corporation," http://www.sony.net/.

[Son05] Sony Computer Entertainment Inc., "TO LAUNCH ITS NEXT GENERATION COMPUTER ENTERTAINMENT SYSTEM,Playstation 3 - Unrivaled Performance with the Introduction of Cell Processor and Many Advanced Technologies, and Backwards compatible with PlayStation and PlayStation 2," Electronic Entertainment Expo (E3), Los Angeles, California, USA, May 2005, http://www.scei.co.jp/corporate/release/pdf/050517e.pdf.

[Son07] Sony Computer Entertainment Inc., "NEW PLAYSTATION 3 (CECHH00 SERIES) COMES IN TWO COLOR VARIATIONS AT A NEW PRICE," Press Release, November 2007, http://www.scei.co.jp/corporate/release/071009ae.html.

[Sta] "Stanisław Jerzy Lec," http://en.wikipedia.org/wiki/Stanislaw_Jerzy_Lec.

[Str06] A. Strandlie, "Track reconstruction in the LHC experiments," 1st LHC Detector Alignment Workshop, CERN, Geneva, Switzerland, September 2006, http://cdsweb.cern.ch/record/1047105?ln=no.

[Str09] B. Strauss, "Projektpraktikum - Sony Playstation III performance analysis of a Hough transform based tracking algorithm for the CBM experiment," December 2009.

[Sup] "Super Proton Synchrotron," http://en.wikipedia.org/wiki/Super_Proton_Synchrotron.

[Syna] "Synplicity," http://www.synplicity.com/.

[Synb] "Synplify Pro," http://www.synplicity.com/products/synplifypro/.

[Syn00] *Synplify - Synplicity Synthesis Reference Manual*, Synplicity, Inc., 935 Stewart Drive, Sunnyvale, CA 94086, May 2000, http://www.synplicity.com/literature/pdf/syn_ref.pdf.

[Syn02] *Synplify - Synplicity Reference Manual*, Synplicity, Inc., 935 Stewart Drive, Sunnyvale, CA 94086, April 2002, http://inst.eecs.berkeley.edu/ cs150/Documents/-SynplifyReference.pdf.

[Ter] "Terra Soft – Linux for Cell, PlayStation PS3, QS20, QS21, QS22, IBM System p, Mercury Cell, and Apple PowerPC." http://www.terrasoftsolutions.com/products/ydl/.

[Thea] "The CbmTask," http://cbm-wiki.gsi.de/cgi-bin/view/-CbmRoot/CbmTask.

[Theb] "The Cross-platform Make system CMake," http://www.-cmake.org/HTML/Index.html.

[Thec] "The HERA-B Experiment," http://www-hera-b.desy.de/-general/info/.

[Thed] "The KOPIO Experiment," http://www.bnl.gov/rsvp/KO-PIO.htm.

[Thee] "The NA49 experiment," http://root.cern.ch/root/html/-examples/na49.html.

[Thef] "The NA60 experiment," http://na60.cern.ch/www.

[Theg] "The ROOT framework," http://root.cern.ch/.

[Theh] "The UrQMD Collaboration," http://th.physik.uni-frank-furt.de/ urqmd/.

[Thei] "The version control system Subversion," http://-subversion.tigris.org/.

[TMP06] P. Tipler, G. Mosca, and D. Pelte, *Physik. Für Wissenschaftler und Ingenieure*, 2nd ed. Spektrum Akademischer Verlag, 2006.

[Tos] "Toshiba," http://en.wikipedia.org/wiki/Toshiba.

[Tra06] C. Traub, "Simulation und Implementierung eines Kalman-Filters zur Spurrekonstruktion auf einem FPGA," Diploma thesis, University of Mannheim, Department of Computer Engineering V, July 2006.

[Uni] "Universal linear accelerator," http://en.wikipedia.org/-wiki/Universal_linear_accelerator.

[Us.] "Us.Wii.com – The Global Wii Experience Website in English," http://us.wii.com/.

[VFJ01] F. Vargas, R. Fagundes, and D. Junior, "A FPGA-based Viterbi algorithm implementation for speechrecognition systems," *Proceedings of the IEEE International Conference on Acoustics, Speech, and Signal Processing (ICASSP 01)*, vol. 2, pp. 1217–1220, May 2001, http://ieeexplore.ieee.org/xpls/abs_-all.jsp?arnumber=941143.

[Vira] "Virtex-4 Configuration Guide," http://www.xilinx.com/-support/documentation/user_guides/ug071.pdf.

[Virb] "Virtex-4 Family Overview," http://www.xilinx.com/support/documentation/data_sheets/ds112.pdf.

[Virc] "Virtex-4 User Guide," http://www.xilinx.com/support/documentation/user_guides/ug070.pdf.

[Vird] "Virtex-5 Family Overview," http://www.xilinx.com/support/documentation/data_sheets/ds100.pdf.

[Vire] "Virtex-5 FPGA Devices," http://www.xilinx.com/online-store/silicon/online_store_v5.htm.

[Virf] "Virtex-5 FPGA System Power Design Considerations," http://www.xilinx.com/support/documentation/white_-papers/wp285.pdf.

[Virg] "Virtex-II Pro and Virtex-II Pro X Platform FPGAs: Complete Data Sheet," http://www.xilinx.com/support/documentation/data_sheets/ds083.pdf.

[Vis] "Visual Studio .NET 2003," http://msdn2.microsoft.com/-en-us/vstudio/Aa700867.aspx.

[Vog95] H. Vogel, *Gerthsen Physik*, 18th ed. Springer Verlag, 1995.

[Wal85] R. Wallace, "A modified Hough transform for lines," in *Proceedings IEEE Computer Society Conference on Computer Vision and Pattern Recognition, San Francisco, California, USA*, 1985, pp. 665–667.

[Wel] "Welcome to the FOPI experiment," `http://www-fopi.gsi.de`.

[Wen99] L. Wenzel, "Digitale Signalverarbeitung ist keine Hexerei, Teil 12," *Elektronik*, vol. 48, no. 19, pp. 60–64, 1999, `http://www.elektronikschule.de/ krausg/ DSP/Dsp_CD/DSP_Application_Notes_and_Books_and_Tutorials/ DSP-Artikelreihe%20(Elektronik)/Teil%2012.pdf`.

[WGP04] T. Wollinger, J. Guajardo, and C. Paar, "Security on FPGAs: State of the Art Implementations and Attacks," *ACM Transactions on Embedded Computing Systems (TECS)*, vol. 3, no. 3, pp. 534–574, August 2004, `http://portal.acm.org/citation.cfm?id=1015047.1015052`.

[Xbo] "Xbox.com | Xbox 360," `http://www.xbox.com/en-US/hardware/?WT.svl=nav`.

[Xila] "Xilinx," `http://www.xilinx.com/`.

[Xilb] "Xilinx," `http://www.xilinx.com/products/silicon_solutions/`.

[Xil05] Xilinx and Birger Engineering, "Xilinx and Birger Engineering," *Embedded magazine*, no. 2, pp. 38–39, September 2005, `http://www.xilinx.com/publications/magazines/emb_02/index.htm`.

[XO92] L. Xu and E. Oja, "Further developments on RHT: basic mechanisms, algorithms, andcomputational complexities," *11th IAPR International Conference on Pattern Recognition*, vol. 1, pp. 125–128, 1992, iSBN:0-8186-2910-X.

[XO93] L. Xu and E. Oja, "Randomized Hough transform (RHT) : basic mechanisms, algorithms, and computational complexities," *Computer Vision Graphics and Image Processing. Image understanding*, vol. 57, no. 2, pp. 131–154, 1993.

[XOK90] L. Xu, E. Oja, and P. Kultanen, "A new curve detection method : randomized Hough transform (RHT)," *Pattern Recognition Letters*, vol. 11, no. 5, pp. 331–338, 1990.

[XV94a] C. Xu and S. Velastin, "A comparison between the standard Hough transform and the Mahalanobis distance Hough transform," *Proceedings of the third European conference on Computer vision*, vol. 1, pp. 95–100, 1994, springer-Verlag New York, Inc.

[XV94b] C. Xu and S. Velastin, "Line and circle finding by the weighted Mahalanobis distance transform and extended Kalman filtering," *Symposium Proceedings of IEEE International Symposium on Industrial Electronics*, pp. 258–263, 1994.

[Yep96] P. Yepes, "A fast track pattern recognition ," *Nuclear Instruments and Methods in Physics Research Section A: Accelerators, Spectrometers, Detectors and Associated Equipment*, vol. 380, no. 3, pp. 582–585, October 1996, `http://www.sciencedirect.com/-science?_ob=ArticleURL&_udi=B6TJM-3VT9H0F-D-&_user=2717328&_coverDate=10%2F11%2F1996&_fmt=-full&_orig=search&_cdi=5314&view=c&_acct=C000056831-&_version=1&_urlVersion=0&_userid=2717328&md5=-a61d92d87f24b0a77b76bc23310ce0c4&ref=full`.

[YK07] R. Yapa and H. Koichi, "A connected component labeling algorithm for grayscale images and application of the algorithm on mammograms," *Proceedings of the 2007 Association for Computing Machinery (ACM) symposium on Applied computing*, pp. 146–152, 2007.

[YWXT04] X. Yu, H. Wai Leong, C. Xu, and Q. Tian, "A robust Hough-based algorithm for partial ellipse detection in broadcast soccer video," *2004. ICME '04. 2004 IEEE International Conference on Multimedia and Expo*, vol. 3, pp. 1555–1558, 2004, `http://ieeexplore.ieee.org/xpls/abs_all.jsp?ar-number=1394544`.

Curriculum Vitae

Dipl.-Inf. Christian Alexander Steinle
Contact: Christian Steinle @ XING

Birthday, Place of birth	08.02.1979 in Heidelberg
Family status	Unmarried

Employment

November 2010 - today	UniCon Software GmbH, Field: OS Development for Thin Clients (X-Server for eLux)

Education

February 2004 - January 2012	Ph. D. in Natural Science, Group Massively Parallel Processing at the University of Heidelberg
October 1998 - December 2003	Academic studies obtaining the graduate technical computer scientist (Major: HWD and VLSI-Design) at the University of Mannheim
August 1988 - June 1998	General qualification of university entrance at the Copernicus-Gymnasium in Philippsburg

Course-related activities

January 2003	- March 2003	Internship: Heidelberger Druckmaschinen AG, Heidelberg (S-RD-CP3 Drive Systems), Field: Embedded Systems programming
October 2002	- December 2002	Internship: Bosch, Schwieberdingen (GS-EC/EES1), Field: PC front-end programming for an embedded system
August 2001	- Juli 2002	Student assistant in group "FPGA-Processors", Uni Mannheim, Field: Digital Image processing
Mai 2001	- Juli 2001	Internship in group "FPGA-Processors", Uni Mannheim, Field: Digital Image processing
February 2001	- April 2001	Student assistant in group "FPGA-Processors", Uni Mannheim, Field: Digital Image processing

Other activities

April 2006	- September 2010	Part-time job: WFE-Warenautomaten GmbH & CoKG, Mannheim, Field: General IT manager & consultant
March 1999	- January 2001	Part-time job: Schülerforum, Wiesental, Field: Coaching in physics, mathematics and electrical engineering
June 1998		Vacation job: TEELA Personalservice, Graben-Neudorf, Field: Unskilled labor
August 1997		Vacation job: SEW-Eurodrive, Graben-Neudorf, Field: Storekeeper
March 1997	- March 2004	Alternative military service: DRK Kirrlach, Rank: Paramedic

Publications

2009	Presentation on the 13'th CBM Collaboration Meeting: L1 Hough Tracking on a Sony Playstation III
2009	CBM Progress Report: Results of a Hough Tracker implementation for CBM
2008	Presentation on the 12'th CBM Collaboration Meeting: L1 Tracking - Status
2008	CBM Progress Report: Implementation of a Hough Tracker for CBM
2007	Presentation on the 10'th CBM Collaboration Meeting: L1 Tracking - Status CBMROOT and Realisation
2007	Presentation on the 9'th CBM Collaboration Meeting: L1 Tracking - Status Hough Tracker
2007	Presentation on the DPG - Meeting: Tracking im Silicon Tracker System des CBM Experiments mittels Hough Transformation
2007	CBM Progress Report: Implementation of a Hough Tracker for CBM
2006	Presentation on the 8'th CBM Collaboration Meeting: L1 Tracking - Status Hough Tracker
2006	Presentation on the DPG - Meeting: Tracking im Silicon Tracker System des CBM Experiments mittels Hough Transformation
2006	GSI Scientific Report: Hardware Implementation of a Hough Tracker for CBM
2006	CBM Progress Report: Implementation of a Hough Tracker for CBM
2005	Presentation on the 5'th CBM Collaboration Meeting: Track Finding with Hough Transform
2005	Paper on Realtime conference: Tracking in the Silicon Tracker System of the CBM Experiment using Hough Transform
2005	CBM Technical Status Report: Hough Transform